Seventh Edition

NDSU

PRECALCULUS II Trigonometry

Customized Version of
"Precalculus Functions and Graphs, 7 th edition"
by Mustafa Munem and James Yizze

Kendall Hunt
publishing company

Kendall Hunt
publishing company

www.kendallhunt.com
Send all inquiries to:
4050 Westmark Drive
Dubuque, IA 52004-1840

ISBN 978-1-5249-2802-5

Printed in the United States of America
10 9 8 7 6 5 4 3 2 1

Contents

CHAPTER 1

Trigonometric Functions

Chapter Contents

Many applications of trigonometric functions involve periodic or cyclic phenomena. For instance, tides cause ocean, sea, and lake currents to flow into and out of coastlines with varying depths that repeat over intervals of time. Trigonometric functions enable us to examine such situations. Suppose that during a 24-hour day, the average depth d (in feet) of water in a certain location, t hours after midnight, is given by the model

$$d = 8 + 4\cos\left(\frac{4\pi}{25}t\right).$$

This model is analyzed in Example 8 on page 55.

Originally, trigonometry was limited to the numerical solutions of triangles and their applications, first in astronomy and later in surveying and navigation.

Today we recognize two points of view in the investigation of the trigonometric functions. In one approach, the unit circle is used to define the trigonometric functions for real numbers. These functions provide mathematical models for periodic phenomena encountered in electronics, engineering, the physical sciences, and the life sciences—phenomena such as sound waves, electromagnetic waves, light waves, business cycles, and biological systems. In the other approach, the trigonometric functions establish relationships between angles and sides of triangles that are used routinely in calculations made by surveyors, engineers, and navigators. We shall explore both approaches and examine the relationships between these two types of trigonometric functions and their properties.

Objectives

1. Locate Points on the Unit Circle
2. Describe and Measure Angles
3. Converting Angle Measurements
4. Find Arc Lengths and Areas of Sectors
5. Solve Applied Problems-Circular Motion

Figure 1

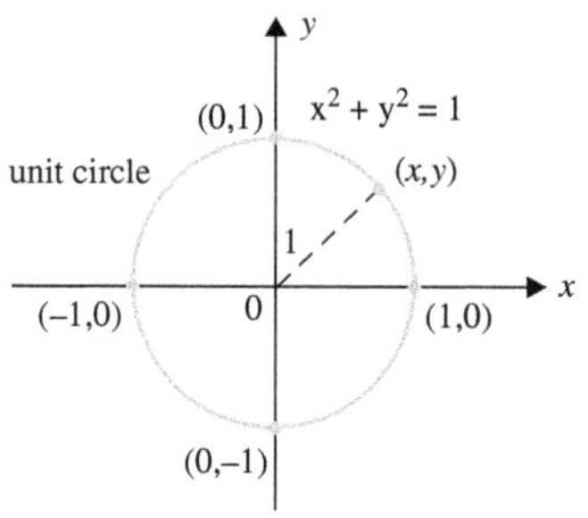

1.1 Preliminary Trigonometric Concepts

We begin by establishing some preliminary concepts necessary for the study of the trigonometric functions.

Locating Points on the Unit Circle

The graph of the equation

$$x^2 + y^2 = 1$$

is called the **unit circle.** It is a circle of radius 1 whose center is at the origin O of a rectangular (Cartesian) coordinate system (Figure 1).

By referring to Figure 2, it is possible to associate each real number with the coordinates of a point on the unit circle in the following way. First, assume that a number line L has the same scale unit as the one used for the unit circle. Next, place the number 0 on the real number line L so that it coincides with the point (1, 0) on the unit circle. Then, the line L is "wrapped around" the circle, either in a *counterclockwise sense* (using the positive part of the real line L) or in a *clockwise sense* (using the negative part of the real line L). Thus, if $t_1 > 0$, it is associated with a point on the unit circle by moving $|t_1|$ units counterclockwise along the circumference of the circle, starting at the point (1, 0) and ending at, say, (x_1, y_1), as illustrated in Figure 2a. If $t_2 < 0$, it is associated with a point on the circle by moving $|t_2|$ units clockwise along the circumference of the circle, starting at the point (1, 0) and ending at, say, (x_2, y_2), as illustrated in Figure 2b.

Figure 2

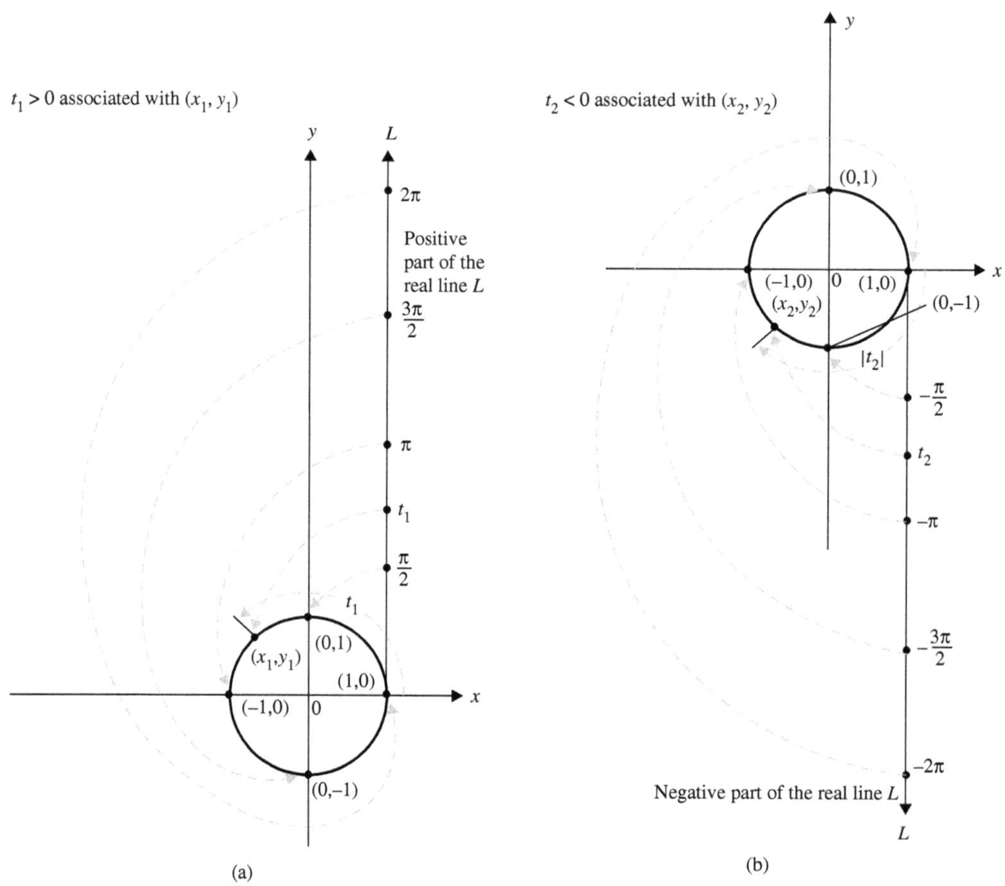

Since the circumference C of any circle is given by $C = 2\pi r$, where r represents the length of the radius of the circle, it follows that the circumference of the unit circle is 2π. If 3.14 is used as an approximation for π, the circumference of the unit circle is approximately 6.28 units. Figure 3 displays some points along the circumference of the unit circle that are obtained by the wrapping process described in Table 1.

Figure 3

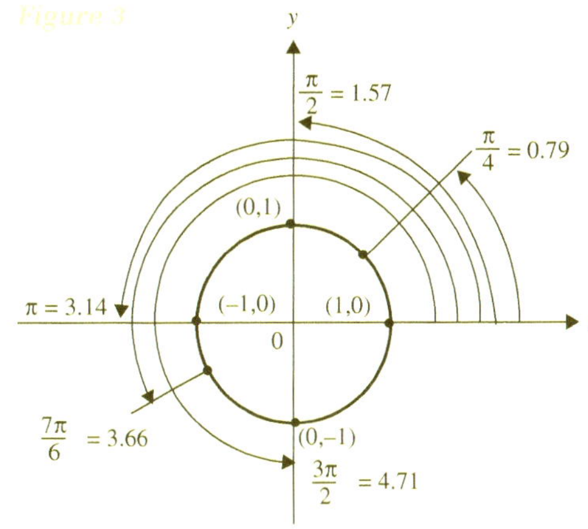

TABLE 1

Start at (1, 0) and Move Counterclockwise	Actual Distance	Approximate Distance ($\pi \approx 3.14$)
One-eighth of the way around the circumference	$\frac{1}{8}(2\pi) = \frac{\pi}{4}$	0.79
One-fourth of the way around the circumference	$\frac{1}{4}(2\pi) = \frac{\pi}{2}$	1.57
One-half of the way around the circumference	$\frac{1}{2}(2\pi) = \pi$	3.14
Seven-twelfths of the way around the circumference	$\frac{7}{12}(2\pi) = \frac{7\pi}{6}$	3.66
Three-fourths of the way around the circumference	$\frac{3}{4}(2\pi) = \frac{3\pi}{2}$	4.71

As a result of the wrapping process just described, it is possible to associate each real number t with a point $P = (x, y)$ on the unit circle, and we use function notation to denote this association as:

$$P(t) = (x, y).$$

Table 2 lists the designation of the points labeled in Figure 4. Thus, we observe that 0 is associated with the point (1, 0); $\pi/2$ is associated with the point (0, 1); π is associated with the point $(-1, 0)$; $3\pi/2$ is associated with the point $(0, -1)$; $-\pi/2$ is associated with the point $(0, -1)$; and $-\pi$ is associated with the point $(-1, 0)$.

TABLE 2

$P(t)$	$P(0)$	$P\left(\frac{\pi}{2}\right)$	$P(\pi)$	$P\left(\frac{3\pi}{2}\right)$	$P\left(-\frac{\pi}{2}\right)$	$P(-\pi)$
(x, y)	(1, 0)	(0, 1)	$(-1, 0)$	$(0, -1)$	$(0, -1)$	$(-1, 0)$

Figure 4

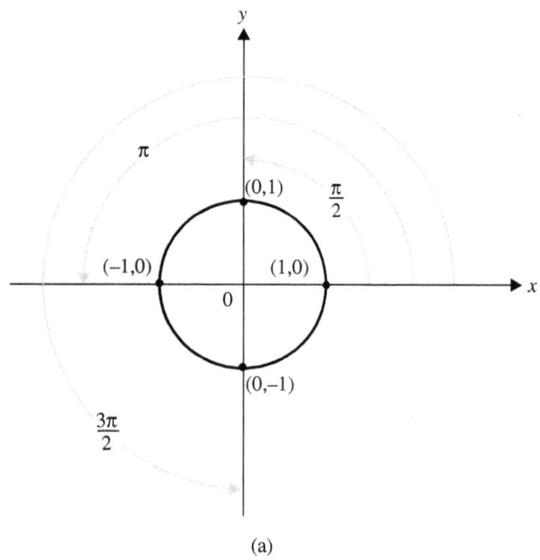

For certain values of t it is possible to find the exact values of the coordinates of $P(t) = (x, y)$ by using some of the analytical tools already studied.

For instance, the coordinates of $P(\pi/4)$ can be found as follows. Let $P(\pi/4) = (a, b)$. Since

$$P\left(\frac{\pi}{4}\right) = P\left(\frac{1}{2} \cdot \frac{\pi}{2}\right),$$

Figure 5

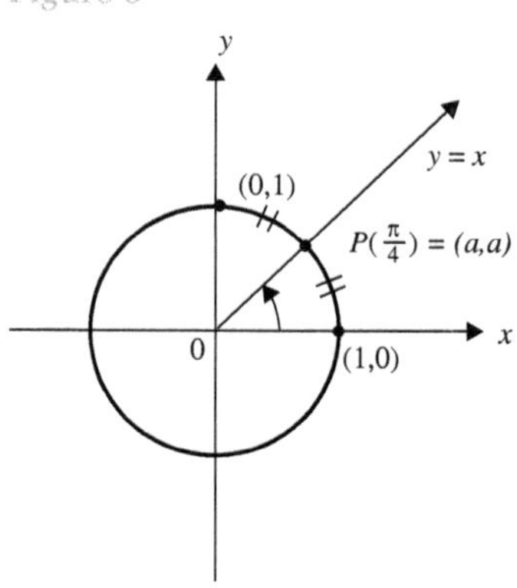

$P(\pi/4)$ is the midpoint of the arc joining the points (1, 0) and (0, 1) on the unit circle (Figure 5). Thus, $P(\pi/4)$ must lie on the line $y = x$, so that $a = b$; that is,

$$P\left(\frac{\pi}{4}\right) = (a, a).$$

Since the coordinates of any point on the unit circle satisfy the equation $x^2 + y^2 = 1$, we have

$$a^2 + a^2 = 1$$
$$2a^2 = 1$$
$$a^2 = \frac{1}{2}.$$

Because $a > 0$ for a point in quadrant I,

$$a = \sqrt{\frac{1}{2}} = \frac{1}{\sqrt{2}} = \frac{\sqrt{2}}{2}$$

so that

$$P\left(\frac{\pi}{4}\right) = \left(\frac{\sqrt{2}}{2}, \frac{\sqrt{2}}{2}\right).$$

Similar arguments can be used to find that the coordinates of $P(\pi/3)$ and $P(\pi/6)$ are, respectively,

$$P\left(\frac{\pi}{3}\right) = \left(\frac{1}{2}, \frac{\sqrt{3}}{2}\right) \quad \text{and} \quad P\left(\frac{\pi}{6}\right) = \left(\frac{\sqrt{3}}{2}, \frac{1}{2}\right) \quad \text{(Problem 54).}$$

The results from the preceding discussion are summarized in Table 3 and displayed in Figure 6.

TABLE 3

$P(t)$	$P\left(\frac{\pi}{6}\right)$	$P\left(\frac{\pi}{4}\right)$	$P\left(\frac{\pi}{3}\right)$
(x, y)	$\left(\frac{\sqrt{3}}{2}, \frac{1}{2}\right)$	$\left(\frac{\sqrt{2}}{2}, \frac{\sqrt{2}}{2}\right)$	$\left(\frac{1}{2}, \frac{\sqrt{3}}{2}\right)$

Figure 6

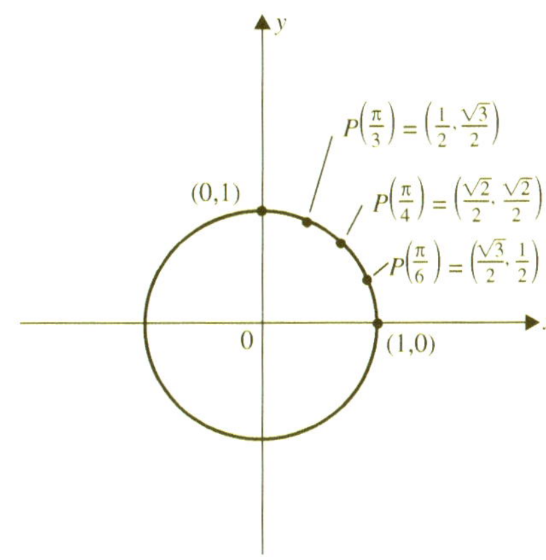

For other values of t, the tools that we have developed so far are insufficient to determine the exact location of $P(t)$. However, it is often helpful to find the quadrant containing the point.

EXAMPLE 1

Determining the Quadrant Where $P(t)$ Lies

Find the quadrant containing the given point (use $\pi = 3.14$).

(a) $P(2)$ (b) $P(-1)$

Figure 7

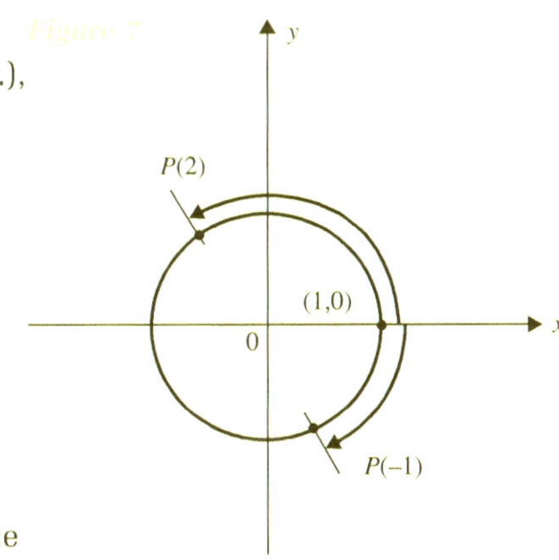

Solution

(a) Since $\pi = 3.14$ and $\pi/2 = 1.57$ (approx.), then

$$\frac{\pi}{2} < 2 < \pi$$

so that $P(2)$ is in quadrant II (Figure 7).

(b) Since $-\pi/2 = -1.57$ (approx.),

$$-\frac{\pi}{2} < -1 < 0.$$

Later, we shall learn how to find the coordinates for points such as P(2) and P(−1).

We locate $P(-1)$ by moving a distance $|-1| = 1$ in the clockwise direction from the point (1, 0) so that $P(-1)$ lies in quadrant IV (Figure 7).

The wrapping process enables us to associate real numbers with points on the unit circle. This concept provides the basis for the modern viewpoint of the study of trigonometry, which will be introduced in Section 1.2. Before introducing the trigonometric functions of angles, we need to review angles and their measurements.

Describing and Measuring Angles

In plane geometry, recall that an **angle** is determined by rotating a ray about its endpoint (the **vertex** of the angle) from some initial position (the **initial side** of the angle) to a terminal position (the **terminal side** of the angle) (Figure 8a). In one scheme that is used to describe an angle, the middle letter represents the vertex, the first letter represents a point on the initial side, and the third letter represents a point on the terminal side. Thus we denote the angle in Figure 8a as $\angle QPR$. Angles are also denoted by lowercase Greek letters such as α in Figure 8b (α is the Greek letter *alpha*). Other Greek letters often used are β (*beta*) (Figure 8c), γ (*gamma*), θ (*theta*), and ϕ (*phi*).

Figure 8

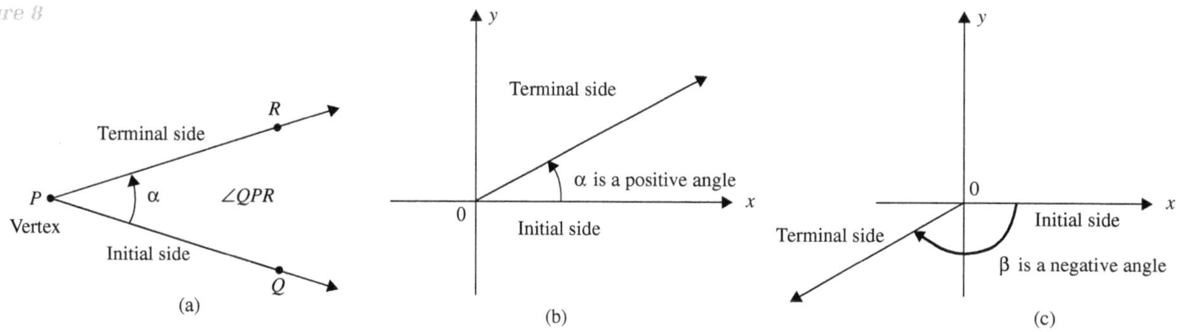

An angle is in **standard position** if it is placed in a rectangular coordinate system with its vertex at the origin O and with the initial side coinciding with the positive x axis. If the angle is formed by a counterclockwise rotation, the angle is considered to be **positive,** whereas if the angle is formed by a clockwise rotation, the angle is **negative.** Figure 8b displays positive angle α in standard position, and Figure 8c displays negative angle β in standard position.

Normally we measure an angle by using *degree* or *radian* measurements.

Definition
One Degree

An angle of **one degree** (1°) is an angle formed by $\frac{1}{360}$ of a complete counterclockwise revolution.

Figure 9 shows angles in standard position measured in degrees. We refer to an angle formed by one-half of a complete counterclockwise revolution as a **straight angle;** it has a measure of $\frac{1}{2}(360°) = 180°$ (Figure 9a). An angle of one-quarter of a counterclockwise revolution is a **right angle;** it has a measure of $\frac{1}{4}(360°) = 90°$ (Figure 9b). An angle is **acute** if its degree measure is between 0° and 90°, for example, 70° (Figure 9c). If an angle measures between 90° and 180°, it is **obtuse,** for example, 120° (Figure 9d).

Figure 9

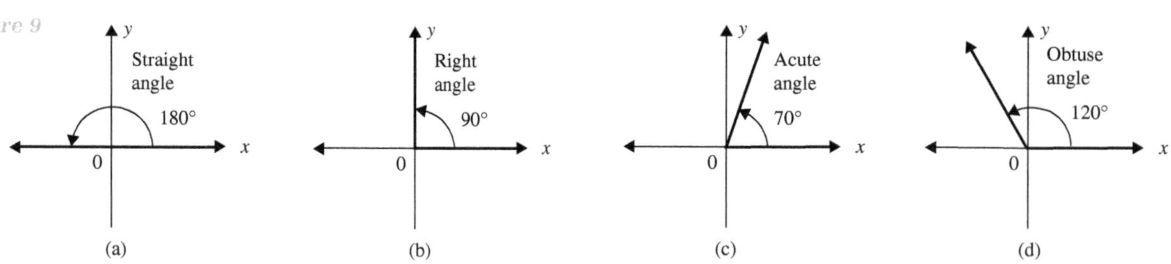

If the terminal side of an angle in standard position lies on either the x axis or the y axis, then the angle is called **quadrantal.** For example, $-360°$, $-270°$, $-180°$, $0°$, $90°$, $180°$, $270°$, and $360°$ are quadrantal angles. Figures 9a and 9b display quadrantals $180°$ and $90°$, respectively.

Angles in standard position that have the same terminal sides are called **coterminal** angles. For example, the three angles

Figure 10

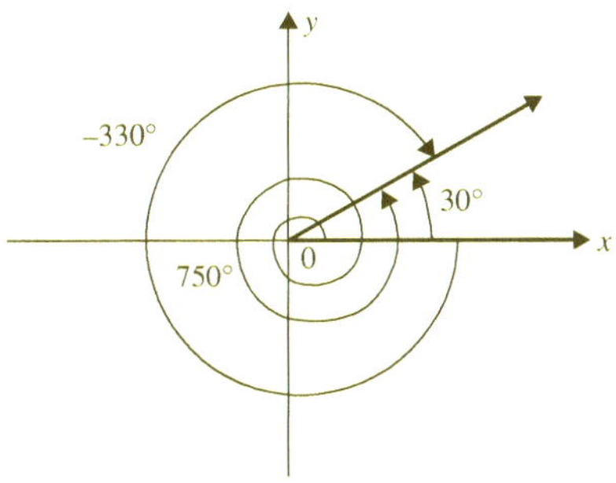

$$30°, \ -330°, \text{ and } 750°$$

are coterminal (Figure 10).

Although parts of a degree can be expressed as a decimal, such parts are sometimes given in *minutes* and *seconds.* A degree can be divided into 60 equal parts called **minutes** (′); a minute can be divided into 60 equal parts called **seconds** (″). Using these relationships, we have

$$1' = \left(\frac{1}{60}\right)^\circ, \quad 1'' = \left(\frac{1}{60}\right)', \quad 1'' = \left(\frac{1}{3600}\right)^\circ, \quad 1^\circ = 60', \quad \text{and} \quad 1' = 60''.$$

Some calculators have keys that automatically convert decimal degree measures to degrees, minutes, and seconds and vice versa. The next example shows how such conversions can be made without such special keys.

EXAMPLE 2

Converting Degree Measures

(a) Express $37.45°$ in terms of degrees, minutes, and seconds.

(b) Express $23°17'37''$ in decimal degree measure to four decimal places.

Solution

(a) $37.45° = 37° + 0.45°$; and $0.45° = (0.45)(60') = 27'$.
Therefore, $37.45° = 37°27'0''$.

(b) $$23^\circ 17' 37'' = \left(23 + \frac{17}{60} + \frac{37}{3600}\right)^\circ$$
$$= 23.2936^\circ \text{ (approx.)}.$$

Another frequently used unit of angle measurement is a *radian.*

Radian Measure Relative to the Unit Circle

> One **radian** is the measure of a positive angle that intercepts an arc of length 1 on a circle of radius 1 (Figure 11).

Figure 11

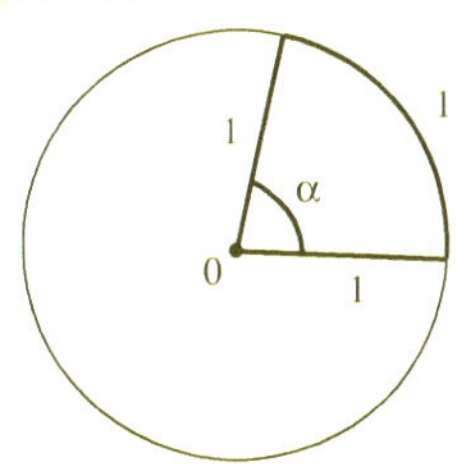

α is an angle of 1 radian

An angle of radian measure t in standard position has its initial side at the positive x axis and its terminal side at the ray containing the origin and the point $P(t)$, which is located through the use of the wrapping scheme described earlier. For example, an angle θ of one radian is generated by a counterclockwise rotation in which the point of intersection of the rotating ray with the unit circle travels 1 unit (Figure 12a). Similarly, an angle θ of radian measure $5\pi/6$ is generated by a counterclockwise rotation in which the point of intersection of the rotating ray with the unit circle travels $5\pi/6$ units (Figure 12b). The angle θ with radian measure -8.9 is generated by a clockwise rotation in which the point of intersection of the rotating ray with the unit circle travels 8.9 units (Figure 12c).

Figure 12

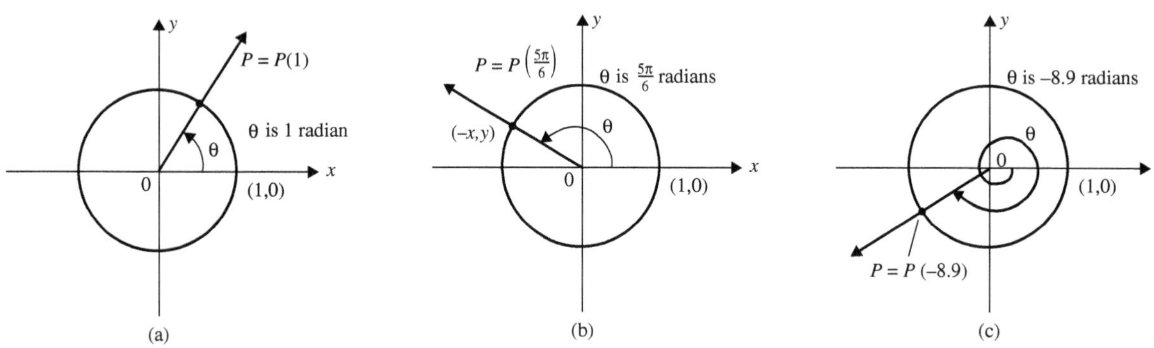

Notice that, given any angle θ, we can associate with it a real number t (the radian measure of θ). Conversely, given any real number t, we can associate with it the angle θ in standard position with terminal side containing the origin O and point $P(t)$.

Figures 13a and b show the two angles in standard position designated by θ when t is $\pi/4$ and $-5\pi/6$, respectively.

Figure 13

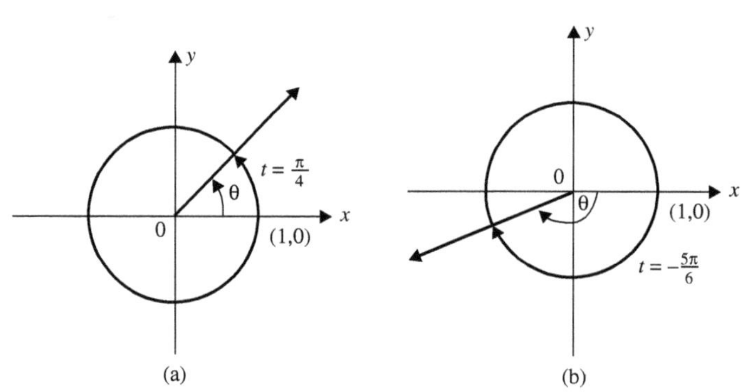

Figure 14

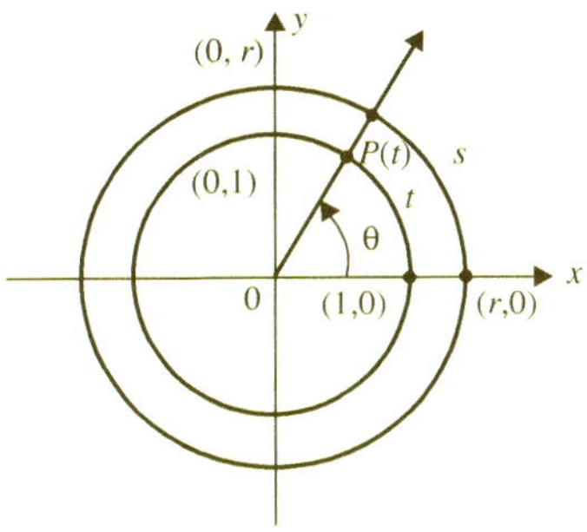

A **central angle** of a circle is an angle that has its vertex at the center of the circle. Figure 14 shows a central angle θ that subtends an arc of length t units on the unit circle and an arc of length s units on a concentric circle of radius r. We know that the radian measure of θ is t. From plane geometry, we know that the ratio of the arc lengths is the same as the ratio of the corresponding radii. That is,

$$\frac{t}{s} = \frac{1}{r}$$

so that we get the following results.

Radian Measure Relative to Any Circle

$$t = \frac{s}{r}.$$

Figure 15

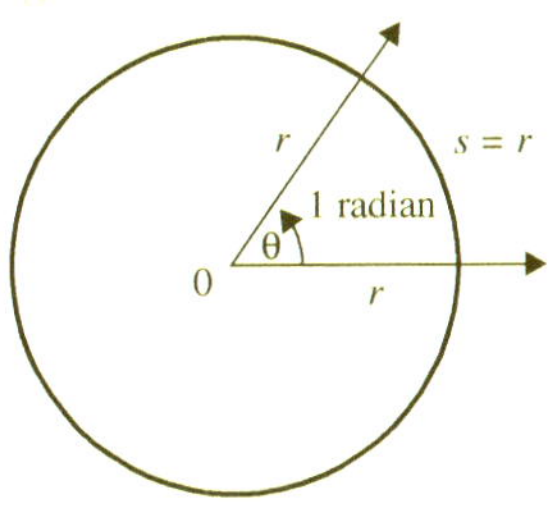

Notice that when $s = r$, the radian measure of θ equals 1; that is,

1 radian is the measure of a central angle that intercepts an arc of the circle equal in length to the radius of the circle (Figure 15).

Converting Angle Measurements

Sometimes we need to convert an angle measure from degrees to radians or vice versa. Table 4 displays corresponding degree and radian measures for some angles.

TABLE 4

Degree Measure (D)	Corresponding Radian Measure (R)
360°	2π
180°	π
90°	$\frac{\pi}{2}$

Graphers have a feature that automatically converts radian measure to degree measure and vice versa.

On the basis of the results in Table 4, we have the following rules for converting between radian measures and degree measures.

If an angle has radian measure $R \neq 0$ and degree measure D, then the following ratio holds:

Formulas Angle Measurement Conversions

$$\frac{D}{R} = \frac{180}{\pi} \quad \text{or} \quad \frac{R}{D} = \frac{\pi}{180}$$

Notice that an angle of 1 radian has a degree measure given by $(180/\pi)°$, or approximately 57.2958° or 57°17′45″.

EXAMPLE 3 **Converting Degrees to Radians**

Convert each degree measure to radian measure.

(a) $-150°$ (b) $26.85°$, to four decimal places

Solution Since $R = (\pi/180)D$, there is $\pi/180$ radian in each degree, so that

(a) $-150°$ corresponds to $\frac{\pi}{180}(-150) = -\frac{5\pi}{6}$ radians.

(b) $26.85°$ corresponds to $\frac{\pi}{180}(26.85) = 0.4686$ radian (approx.).

Finding Arc Lengths and Areas of Sectors

Figure 16

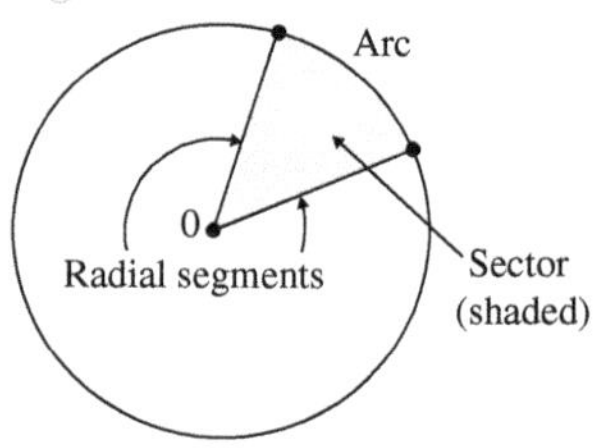

As displayed in Figure 16, a **sector of a circle** is a region inside the circle bounded by an arc and radial line segments formed by the center of the circle 0 and the end points of the arc. Our objective is to derive formulas for the arc lengths and areas of sectors.

Earlier we established that if a central angle θ of a circle of radius r subtends an arc of length s, then the radian measure t of θ is given by the formula $t = \frac{s}{r}$.

By solving the formula for s, we obtain the following formula.

Formula
Arc Length

> The **arc length of a circular sector of a circle** with radius r and central angle θ is given by
>
> $$s = rt,$$
>
> where t is the measure of θ expressed in radians.

The entire interior of a circle is also a sector.

From geometry, the ratio $\dfrac{\text{Area of a Sector of a circle}}{\text{Radian measure of its central angle}}$ is constant.

Thus, if we have a sector formed from a circle of radius r with a central angle that has radian measure t (Figure 17), it follows that

Figure 17

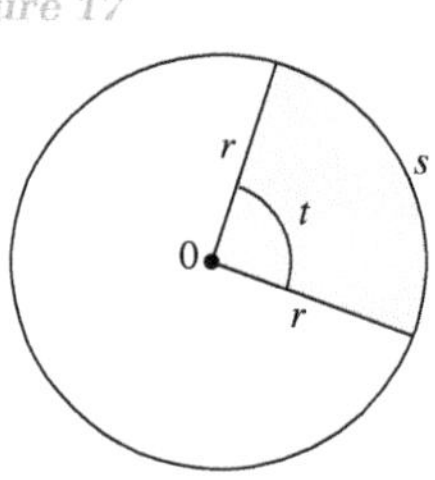

$$\frac{\text{Area of Sector}}{t} = \frac{\text{Area of Circle}}{2\pi} = \frac{\pi r^2}{2\pi} = \frac{1}{2}r^2 \quad \text{or} \quad \frac{\text{Area of Sector}}{t} = \frac{1}{2}r^2$$

so that

Area of Sector $= \frac{1}{2}r^2t$. Since $s = rt$, we can rewrite the latter formula as

$$\text{Area of Sector} = \frac{1}{2}r \cdot (rt) = \frac{1}{2}rs.$$

Thus we have the following general result.

Formulas
Area of a Sector

> If A represents the **area of a sector of a circle** of radius r formed by a central angle θ of radian measure t and a subtended arc length s, then
>
> $$A = \frac{1}{2}r^2t \quad \text{or} \quad A = \frac{1}{2}rs.$$

EXAMPLE 4

Finding an Arc Length and Area of a Sector

Given a circle of radius 6 centimeters and a central angle of $\theta = 240°$ (Figure 18),

(a) Find the length s of the arc intercepted on the circle.

(b) Find the area of the sector formed by the angle.

Solution

Figure 18

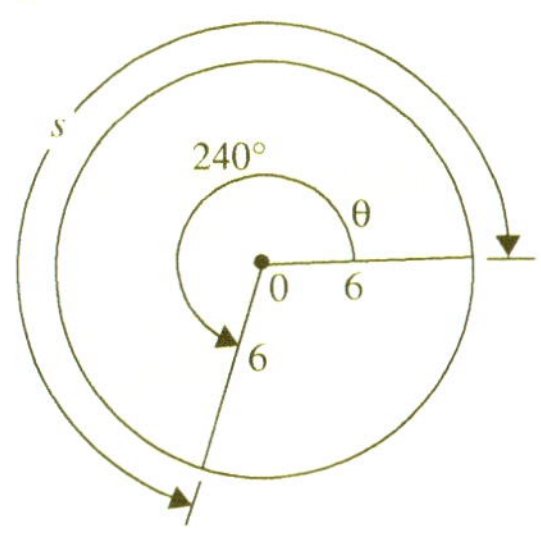

(a) We first convert 240° to radian measure t to get

$$t = \frac{\pi}{180}D = \left(\frac{\pi}{180}\right)(240) = \frac{4\pi}{3} \text{ radians.}$$

So $$s = rt = 6\left(\frac{4\pi}{3}\right) = 8\pi.$$

Therefore, the arc length is 8π or about 25.13 centimeters.

(b) To find the area of the sector we can proceed in either of the following ways:

Using the Result from Part (a)

Substituting our result from part (a) into the formula

$$A = \frac{1}{2}rs$$

we get

$$A = \frac{1}{2}(6)(8\pi) = 24\pi.$$

Alternate Solution

After converting 240° to radian measure $\frac{4\pi}{3}$, we substitute into the formula

$$A = \frac{1}{2}r^2t$$

to get

$$A = \frac{1}{2}(6)^2\left(\frac{4\pi}{3}\right) = \frac{1}{2}(36)\left(\frac{4\pi}{3}\right) = 24\pi.$$

Either way, the area is 24π or about 75.40 square centimeters.

Solving Applied Problems—Circular Motion

Figure 19

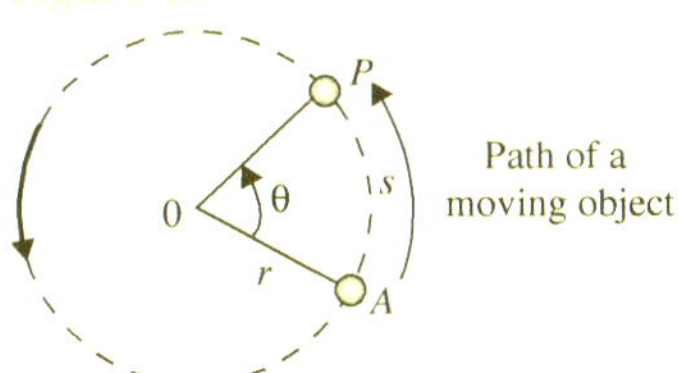

Angles measured in radians allow us to study the motion of an object moving at a constant speed around a circle of radius r. Suppose that a moving object starts at point A and after T units of time reaches point P (Figure 19). If the arc $\widehat{AP}$ has length s, then the object has moved s units of distance in T units of time. Since *distance divided by time equals speed*, we represent the *linear speed* v of the moving object as follows.

Formula
Linear Speed

$$v = \frac{s}{T}.$$

The ratio that measures the rate of change of the angle θ, where θ is measured in radians, with respect to T units of time is called the **angular speed ω** (Greek letter omega) and is given by the following formula.

Formula

Angular Speed

$$\omega = \frac{\theta}{T}.$$

Since the arc length $s = r\theta$, it follows that

$$v \text{ (linear speed)} = \frac{s}{T} = \frac{r\theta}{T} = r\left(\frac{\theta}{T}\right) = r\omega.$$

So the relationship between the linear speed v and angular speed ω is given by

$$v = r\omega \quad \text{or} \quad \omega = \frac{v}{r}.$$

In other words, the linear speed of an object moving along a circular path of radius r is the product of the angular speed and the radius.

EXAMPLE 5

Finding the Angular Speed of a Flywheel

A belt passes over the rim of a flywheel with radius 18 centimeters (Figure 20).

(a) Find the angular speed of a point on the rim of the wheel if the belt drives the wheel at a linear speed of $v = 576$ centimeters per second.

(b) Find the angular speed if the belt drives the wheel at a rate of 12 rotations every 2 seconds. Round off the answer to two decimal places.

Solution

(a) Since the angular speed is given by

$$\omega = \frac{v}{r}, \text{ where } v = 576 \text{ and } r = 18,$$

it follows that

$$\omega = \frac{576}{18} = 32 \text{ radians per second.}$$

Figure 20

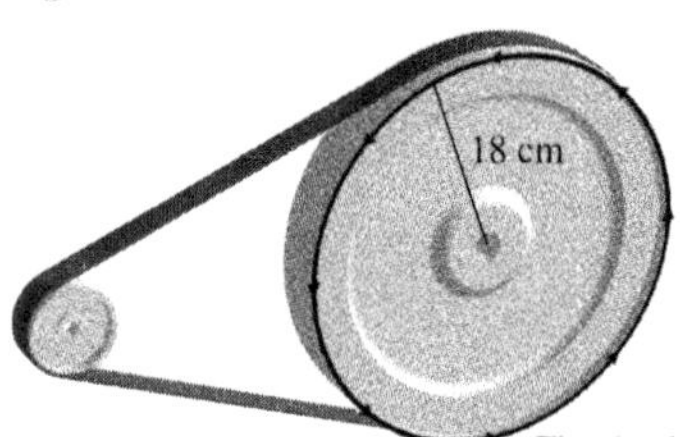

(b) Since the belt drives a point on the rim of the wheel 12 times around the circle in 2 seconds, a point on the rim of the wheel rotates $12(2\pi) = 24\pi$ radians in 2 seconds. Therefore, the angular speed of the belt is

$$\omega = \frac{\theta}{T} = \frac{24\pi}{2} = 12\pi, \text{ or approximately 37.70 radians per second.}$$

PROBLEM SET 1.1

Mastering the Concepts

In problems 1 and 2, find the value of t if the point $P(t)$ is located by starting at point (1, 0) and moving around the circumference of the unit circle as specified.

1. (a) One-third of the way, clockwise
(b) Five-fourths of the way, counterclockwise

2. (a) Three-eighths of the way, clockwise
(b) One-sixth of the way, counterclockwise

3. Use Figure 6 on page 5 and symmetry to find the coordinates of each point.
(a) $P\left(-\frac{\pi}{3}\right)$ (b) $P\left(-\frac{\pi}{4}\right)$ (c) $P\left(-\frac{\pi}{6}\right)$

4. Find the coordinates of each point.
(a) $P(4\pi)$ (b) $P\left(\frac{5\pi}{2}\right)$ (c) $P\left(-\frac{5\pi}{2}\right)$

In problems 5–8, display each point on the unit circle and find the quadrant (if any) containing each point.

5. (a) $P\left(\frac{\pi}{7}\right)$
(b) $P\left(\frac{-3\pi}{5}\right)$
(c) $P\left(\frac{-2\pi}{5}\right)$
(d) $P\left(\frac{17\pi}{8}\right)$

6. (a) $P\left(\frac{17\pi}{6}\right)$
(b) $P\left(\frac{-8\pi}{7}\right)$
(c) $P\left(\frac{31\pi}{8}\right)$
(d) $P\left(\frac{-9\pi}{5}\right)$

7. (a) $P(6)$
(b) $P(1.4)$
(c) $P(3.6)$
(d) $P(-5.7)$

8. (a) $P(-11.2)$
(b) $P(23.5)$
(c) $P(-3.1)$
(d) $P(-13.4)$

In problems 9–14, locate each angle in standard position. Specify its quadrant location, and sketch two angles coterminal with the given angle.

9. (a) $\frac{3\pi}{4}$
(b) $-\frac{\pi}{6}$

10. (a) $-\frac{5\pi}{12}$
(b) $\frac{7\pi}{6}$

11. (a) $-45°$
(b) $120°$

12. (a) $210°$
(b) $105°$

13. (a) $\frac{17\pi}{3}$
(b) $-700°$

14. (a) $-\frac{11\pi}{3}$
(b) $538°$

In problems 15 and 16, express each angle measure in terms of degrees, minutes, and seconds.

15. (a) $16.31°$
(b) $-87.81°$
(c) $-156.63°$
(d) $89.74°$

16. (a) $64.14°$
(b) $-12.16°$
(c) $-213.68°$
(d) $463.09°$

In problems 17 and 18, express each angle in terms of decimal degrees. Round off each answer to two decimal places.

17. (a) $38°18'$
(b) $65°11'23''$
(c) $-141°28'15''$
(d) $244°46'15''$

18. (a) $35°41'$
(b) $-48°15'25''$
(c) $1°1'1''$
(d) $-10°10'10''$

In problems 19–24, convert each angle measure to radians. Express the answer in terms of π and also as a decimal rounded off to two decimal places.

19. (a) $75°$
(b) $-135°$

20. (a) $240°$
(b) $-7.5°$

21. (a) $-95°$
(b) $444°$

22. (a) $-220°$
(b) $330°$

23. (a) $67.5°$
(b) $30°30'36''$

24. (a) $17.45°$
(b) $115°13'44''$

In problems 25–30, convert each radian measure to degrees. Round off each answer to two decimal places.

25. (a) $\frac{2\pi}{3}$
(b) $\frac{43\pi}{6}$

26. (a) $\frac{11\pi}{6}$
(b) $\frac{7\pi}{18}$

27. (a) $\frac{7\pi}{12}$
(b) $-\frac{4\pi}{9}$

28. (a) $-\frac{3\pi}{8}$
(b) $-\frac{\pi}{14}$

29. (a) 5
(b) -2.3

30. (a) -1
(b) 4.6

In problems 31–36, s represents the arc length corresponding to a central angle θ on a circle of radius r.

(a) Find the missing measurement to two decimal places.
(b) Sketch the sector that is formed in each situation and find its area. Round off the answer to two decimal places.

31. $r = 7$ inches, $\theta = \dfrac{3\pi}{14}$, $s = ?$

32. $s = 6$ feet, $\theta = \dfrac{\pi}{7}$, $r = ?$

33. $r = 1.8$ meters, $\theta = 217°$, $s = ?$

34. $r = 11$ centimeters, $\theta = \dfrac{7\pi}{5}$, $s = ?$

35. $s = 5$ yards $\theta = \dfrac{5\pi}{3}$, $r = ?$

36. $r = 7$ feet, $\theta = 110°$, $s = ?$

Applying the Concepts

In problems 37–48, round off the results to two decimal places.

37. Nautical Mile: A nautical mile is the arc length intersected on the surface of the Earth by a central angle of 1 minute (Figure 21). Assume that the radius of the Earth is 3960 miles.

(a) How many miles are there in 1 nautical mile?

(b) How many feet are there in 1 nautical mile?

Figure 21

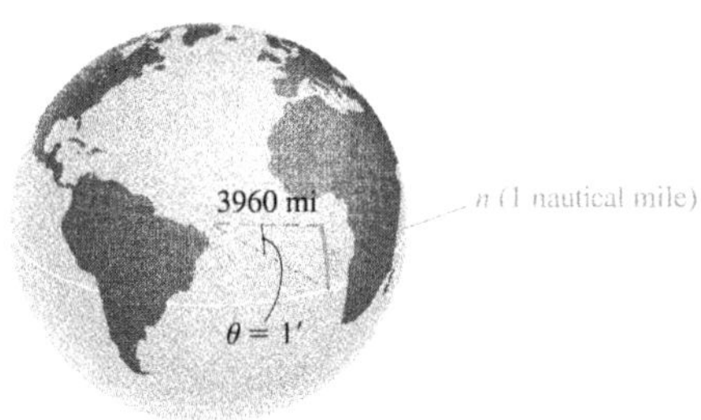

38. Pendulum Swing: Find the length of a pendulum if the tip of the pendulum traces an arc of 6 feet and the measure of the angle that this arc subtends is $\pi/5$ radian.

39. Child's Swing: The length of each chain supporting the seat of a child's swing is 8 feet. When the swing moves from its extreme forward position to its extreme backward position, the radian measure of the angle swept out by the chains is $(5\pi)/6$. How far does the seat travel in one trip between these extreme points; that is, what is the length of the arc generated by the seat through one swing between the two extreme points?

40. Revolving Door: A revolving door rotates through an angle of 2.64 radians before it is stopped. If one half of the door is 4 feet wide, through what distance does the edge of the door move?

41. Windshield Wiper: A windshield wiper of a car is 48 centimeters long, and it rotates at an angular speed of 12 degrees per second. Find the linear speed of the tip of the wiper.

42. Central Angle of a Pulley: A pulley of radius 25 centimeters uses a belt to drive another pulley of radius 15 centimeters. Find the radian measure of the angle θ_2 through which the smaller pulley turns as the larger pulley makes an angle θ_1 of 140° (Figure 22).

Figure 22

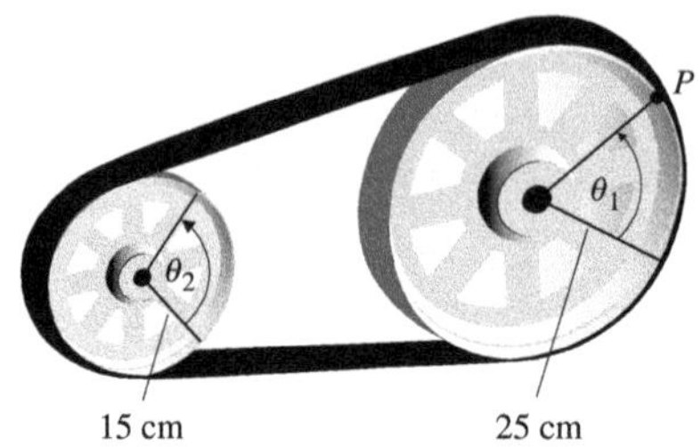

43. Tire Rotations: A car with tires that are 22 inches in diameter travels at 55 miles per hour.

(a) Find the angular speed of each tire in radians per minute.

(b) How many rotations per minute does each tire make?

44. Satellite Orbit: A satellite is orbiting the Earth in a circular orbit with a radius of 7680 kilometers. If it makes $\dfrac{3}{4}$ of a revolution every hour, find:

(a) Its angular speed

(b) Its linear speed

45. Bicycle Sprocket Wheels: A bicycle's small sprocket wheel of radius 2.75 inches is connected by a chain to a large sprocket wheel of radius 4.50 inches (Figure 23). The small sprocket wheel turns at 50 rotations per minute. Find:

(a) The linear speed (in inches per minute) of the chain connecting the sprocket wheels.

(b) The number of rotations per minute of the large sprocket wheel.

Figure 23

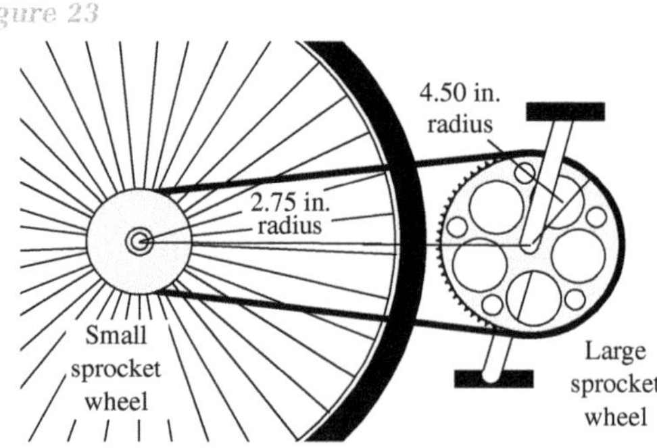

46. **The Earth's Rotation:** The Earth, which is approximately 93,000,000 miles from the sun, revolves about the sun in a nearly circular orbit in approximately 365 days. Find the linear speed (in miles per day) of the Earth in its orbit.
47. **Waterwheel:** A point on the rim of a circular waterwheel of radius 8 feet makes one complete rotation every 10 seconds. Find the angular and linear speeds of the circular motion of the moving point.
48. **Steamboat:** As the paddlewheel with radius 9 feet of a steamboat turns, a point on the end of the paddle blade makes one complete rotation every 8 seconds. Determine the angular and linear speeds of the circular motion of the point on the paddle blade.

Developing and Extending the Concepts

49. Figure 24 shows three circles with the same center, a common central angle θ, and radii r_1, r_2, and r_3, where $r_1 < r_2 < r_3$. A student makes the following assertion: Since the radian measure of a central angle of a circle depends on the radius of the circle, it follows that the circle with a larger radius has a bigger central angle. Explain why this assertion is false.

Figure 24

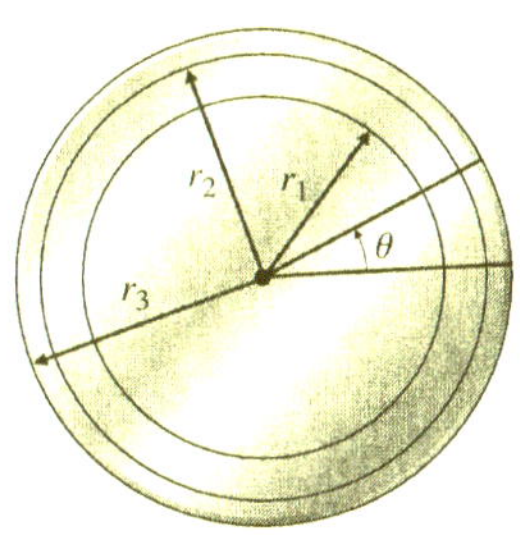

50. The following assertion is made: Suppose that angle θ_1 has radian measure t and angle θ_2 has radian measure $t + 2\pi$. When both angles are placed in standard position, they have the same terminal side. It follows that

$$\theta_1 = \theta_2.$$

Explain why this assertion is wrong.

51. Which angle is larger, angle $\alpha = 1$ or angle $\beta = 1°$? Explain your answer.
52. Suppose that you ride on a merry-go-round at a carnival. You realize that you go faster when you sit near the outside than when you sit near the inside, even though the merry-go-round travels at a constant angular speed (Figure 25). Explain this phenomenon.

Figure 25

53. What is the radian measure of the smaller of the positive angles between the hands of a clock at 11:30?
54. Prove that

(a) $P\left(\frac{\pi}{3}\right) = \left(\frac{1}{2}, \frac{\sqrt{3}}{2}\right)$ *Hint:* Let $A = (1, 0)$, $B = P\left(\frac{\pi}{3}\right) = (a, b)$, and $C = P\left(\frac{2\pi}{3}\right) = (-a, b)$.

Since the lengths of arcs $\overset{\frown}{AB}$ and $\overset{\frown}{BC}$ are equal (each is of length $\frac{\pi}{3}$) (Figure 26), it follows from geometry that the chords $\overline{AB}$ and $\overline{BC}$ are equal in length, so that

Figure 26

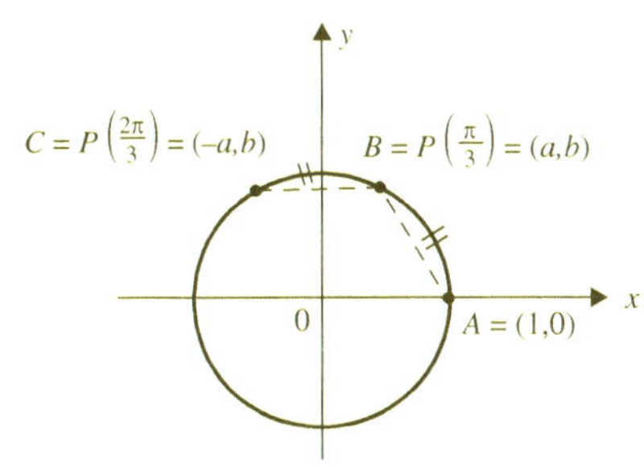

$$|\overline{AB}| = |\overline{BC}| \quad \text{and thus} \quad |\overline{AB}|^2 = |\overline{BC}|^2.$$

Now use the distance formula.

(b) $P\left(\frac{\pi}{6}\right) = \left(\frac{\sqrt{3}}{2}, \frac{1}{2}\right)$

Interlude: Practice Problems 1.1, Part I

1. Determine how far the tip of the minute hand on a clock moves over the course of 42 minutes if the hand is 8 cm long. Give the result accurate to two decimal places.

2. A satellite following a circular orbit around the earth orbits 600 miles above the earth's surface. Calculate the radian measure (accurate to two decimal places) of the angle through which the satellite moves, measured at the center of the earth, as it travels 1000 miles along its orbit. Assume that the radius of the earth is 3960 miles.

3. A total solar eclipse occurs when the moon passes between the earth and the sun and the angle subtended by the diameter of the moon is at least as large as the angle subtended by the diameter of the sun (see Figure 27).

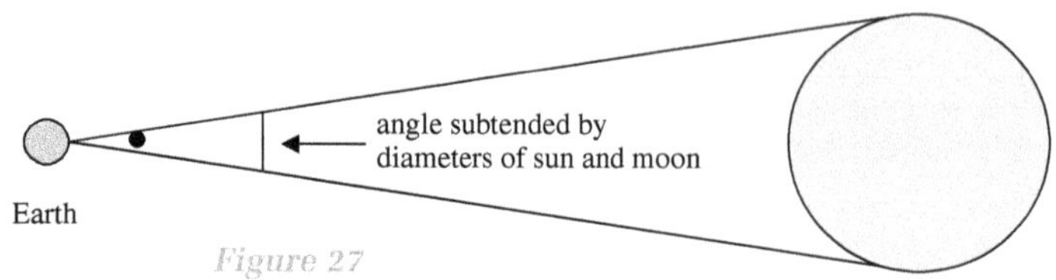

Figure 27

(a) The distances between the earth and moon and the earth and sun change with time. For a total solar eclipse to occur, would it be better for the sun to be as far away from earth as possible and the moon as close to earth as possible, or vice versa? Explain.

(b) The diameters of the sun and moon are, respectively, about 864,000 miles and 2160 miles. Find the maximum distance (to the nearest mile) that the moon can be from the earth for a total eclipse to occur when the sun is at its maximum distance, 94,500,000 miles, from the earth.

Interlude: Practice Problems 1.1, Part II

1. On a certain bicycle, the rear wheel has a radius of 14 inches. The sprocket attached to the wheel has a radius of 2 inches and is connected by the bike chain to the pedal sprocket, which has a radius of 4 inches. Suppose the cyclist is pedaling at a rate of 1 revolution per second.

   (a) Calculate the angular and linear speeds of the pedal sprocket, and label each appropriately.

   (b) Calculate the angular and linear speeds of the wheel sprocket, and label each appropriately.

   (c) Estimate the speed of the bike, in feet per second and in miles per hour. Give both estimates accurate to two decimal places.

2. Suppose that two poles that are perfectly vertical (meaning each pole points directly towards the center of the earth) are erected 300 miles apart and that if extended their ends would meet at the center of the earth at an angle of about 4.31878°. Estimate the radius of the earth. Round your result to three decimal places.

3. Three gears of radius 10 cm, 20 cm, and 30 cm are aligned in order of size. Suppose that the largest gear completes 5 revolutions.

(a) Determine the number of revolutions completed by both of the smaller gears.

(b) If the largest gear is rotating at a rate of 5 revolutions per second, determine the linear and angular velocity of the smallest gear.

Objectives

1. Define Trigonometric Functions of Real Numbers—Circular Functions
2. Evaluate Trigonometric Functions of Special Values
3. Use Symmetry to Evaluate Trigonometric Functions
4. Approximate Trigonometric Function Values
5. Solve Applied Problems

1.2 Trigonometric Functions of Real Numbers—Circular Functions

The coordinates (x, y) of point $P(t)$ on the unit circle (Figure 1) are used to define the six *trigonometric functions* or *circular functions* of t. These functions are referred to and are abbreviated as follows: (We may consider the point $P(t)$ in *any* quadrant.)

Name of Trigonometric Function	Abbreviation of the Function at t
sine	$\sin t$
cosine	$\cos t$
tangent	$\tan t$
cotangent	$\cot t$
secant	$\sec t$
cosecant	$\csc t$

Figure 1

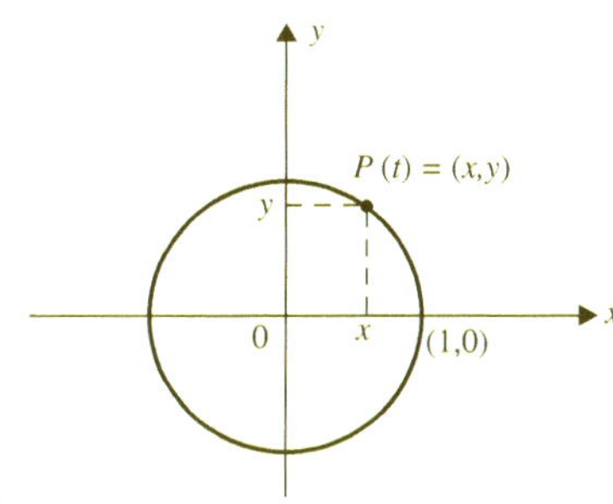

Defining Trigonometric Functions of Real Numbers—Circular Functions

The six trigonometric functions, which have domains that consist of real numbers, are defined as follows:

Definition

Trigonometric Functions of a Real Number—Circular Functions

Let t be a real number and let $P(t) = (x, y)$ be the point on the unit circle associated with t (Figure 1). Then

$$\sin t = y \qquad \csc t = \frac{1}{y},\quad y \neq 0$$

$$\cos t = x \qquad \sec t = \frac{1}{x},\quad x \neq 0$$

$$\tan t = \frac{y}{x},\quad x \neq 0 \qquad \cot t = \frac{x}{y},\quad y \neq 0.$$

EXAMPLE 1

Using the Definition of Circular Functions

Use the definition to determine the values of the six circular functions of t when $P(t) = \left(-\frac{4}{5}, \frac{3}{5}\right)$.

Solution

Note that $\left(-\frac{4}{5}\right)^2 + \left(\frac{3}{5}\right)^2 = 1$, so that $P(t) = \left(-\frac{4}{5}, \frac{3}{5}\right)$ lies on the unit circle. It follows from the definition that

$$\sin t = y = \frac{3}{5} \qquad \csc t = \frac{1}{y} = \frac{1}{\frac{3}{5}} = \frac{5}{3}$$

$$\cos t = x = -\frac{4}{5} \qquad \sec t = \frac{1}{x} = \frac{1}{-\frac{4}{5}} = -\frac{5}{4}$$

$$\tan t = \frac{y}{x} = \frac{\frac{3}{5}}{-\frac{4}{5}} = -\frac{3}{4} \qquad \cot t = \frac{x}{y} = \frac{-\frac{4}{5}}{\frac{3}{5}} = -\frac{4}{3}.$$

Evaluating Trigonometric Functions of Special Values

When $P(t)$ is a point on the coordinate axes, then the trigonometric values of t are found as illustrated in the next example.

EXAMPLE 2 **Evaluating Circular Functions**

Determine the values of the six trigonometric functions of t.

(a) $t = 0$ (b) $t = \frac{\pi}{2}$

Solution (a) Since $P(0) = (1, 0)$ (Figure 2), it follows from the definition that the values are

Figure 2

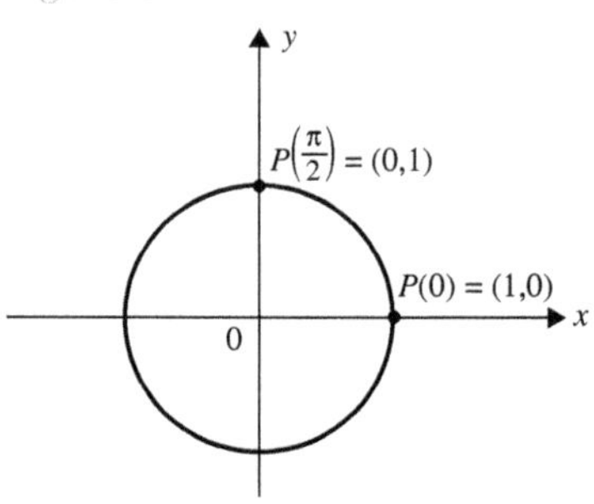

$$\sin 0 = y = 0 \qquad \csc 0 = \frac{1}{y} = \frac{1}{0} \text{ (undefined)}$$

$$\cos 0 = x = 1 \qquad \sec 0 = \frac{1}{x} = \frac{1}{1} = 1$$

$$\tan 0 = \frac{y}{x} = \frac{0}{1} = 0 \qquad \cot 0 = \frac{x}{y} = \frac{1}{0} \text{ (undefined).}$$

(b) $P\left(\frac{\pi}{2}\right) = (0, 1)$ (Figure 2), so that from the definition it follows that the values are

$$\sin\frac{\pi}{2} = y = 1 \qquad \csc\frac{\pi}{2} = \frac{1}{y} = \frac{1}{1} = 1$$

$$\cos\frac{\pi}{2} = x = 0 \qquad \sec\frac{\pi}{2} = \frac{1}{x} = \frac{1}{0} \text{ (undefined)}$$

$$\tan\frac{\pi}{2} = \frac{y}{x} = \frac{1}{0} \text{ (undefined)} \qquad \cot\frac{\pi}{2} = \frac{x}{y} = \frac{0}{1} = 0.$$

For *special values* of t, $\frac{\pi}{4}$, $\frac{\pi}{3}$, and $\frac{\pi}{6}$, we can find the trigonometric values by using results established in Section 1.1.

EXAMPLE 3 **Evaluating Circular Functions for $\frac{\pi}{4}$, $\frac{\pi}{3}$, and $\frac{\pi}{6}$**

Determine the six trigonometric values of t.

(a) $t = \frac{\pi}{4}$ (b) $t = \frac{\pi}{3}$ (c) $t = \frac{\pi}{6}$

Solution (a) In Section 1.1, we proved that $P\left(\frac{\pi}{4}\right) = \left(\frac{\sqrt{2}}{2}, \frac{\sqrt{2}}{2}\right)$ (Figure 3). It follows from the definition that

Figure 3

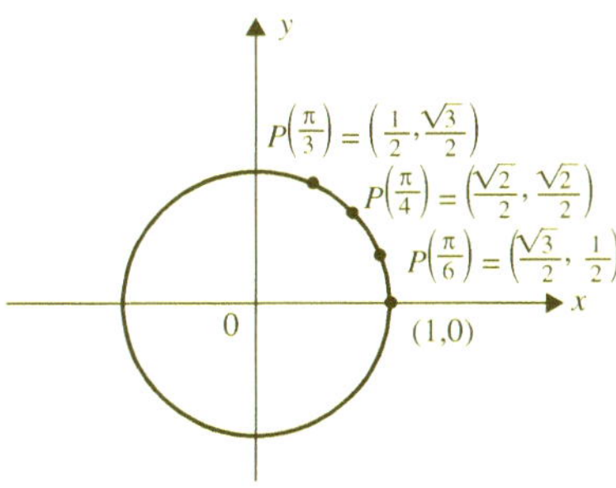

$$\sin\frac{\pi}{4} = y = \frac{\sqrt{2}}{2} \qquad \csc\frac{\pi}{4} = \frac{1}{y} = \frac{1}{\frac{\sqrt{2}}{2}} = \frac{2}{\sqrt{2}} = \sqrt{2}$$

$$\cos\frac{\pi}{4} = x = \frac{\sqrt{2}}{2} \qquad \sec\frac{\pi}{4} = \frac{1}{x} = \frac{1}{\frac{\sqrt{2}}{2}} = \frac{2}{\sqrt{2}} = \sqrt{2}$$

$$\tan\frac{\pi}{4} = \frac{y}{x} = \frac{\frac{\sqrt{2}}{2}}{\frac{\sqrt{2}}{2}} = 1 \qquad \cot\frac{\pi}{4} = \frac{x}{y} = \frac{\frac{\sqrt{2}}{2}}{\frac{\sqrt{2}}{2}} = 1.$$

(b) Since $P\left(\frac{\pi}{3}\right) = \left(\frac{1}{2}, \frac{\sqrt{3}}{2}\right)$ (Figure 3), it follows that

$$\sin\frac{\pi}{3} = y = \frac{\sqrt{3}}{2} \qquad \csc\frac{\pi}{3} = \frac{1}{y} = \frac{1}{\frac{\sqrt{3}}{2}} = \frac{2}{\sqrt{3}} = \frac{2\sqrt{3}}{3}$$

$$\cos\frac{\pi}{3} = x = \frac{1}{2} \qquad \sec\frac{\pi}{3} = \frac{1}{x} = \frac{1}{\frac{1}{2}} = 2$$

$$\tan\frac{\pi}{3} = \frac{y}{x} = \frac{\frac{\sqrt{3}}{2}}{\frac{1}{2}} = \sqrt{3} \qquad \cot\frac{\pi}{3} = \frac{x}{y} = \frac{\frac{1}{2}}{\frac{\sqrt{3}}{2}} = \frac{1}{\sqrt{3}} = \frac{\sqrt{3}}{3}.$$

(c) Since $P\left(\frac{\pi}{6}\right) = \left(\frac{\sqrt{3}}{2}, \frac{1}{2}\right)$ (Figure 3), it follows that

$$\sin\frac{\pi}{6} = y = \frac{1}{2} \qquad \csc\frac{\pi}{6} = \frac{1}{y} = \frac{1}{1/2} = 2$$

$$\cos\frac{\pi}{6} = x = \frac{\sqrt{3}}{2} \qquad \sec\frac{\pi}{6} = \frac{1}{x} = \frac{1}{\sqrt{3}/2} = \frac{2}{\sqrt{3}} = \frac{2\sqrt{3}}{3}$$

$$\tan\frac{\pi}{6} = \frac{y}{x} = \frac{\frac{1}{2}}{\frac{\sqrt{3}}{2}} = \frac{\sqrt{3}}{3} \qquad \cot\frac{\pi}{6} = \frac{x}{y} = \frac{\frac{\sqrt{3}}{2}}{\frac{1}{2}} = \sqrt{3}.$$

Table 1 lists a summary of the trigonometric function values for $\frac{\pi}{6}, \frac{\pi}{4}$ and $\frac{\pi}{3}$.

TABLE 1 Trigonometric Values for Special Real Numbers

Real Number t	$\sin t$	$\cos t$	$\tan t$	$\csc t$	$\sec t$	$\cot t$
$\frac{\pi}{6}$	$\frac{1}{2}$	$\frac{\sqrt{3}}{2}$	$\frac{\sqrt{3}}{3}$	2	$\frac{2\sqrt{3}}{3}$	$\sqrt{3}$
$\frac{\pi}{4}$	$\frac{\sqrt{2}}{2}$	$\frac{\sqrt{2}}{2}$	1	$\sqrt{2}$	$\sqrt{2}$	1
$\frac{\pi}{3}$	$\frac{\sqrt{3}}{2}$	$\frac{1}{2}$	$\sqrt{3}$	$\frac{2\sqrt{3}}{3}$	2	$\frac{\sqrt{3}}{3}$

Using Symmetry to Evaluate Trigonometric Functions

Figure 4

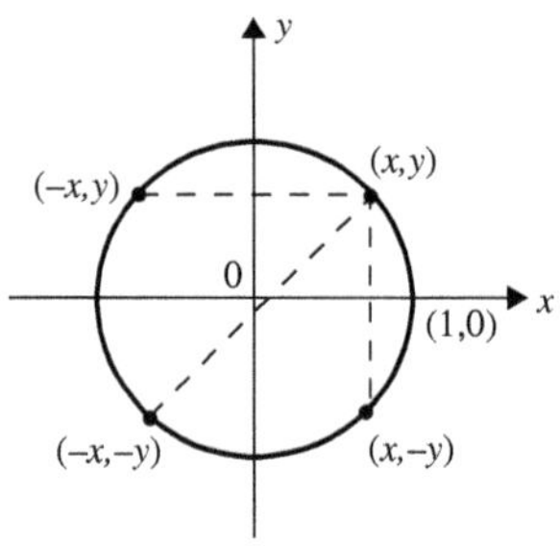

Using the symmetry of the unit circle and special values we can find values of the trigonometric functions for other values of t when $P(t)$ is in quadrants other than quadrant I. Because of the symmetry of the unit circle $x^2 + y^2 = 1$, it follows that the reflection of any point on the circle across the x axis, y axis, and the origin results in another point on the circle. Thus, if (x, y) is a point on the unit circle in quadrant I, then $(x, -y)$, $(-x, y)$, and $(-x, -y)$ are also on the unit circle (Figure 4).

We know that $P(\pi/6) = (\sqrt{3}/2, 1/2)$. It follows that reflections of this point across the x axis, the y axis, and the origin are given by the points $P(11\pi/6) = (\sqrt{3}/2, -1/2)$, $P(5\pi/6) = (-\sqrt{3}/2, 1/2)$, and $P(7\pi/6) = (-\sqrt{3}/2, -1/2)$, respectively (Figure 5). With this information, the values of the trigonometric functions for $t = 5\pi/6$, $7\pi/6$, and $11\pi/6$ can be obtained by using the definition given earlier. For instance,

Figure 5

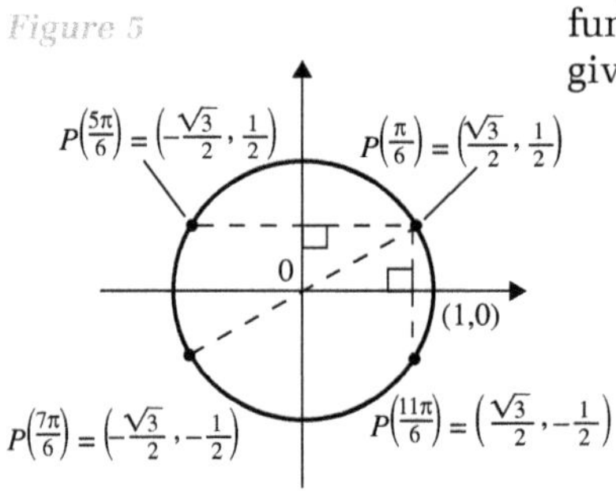

$$\sin\frac{5\pi}{6} = y = \frac{1}{2}, \cos\frac{7\pi}{6} = x = -\frac{\sqrt{3}}{2},$$

$$\tan\frac{11\pi}{6} = \frac{y}{x} = \frac{-1/2}{\sqrt{3}/2} = -\frac{1}{\sqrt{3}} = -\frac{\sqrt{3}}{3}.$$

The coordinates of the points on the unit circle corresponding to the values between 0 and 2π are listed in Appendix II.

EXAMPLE 4 **Evaluating Trigonometric Functions on a Unit Circle**

Use Figure 5 to find each value.

(a) $\sin\left(-\frac{\pi}{6}\right)$ (b) $\csc\left(-\frac{\pi}{6}\right)$

Solution We locate $P(-\pi/6)$ by moving clockwise from the point $(1, 0)$ a distance of $|-\pi/6|$. We see that $P(-\pi/6)$ is in quadrant IV, and by using symmetry, we note that it coincides with $P(11\pi/6)$ (Figure 5), so that

$$P\left(-\frac{\pi}{6}\right) = P\left(\frac{11\pi}{6}\right) = \left(\frac{\sqrt{3}}{2}, -\frac{1}{2}\right).$$

Using the definition, we have

(a) $\sin\left(-\frac{\pi}{6}\right) = y = -\frac{1}{2}$ (b) $\csc\left(-\frac{\pi}{6}\right) = \frac{1}{y} = \frac{1}{-\frac{1}{2}} = -2.$

We can use a similar argument to find the coordinates for $P(t)$ and the values of the trigonometric functions if $t = 3\pi/4, 5\pi/4$, and $7\pi/4$ (Figure 6a), or if $t = 2\pi/3$, $4\pi/3$, and $5\pi/3$ (Figure 6b).

Figure 6

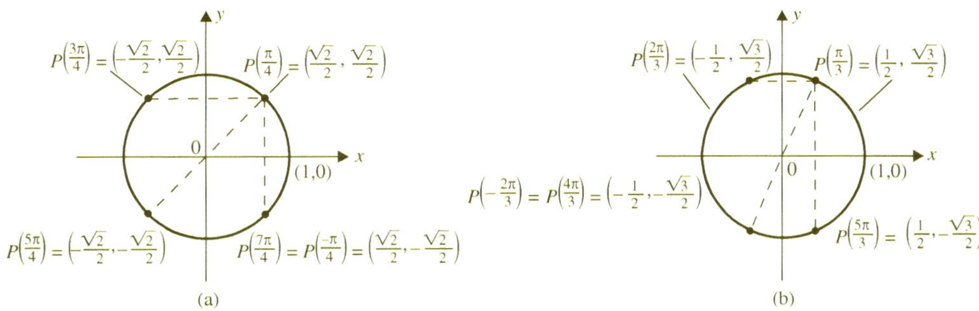

Thus,

$$\sin\left(\frac{3\pi}{4}\right) = y = \frac{\sqrt{2}}{2}$$

and

$$\cos\left(\frac{-\pi}{4}\right) = x = \frac{\sqrt{2}}{2}.$$

Also,

$$\tan\left(\frac{5\pi}{3}\right) = \frac{y}{x} = \frac{\left(-\frac{\sqrt{3}}{2}\right)}{\left(\frac{1}{2}\right)} = -\sqrt{3}$$

and

$$\cot\left(-\frac{2\pi}{3}\right) = \cot\left(\frac{4\pi}{3}\right) = \frac{x}{y} = \frac{\left(-\frac{1}{2}\right)}{\left(\frac{-\sqrt{3}}{2}\right)} = \frac{1}{\sqrt{3}} \text{ or } \frac{\sqrt{3}}{3}.$$

Approximating Trigonometric Function Values

We can approximate values of the trigonometric functions of a given real number by using a calculator. In order to perform such calculation for real numbers, the calculator must be set in *radian* mode.

Most calculators have keys for only the sine, cosine, and tangent functions. To evaluate the remaining three trigonometric functions on such calculators, we make use of the following reciprocal relationships.

Reciprocal Relationships

For a real number t, the following relationships hold:

(i) $\csc t = \dfrac{1}{\sin t}$ (ii) $\sec t = \dfrac{1}{\cos t}$ (iii) $\cot t = \dfrac{1}{\tan t}$

These relationships follow directly from the definitions of the trigonometric functions. By referring to Figure 7, we observe that

Figure 7

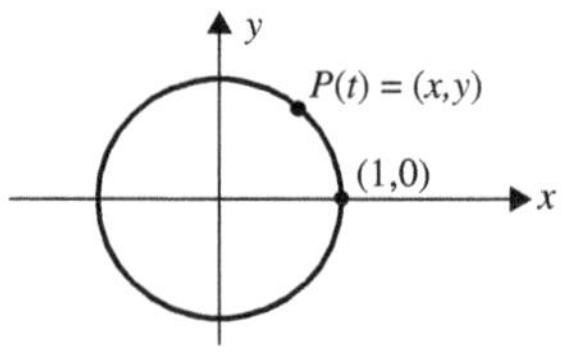

$$\csc t = \frac{1}{y} = \frac{1}{\sin t}$$

$$\sec t = \frac{1}{x} = \frac{1}{\cos t}$$

and

$$\cot t = \frac{x}{y} = \frac{1}{\left(\frac{x}{y}\right)} = \frac{1}{\tan t}.$$

The keys marked sin⁻¹, cos⁻¹, and tan⁻¹ on a calculator do not represent reciprocals. They are used to calculate inverse function values.

Thus the values of the cosecant, secant, and cotangent functions are found by using the reciprocal key 1/x or x⁻¹ along with the function keys for the sine, cosine, and tangent, respectively.

Before using a calculator to find function values of a real number, we must be sure that the calculator is in radian mode.

EXAMPLE 5

Using a Calculator to Approximate Values of Trigonometric Functions

Use a calculator to approximate each value to four decimal places.

(a) $\sin 1$ (b) $\cos 60$ (c) $\cot \frac{\pi}{8}$ (d) $\csc(-4.15)$

Solution First we set the calculator in radian mode. Then we proceed as follows.

(a) $\sin 1 = 0.8415$ (approx.)

(b) $\cos 60 = -0.9524$ (approx.)

(c) $\cot \frac{\pi}{8} = \frac{1}{\tan \frac{\pi}{8}} = 2.4142$ (approx.)

(d) $\csc(-4.15) = \frac{1}{\sin(-4.15)} = 1.1820$ (approx.)

From the definitions of the trigonometric functions we know that if the point $P(t) = (x, y)$, then $\cos t = x$ and $\sin t = y$. By substitution it follows that

$$\boxed{P(t) = (\cos t, \sin t).}$$

For instance,

$$P\left(\frac{\pi}{4}\right) = \left(\cos \frac{\pi}{4}, \sin \frac{\pi}{4}\right) = \left(\frac{\sqrt{2}}{2}, \frac{\sqrt{2}}{2}\right)$$

$$P\left(\frac{5\pi}{6}\right) = \left(\cos \frac{5\pi}{6}, \sin \frac{5\pi}{6}\right) = \left(\frac{-\sqrt{3}}{2}, \frac{1}{2}\right)$$

and

$$P(-0.725) = (\cos(-0.725), \sin(-0.725))$$
$$= (0.7485, -0.6631) \text{ (approx.)}.$$

Solving Applied Problems

EXAMPLE 6

Solving a Motion Problem

Figure 8a shows the shaft of a horizontal piston attached to the rim of a wheel at point A. Suppose that the radius of the wheel is 1 foot and assume that the shaft begins from a horizontal position. Determine how far the piston head moves when A moves along the rim of the wheel.

(a) 0.7 foot (b) 0.9 foot

Round off each answer to two decimal places.

Figure 8a

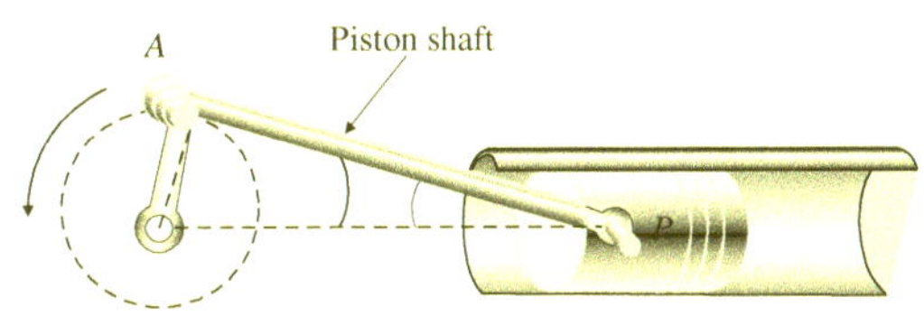

Solution Figure 8b shows a schematic diagram of the piston. Assume that A is a point on the unit circle that starts at (1, 0).

Figure 8b

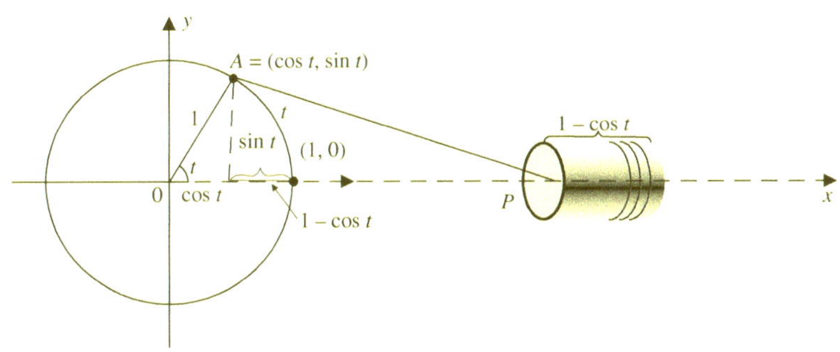

As the wheel rotates counterclockwise, the coordinates of A are given by

$$A = (\cos t, \sin t)$$

where t (in feet) is the length of the circular arc. When A has moved t feet, its x coordinate is $\cos t$, so the piston's head has moved $(1 - \cos t)$ feet.

Consequently,

(a) if $t = 0.7$ foot, then $1 - \cos t = 1 - \cos 0.7 = 0.24$ foot (approx.);

(b) if $t = 0.9$ foot, then $1 - \cos t = 1 - \cos 0.9 = 0.38$ foot (approx.).

PROBLEM SET 1.2

Mastering the Concepts

In problems 1–6, show that each point $P(t)$ lies on the unit circle, and then find the values of the trigonometric functions of t.

1. $P(t) = \left(\frac{-\sqrt{3}}{2}, \frac{1}{2}\right)$

2. $P(t) = \left(\frac{-4}{5}, \frac{-3}{5}\right)$

3. $P(t) = \left(\frac{5}{13}, -\frac{12}{13}\right)$

4. $P(t) = \left(\frac{1}{\sqrt{10}}, -\frac{3}{\sqrt{10}}\right)$

5. $P(t) = \left(\frac{-3}{\sqrt{13}}, \frac{2}{\sqrt{13}}\right)$

6. $P(t) = \left(\frac{-4}{\sqrt{17}}, \frac{-1}{\sqrt{17}}\right)$

In problems 7–10, determine the coordinates of $P(t)$. Then find the exact values of the trigonometric functions of t.

7. (a) $t = -\frac{\pi}{2}$ (b) $t = \frac{3\pi}{2}$

8. (a) $t = \pi$ (b) $t = \frac{5\pi}{2}$

9. (a) $t = -2\pi$ (b) $t = -\frac{5\pi}{2}$

10. (a) $t = \frac{-7\pi}{2}$ (b) $t = -\frac{3\pi}{2}$

In problems 11–22, use the coordinates of $P\left(\frac{\pi}{6}\right)$, $P\left(\frac{\pi}{4}\right)$, and $P\left(\frac{\pi}{3}\right)$ (Figure 3), along with symmetry, to find the exact value of each expression.

11. $\sin\frac{3\pi}{4}$

12. $\sin\left(-\frac{3\pi}{4}\right)$

13. $\cos\frac{4\pi}{3}$

14. $\cos\frac{7\pi}{4}$

15. $\sin\frac{11\pi}{6}$

16. $\sin\frac{5\pi}{3}$

17. $\cos\left(-\frac{5\pi}{6}\right)$

18. $\cos\left(-\frac{7\pi}{4}\right)$

19. $\sin\left(-\frac{\pi}{4}\right)$

20. $\sin\left(-\frac{7\pi}{4}\right)$

21. $\cos\frac{5\pi}{4}$

22. $\cos\left(-\frac{5\pi}{4}\right)$

In problems 23–34, use a calculator to approximate each value to four decimal places.

23. $\sin 3$

24. $\cos\sqrt{2}$

25. $\cos(-5)$

26. $\sin(-17)$

27. $\tan 1$

28. $\cot 4$

29. $\cot(-1)$

30. $\sec 23$

31. $\csc(5.81)$

32. $\csc\left(\frac{7}{\pi}\right)$

33. $\sec\left(\frac{\pi}{7}\right)$

34. $\tan(\cos(-3.5))$

In problems 35 and 36, use $P(t) = (x, y) = (\cos t, \sin t)$ to find the coordinates of each point. Round off the answers to four decimal places.

35. (a) $P(2.7)$
(b) $P(-4.18)$
(c) $P(-3.58)$
(d) $P(5.73)$

36. (a) $P(-0.432)$
(b) $P(5.129)$
(c) $P(0.572)$
(d) $P(-0.636)$

Applying the Concepts

37. **Piston Movement:** Figure 9 shows a typical piston engine, where θ is the angle between the crankshaft $\overline{OA}$ and the connecting rod $\overline{AP}$. It can be shown that the distance $d = |\overline{OP}|$ is given by

$$d = 4\sqrt{4 - 2\cos\theta}.$$

Find the distance d (in inches), when $\theta = \pi/6$, $\pi/5$, or $\pi/4$. Round off the answers to two decimal places.

Figure 9

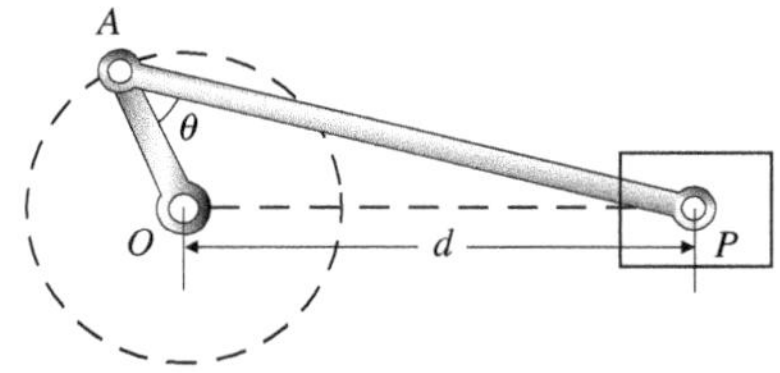

38. **Speed of Rolling Pulley:** If a pulley were spun in place, then a point P on the rim of the pulley would travel in a circular motion around its axle. Suppose that a pulley is moving at a constant speed v (in feet per second), then the forward speed s (in feet per second) of a fixed point on the pulley is given by the equation

$$s = v(1 + \cos\theta)$$

where θ is the angle shown in Figure 10. Find the forward speed s of a point P on the pulley moving at a rate of 45 feet per second, when $\theta = \pi/6$, $\pi/5$, or $\pi/4$. Round off the answers to two decimal places.

Figure 10

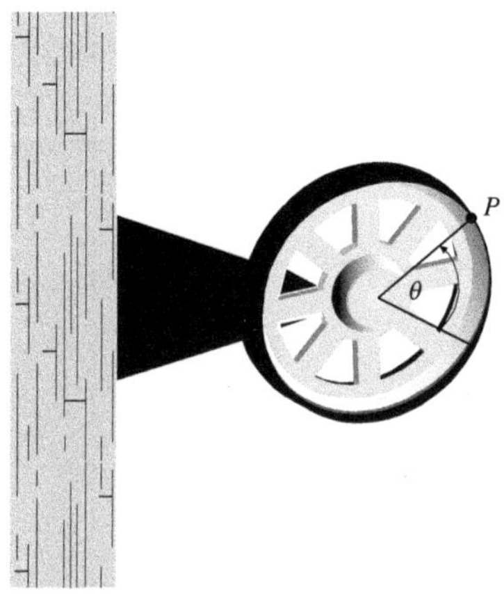

Developing and Extending the Concepts

In problems 39 and 40, find the values of the trigonometric functions of t under the given conditions.

39. $P(t)$ is in quadrant III and lies on the line $y = 3x$.

40. $P(t)$ is in quadrant IV and lies on the line $3x + 2y = 0$.

41. Explain why $(\cos t)^2$ and $\cos t^2$ are different. Give examples to support your argument.

42. The function

$$f(t) = \frac{\sin t}{t}$$

is used in calculus. For parts (a)–(d), find each value and round off each answer to four decimal places.

(a) $f(0.1)$ (b) $f(0.01)$

(c) $f(0.001)$ (d) $f(0.0001)$

(e) Discuss the behavior of f as $t \to 0^+$.

43. In calculus it is shown that $\sin t$ can be approximated by the polynomial function

$$g(t) = t - \frac{t^3}{6} + \frac{t^5}{120} \quad \text{where } -1 \le t \le 1.$$

Compare $\sin 0.1$ and $g(0.1)$, $\sin 0.2$ and $g(0.2)$, $\sin(-0.3)$ and $g(-0.3)$, and $\sin(-1)$ and $g(-1)$. Round off the values to four decimal places.

Interlude: Practice Problems 1.2

1. On the unit circle, sketch two angles in standard position, that are not coterminal, of measures θ and α such that $\sec\theta = \sec\alpha = \frac{5}{2}$. Label the points of intersection of the terminal sides of the angles and the unit circle with exact coordinates.

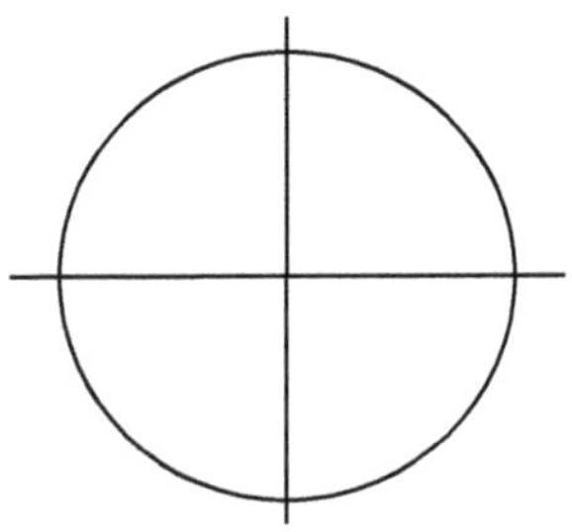

2. Suppose that $\frac{3\pi}{2} < t < 2\pi$ and $\cos t = q$. Express the values of the other five trigonometric functions at t in terms of q (fill in the blanks).

$\sin t =$ ______________________________

$\tan t =$ ______________________________

$\sec t =$ ______________________________

$\csc t =$ ______________________________

$\cot t =$ ______________________________

3. Suppose that $\csc t = \sqrt{7}$ and $-\pi < t < 0$. Calculate the exact values of the other five trigonometric functions at t (fill in the blanks).

____

____

____

$\sin t =$ ______________________

$\cos t =$ ______________________

$\tan t =$ ______________________

$\sec t =$ ______________________

$\cot t =$ ______________________

4. Suppose that $\cos(-t) = 0.1$ and $0 < t < \frac{\pi}{2}$. Calculate the values of the other five trigonometric functions at t (fill in the blanks).

____

____

____

$\sin t =$ ______________________

$\tan t =$ ______________________

$\csc t =$ ______________________

$\sec t =$ ______________________

$\cot t =$ ______________________

Objectives

1. Determine the Signs of Trigonometric Function Values
2. Establish the Quotient Identities
3. Establish the Even-Odd and Pythagorean Identities
4. Establish the Periodicity of Trigonometric Functions
5. Solve Applied Problems

1.3 Properties of the Trigonometric Functions

We are able to derive several properties of the trigonometric functions by examining their definitions.

Determining the Signs of Trigonometric Function Values

The algebraic sign of a trigonometric function value depends on the quadrant in which $P(t)$ lies.

For instance, as Figure 1 shows, $\sin t$ is positive for $P(t)$ in quadrants I and II, and $\sin t$ is negative in quadrants III and IV. Similarly, we can determine the algebraic signs of the other five trigonometric functions depending on the quadrant where $P(t)$ lies. The results are listed in Table 1.

Figure 1

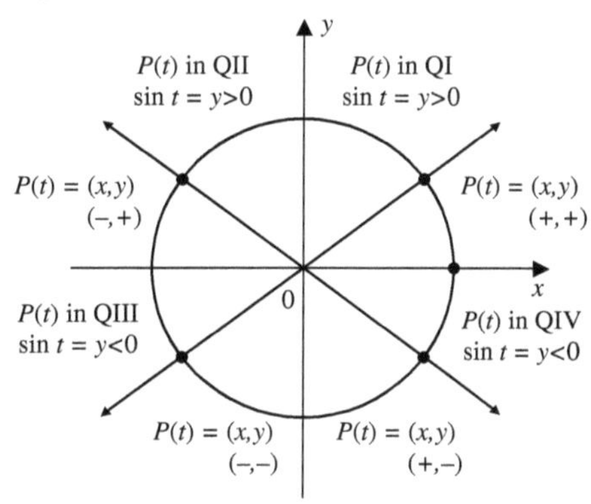

TABLE 1 Signs of the Values of the Trigonometric Functions of t

Quadrant:	Positive Values	Negative Values
I	All	None
II	$\sin t$, $\csc t$	$\cos t$, $\sec t$, $\tan t$, $\cot t$
III	$\tan t$, $\cot t$	$\sin t$, $\cos t$, $\csc t$, $\sec t$
IV	$\cos t$, $\sec t$	$\sin t$, $\csc t$, $\tan t$, $\cot t$

EXAMPLE 1 **Finding the Quadrant Location of $P(t)$**

Find the quadrant in which $P(t)$ lies if

$$\cot t > 0 \quad \text{and} \quad \cos t < 0.$$

Solution Using Table 1, we find that $\cot t > 0$ if $P(t)$ is in quadrant I or III, and $\cos t < 0$ if $P(t)$ is in quadrant II or III. For both conditions to be satisfied, $P(t)$ must be in quadrant III.

Establishing the Quotient Identities

An **identity** is an equation that is true for all values of the variable for which both sides of the equation are defined. In Section 1.2 we established the reciprocal identities. Upon examining Figure 2, we discover that

Figure 2

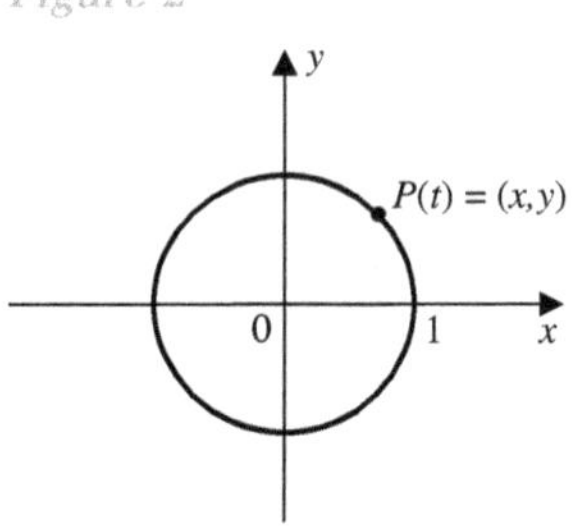

$$\tan t = \frac{y}{x} = \frac{\sin t}{\cos t}$$

and

$$\cot t = \frac{x}{y} = \frac{\cos t}{\sin t}.$$

Thus we are led to the following identities.

Quotient Identities

$$\tan t = \frac{\sin t}{\cos t} \quad \text{and} \quad \cot t = \frac{\cos t}{\sin t}$$

EXAMPLE 2 **Using the Quotient Identities**

Suppose $\sin t = \frac{3}{5}$ and $\cos t = \frac{-4}{5}$.
Find $\tan t$ and $\cot t$.

Solution By the quotient identities, it follows that

$$\tan t = \frac{\sin t}{\cos t} = \frac{(3/5)}{(-4/5)} = -\frac{3}{4}$$

and

$$\cot t = \frac{\cos t}{\sin t} = \frac{(-4/5)}{(3/5)} = -\frac{4}{3}.$$

Establishing Even-Odd and Pythagorean Identities

By using the definitions of the trigonometric functions we can derive additional identities classified as the *even-odd* identities and the *Pythagorean* identities.

Suppose that

$$P(t) = (x, y).$$

Figure 3

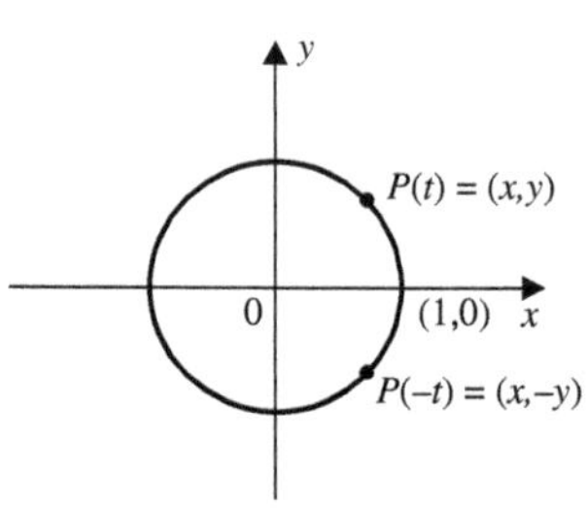

By symmetry, $P(-t) = (x, -y)$ (Figure 3).

Using the definitions of the trigonometric functions, we have

$$\cos t = x \quad \text{and} \quad \cos(-t) = x.$$

Also,

$$\sin t = y \quad \text{and} \quad \sin(-t) = -y$$

so that

$$\sin(-t) = -y = -\sin t.$$

It follows that

$$\cos(-t) = \cos t \quad \text{and} \quad \sin(-t) = -\sin t.$$

From the quotient identity, we get

$$\tan(-t) = \frac{\sin(-t)}{\cos(-t)} = \frac{-\sin t}{\cos t} = -\tan t.$$

So $\tan(-t) = -\tan t$.

Similar arguments can be used to establish identities for the cosecant, secant, and cotangent functions. Thus we are led to the following formulas, which are called the *even-odd* identities.

Even-Odd Identities

Function	Identity	Even or Odd
sine	$\sin(-t) = -\sin t$	odd
cosine	$\cos(-t) = \cos t$	even
tangent	$\tan(-t) = -\tan t$	odd
cosecant	$\csc(-t) = -\csc t$	odd
secant	$\sec(-t) = \sec t$	even
cotangent	$\cot(-t) = -\cot t$	odd

EXAMPLE 3 **Using the Even-Odd Identities**

Suppose that $\sin t = -3/5$ and $\cos t = 4/5$. Find

(a) $\tan(-t)$ (b) $\sec(-t)$ (c) $\csc(-t)$

Solution

(a) $$\tan(-t) = -\tan t = -\frac{\sin t}{\cos t} = -\frac{\left(-\frac{3}{5}\right)}{\left(\frac{4}{5}\right)} = -\left(\frac{-3}{4}\right) = \frac{3}{4}$$

(b) $$\sec(-t) = \sec t = \frac{1}{\cos t} = \frac{1}{\left(\frac{4}{5}\right)} = \frac{5}{4}$$

(c) $$\csc(-t) = -\csc t = -\frac{1}{\sin t} = -\frac{1}{\left(-\frac{3}{5}\right)} = -\left(\frac{5}{-3}\right) = \frac{5}{3}$$

Figure 4

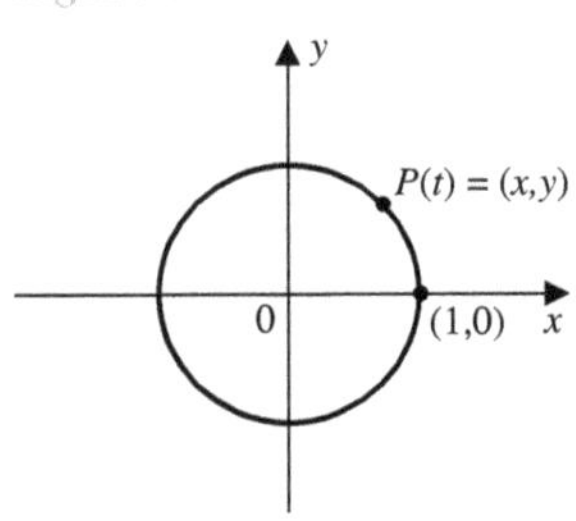

To derive the Pythagorean identities, we begin by considering any $P(t) = (x, y)$ on the unit circle as illustrated in Figure 4.

By definition, we know that

$$\cos t = x \quad \text{and} \quad \sin t = y.$$

After adding the squares of these two equations to each other, we get

$$(\cos t)^2 + (\sin t)^2 = x^2 + y^2 = 1$$

because (x, y) is on the unit circle.

So we get the identity

$$(\cos t)^2 + (\sin t)^2 = 1.$$

It is customary to write powers of trigonometric expressions such as $(\sin t)^n$ and $(\cos t)^n$, respectively, as

$$(\sin t)^n = \sin^n t \quad \text{and} \quad (\cos t)^n = \cos^n t$$

except when $n = -1$.

So we write

$$\cos^2 t = (\cos t)^2 \quad \text{and} \quad \sin^2 t = (\sin t)^2.$$

Thus the identity

$$(\cos t)^2 + (\sin t)^2 = 1 \quad \text{is written as}$$
$$\cos^2 t + \sin^2 t = 1.$$

We use this latter identity to prove the other two Pythagorean identities. If we divide both sides of the identity by $\cos^2 t$ we get

$$\frac{\cos^2 t}{\cos^2 t} + \frac{\sin^2 t}{\cos^2 t} = \frac{1}{\cos^2 t}$$
$$1 + \left(\frac{\sin t}{\cos t}\right)^2 = \left(\frac{1}{\cos t}\right)^2$$
$$1 + \tan^2 t = \sec^2 t.$$

Similarly, if we divide both sides of the identity $\cos^2 t + \sin^2 t = 1$ by $\sin^2 t$ and then simplify the result, we get

$$\cot^2 t + 1 = \csc^2 t.$$

Thus we are led to the following results:

Pythagorean Identities

(i) $\cos^2 t + \sin^2 t = 1$	(ii) $1 + \tan^2 t = \sec^2 t$	(iii) $1 + \cot^2 t = \csc^2 t$

The identities can be used to compute values of the trigonometric functions as illustrated in the following examples.

EXAMPLE 4

Evaluating Functions by Using Identities

Use identities to find the exact values of the other five trigonometric functions of t if

$$\sin t = \frac{8}{17} \quad \text{and} \quad \cos t < 0.$$

Solution Since $\sin t$ is positive and $\cos t$ is negative, it follows that $P(t)$ is in quadrant II.

Using the Pythagorean identity $\cos^2 t + \sin^2 t = 1$, we have

$$\cos^2 t = 1 - \sin^2 t.$$

Thus

$$\cos t = \pm\sqrt{1 - \sin^2 t}.$$

However, we are given that $\cos t$ is negative, so

$$\cos t = -\sqrt{1 - \left(\frac{8}{17}\right)^2} = -\sqrt{1 - \frac{64}{289}} = -\sqrt{\frac{225}{289}} = -\frac{15}{17}.$$

Next we use the quotient and reciprocal identities to get

$$\tan t = \frac{\sin t}{\cos t} = \frac{\left(\frac{8}{17}\right)}{\left(-\frac{15}{17}\right)} = \frac{-8}{15} \qquad \cot t = \frac{\cos t}{\sin t} = \frac{\left(-\frac{15}{17}\right)}{\left(\frac{8}{17}\right)} = \frac{-15}{8}$$

$$\sec t = \frac{1}{\cos t} = \frac{1}{\left(-\frac{15}{17}\right)} = \frac{-17}{15} \qquad \csc t = \frac{1}{\sin t} = \frac{1}{\left(\frac{8}{17}\right)} = \frac{17}{8}.$$

Establishing the Periodicity of Trigonometric Functions

Because the unit circle has circumference 2π, we notice that if we add any integer multiple of 2π to t or subtract any integer multiple of 2π from t, the resulting point is at the same location as $P(t) = (x, y)$, as illustrated in Figure 5, that is,

Figure 5

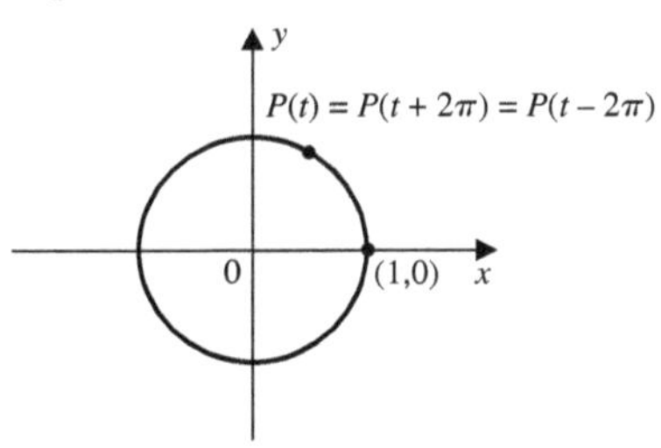

$$P(t + k \cdot 2\pi) = P(t),\ k \text{ is any integer.}$$

Correspondingly,

$$\sin t = y \quad \text{and} \quad \sin(t + k \cdot 2\pi) = y,\ k \text{ is any integer.}$$

so that

$$\boxed{\sin(t + k \cdot 2\pi) = \sin t, \text{ for any integer } k.}$$

In other words, the trigonometric function values of the sine function repeat every 2π units.

EXAMPLE 5 **Using Repetitive Behavior of Trigonometric Functions**

Use the repetitive behavior of the sine function to find the *exact* value of

$$\sin \frac{13\pi}{6}.$$

Solution Since

$$\frac{13\pi}{6} = \frac{\pi}{6} + \frac{12\pi}{6} = \frac{\pi}{6} + 2\pi \text{ (Figure 6)},$$

Figure 6

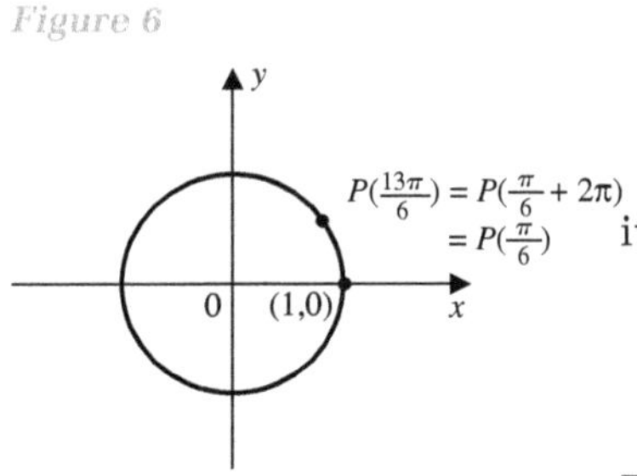

it follows that

$$\sin \frac{13\pi}{6} = \sin\left(\frac{\pi}{6} + 2\pi\right) = \sin \frac{\pi}{6} = \frac{1}{2}.$$

Functions with this kind of repetitive behavior are called *periodic functions*.

In general, we have the following results.

Property
Periodicity of Trigonometric Functions

$$\sin t = \sin(t + k \cdot 2\pi),$$
$$\cos t = \cos(t + k \cdot 2\pi),$$
$$\tan t = \tan(t + k \cdot 2\pi),$$
$$\cot t = \cot(t + k \cdot 2\pi),$$
$$\sec t = \sec(t + k \cdot 2\pi),$$
$$\text{and} \quad \csc t = \csc(t + k \cdot 2\pi),$$
where k is any integer.

It can be shown that 2π is the *smallest positive period,* called the **fundamental period,** for the sine, cosine, cosecant, and secant functions.

However the tangent and cotangent functions have a fundamental period of π. To understand why this is the case, let us suppose that t is a given real number and

$$P(t) = (x, y).$$

Figure 7

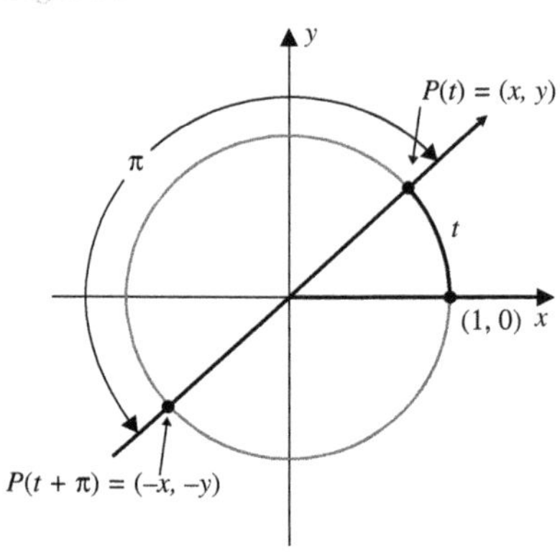

By symmetry,

$$P(t + \pi) = (-x, -y) \text{ (Figure 7).}$$

From the definition of the tangent function, we have

$$\tan(t + \pi) = \frac{-y}{-x} = \frac{y}{x} \quad \text{and} \quad \tan t = \frac{y}{x}.$$

Thus

$$\tan(t + \pi) = \tan t.$$

It follows that

$$\boxed{\tan(t + k\pi) = \tan t \quad \text{where } k \text{ is any integer.}}$$

So the tangent function has a period of π, that is, the tangent function values repeat every π units.

For example,

$$\tan\frac{10\pi}{3} = \tan\left(\frac{\pi}{3} + \frac{9\pi}{3}\right) = \tan\left(\frac{\pi}{3} + 3\pi\right) = \tan\frac{\pi}{3} = \sqrt{3}.$$

Similarly, it can be shown that the cotangent also has a period of π, that is,

$$\boxed{\cot(t + k\pi) = \cot t, \text{ where } k \text{ is any integer.}}$$

Solving Applied Problems

In calculus, algebraic expressions of the form

$$\sqrt{a^2 - u^2},\ \sqrt{u^2 + a^2},\quad \text{and}\quad \sqrt{u^2 - a^2},$$

where a is a positive constant, can be converted to more useful forms by making appropriate substitutions involving trigonometric functions. This technique is called **trigonometric substitution.** The idea is to rewrite radical expressions as trigonometric expressions containing no radical, as illustrated in the next example.

EXAMPLE 6 **Converting a Radical to a Trigonometric Form**

Use the substitution $u = 5 \sin t$, where t is a real number such that $P(t)$ is in quadrant I, to rewrite

$$\sqrt{25 - u^2}$$

as a trigonometric expression containing no radical.

Solution After rewriting the Pythagorean identity

$$\cos^2 t + \sin^2 t = 1$$

as

$$\cos^2 t = 1 - \sin^2 t,$$

we proceed as follows:

$$\begin{aligned}
\sqrt{25 - u^2} &= \sqrt{25 - (5 \sin t)^2} && \text{Substitute } u = 5 \sin t \\
&= \sqrt{25 - 25 \sin^2 t} && \text{Multiply} \\
&= \sqrt{25(1 - \sin^2 t)} && \text{Factor} \\
&= 5\sqrt{1 - \sin^2 t} && \text{Simplify} \\
&= 5\sqrt{\cos^2 t} && \text{Rewrite of Identity} \\
&= 5 \cos t && \cos t > 0 \text{ since } P(t) \text{ is in quadrant I}
\end{aligned}$$

PROBLEM SET 1.3

Mastering the Concepts

In problems 1–8, use the reciprocal and quotient relationships to find the other four trigonometric function values of t under the given conditions.

1. $\cos t = \frac{12}{13}$ and $\sin t = \frac{5}{13}$
2. $\cos t = \frac{2}{5}$ and $\sin t = \frac{\sqrt{21}}{5}$
3. $\tan t = \frac{12}{5}$ and $\cos t = \frac{-5}{13}$
4. $\cot t = \frac{-2}{\sqrt{21}}$ and $\cos t = \frac{2}{5}$
5. $\sin t = \cos t = \frac{-1}{\sqrt{2}}$
6. $\tan t = 1$ and $\sin t = \frac{-1}{\sqrt{2}}$
7. $\sin t = 0$ and $\cos t = 1$
8. $\cos t = 0$ and $\sin t = -1$

In problems 9–12, indicate the quadrant in which $P(t)$ lies.

9. (a) $\cos t > 0$ and $\sin t < 0$
(b) $\sec t > 0$ and $\cot t < 0$

10. (a) $\sin t > 0$ and $\cos t < 0$
(b) $\sin t > 0$ and $\sec t < 0$

11. (a) $\tan t < 0$ and $\cos t < 0$
(b) $\tan t < 0$ and $\cos t > 0$

12. (a) $\cos t < 0$ and $\cot t < 0$
(b) $\sec t < 0$ and $\cot t > 0$

In problems 13–16, suppose that $\sin t = -12/13$ and $\cos t = -5/13$. Find each value.

13. $\csc(-t)$ **14.** $\tan(-t)$
15. $\cot(-t)$ **16.** $\sec(-t)$

In problems 17 and 18, verify that the equation is true for the given value of t by rounding off the values to four decimal places.

17. (a) $f(-t) = -f(t)$, where $f(t) = \sin t$ and $t = 3.752$
(b) $f(-t) = f(t)$, where $f(t) = \cos t$ and $t = 5.384$

18. (a) $f(-t) = -f(t)$, where $f(t) = \tan t$ and $t = 4.138$
(b) $f(-t) = f(t)$, where $f(t) = \sec t$ and $t = 2.874$

In problems 19–26, use identities to find the exact values of the other five trigonometric functions.

19. $\sin t = \frac{15}{17}$; t is in quadrant I

20. $\cos t = \frac{5}{13}$; $\tan t > 0$

21. $\cos t = \frac{-4}{5}$; t is in quadrant II

22. $\sin t = \frac{-7}{25}$; $\cot t < 0$

23. $\csc t = \frac{17}{8}$; t is in quadrant II

24. $\tan t = -3$; $\cos t > 0$

25. $\sec t = -\frac{5}{3}$; $\sin t < 0$

26. $\cot t = \frac{3}{4}$; $\sin t < 0$

In problems 27–30, use identities to find the approximate value of the other five trigonometric functions. Round off each answer to two decimal places.

27. $\sin t = 0.57$; $\tan t < 0$

28. $\cos t = -0.83$; $\cot t > 0$

29. $\tan t = -1.53$; $\sin t > 0$

30. $\cot t = 3.41$; $\csc t > 0$

In problems 31–40, use periodicity to find the exact values.

31. $\cos\left(\frac{5\pi}{6} + 2\pi\right)$ **32.** $\sin\left(\frac{\pi}{4} - 2\pi\right)$

33. $\sin\left(\frac{2\pi}{3} + 4\pi\right)$ **34.** $\cos\left(\frac{11\pi}{6} - 6\pi\right)$

35. $\cot\left(\frac{41\pi}{4}\right)$ **36.** $\tan 7\pi$

37. $\csc\left(\frac{-43\pi}{4}\right)$ **38.** $\sec\left(2\pi - \frac{\pi}{6}\right)$

39. $\tan\left(\frac{-65\pi}{6}\right)$ **40.** $\cot\left(\frac{-15\pi}{2}\right)$

Applying the Concepts

In problems 41–46, rewrite each algebraic expression as a trigonometric expression containing no radical by using the given trigonometric substitution. Assume that t is a real number such that $P(t)$ is in quadrant I.

41. $\sqrt{4 - u^2}$; $u = 2 \sin t$

42. $\sqrt{u^2 - 4}$; $u = 2 \sec t$

43. $\sqrt{u^2 + 9}$; $u = 3 \tan t$

44. $\sqrt{9u^2 + 4}$; $u = \frac{2}{3} \tan t$

45. $\sqrt{4 - 9u^2}$; $u = \frac{2}{3} \sin t$

46. $\sqrt{u^2 - 81}$; $u = 9 \sec t$

Developing and Extending the Concepts

47. Show that the following even-odd identities are true:
(a) $\cot(-t) = -\cot t$
(b) $\sec(-t) = \sec t$
(c) $\csc(-t) = -\csc t$

48. If $f(t) = \cos t$, where t is a real number, explain why $f(t) = f(t + 2\pi) = f(t - 2\pi)$.

49. Use the Pythagorean identity

$$\cos^2 t + \sin^2 t = 1$$

to prove that

$$1 + \cot^2 t = \csc^2 t.$$

50. Explain why it is *impossible* for $\sin t > 1$ and $|\csc t| < 1$ for any real number t.

Interlude: Practice Problems 1.3

1. Suppose that $0 < t < \frac{\pi}{2}$ and $\cos t = q$. Evaluate the following in terms of q (fill in the blanks).

(a) $\cos(2\pi - t) =$ ____________________

(b) $\sec(-t) =$ ____________________

(c) $\tan(2\pi - t) =$ ____________________

(d) $\csc(t + \pi) =$ ____________________

(e) $\cot(t - \pi) =$ ____________________

(f) $\cos(3\pi - t) =$ ____________________

2. Assume that $\sin\left(\frac{\pi}{10}\right) = \frac{\sqrt{5}-1}{4}$. Calculate the exact values for the following (fill in the blanks).

____

____

____

(a) $\cos\left(\frac{\pi}{10}\right) =$ ____________________

(b) $\sin\left(\frac{31\pi}{10}\right) =$ ____________________

(c) $\tan\left(\frac{19\pi}{10}\right) =$ ____________________

(d) $\sec\left(-\frac{49\pi}{10}\right) =$ ____________________

(e) $\cot\left(-\frac{11\pi}{10}\right) =$ ____________________

Objectives

1. Determine the Domain and Range of the Sine and Cosine Functions
2. Graph the Sine and Cosine Functions
3. Transformations of the Sine and Cosine Graphs
4. Solve Applied Problems
5. Model Simple Harmonic Motion

1.4 Graphs of Sine and Cosine Functions

Now that we have established the trigonometric functions for real numbers, we turn our attention to graphing them. We begin with the sine and cosine functions. They are used extensively in applied mathematics.

Determining the Domain and Range of the Sine and Cosine Functions

By analyzing the definitions of the sine and cosine functions we can determine their domains and ranges. Later in Section 1.5 we will do the same for the other four trigonometric functions.

Assume that $P(t) = (x, y)$ is a point on the unit circle (Figure 1), then

Figure 1

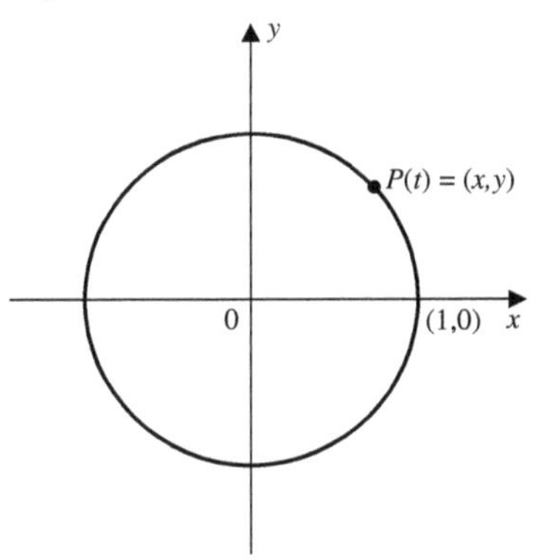

$$x = \cos t \quad \text{and} \quad y = \sin t.$$

Since (x, y) is on the unit circle, it follows that *no matter what value* of t is chosen we have

$$-1 \le x \le 1 \quad \text{and} \quad -1 \le y \le 1.$$

Consequently,

$$-1 \le \cos t \le 1 \quad \text{and} \quad -1 \le \sin t \le 1.$$

Thus we have the following result.

Domain and Range
Sine and Cosine

For $f(t) = \sin t$ and $g(t) = \cos t$, the domain consists of all real numbers, and the range of each function is the interval $[-1, 1]$.

EXAMPLE 1 **Finding the Range of a Sine Function**

Use $-1 \le \sin t \le 1$ to find the range of $f(t) = 2 \sin t + 5$.

Solution To find the range of f we proceed as follows:

$-1 \le \sin t \le 1$	Sine has range $[-1, 1]$
$-2 \le 2 \sin t \le 2$	Multiply by 2
$3 \le 2 \sin t + 5 \le 7$	Add 5

Thus $3 \le f(t) \le 7$, that is, the range of f is the interval $[3, 7]$.

Graphing the Sine and Cosine Functions

1. The Sine Graph

To graph the sine function, we recall two important facts:

(i) If t is a real number and $P(t) = (x, y)$ is the associated point on the unit circle, then $\sin t = y$ (Figure 2a).

(ii) As $P(t)$ moves one complete revolution counterclockwise around the unit circle, starting at $P(0) = (1, 0)$ and ending at $P(2\pi) = (1, 0)$, correspondingly, t increases from 0 to 2π along the horizontal axis in a Cartesian system (Figure 2b).

Figure 2

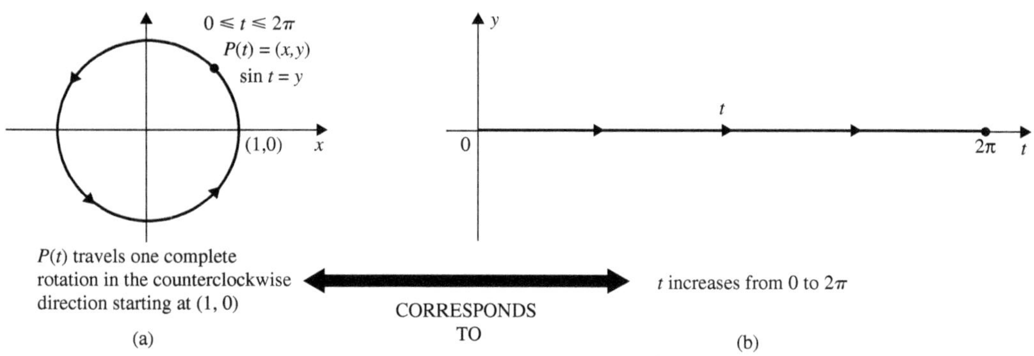

Figure 3a shows the locations of several specific points for $P(t) = (x, y)$. Since $y = \sin t$, we can obtain a rough sketch of the graph of $f(t) = \sin t$ by plotting the points (t, y) in a Cartesian system and then connecting the points with a smooth graph (Figure 3b). The graph that results when t increases from 0 to 2π, corresponding to $P(t)$ moving one counterclockwise rotation, is called **one cycle** of the sine function.

Figure 3

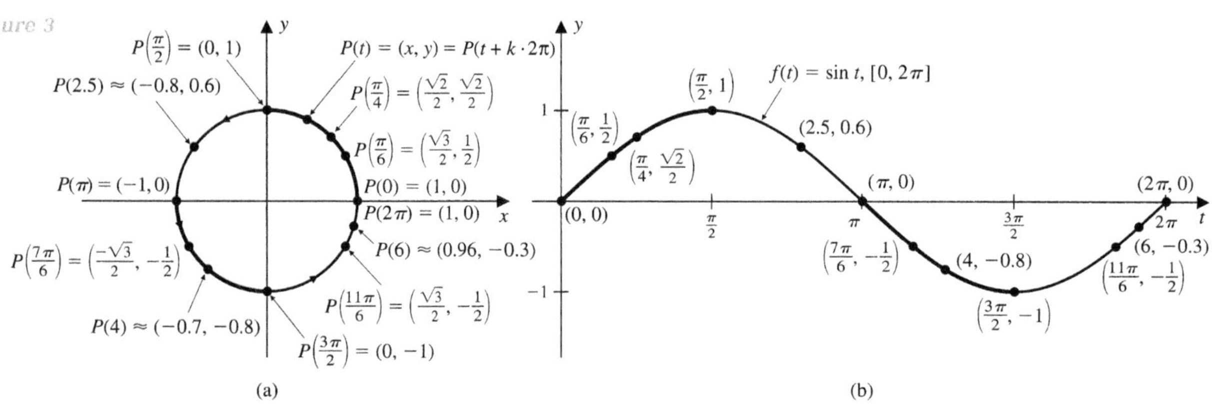

The graph in Figure 3b reveals *five key points* for the sine cycle, that is, locations of maximum and minimum points and of x intercepts. There is a maximum point at $(\pi/2, 1)$, a minimum at $(3\pi/2, -1)$, and x intercepts at $(0, 0)$, $(\pi, 0)$ and $(2\pi, 0)$.

When graphing trigonometric functions in the Cartesian plane, it is customary to use x as the independent variable instead of t, and y as the dependent variable.

Because the sine has period 2π, each time t increases or decreases by 2π units, $P(t)$ traverses the unit circle again, and correspondingly, other sine cycles are generated. So we obtain a complete graph of $y = \sin x$ by repeating the cycle in Figure 3b to the right and to the left every 2π units (Figure 4).

Upon reading the sine graph we are led to the properties listed in Table 1. Notice that the graph confirms properties already developed earlier for the sine function.

Figure 4

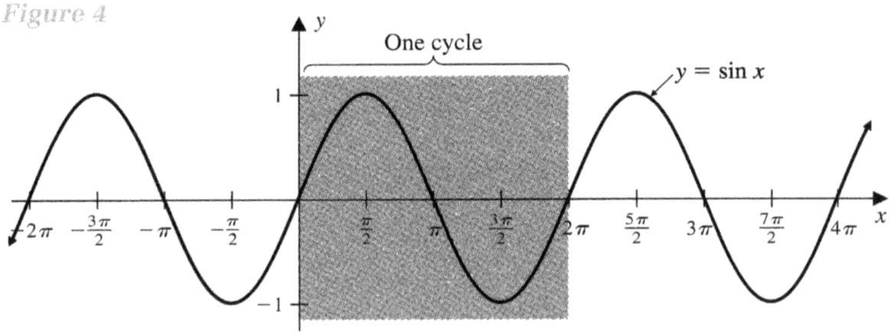

TABLE 1 Properties of the Sine Function

Property	$y = \sin x$
Domain	All real numbers
Range	Interval $[-1, 1]$; that is; $-1 \le \sin x \le 1$
Period	2π
Symmetry	Symmetric with respect to the origin
Even or odd	The function is *odd,* that is, $\sin(-x) = -\sin x$

2. The Cosine Graph

As with the sine function, we sketch one cycle of the graph of the cosine function $y = \cos x$ in the interval $[0, 2\pi]$ by first plotting some points (Table 2) and then connecting them with a smooth graph (Figure 5). The five key points are listed in Table 3.

TABLE 2

x	$\cos x$
0	1
$\frac{\pi}{3}$	$\frac{1}{2}$
$\frac{\pi}{2}$	0
π	-1
$\frac{4\pi}{3}$	$-\frac{1}{2}$
$\frac{3\pi}{2}$	0
2π	1

Figure 5

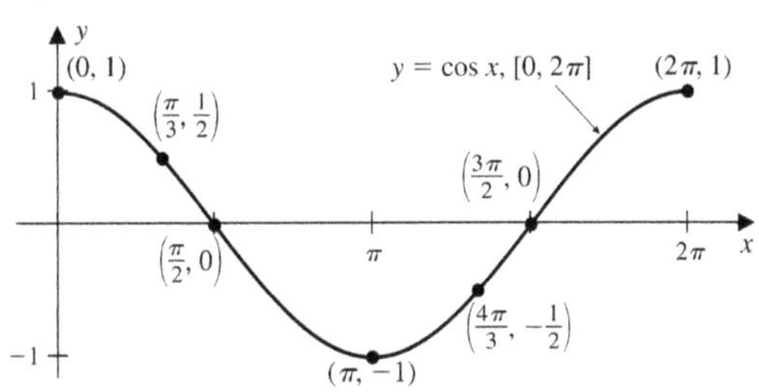

TABLE 3 Key Points for $y = \cos x$, $[0, 2\pi]$

Maximums	$(0, 1)$ and $(2\pi, 1)$
Minimum	$(\pi, -1)$
x-intercepts	$\left(\frac{\pi}{2}, 0\right)$ and $\left(\frac{3\pi}{2}, 0\right)$

Because the cosine has period 2π, we repeat the cycle in Figure 5 to the right and to the left every 2π units, to get the complete graph of $y = \cos x$ (Figure 6). From the graph, we observe the properties of the cosine function listed in Table 4. Once again, the graph confirms properties developed earlier for the cosine function.

Figure 6

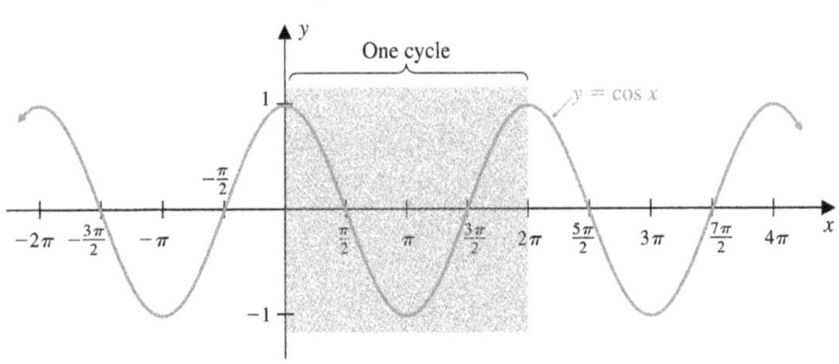

TABLE 4 Properties of the Cosine Function

Property	$y = \cos x$
Domain	All real numbers
Range	Interval $[-1, 1]$; that is, $-1 \le \cos x \le 1$
Period	2π
Symmetry	Symmetric with respect to the y axis
Even or odd	The function is *even,* that is, $\cos(-x) = \cos x$

Graphs of one cycle of $y = \sin x$ (Figure 3b) and of $y = \cos x$ (Figure 5) including their key points should be learned.

Transformations of the Sine and Cosine Graphs

Our objective here is to use transformations of the graphs of $y = \sin x$ and $y = \cos x$ to obtain graphs of functions of the more general forms.

$$F(x) = a \sin(kx + b) + c \quad \text{and} \quad G(x) = a \cos(kx + b) + c$$

where a, k, b and c are constants.

1. Amplitude: Vertical Scaling and Vertical Shifting

The next example illustrates the applications of vertical scaling and vertical shifting to the sine and cosine functions.

EXAMPLE 2 **Using Vertical Scaling and Vertical Shifting to Graph**

Use transformations of the sine and cosine graphs to sketch the graph of each function. Also determine the range and period.

(a) $F(x) = 5 \sin x$ (b) $G(x) = \frac{1}{5} \cos x + 1$

Solution

(a) First we sketch the graph of $y = \sin x$; then we multiply each ordinate by 5. Geometrically, this means that the graph of $y = \sin x$ is vertically stretched by a factor of 5 to obtain the graph of $F(x) = 5 \sin x$ (Figure 7a). From the graph, we see that the range of F is the interval $[-5, 5]$ and the period is 2π.

(b) We start by graphing $y = \cos x$.

Next the graph of $g(x) = (1/5) \cos x$ is obtained by vertically shrinking the graph of $y = \cos x$ by a factor of $(1/5)$ (Figure 7b). Finally, we get the graph of

$$G(x) = \left(\frac{1}{5}\right)\cos x + 1$$

by shifting the graph of g up 1 unit (Figure 7b).

The range of G is the interval

$$\left[-\frac{1}{5} + 1, \frac{1}{5} + 1\right] \quad \text{or} \quad \left[\frac{4}{5}, \frac{6}{5}\right],$$

and the period is 2π.

Figure 7

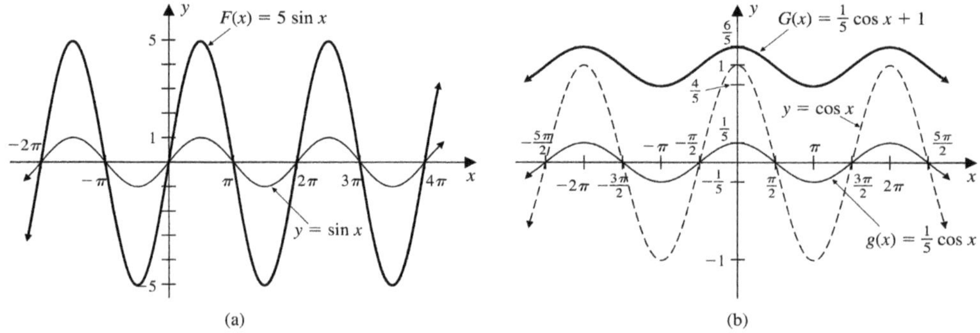

Example 2 leads us to the following generalizations.

Property

Amplitude—Vertical Scaling and Vertical Shifting

(i) The graph of

$$f(x) = a \sin x \quad \text{or} \quad g(x) = a \cos x, \quad \text{where } a \neq 0$$

is obtained from the graph of

$$y = \sin x \quad \text{or} \quad y = \cos x,$$

respectively, by *stretching* or *compressing vertically* by a factor of $|a|$, depending on whether

$$|a| > 1 \quad \text{or} \quad |a| < 1,$$

respectively. If $a < 0$ we also *reflect* the resulting graph across the x axis.

(ii) The graph of

$$F(x) = a \sin x + c \quad \text{or} \quad G(x) = a \cos x + c$$

is obtained, respectively, by *vertically shifting* the graph of

$$f(x) = a \sin x \quad \text{or} \quad g(x) = a \cos x$$

$|c|$ units,

$$\text{upward if } c > 0$$

and

$$\text{downward if } c < 0.$$

In either case, the value of $|a|$ is called the **amplitude** of the function.

For example, the amplitudes of

$$F(x) = 5 \sin x,\ G(x) = \frac{1}{5} \cos x + 1, \quad \text{and} \quad H(x) = -3 \cos x + 5$$

are, respectively, $|5| = 5$, $\left|\frac{1}{5}\right| = \frac{1}{5}$, and $|-3| = 3$.

2. Period: Horizontal Scaling

Next we consider the graphs of functions of the forms

$$f(x) = \sin kx \text{ and } g(x) = \cos kx, \text{ where } k \text{ is a constant.}$$

To better understand the effect that k has on the graphs of $y = \sin x$ and $y = \cos x$, let us consider the following example in which we compare the graphs of $f(x) = \sin 2x$ and $y = \sin x$.

EXAMPLE 3

Comparing Periods

Sketch the graph of the function

$$f(x) = \sin 2x$$

by using the values of x in interval $[0, \pi]$. On the same coordinate system, display the graph of $y = \sin x$ in interval $[0, 2\pi]$. Compare the periods of the two functions.

Solution Both graphs have the same amplitude, 1, but how do their periods compare? Table 5 lists values of x in interval $[0, \pi]$ along with corresponding values of $2x$ and $\sin 2x$. By using the table entries we plot points and then connect them with a smooth curve to get the graph of f (Figure 8). This graph appears to be a sine cycle with period π.

To confirm this observation, we know that

$$y = \sin \blacksquare \text{ completes one cycle when } 0 \le \blacksquare \le 2\pi.$$

Correspondingly,

$$f(x) = \sin 2x \text{ completes one cycle}$$
$$\text{when } 0 \le 2x \le 2\pi.$$

That is,

$$\text{when } 0 \le x \le \pi.$$

So the period is π.

TABLE 5

x	$2x$	$f(x) = \sin 2x$
0	0	0
$\frac{\pi}{6}$	$\frac{\pi}{3}$	$\frac{\sqrt{3}}{2}$
$\frac{\pi}{4}$	$\frac{\pi}{2}$	1
$\frac{\pi}{2}$	π	0
$\frac{7\pi}{12}$	$\frac{7\pi}{6}$	$-\frac{1}{2}$
$\frac{3\pi}{4}$	$\frac{3\pi}{2}$	-1
π	2π	0

Figure 8

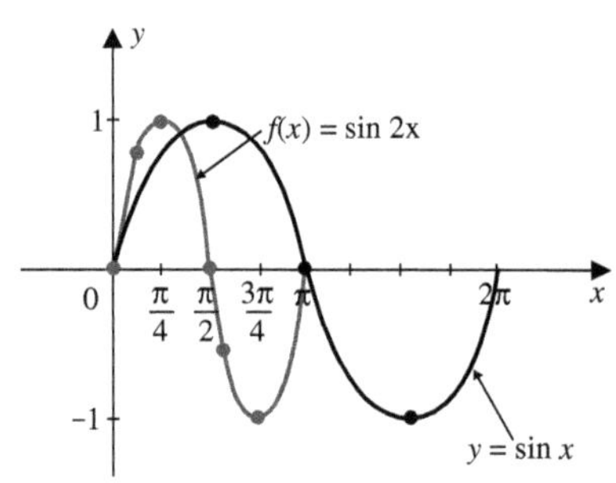

The graph of $y = \sin x$ in interval $[0, 2\pi]$ is also shown in Figure 8. The period of this function is 2π. Upon comparing the periods of the two functions, we discover that by replacing x with $2x$ in the equation $y = \sin x$, we obtain a function f whose graph is the sine curve that is horizontally compressed by one-half.

Our analysis in Example 3 is generalized as follows:
Given

$$f(x) = \sin kx, \quad \text{where } k > 0,$$

then one cycle of the graph of f occurs if

$$0 \le kx \le 2\pi,$$

that is, if

$$0 \le x \le \frac{2\pi}{k}.$$

A similar analysis applies to

$$g(x) = \cos kx, \quad \text{for} \quad k > 0.$$

Thus we have the following general property:

Property

Period—Horizontal Scaling

The graph of

$$f(x) = \sin kx \quad \text{or} \quad g(x) = \cos kx, \quad k > 0,$$

can be obtained by *horizontally scaling* the graph of

$$y = \sin x \quad \text{or} \quad y = \cos x,$$

respectively, by a factor of $1/k$. The **period** of f or g is $2\pi/k$ so that one cycle of the graph occurs in the interval $[0, 2\pi/k]$.

EXAMPLE 4 **Using Horizontal Scaling**

Use horizontal scaling to graph one cycle of $G(x) = \cos \frac{x}{2}$. Indicate the period of G and locate the key points.

Solution Using the above property, we conclude that the given function

$$G(x) = \cos \frac{x}{2} = \cos \frac{1}{2} x$$

has a period given by

$$\frac{2\pi}{\left(\frac{1}{2}\right)} = 4\pi.$$

This means that one cycle of the graph of G can be obtained by stretching the cosine cycle horizontally by a factor of $\frac{1}{\left(\frac{1}{2}\right)} = 2$ over interval $[0, 4\pi]$ (Figure 9).

The key points, namely, the maximums, the minimum, and the x intercepts, are located by proportionally scaling their locations horizontally. They are shown in Figure 9.

Figure 9

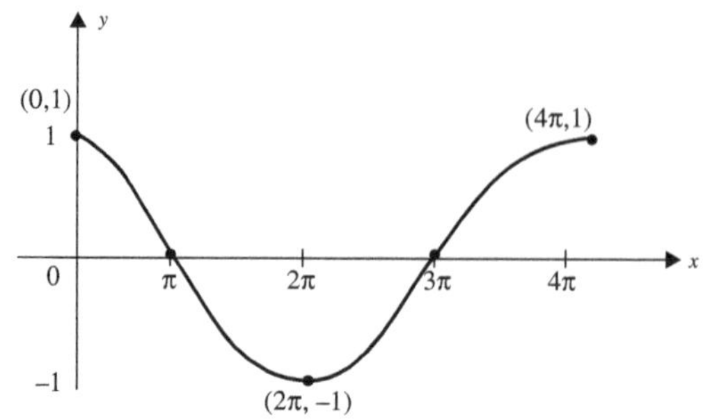

Notice that Example 3 illustrates that if $k > 1$, the cycle is *compressed* horizontally. On the other hand, if $0 < k < 1$, as was the case in Example 4, then the cycle is *stretched* horizontally.

There is no loss in generality by having k positive in the above property. If k is negative, we can rewrite the function using the even-odd properties in such a way that the coefficient of x turns out to be positive. For instance,

$f(x) = \sin(-3x) = -\sin 3x$ (Graph $y = \sin 3x$ and then reflect it across the x axis.)

and

$g(x) = \cos(-3x) = \cos 3x.$ (Same graph as the one for $y = \cos 3x$.)

3. Phase Shift: Horizontal Shifting

Now we consider graphs of functions of the forms

$$f(x) = \sin(kx + b) \text{ and } g(x) = \cos(kx + b)$$

where $k > 0$ and $b \neq 0$.

First, let us consider the situation when $k = 1$.

Recall that the graph of a function

$$y = f(x - h)$$

can be obtained by horizontally shifting the graph of $y = f(x)$ to the right if $h > 0$ and to the left if $h < 0$.

For instance, the graph of

$$f(x) = \sin\left(x + \frac{\pi}{6}\right) = \sin\left[x - \left(-\frac{\pi}{6}\right)\right]$$

Figure 10

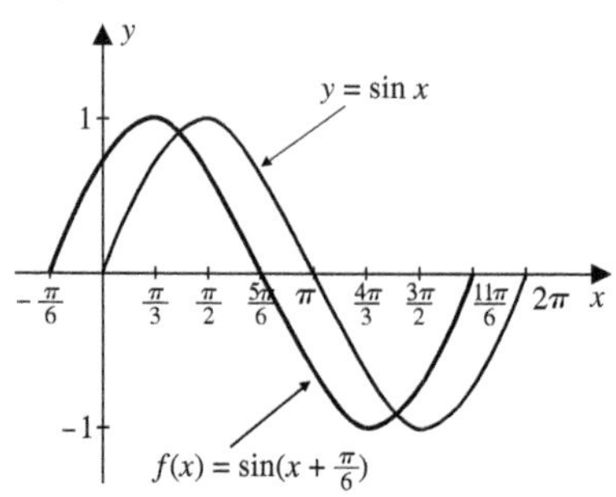

can be obtained by horizontally shifting the graph of $y = \sin x$, $\frac{\pi}{6}$ unit to the left. Figure 10 shows the result of shifting one cycle.

In a sense, $-\pi/6$ is the starting position of the sine cycle for f. Notice that the x values of the five key points of the graph of f are obtained by subtracting $\pi/6$ from the x values of the five key points of $y = \sin x$ to obtain

$$-\frac{\pi}{6}, \frac{\pi}{2} - \frac{\pi}{6} = \frac{\pi}{3}, \pi - \frac{\pi}{6} = \frac{5\pi}{6}, \frac{3\pi}{2} - \frac{\pi}{6} = \frac{4\pi}{3}, \text{ and } 2\pi - \frac{\pi}{6} = \frac{11\pi}{6}.$$

To help us understand how the graph of either

$$y = \sin x \quad \text{or} \quad y = \cos x$$

is affected when replacing

$$x \quad \text{with} \quad kx + b,$$

where $k > 0$, but $k \neq 1$, let us consider the following example.

EXAMPLE 5

Shifting and Scaling a Graph Horizontally

Given

$$f(x) = \sin\left(2x + \frac{\pi}{2}\right)$$

(a) Identify a starting and ending position of one cycle of the graph of f. Find the period of f.

(b) Graph one cycle by using the results in part (a)

Solution

(a) We know that $y = \sin x$ completes one cycle when $0 \le x \le 2\pi$. Correspondingly, $f(x) = \sin[2x + (\pi/2)]$ completes one cycle when

$$0 \le 2x + \frac{\pi}{2} \le 2\pi$$

that is, when

$$-\frac{\pi}{2} \le 2x \le 2\pi - \frac{\pi}{2} \qquad \text{Subtract } \frac{\pi}{2} \text{ from each side}$$

$$-\frac{\pi}{4} \le x \le \pi - \frac{\pi}{4} \qquad \text{Divide by 2}$$

$$-\frac{\pi}{4} \le x \le \frac{3\pi}{4} \qquad \text{Simplify}$$

↑ Starting position of one cycle ↑ Ending position of one cycle

Here, the starting position is the solution of $2x + \frac{\pi}{2} = 0.$

The length of the interval $\left[-\frac{\pi}{4}, \frac{3\pi}{4}\right]$ is given by

$$\frac{3\pi}{4} - \left(-\frac{\pi}{4}\right) = \frac{4\pi}{4} = \pi$$

which is the same as

$$\frac{2\pi}{k} = \frac{2\pi}{2} = \pi.$$

Consequently, the period of f is π.

(b) Since the graph of f generates one cycle on interval $\left[-\frac{\pi}{4}, \frac{3\pi}{4}\right]$, which has length π, its graph can be obtained by shifting the graph of $y = \sin x$ (which has period 2π) horizontally left $\frac{\pi}{4}$ unit and then compressing it by a factor of $\frac{1}{k} = \frac{1}{2}$ (Figure 11).

Figure 11

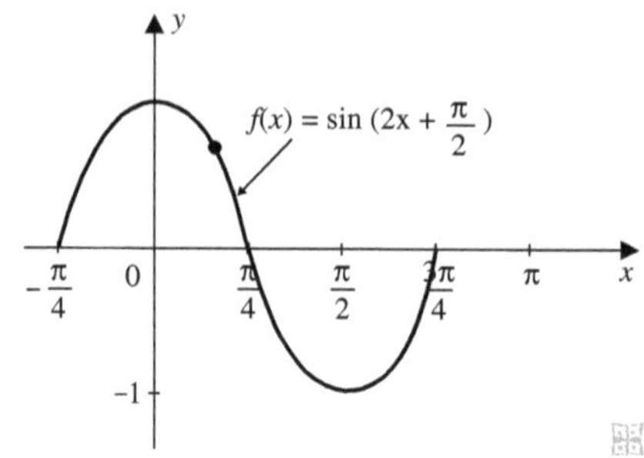

We generalize the results of Example 5 as follows.

Property
Phase Shift and Period

The graph of

$$f(x) = \sin(kx + b) \quad \text{or} \quad g(x) = \cos(kx + b), \quad \text{where} \quad k > 0$$

can be obtained by horizontally shifting the graph of

$$y = \sin x \quad \text{or} \quad y = \cos x,$$

respectively, by $|-b/k|$ units to the left if $-b/k < 0$ and to the right if $-b/k > 0$, and then horizontally scaling it by a factor of $\frac{1}{k}$. The resulting function has a period of $\frac{2\pi}{k}$.

The number $-b/k$ is called the **phase shift** of the function.

The phase shift can be found by setting $kx + b = 0$ and solving for x to get $x = -b/k$.

In a sense, the phase shift is the starting position of one cycle.

For example,

the function $f(x) = \sin\left(2x + \frac{\pi}{2}\right)$ has phase shift

$$\frac{-b}{k} = \frac{\left(-\frac{\pi}{2}\right)}{2} = -\frac{\pi}{4} \quad \text{and period} \quad \frac{2\pi}{2} = \pi;$$

the function $g(x) = 4\cos\left(\frac{1}{3}x - 2\right) = 4\cos\left(\frac{1}{3}x + (-2)\right)$ has phase shift

$$\frac{-b}{k} = \frac{-(-2)}{\left(\frac{1}{3}\right)} = 6 \quad \text{and} \quad \text{period} \; \frac{2\pi}{\left(\frac{1}{3}\right)} = 6\pi.$$

Based on our previous work, let us summarize the information about sine and cosine graphs that can be extracted from function equations of the forms

$$F(x) = a\sin(kx + b) \quad \text{and} \quad G(x) = a\cos(kx + b).$$

Summary
Amplitude, Period, and Phase Shift Properties

Graphs of functions of the form

$F(x) = a\sin(kx + b)$ or $G(x) = a\cos(kx + b)$, for $a \neq 0$ and $k > 0$,

have the following properties:

(i) The amplitude is $|a|$ (ii) The period is $\dfrac{2\pi}{k}$

(iii) The phase shift is $\dfrac{-b}{k}$

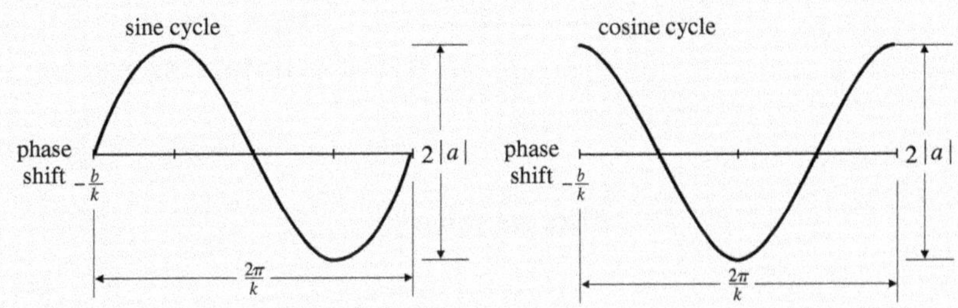

EXAMPLE 6

Using Properties to Sketch a Cosine Graph

Identify and use the phase shift, period, and amplitude to graph one cycle of the function $f(x) = 2\cos(3x - 1)$. Locate the key points, that is, the locations of minimum points, maximum points, and x intercepts.

Solution First we rewrite the given function as

$$f(x) = 2\cos(3x - 1) = 2\cos[3x + (-1)].$$

Thus

$$a = 2, \quad k = 3, \quad \text{and} \quad b = -1$$

so that

(i) the amplitude is $|a| = |2| = 2$;

(ii) the period is $\dfrac{2\pi}{k} = \dfrac{2\pi}{3}$;

(iii) the phase shift is $\dfrac{-b}{k} = \dfrac{-(-1)}{3} = \dfrac{1}{3}$.

Figure 12

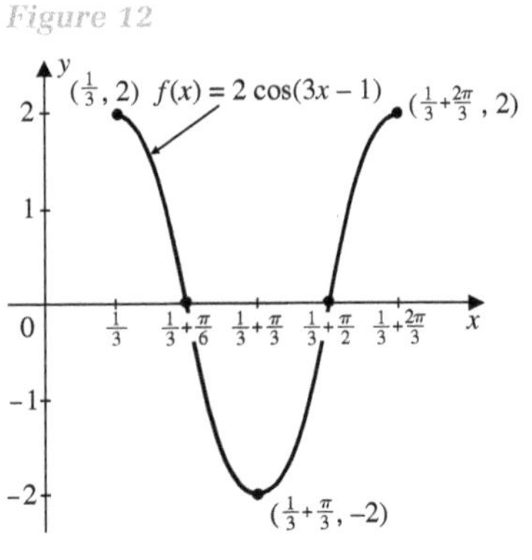

Consequently, one cycle of the graph of f, which is a cosine curve, starts at $x = 1/3$, covers an interval of length $(2\pi)/3$, to end at $x = (1/3) + (2\pi)/3$, and is vertically stretched by a factor of 2 (Figure 12).

The x values of the five key points are found by setting

$$3x - 1 = 0, \quad \frac{\pi}{2}, \quad \pi, \quad \frac{3\pi}{2}, \quad \text{and} \quad 2\pi$$

to get

$$x = \frac{1}{3}, \quad \frac{1}{3} + \frac{\pi}{6}, \quad \frac{1}{3} + \frac{\pi}{3}, \quad \frac{1}{3} + \frac{\pi}{2}, \quad \text{and} \quad \frac{1}{3} + \frac{2\pi}{3}.$$

The locations of the key points are shown in Figure 12.

It is not necessary to memorize the formulas for the period and phase shift for the functions

$$F(x) = \sin(kx + b) \quad \text{and} \quad G(x) = \cos(kx + b).$$

They are easily derived as follows:

> Solve $kx + b = 0$ and $kx + b = 2\pi$ for x to get, respectively, $x = \frac{-b}{k}$ (phase shift) and $x = -\frac{b}{k} + \frac{2\pi}{k}$ (phase shift plus period).

For example, if $f(x) = \sin(7x - 5)$, then we solve

$$7x - 5 = 0 \text{ to get } x = \frac{5}{7}, \text{ which is the phase shift.}$$

Also, we solve

$$7x - 5 = 2\pi \text{ to get } x = \frac{5}{7} + \frac{2\pi}{7} \text{ (phase shift plus period)}$$

so that the period of f is $\frac{2\pi}{7}$.

The graphs of sine and cosine function are called **sine waves** or **sinusoidal curves.** The characteristics of the graph of such a curve can be used to find its equation in situations such as the one illustrated in the next example.

EXAMPLE 7

Finding an Equation for a Sine Wave

Given the graph of one cycle of the function $y = a \sin(kx + b)$ in Figure 13, find values for $a > 0$, $k > 0$, and b. Also find the amplitude, period, and phase shift.

Solution

Figure 13

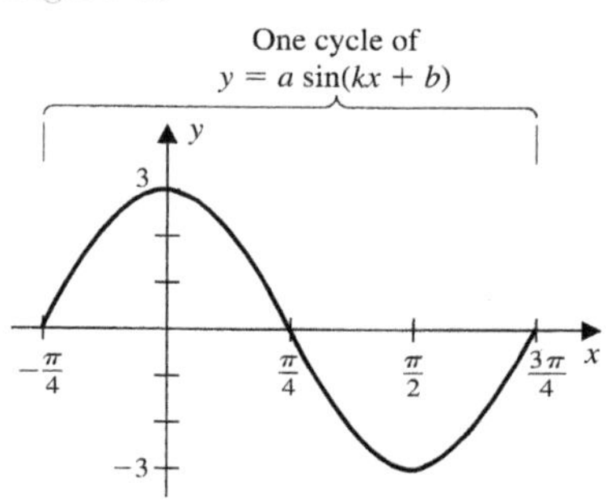

The graph shows that the amplitude is 3; that is, $a = 3$. The period, which is the length of the interval containing the cycle, is given by

$$\frac{3\pi}{4} - \left(\frac{-\pi}{4}\right) = \pi$$

so

$$\frac{(2\pi)}{k} = \pi, \quad \text{or} \quad k = 2.$$

Since the phase shift is shown to be $-\pi/4$, it follows that $-b/k = -b/2 = -\pi/4$, or $b = \pi/2$. Thus the equation is

$$y = 3\sin\left(2x + \frac{\pi}{2}\right).$$

Functions such as

$$f(x) = \sin 2x + \cos x,\ g(x) = 5 \sin 3x \cos 2x, \text{ and } h(x) = \sin^2 4x$$

which are *not* of the form

$$F(x) = a \sin(kx + b) \quad \text{or} \quad G(x) = a \cos(kx + b)$$

are difficult to graph by hand.

For instance, the graphs of f, g, and h, given on the interval $[0, 2\pi]$ produced by a grapher set in radian mode (for real numbers), are shown in Figures 14a, 14b, and 14c, respectively.

Figure 14

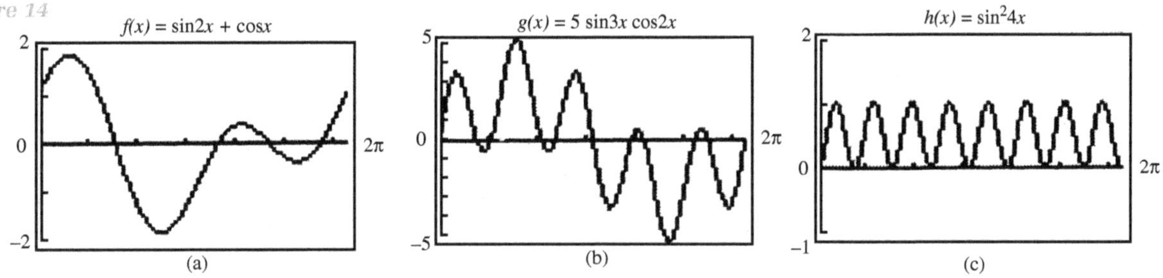

(a) (b) (c)

Solving Applied Problems

When sine and cosine functions are used to model periodic phenomena, their graphs help us to interpret the situation.

EXAMPLE 8

Modeling a Tide Problem

© JupiterImages, Inc.

Tides cause ocean, sea, and lake currents to flow into and out of harbors, canals, and coastlines with varying depths. Suppose for a certain location, the depth of water d (in feet) varies with time t (in hours) according to the model

$$d(t) = 8 + 4\cos\left(\frac{4\pi}{25}t\right)$$

where $t = 0$ represents midnight.

(a) Use this model to predict the depth of water at the start, at 2:00 AM and at 6:00 AM. Round off the answers to two decimal places.

(b) Sketch the graph of this model over a 24-hour day.

(c) Find the amplitude and period. Interpret these values.

(d) Find the minimum and maximum depth of water and determine when they occur. Interpret your answers.

Solution (a) At the start, $t = 0$, so the depth of water is given by

$$\begin{aligned} d(0) &= 8 + 4\cos 0 \\ &= 8 + 4 = 12 \text{ ft.} \end{aligned}$$

At 2:00 AM, $t = 2$, so we have

$$\begin{aligned} d(2) &= 8 + 4\cos\left(\frac{8\pi}{25}\right) \\ &= 10.14 \text{ ft (approx.).} \end{aligned}$$

At 6:00 AM, $t = 6$, so we have

$$\begin{aligned} d(6) &= 8 + 4\cos\left(\frac{24\pi}{25}\right) \\ &= 4.03 \text{ ft (approx.).} \end{aligned}$$

(b) Figure 15 shows the graph of this model over a 24-hour day.

Figure 15

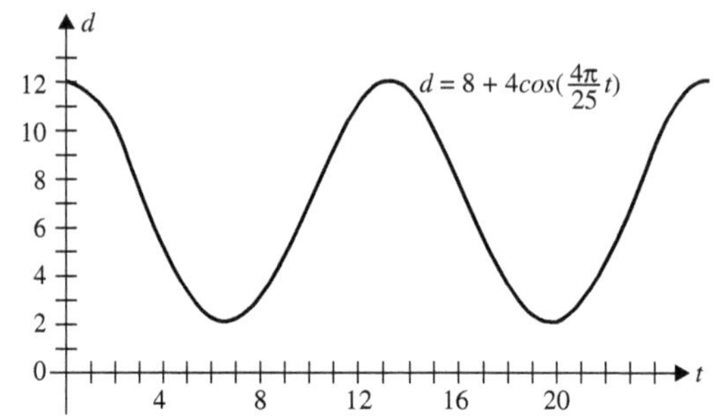

(c) The amplitude is 4, and the period is given by

$$\frac{2\pi}{\left(\frac{4\pi}{25}\right)} = \frac{25}{2} = 12.5 \text{ hours.}$$

The amplitude of 4 indicates that the depth varies no more than 4 feet from its average of 8 feet. Thus the water level rises to a depth of no more than $8 + 4 = 12$ feet and no less than $8 - 4 = 4$ feet. This behavior repeats every 12.5 hours.

(d) The minimum depth d occurs when the cosine is minimum, that is, when the cosine is -1. So the minimum depth is

$$8 + 4(-1) = 4 \text{ feet}$$

which occurs when

$$(4\pi/25)t = \pi \text{ and } (4\pi/25)t = 3\pi;$$

that is, when $t = \pi \cdot (25/4\pi) = 25/4 = 6.25$ and $t = 3\pi \cdot (25/4\pi) = 18.75$. This happens at 6:15 AM and then again at 6:45 PM.

The maximum depth is $8 + 4(1) = 12$ feet, when the cosine is at its maximum value of 1. This occurs when

$$(4\pi/25)t = 0, \text{ and when } (4\pi/25)t = 2\pi;$$

that is, when $t = 0$ and when $t = 2\pi \cdot (25/4\pi) = (25/2) = 12.5$.

That is, the maximum depth of 12 feet happens at the start (midnight) and again at 12:30 PM.

At times the least-squares method for finding a regression equation that best fits given data results in a sine function of the form

$$y = a\sin(kx + b) + c.$$

EXAMPLE 9

Modeling Monthly Temperatures

The data in the given table lists average monthly morning temperatures in degrees Fahrenheit in a southern city for certain months of the year.

(a) Plot the data in the table on a scattergram. Use the horizontal axis for the t values and the vertical axis for the F values.

(b) Use a grapher to find a sine function of the form

$$F = a\sin(kt + b) + c$$

that best fits the data. Round off a, b, c, and k to three decimal places.

(c) Graph the function in part (b) of best fit on the same coordinate system as the scattergram.

(d) Predict the average monthly morning temperature in the months of

March, June, and October.

Round off the answers to one decimal place.

Month	t	Average Morning Monthly Temperature F
Jan	1	43.2
Feb	2	47.9
April	4	58.8
May	5	67.2
July	7	75.3
Aug	8	72.4
Nov	11	46.5
Dec	12	44.1

Solution

(a) By using the horizontal axis to represent the t values and the vertical axis to represent the F values, we get the scattergram shown in Figure 16a.

(b) Upon using a grapher to produce the regression equation and after rounding off to three decimal places we get the model

$$F = 15.696\sin(0.574t - 2.305) + 59.638.$$

(c) The graph of this equation, along with the scattergram for the given data, is shown in Figure 16b.

Figure 16

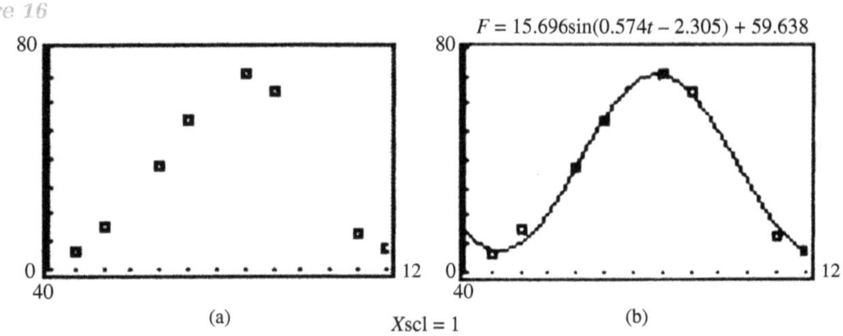

Xscl = 1
Yscl = 5

(d) To predict the average monthly morning temperature from this model for the month of March, we substitute $t = 3$ into the regression equation to get $51.0°F$ (approx.).
Similarly, for June, $t = 6$ so that $F = 73.9°$ (approx.), and for October, $t = 10$ so that $F = 55.1°$ (approx.).

Modeling Simple Harmonic Motion

In the real world there are many phenomena that involve oscillation or vibration in a uniform manner, repeating periodically in definite intervals of time. These phenomena can be modeled by *sinusoidal curves,* also called *simple harmonic curves,* which are the graphs of sine and cosine functions. Examples include alternating electrical current, sound waves, light waves, pendulums, and mass-spring systems. A mathematical model for a quantity y that is oscillating in a simple harmonic manner is given by either of the following equations:

$$\boxed{y = a \cos \omega t \quad \text{or} \quad y = a \sin \omega t}$$

where t represents elapsed time and $\omega > 0$ is called the **angular frequency.**

The **period** $\boldsymbol{T}$ for either $y = a \cos \omega t$ or $y = a \sin \omega t$ is given by

$$\boxed{T = \frac{2\pi}{\omega}.}$$

The **frequency** $\boldsymbol{f}$ of a simple harmonic motion is the number of oscillations per unit of time, so

$$\boxed{f = \frac{1}{T} = \frac{\omega}{2\pi}.}$$

EXAMPLE 10 **Finding the Frequency of a Simple Harmonic Model**

Suppose that alternating electric current is described by the simple harmonic model given by

$$y = 14.1 \sin 120\pi t$$

where y (in amperes) is the current at time t (in seconds). Find the amplitude, period, and frequency of the current.

Solution In this situation, $y = a \sin \omega t$, where $a = 14.1$ and $\omega = 120\pi$. So the amplitude of the current is 14.1 and the period T is given by

$$T = \frac{2\pi}{\omega} = \frac{2\pi}{120\pi} = \frac{1}{60}$$

that is, the period is $\frac{1}{60}$ seconds. The frequency f of the current is given by

$$f = \frac{1}{T} = \frac{1}{\left(\frac{1}{60}\right)} = 60.$$

Therefore, the frequency is 60 cycles per second, or 60 hertz (Hz). ■

If a weight suspended from a spring is pulled down a distance $|a|$ and then released, it oscillates vertically in periodic motion. To model the situation, we let y (say, in centimeters) represent the *directed distance* of the weight from its rest position after time t (say, in seconds). We consider y to be a function of t. If the values of y are plotted for specific values of t, and if friction is neglected, then the resulting graph (Figure 17) has an equation of one of the following forms:

$$y = a \cos\sqrt{\frac{k}{m}}\,t \quad \text{or} \quad y = a \sin\sqrt{\frac{k}{m}}\,t$$

Figure 17

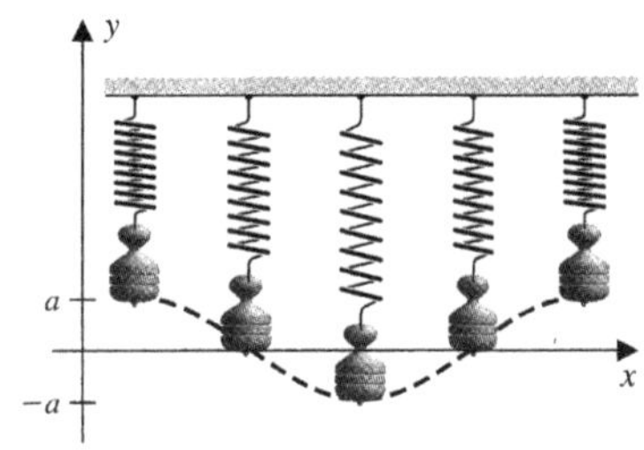

Here, k, the **stiffness coefficient,** is a constant associated with the particular spring, and m is the **mass** of the weight.

The variable y is also referred to as the **displacement.** A positive value of y indicates that it is above its rest position, and a negative value of y indicates that it is below its rest position. The farther the weight is pulled down before it is released, the greater the amplitude of motion will be. Furthermore, the stiffer the spring, the more rapidly the weight will oscillate, and thus the smaller the time period of repeating motion will be. Since the period is given by $T = 2\pi/\omega$, where $\omega = \sqrt{k/m}$, it takes $2\pi/\omega$ units of time to go through one complete oscillation.

EXAMPLE 11

Finding the Equation of Motion of an Object Suspended from a Spring

An object of mass 4 grams suspended from a spring is pulled down 2 centimeters and released at time $t = 0$ so that the object oscillates vertically.

(a) Determine which equation,

$$y = a \cos\sqrt{\frac{k}{m}}\,t \quad \text{or} \quad y = a \sin\sqrt{\frac{k}{m}}\,t,$$

is better suited to model this situation.

(b) Write an equation of motion if the stiffness coefficient k is 64 and t is the number of elapsed seconds.

(c) Find the period, frequency, and amplitude of the motion.

(d) Give a physical interpretation of the answers in part (c).

(e) Graph three cycles of the displacement y as a function of t.

Solution (a) First, we compare the two equations available to model the situation:

Equation	$y = a \cos \sqrt{\dfrac{k}{m}}\, t$	$y = a \sin \sqrt{\dfrac{k}{m}}\, t$
Value of Equation When $t = 0$	$y = a \cos 0 = a$	$y = a \sin 0 = 0$

For the given situation, the displacement is 2 centimeters below the resting position at the start of the motion; that is, when $t = 0$, $y = -2$. So the equation $y = a \cos \sqrt{k/m}\, t$, with $a = -2$, is a more suitable model.

(b) Using the equation $y = a \cos \sqrt{k/m}\, t$ with $a = -2$, $k = 64$, and $m = 4$, we get the motion equation

$$y = -2 \cos \sqrt{\frac{64}{4}}\, t = -2 \cos 4t.$$

(c) The period $T = 2\pi/\omega = 2\pi/4 = \pi/2$. The frequency $f = 1/T = 2/\pi$ cycles per second, and the amplitude is $|a| = |-2| = 2$ centimeters.

(d) Since the period is $\pi/2$, it takes about 1.57 seconds for one complete oscillation. An amplitude of 2 means the highest and lowest positions are 2 centimeters above and below the resting position. A frequency of $2/\pi$ indicates that it makes about 0.64 oscillation per second.

(e) The graph in Figure 18 shows three cycles of this situation.

Figure 18

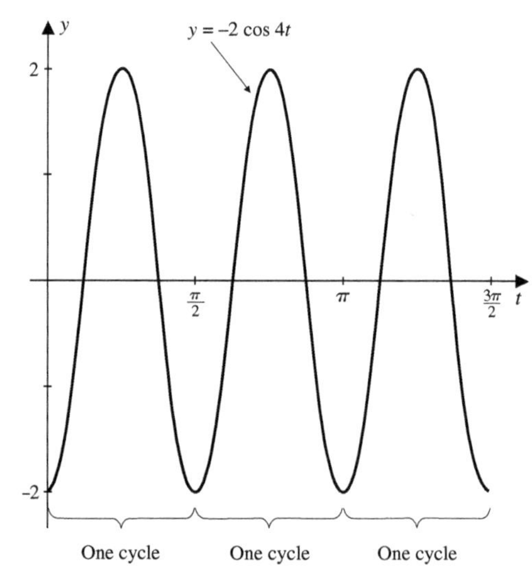

PROBLEM SET 1.4

Mastering the Concepts

In problems 1 and 2, find the range of each function. Express the answer in interval notation.

1. (a) $f(\theta) = 3 \sin \theta$

(b) $h(\theta) = 4 - \frac{1}{3} \cos \theta$

(c) $f(\theta) = \frac{2 \sin \theta + 5}{3}$

2. (a) $g(\theta) = -\frac{1}{2} \cos \theta$

(b) $g(\theta) = \frac{1}{\cos \theta}$

(c) $g(\theta) = \frac{3 + 2 \cos \theta}{-5}$

In problems 3–18, use the graph of the sine or cosine function and transformations to sketch the graph of each function. Indicate the range, amplitude, and period of each function.

3. (a) $y = 3 \cos x$
(b) $y = -3 \cos x$

4. (a) $y = -5 \sin x$
(b) $y = \frac{4}{3} \sin x$

5. (a) $y = \frac{2}{3} \cos(-x)$
(b) $y = \frac{1}{2} \cos x$

6. (a) $y = -2 \cos x$
(b) $y = 2 \cos(-x)$

7. (a) $y = \cos 2x$
(b) $y = \cos \frac{x}{2}$

8. (a) $y = \sin 3x$
(b) $y = \sin \frac{x}{3}$

9. (a) $y = \cos \frac{\pi x}{3}$
(b) $y = \cos(-2\pi x)$

10. (a) $y = \sin \frac{\pi x}{3}$
(b) $y = -\sin \frac{\pi x}{2}$

11. (a) $y = \frac{5}{3} \sin (-6x)$
(b) $y = 5 \sin 10x$

12. (a) $y = 3 \cos \left(\frac{-\pi x}{6}\right)$
(b) $y = -3 \cos 5x$

13. (a) $y = -3 \cos \frac{\pi x}{2}$
(b) $y = \pi \cos\left(\frac{-5x}{2}\right)$

14. (a) $y = -5 \sin \frac{2\pi x}{3}$
(b) $y = -\frac{1}{2} \sin \frac{3x}{2}$

15. (a) $y = 1 + \sin x$
(b) $y = 3 - \sin x$

16. (a) $y = 2 + \cos x$
(b) $y = -\frac{3}{4} + 2 \cos x$

17. (a) $y = 2 + 3 \cos \frac{\pi x}{4}$
(b) $y = -2 - 3 \cos 2\pi x$

18. (a) $y = \frac{1}{2} - \frac{1}{2} \sin \frac{\pi x}{2}$
(b) $y = -2 - \sin \frac{\pi x}{6}$

19. Match each function with one of the graphs shown in Figure 19.

$f(x) = 0.5 \sin x$; $g(x) = \sin x$; $h(x) = 2 \sin x$

Figure 19

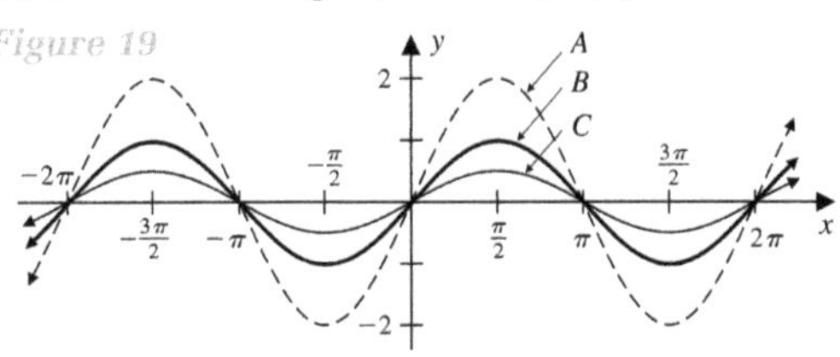

20. Match each function with one of the graphs shown in Figure 20.

$f(x) = 2 \sin x + 1$; $g(x) = \sin x + 3$;
$h(x) = 0.5 \sin x - 2$

Figure 20

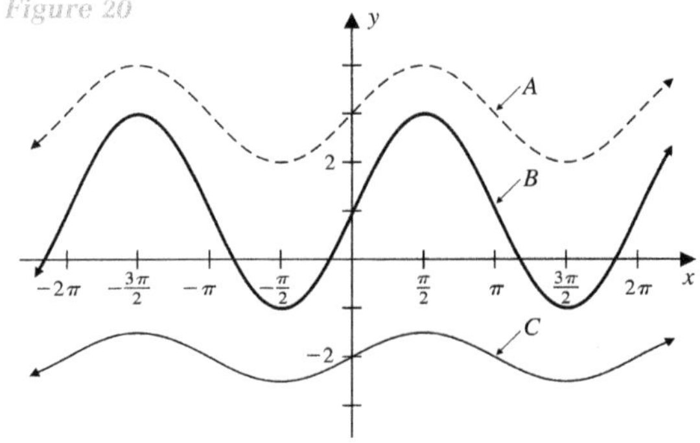

In problems 21–36, identify and use the phase shift, period, and amplitude to graph one cycle of each function.

21. $y = 2\sin\left(x - \frac{\pi}{4}\right)$

22. $y = -3\sin\left(x + \frac{\pi}{3}\right)$

23. $y = 4\cos\left(x + \frac{\pi}{6}\right)$

24. $y = 2\cos\left(x - \frac{\pi}{3}\right)$

25. $y = -2\sin(2x - \pi)$

26. $y = -2\cos(3x - \pi)$

27. $y = 2\sin(\pi x - \pi)$

28. $y = 3\cos(\pi x - \pi)$

29. $y = -3\cos(3x + 2)$

30. $y = -3\sin(4x - 1)$

31. $y = 2\cos\left(\frac{\pi x}{2} - \frac{1}{4}\right)$

32. $y = -4\sin\left(\frac{\pi x}{3} + \frac{1}{3}\right)$

33. $y = -2\sin(5 - 3x)$

34. $y = -2\cos(3 - 2x)$

35. $y = 2 + 3\cos(2x - 2\pi)$

36. $y = -2 - 3\sin(3x + \pi)$

In problems 37 and 38, each figure shows the graph of one cycle of the function

$$y = a\sin(kx + b).$$

Find values for $a > 0$, $k > 0$, and b, and write the resulting equation. Also find the amplitude, period, and phase shift.

37. (a)

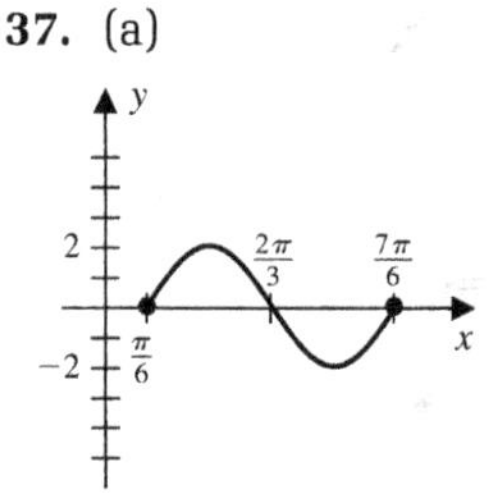

(b)

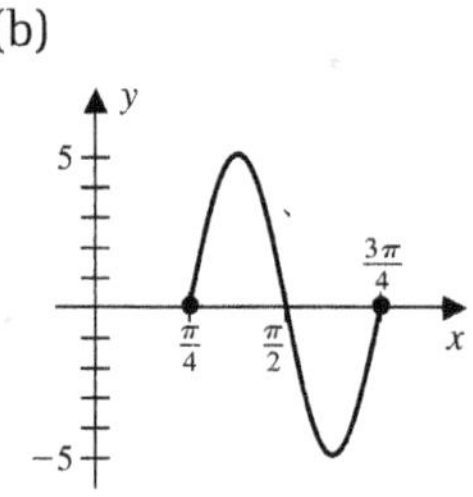

38. (a)

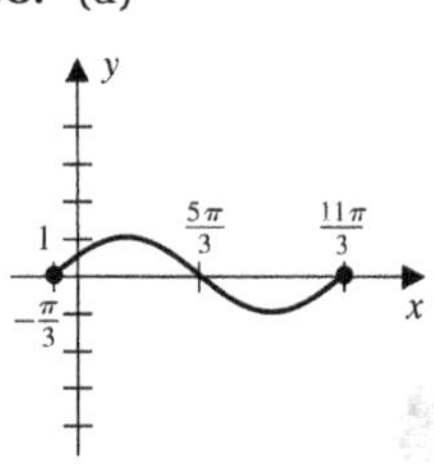

(b)

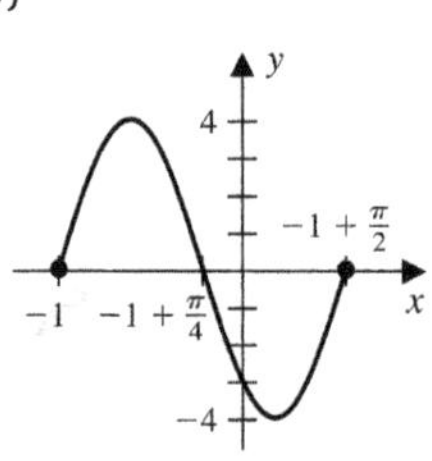

39. Write an equation for a sine function with the following properties:

$$\text{amplitude} = 3, \quad \text{period} = \frac{\pi}{3}, \quad \text{phase shift} = 0$$

Then graph one cycle.

40. Write an equation for a cosine function with the following properties:

$$\text{amplitude} = 2, \quad \text{period} = \frac{\pi}{6}, \quad \text{phase shift} = \frac{\pi}{2}$$

Then graph one cycle.

Applying the Concepts

In problems 41–48, round off the answers to two decimal places.

41. Blood Pressure: The blood pressure of a person is given by the model

$$y = 110 + 30\cos\frac{12\pi t}{5}$$

where t is in seconds and y is in millimeters of mercury (mm Hg).

(a) Sketch the graph of this function for $0 \leq t \leq 2.5$.

(b) Determine the period of the function.

(c) Use the graph to determine the maximum and minimum blood pressure of the person for the given interval.

42. Electrical Circuit: The electricity supplied to homes is called alternating current (ac), and the current I (in amperes) is related to time t (in seconds) by the equation

$$I = 5\cos 120\pi t.$$

The voltage V (in volts) that causes this current to flow is related to time t (in seconds) by the equation

$$V = 170\cos[120\pi(t + 0.031)].$$

(a) Sketch the graphs of the current and voltage functions for $0 \leq t \leq 0.004$.

(b) Use the graphs to estimate the voltage at the time the current is a maximum.

(c) Use the graphs to estimate the current at the time the voltage is a maximum.

(d) Predict the first time when the voltage reaches 160 volts.

43. Height of a Point on a Waterwheel: Figure 21 shows a rotating waterwheel. At time t (in seconds) the height h (in feet) of a point P on the rim of the wheel above the surface of the water is given by the model

$$h(t) = 7 + 8\cos\left(\frac{\pi t}{5} - \frac{2}{5}\right).$$

Figure 21

(a) Find the height of the point P on the wheel after 2 seconds, after 5 seconds, and after 10 seconds.

(b) How long does it take for the point P on the wheel to go around once?

(c) Sketch the graph of h for two cycles.

(d) Find the minimum and maximum heights of P on the wheel.

44. Sound Wave: The difference D (in dynes per square centimeter) between the atmospheric pressure and the air pressure at the eardrum produced by a sound wave at t seconds is given by the function

$$D(t) = 4\sin\left(100\pi t + \frac{\pi}{4}\right).$$

(a) Sketch the graph of this function for $0 \le t \le 0.18$.

(b) Find the period and amplitude.

(c) Find the minimum and maximum values of $D(t)$.

45. Seasonal Business Profit: The profit P (in thousands of dollars) per month from the sales of scuba equipment for 2 years beginning in January 1, 2004, is given by the function

$$P(t) = 900 - 700\cos\frac{\pi t}{6}$$

where t is the elapsed time (in months) and $t = 0$ corresponds to January 1, 2004.

(a) Graph the function for $0 \le t \le 24$.

(b) Find the amplitude and period, and interpret them.

(c) Determine the profit at 3 months, 9 months, 1 year, 18 months, and 2 years.

(d) What is the maximum profit, and when does it occur?

46. Tides: During a 24-hour day, the mathematical model

$$d = 3\cos 0.52t + 9$$

is used to predict the tide. That is, it predicts how much above mean sea level the tide will raise or lower the water depth d (in feet) in a certain harbor, t hours after midnight.

(a) Sketch the graph of this function over a 24-hour day.

(b) Find the amplitude and period, and interpret them.

(c) Find the minimum and maximum depth of water in the harbor, and determine when they occur.

(d) Interpret your answers for part (c) in terms of this tidal wave.

47. Hours of Daylight: Suppose that the hours $h(t)$ of daylight for a certain city are related to the day of the year by the model

$$h(t) = 2.3\sin\left[\frac{2\pi}{365}(t - 80)\right] + 12$$

where t is the day of the year, and $t = 1$ corresponds to January 1.

(a) Sketch the graph of this function for

$$1 \le t \le 365.$$

(b) Find the amplitude and the period.

(c) Determine the maximum and minimum number of hours of daylight and when they occur.

48. Business Profit: A company's profit fluctuates during a 12-month period according to the model

$$P(t) = 30{,}000 - 70{,}000(\sin t + 0.5\cos 0.5t)$$

where $P(t)$ represents the profit (in dollars) after t months.

(a) Find the profit (or loss) at the beginning of the 12-month period, after 5 months, after 8 months, and after 1 year.

(b) Sketch the graph of P for

$$0 \le t \le 12.$$

(c) Interpret the graph to determine when there is a profit.

(d) What are the maximum and minimum profits over the 12-month period, and when do they occur?

49. **Temperature Trend:** The data in the following table lists the average monthly temperature F (in degrees Fahrenheit) for a mountainous region for certain months of the year.

Month	t	Average Monthly Temperature F
Jan	1	24
Feb	2	28.3
April	4	39.2
May	5	47.1
July	7	56.4
Aug	8	57
Oct	10	41.8
Dec	12	29.6

(a) Plot the data in the above table on a scattergram. Use the horizontal axis for the t values and the vertical axis for the F values.

(b) Use a grapher to find a sine function of the form

$$F = a\sin(kt + b) + c$$

that best fits the data. Round off a, k, b, and c to three decimal places.

(c) Graph the function in part (b) of best fit on the same coordinate system as the scattergram.

(d) Predict the average monthly temperatures in the months of March, June, and September. Round off the answers to one decimal place.

50. **Precipitation Trend:** The data in the following table lists the average monthly precipitation P (in inches) in a region for certain months of the year.

(a) Plot the data in the table at the top of the next column on a scattergram. Use the horizontal axis for the t values and the vertical axis for the P values.

Month	t	Average Precipitation P in inches
Jan	1	1.1
Feb	2	0.9
Mar	3	1.7
May	5	3.1
June	6	3.9
Aug	8	4.1
Sept	9	3.2
Nov	11	1.6
Dec	12	1.2

(b) Use a grapher to find a regression model of the form

$$P = a\sin(kt + b) + c$$

that best fits the data. Round off a, k, b, and c to three decimal places.

(c) Graph the function in part (b) of best fit on the same coordinate system as the scattergram.

(d) Predict the precipitation in the months of April, July, and October. Round off the answers to one decimal place.

In problems 51 and 52, a simple harmonic model is described by the given equation, where t is time (in seconds). Determine the amplitude, period, and frequency of each model.

51. (a) $y = \sin 200\pi t$
(b) $y = 2\cos 0.001\pi t$

52. (a) $y = 0.15\sin 792\pi t$
(b) $y = 0.06\cos 120\pi t$

53. **Spring–Weight Model:** An object of mass 30 grams suspended from a spring is pulled down 50 centimeters and released at time $t = 0$ so that the mass oscillates vertically.

(a) Write an equation of motion if the stiffness coefficient $k = 8$ and t is the number of elapsed seconds.

(b) Graph two cycles of the displacement y as a function of t.

(c) Find the amplitude, period, and frequency of the motion and interpret them.

54. **Spring–Weight Model:** A weight suspended from a spring oscillates up and down according to the model

$$y = a\cos\sqrt{\frac{k}{m}}\,t.$$

If the stiffness coefficient of the spring $k = 4000$, how large a mass should be attached so that the spring will oscillate 6.4 times per second?

55. Pendulum Swing Model: A pendulum swings uniformly back and forth through an arc length of 0.05 meter, taking 2.2 seconds to move from the position directly above point A to the position directly above point B (Figure 22). Assume the motion of the pendulum is simple harmonic.

(a) Find the angular frequency ω.

(b) Find the equation for y as a function of t, where y denotes the displacement of the mass at t seconds.

Figure 22

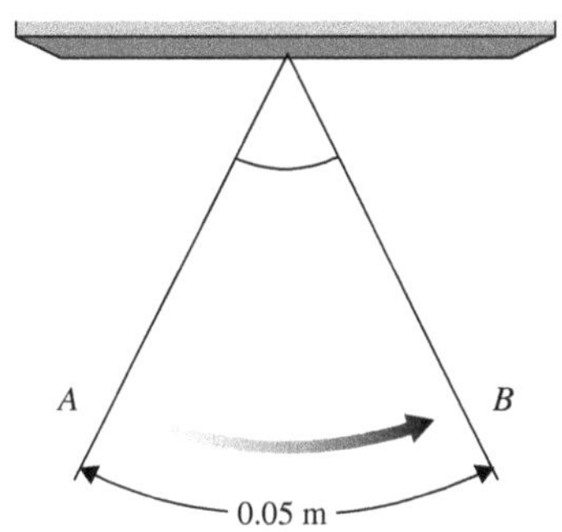

(c) Sketch the graph for three cycles.

56. Grandfather Clock: A grandfather clock has a pendulum of length $L = 0.99$ meter, and it swings through an arc length of 0.25 meter according to the model

$$y = a \sin \omega t.$$

(a) Find the period

$$T = 2\pi\sqrt{\frac{L}{g}},$$

where $g = 9.8$ meters per second.

(b) Find the angular frequency ω.

(c) Determine the equation for the displacement y as a function of t.

(d) Sketch the graph of the equation for three cycles.

Developing and Extending the Concepts

57. Assume

$$f(x) = \sin\left(\frac{\pi}{2} - x\right)$$

and

$$g(x) = \cos x,$$

where $-\pi \le x \le \pi$.

Compare the graphs. Are they the same? Explain.

58. Sketch the graphs of

$$f(x) = [\sin x]^2$$

and

$$g(x) = \left[\cos\left(\frac{\pi}{2} - x\right)\right]^2$$

on the interval $[0, 2\pi]$. How do the graphs compare?

59. Assume $y = \cos |x|$, where $-2\pi \le x \le 2\pi$. Compare this v graph to the graph of $f(x) = \cos x$. Are the two graphs the same or different? Explain.

60. Compare the graph of $f(x) = \sin x$ to the graph of $g(x) = |\sin x|$ for $0 \le x \le 2\pi$. Where are they the same? Where are they different? Explain.

61. $f(x) = \dfrac{\sin x}{x}$

(a) Is $f(0)$ defined?

(b) Use a grapher to graph f, and then use the graph to help determine the behavior of $f(x)$ as $x \to 0$.

(c) Complete the following table to support the conclusion in part (b):

	$x \to 0^-$					$x \to 0^+$			
x	-1	-0.5	-0.01	-0.001	0	0.001	0.01	0.5	1
$f(x) = \dfrac{\sin x}{x}$					—				

62. Use a grapher to graph

$$f(x) = \cos x$$

and

$$g(x) = 1 - \frac{x^2}{2} + \frac{x^4}{24}$$

for the interval

$$[-2.75, 2.75].$$

Compare the values of $f(x)$ and $g(x)$ in this interval.

In problems 63 and 64, use a grapher to sketch the graph of each function, where

$$-2\pi \le x \le 2\pi.$$

If the function is periodic, find its period and its amplitude.

63. $y = \sin x + 2 \cos x$

64. $y = 4 \sin x + 3 \cos x$

Interlude: Practice Problems 1.4

1. Determine an equation of the form $y = c + a\sin(kt + b)$ for a sine curve that models the water depth (in feet) at the end of a pier over the course of several hours, based on the information given below. Be clear how you have established the values of a, b, c, and k.

Time, t	Midnight	2 a.m.	4 a.m.	6 a.m.	8 a.m.	10 a.m.	Noon
Depth, y	10.2	15.2	17.6	15.2	10.2	7.2	9.08

2. Imagine riding an exceptionally large and slow-moving ferris wheel that has a diameter of 500 feet and completes one rotation every 20 minutes (a similar such wheel was scheduled to begin operation in Las Vegas in early 2014). Suppose that t represents the number of minutes you have been riding the wheel, $f(t)$ denotes the height of your car (in feet) above the ground, and that $f(0)= 0$. Assuming that the wheel rotates in a counterclockwise direction at a constant rate, find the rule of $f(t)$.

Objectives

1. Graph the Tangent Function
2. Graph the Cotangent Function
3. Graph the Secant and Cosecant Functions

1.5 Graphs of Other Trigonometric Functions

So far, we have graphed sine and cosine functions. In this section we develop the graphs of the other four trigonometric functions by taking advantage of the fact that they can be written in terms of the sine and cosine functions.

Graphing the Tangent Function

Since the tangent has a period of π, our strategy is to graph $y = \tan x$ on interval $[-\pi/2, \pi/2]$, which has length π, and then repeat the resulting graph in both directions.

By using the graphs of $g(x) = \cos x$ and $f(x) = \sin x$ on interval $[-\pi/2, \pi/2]$ (Figures 1a and 1b), along with the quotient identity $\tan x = \sin x/\cos x$, we can discover three important features:

(1) the domain, (2) vertical asymptotes, and (3) an x intercept.

Figure 1

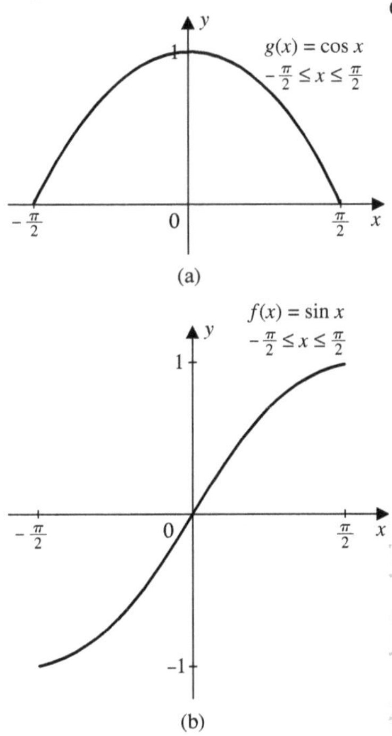

1. **Domain:** From the quotient relationship $\tan x = \sin x/\cos x$, it follows that $y = \tan x$ is undefined for those values of x where $\cos x = 0$. These values, which are the same as the x intercepts of the graph of $g(x) = \cos x$, are $x = -\pi/2$ and $x = \pi/2$ (Figure 1a). Consequently, the domain of $y = \tan x$ consists of all values of x for which $x \neq \pm\pi/2$ on interval $[-\pi/2, \pi/2]$.
2. **Asymptotes:** To help find out what happens to the values of $\tan x$ as x approaches $\pi/2$, we examine the entries in Table 1. They suggest that as x approaches $\pi/2$ from the left, $\tan x$ increases without bound; that is,

$$\tan x \to +\infty \quad \text{as} \quad x \to \left(\frac{\pi}{2}\right)^-.$$

Thus the line $x = \pi/2$ is a vertical asymptote of the graph.

TABLE 1

$x \to \left(\frac{\pi}{2}\right)^-$								
x	0	0.5	1.0	1.5	1.56	1.569	1.57	$\frac{\pi}{2}$
$y = \tan x$	0	0.546	1.557	14.101	92.620	556.691	1255.766	Undefined

Similarly, it can be shown that

$$\tan x \to -\infty \quad \text{as} \quad x \to \left(-\frac{\pi}{2}\right)^+$$

so $x = -\pi/2$ is also a vertical asymptote.

3. **x intercept:** The x intercept, that is, the value of x for which $\tan x = 0$ on interval $[-\pi/2, \pi/2]$, is the same as the x intercept of $f(x) = \sin x$ (Figure 1b) which is $x = 0$.

Figure 2

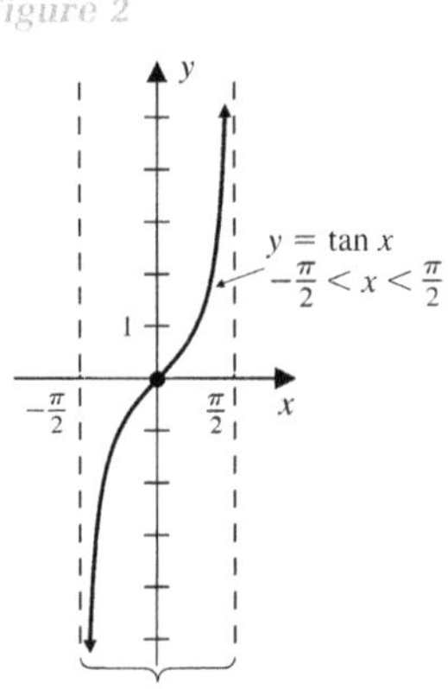

By using this information and plotting a few points, we obtain a sketch of the graph of $y = \tan x$ in the interval $(-\pi/2, \pi/2)$ (Figure 2).

Since the tangent function has a period of π, we repeat the graph in Figure 2 to the right and left every π units to get a complete graph of the tangent function (Figure 3). The periodicity of π enables us to locate all the other x intercepts and vertical asymptotes because they repeat every π units.

Figure 3

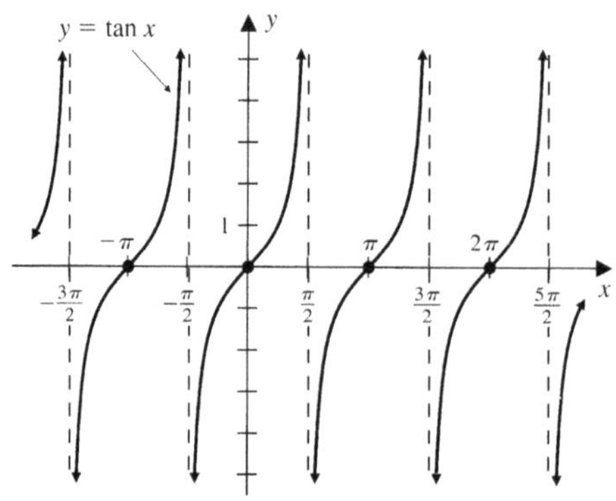

By reading the graph in Figure 3, we are led to the properties of the tangent function listed in Table 2.

TABLE 2 Properties of the Tangent Function

Property	$y = \tan x$
Domain	All real numbers except $(\pi/2) + n\pi$, where n is an integer
Range	All real numbers
Period	π
Symmetry	Symmetric with respect to the origin
Even or odd	The function is odd, that is, $\tan(-x) = -\tan x$
Asymptotes	$x = \pm\pi/2, \pm(3\pi)/2, \pm(5\pi)/2, \ldots, \pm[(\pi/2) + n\pi]$, where n is an integer
x intercepts	$0, \pm\pi, \pm 2\pi, \ldots \pm n\pi$, where n is an integer
Increasing or decreasing behavior	Increases between consecutive asymptotes

By using arguments similar to those used for the sine and cosine functions in Section 1.4, we are led to the following result.

Property

Tangent Period and Phase Shift

The period of $f(x) = a \tan(kx + b)$ is $\frac{\pi}{k}$,

and

its phase shift is $-\frac{b}{k}$, for $k > 0$.

If k is negative, we can use the fact that the tangent is an odd function to rewrite the given function with a positive coefficient of x, as we did for the sine and cosine functions.

For instance,

$$\begin{aligned} f(x) &= \tan(-3x + 1) \\ &= \tan[-(3x - 1)] \\ &= -\tan(3x - 1). \end{aligned}$$

EXAMPLE 1

Graphing a Tangent Function

Sketch the graph of

$$f(x) = \tan\left(x - \frac{\pi}{6}\right).$$

for one period by using a transformation of the graph of $y = \tan x$ in the interval $(-\pi/2, \pi/2)$. Specify the period, phase shift, and vertical asymptotes.

Solution We can shift the graph of $y = \tan x$ (Figure 4a), $\pi/6$ unit to the right to get the graph of f (Figure 4b). After rewriting $f(x)$ in the form $a \tan(kx + b)$, we get

$$f(x) = 1 \tan\left[1x + \left(-\frac{\pi}{6}\right)\right].$$

So that the period is

$\pi/k = \pi/1 = \pi$, which is the same as the period for $y = \tan x$,

and the phase shift is given by

$$\frac{-b}{k} = \frac{-(-\pi/6)}{1} = \frac{\pi}{6}.$$

The left asymptote for $y = \tan x$ shifts from $x = -\pi/2$ to

$$x = -\frac{\pi}{2} + \frac{\pi}{6} = -\frac{\pi}{3}$$

for f.
The right asymptote for $y = \tan x$ shifts from $x = \pi/2$ to

$$x = \frac{\pi}{2} + \frac{\pi}{6} = \frac{2\pi}{3}$$

for f.

Figure 4

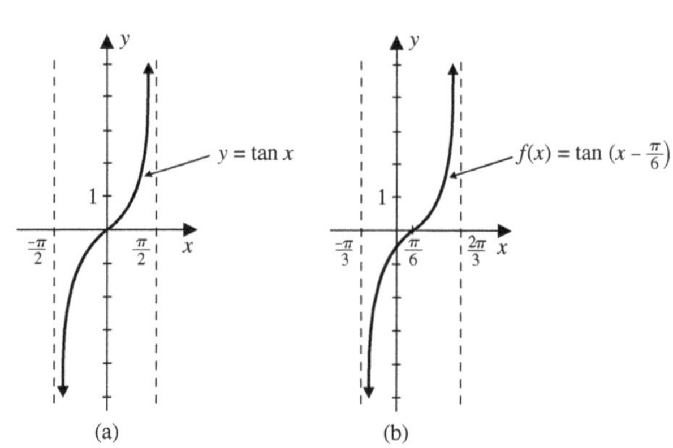

Graphing the Cotangent Function

Figure 5

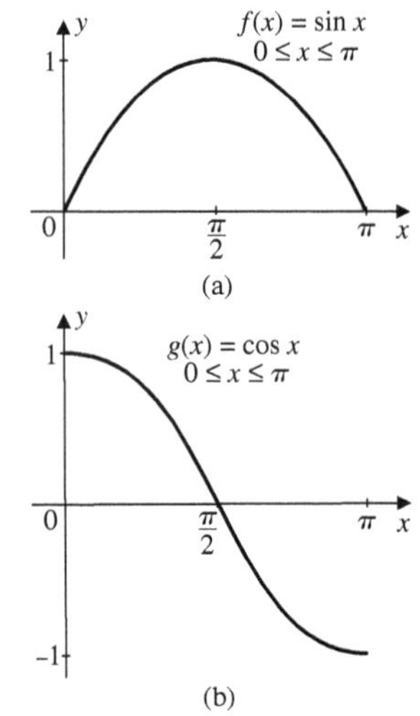

Since the cotangent has period π, we begin by graphing $y = \cot x$ on interval $[0, \pi]$. By using the quotient identity, along with the graphs of $f(x) = \sin x$ (Figure 5a) and $g(x) = \cos x$ (Figure 5b) on the interval $[0, \pi]$, we arrive at the following features.

1. **Domain:** The quotient relationship $\cot x = \cos x/\sin x$ implies that the $\cot x$ is undefined whenever $\sin x = 0$ on interval $[0, \pi]$, that is, when $x = 0$ or $x = \pi$. So the domain consists of all numbers x in interval $[0, \pi]$ except 0 and π.
2. **Asymptote:** By examining the behavior of $y = \cot x$ as $x \to 0^+$, we find that $\cot x \to +\infty$. Also, $\cot x \to -\infty$ as $x \to \pi^-$. So both $x = 0$ and $x = \pi$ are vertical asymptotes.
3. **x-intercept:** On interval $[0, \pi]$, $\cot x = 0$ whenever $\cos x = 0$, that is, if $x = \pi/2$. So $x = \pi/2$ is an x intercept.

By plotting some points and using the above information we obtain the graph of $y = \cot x$ for interval $[0, \pi]$ (Figure 6a).

To obtain the complete graph of $y = \cot x$, we use the fact that the cotangent has period π by repeating the graph in Figure 6a every π units to the right and left (Figure 6b).

Figure 6

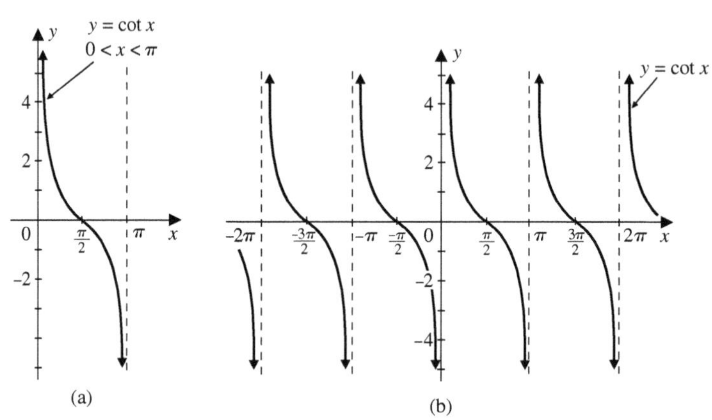

The properties of the cotangent function listed in Table 3 are read off of Figure 6b.

TABLE 3 Properties of the Cotangent Function

Property	$y = \cot x$
Domain	All real numbers except $n\pi$, where n is an integer
Range	All real numbers
Period	π
Symmetry	Symmetric with respect to the origin
Even or odd	The function is odd, that is, $\cot(-x) = -\cot x.$
Asymptotes	$x = 0, \pm\pi, \pm 2\pi, \ldots, \pm n\pi$, where n is an integer
x intercepts	$\pm\frac{\pi}{2}, \pm\frac{3\pi}{2}, \pm\frac{5\pi}{2}, \ldots, \pm\left(\frac{\pi}{2} + n\pi\right)$, where n is an integer
Increasing or decreasing behavior	Decreases between consecutive asymptotes

As in the case for the tangent function, graphs of cotangent functions of the form

$$f(x) = a\cot(kx + b)$$

have the following characteristics:

Property
Cotangent Period and Phase Shift

> The period for the function
>
> $$f(x) = a\cot(kx + b) \text{ is } \frac{\pi}{k}$$
>
> and its phase shift is
>
> $$\frac{-b}{k} \text{ for } k > 0.$$

EXAMPLE 2

Graphing a Cotangent Function

Sketch the graph of one period of the function

$$g(x) = \cot\left(2x - \frac{\pi}{2}\right)$$

by using transformations of the graph of one period of $y = \cot x$ in the interval $(0, \pi)$. Find the period and phase shift, and determine the asymptotes.

Solution We begin by graphing $y = \cot x$ in the interval $(0, \pi)$ (Figure 7a). Since

$$g(x) = \cot\left(2x - \frac{\pi}{2}\right) = 1\cot\left[2x + \left(-\frac{\pi}{2}\right)\right],$$

its period is given by

$$\frac{\pi}{k} = \frac{\pi}{2}$$

which is $\frac{1}{2}$ the period of $y = \cot x$.
Also, the phase shift of g is given by

$$\frac{-b}{k} = \frac{-\left(-\frac{\pi}{2}\right)}{2} = \frac{\pi}{4}.$$

Alternately, the phase shift can be found, as we did for sine and cosine functions, by solving

$$2x - \frac{\pi}{2} = 0 \quad \text{to get} \quad x = \frac{\pi}{4}.$$

It follows that the graph of g can be obtained by shifting the graph of $y = \cot x$ to the right $\pi/4$ unit and then horizontally compressing it by a scale factor of $\frac{1}{2}$ (Figure 7b).

Figure 7

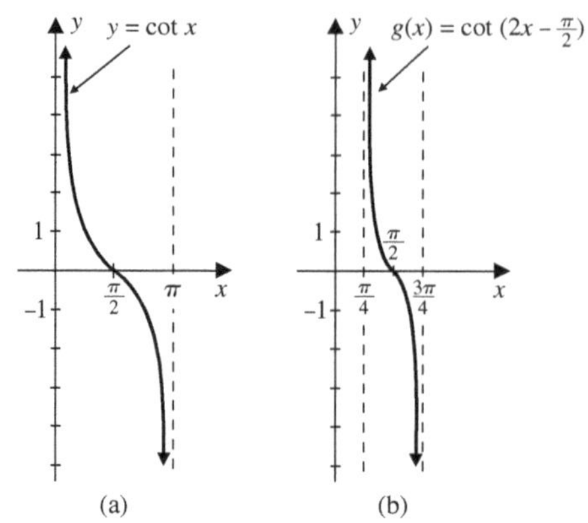

Notice that since the vertical asymptotes of $y = \cot x$ are at

$$x = 0 \text{ and } x = \pi,$$

correspondingly, the vertical asymptotes of g occur when

$$2x - \frac{\pi}{2} = 0 \text{ or } \pi, \text{ that is, when } x = \frac{\pi}{4} \text{ or } \frac{3\pi}{4}.$$

Graphing the Secant and Cosecant Functions

To graph the secant and cosecant functions, we take advantage of what we know about the graphs of the cosine and sine functions.

The graph of $y = \sec x$ can be sketched with the aid of the reciprocal identity

$$\sec x = \frac{1}{\cos x}.$$

To sketch the graph of $y = \sec x$, we first graph the cosine function $y = \cos x$. Vertical asymptotes of $y = \sec x$ occur at the x intercepts of $y = \cos x$, where $\cos x = 0$, that is, when $x = \pi/2 + n\pi$, for any integer of n.

Notice that because of the reciprocal relationship the secant increases where the cosine decreases, and it decreases where the cosine increases (Figure 8).

Figure 8

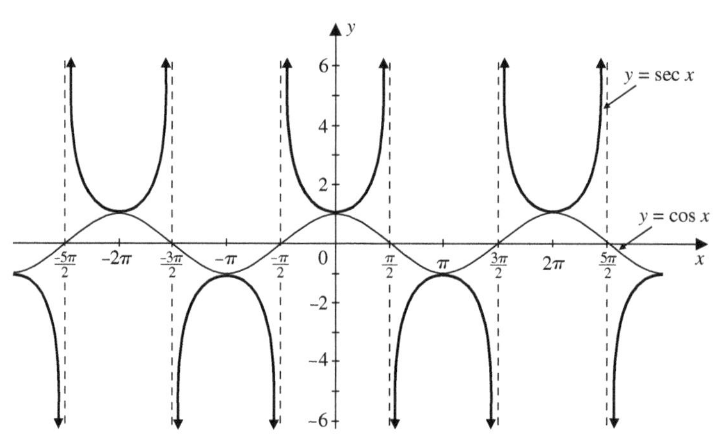

The graph shows that the range of the secant function includes all numbers y, where $y \leq -1$ or $y \geq 1$. In addition, the graph repeats itself every 2π units, confirming the fact that the function $y = \sec x$ is periodic with period 2π.

Similarly, the graph of the cosecant function $y = \csc x$ can be sketched by using

$$\csc x = \frac{1}{\sin x}$$

and the sine graph.

The vertical asymptotes occur when $\sin x = 0$, or when $x = n\pi$, for any integer n. Here the cosecant increases where the sine decreases, and it decreases where the sine increases (Figure 9).

Figure 9

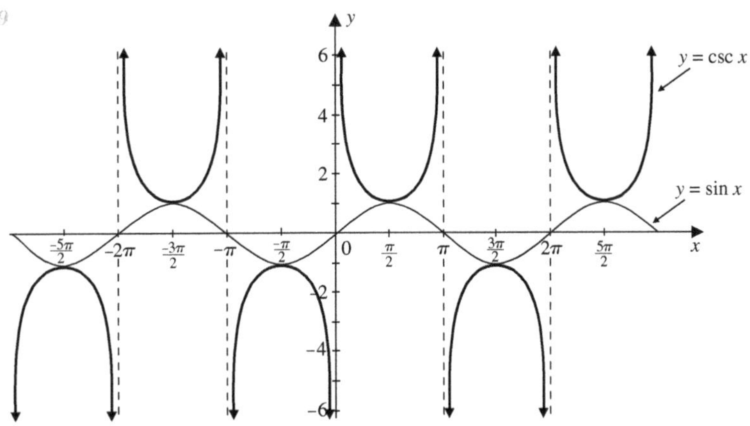

The graph reveals that the range of the cosecant function includes all numbers y, where $y \leq -1$ or $y \geq 1$. The graph repeats every 2π units, confirming that the function $y = \csc x$ is periodic with period 2π.

Table 4 summarizes properties of the secant and cosecant functions, which can be read off of the graphs in Figures 8 and 9.

TABLE 4 Properties of the Secant and Cosecant Functions

Property	$y = \sec x$	$y = \csc x$
Domain	All real numbers except $\pi/2 + n\pi$, where n is an integer	All real numbers except $n\pi$, where n is an integer
Range	All real numbers y such that $y \leq -1$ or $y \geq 1$	All real numbers y such that $y \leq -1$ or $y \geq 1$
Period	2π	2π
Symmetry	Symmetric with respect to the y axis	Symmetric with respect to the origin
Even or odd	The function is *even*, that is, $\sec(-x) = \sec x$	The function is *odd*, that is, $\csc(-x) = -\csc x$
Aymptotes	$x = (\pi/2) + n\pi$, where n is an integer	$x = n\pi$, where n is an integer

As with the sine and cosine functions, we have the following graph features.

Property

Secant and Cosecant, Period and Phase Shift

The functions

$$f(x) = a \sec(kx + b) \text{ and } g(x) = a \csc(kx + b)$$

have period

$$\frac{2\pi}{k}$$

and phase shift

$$\frac{-b}{k} \quad \text{for } k > 0.$$

For instance, the function

$$f(x) = 2\csc\left(3x - \frac{\pi}{4}\right) \text{ has a period of } \frac{2\pi}{3}$$

and a phase shift of $\dfrac{-\left(\dfrac{-\pi}{4}\right)}{3} = \dfrac{\pi}{12}$.

The graphs of the secant and cosecant functions shown in Figures 8 and 9, respectively, can be used to graph other trigonometric functions, as illustrated in the next example.

EXAMPLE 3

Graphing Secant and Cosecant Functions

Use transformations to sketch the graph of one period of each function. Find the period, phase shift, vertical asymptotes, and the range.

(a) $f(x) = 3\csc\dfrac{x}{2}$ (b) $g(x) = \sec(3x + \pi)$

Solution

(a) We will use one period of the graph of $y = \csc x$ in interval $(-\pi, \pi)$ (Figure 9) to sketch one period of the graph of

$$f(x) = 3\csc\frac{x}{2} = 3\csc\left(\frac{1}{2}x\right).$$

In the form

$$f(x) = a\csc(kx + b),$$

$$a = 3, \quad b = 0, \quad \text{and} \quad k = \frac{1}{2}.$$

So the period of f is

$$\frac{2\pi}{k} = \frac{2\pi}{\left(\frac{1}{2}\right)} = 4\pi = 2 \cdot 2\pi.$$

Since $b = 0$, the phase shift is 0.

Thus to get the graph of f from the graph of $y = \csc x$, which has a period of 2π, the latter graph has to be stretched horizontally by a factor of 2.

Since the vertical asymptotes for $y = \csc x$ on interval $[-\pi, \pi]$ occur when

$$x = -\pi, \quad 0, \quad \text{or} \quad \pi,$$

it follows that three vertical asymptotes for f occur when

$$x/2 = -\pi, \quad 0, \quad \text{or} \quad \pi,$$

that is, when

$$x = -2\pi, \quad 0, \quad \text{or} \quad 2\pi.$$

Because $a = 3$, we also need to stretch the graph vertically to get the graph of f. (Figure 10a). Note that the vertical stretch of multiple of 3 causes the range to become all values of y such that $y \le -3$ or $y \ge 3$.

(b) We can obtain one period of the graph of

$$g(x) = \sec(3x + \pi)$$

from one period of the graph of $y = \sec x$ on interval $[0, 2\pi]$ (Figure 8). The period of g is given by

Figure 10

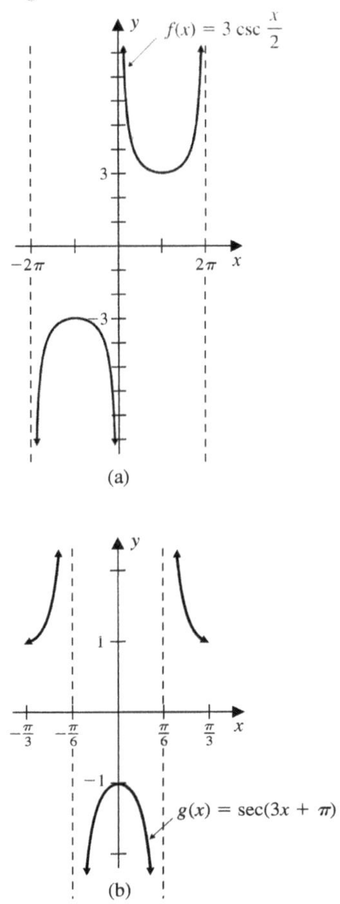

$$\frac{2\pi}{k} = \frac{2\pi}{3} = \frac{1}{3} \cdot 2\pi.$$

So the graph of $y = \sec x$, which has period 2π, must be horizontally shrunk by a multiple of 1/3. Also, we need to shift the graph to the left by $\pi/3$ units because the phase shift is

$$-\frac{b}{k} = -\frac{\pi}{3}.$$

Two vertical asymptotes of $y = \sec x$ occur when

$$x = \frac{\pi}{2} \quad \text{or} \quad \frac{3\pi}{2}.$$

Correspondingly, g has vertical asymptotes when

$$3x + \pi = \frac{\pi}{2} \quad \text{or} \quad \frac{3\pi}{2}$$

that is, when

$$x = -\frac{\pi}{6} \quad \text{or} \quad \frac{\pi}{6}.$$

Figure 10b shows the graph of one period of g. Since there is no vertical scaling ($a = 1$), the range of g includes all values of y such that $y \le -1$ or $y \ge 1$.

Most graphers do not have function keys for the cotangent, secant, and cosecant functions. To graph these functions, we use the TAN, COS, and SIN keys, respectively, and the reciprocal key 1/x or x^{-1}, as we did when we evaluated these functions. For instance the graph of $y = \cot(4x - 3)$ is obtained by graphing

$$y = \frac{1}{\tan(4x - 3)}.$$

We graph $f(x) = 5 \sec 3x$ by graphing

$$f(x) = \frac{5}{\cos 3x}$$

and so forth.

Care must be taken when reading such graphs produced by a grapher. Depending on the window settings and the grapher's features, it may turn out that the graph will appear connected or continuous even though it should not because there are vertical asymptotes. Figure 11 displays such an inaccuracy for the graph of $y = \cot(4x - 3)$.

Figure 11

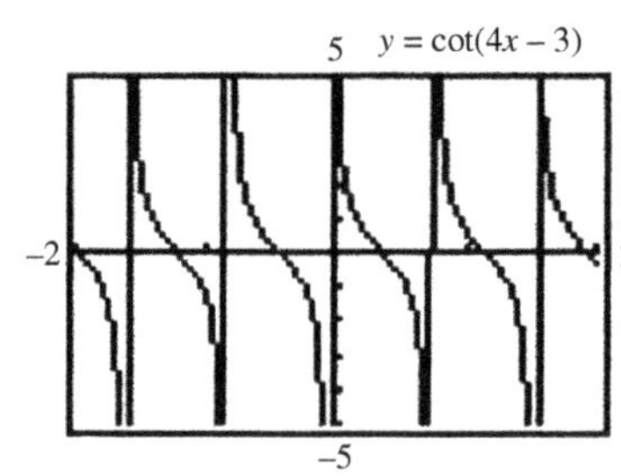

Graphers enable us to gain more insight into the characteristics of mathematical models involving trigonometric functions. For instance, Figure 12 shows the graph of the function $L = 8 \csc t + \sec t$, $0 < t < \pi/2$.

Upon reading the graph of L, we observe there is a minimum or a low point on the graph. By using the MIN-finding feature of a grapher, we find the low point is located at approximately (1.11, 11.18).

Figure 12

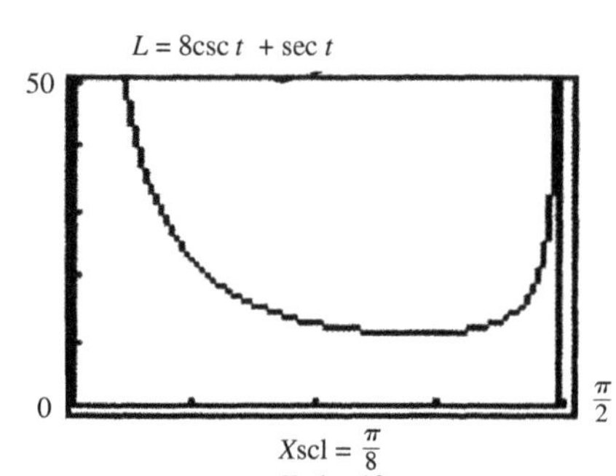

PROBLEM SET 1.5

Mastering the Concepts

In problems 1–24, use transformations of the graphs of the basic trigonometric functions to sketch the graph of one period for each function. Specify the period and phase shift.

1. $y = 3 \tan x$

2. $y = -3 \tan x$

3. $y = -\frac{3}{2} \cot x$

4. $y = -2 \sec x$

5. $y = 5 \csc x$

6. $y = 2 \sec \frac{x}{4}$

7. $y = 1 + 2 \csc x$

8. $y = 2 - \sec x$

9. $y = -2 \cot 4x$

10. $y = 2 \tan \frac{\pi x}{4}$

11. $y = -2 \sec \frac{2x}{3}$

12. $y = -3 \csc \frac{\pi x}{6}$

13. $y = 2 \csc 4x$

14. $y = -3 \sec \pi x$

15. $y = 2 \csc\left(x - \frac{\pi}{6}\right)$

16. $y = 3 \csc\left(x + \frac{\pi}{3}\right)$

17. $y = \tan\left(x - \frac{\pi}{4}\right)$

18. $y = \tan\left(x + \frac{2\pi}{3}\right)$

19. $y = 1.5 \cot\left(x - \frac{\pi}{2}\right)$

20. $y = 3 \cot\left(x + \frac{5\pi}{6}\right)$

21. $y = 3 \sec\left(\frac{x}{4} + \frac{\pi}{4}\right)$

22. $y = 2 \csc\left(3x - \frac{\pi}{2}\right)$

23. $y = -2 \cot\left(\frac{\pi x}{4} - \frac{\pi}{4}\right)$

24. $y = -2 \sec\left(\frac{\pi x}{6} + \frac{\pi}{6}\right)$

25. Use the graphs of the functions to indicate where each function listed in the table is increasing or decreasing.

	As x increases from:			
	0 to $\frac{\pi}{2}$	$\frac{\pi}{2}$ to π	π to $\left(\frac{3\pi}{2}\right)$	$\left(\frac{3\pi}{2}\right)$ to 2π
(a) $y = \tan x$				
(b) $y = \cot x$				
(c) $y = \sec x$				
(d) $y = \csc x$				

26. (a) Explain why $y = \tan x$ is not one-to-one.

(b) Is $f(x) = \tan x$, where $-\pi/2 < x < \pi/2$, one-to-one? Use a graph to support the assertion.

(c) Is $f(x) = \tan x$, where $0 < x < 2\pi$, one-to-one? Use a graph to support the assertion.

In problems 27–30, use the graph of the indicated function to complete each statement.

27. (a) As $x \to \left(\frac{\pi}{4}\right)^+$, $\tan x \to$ _____

(b) As $x \to \left(\frac{\pi}{2}\right)^+$, $\tan x \to$ _____

28. (a) As $x \to 0^+$, $\cot x \to$ _____

(b) As $x \to \left(\frac{\pi}{2}\right)^+$, $\sec x \to$ _____

29. (a) As $x \to \left(\frac{\pi}{2}\right)^-$, $\sec x \to$ _____

(b) As $x \to \left(\frac{\pi}{4}\right)^-$, $\cot x \to$ _____

30. (a) As $x \to \left(\frac{\pi}{6}\right)^-$, $\tan x \to$ _____

(b) As $x \to \pi^+$, $\cot \to$ _____

In problems 31–34, one period of a graph is shown. Find an equation of the given form for the graph.

31. $y = \tan kx$, $k > 0$

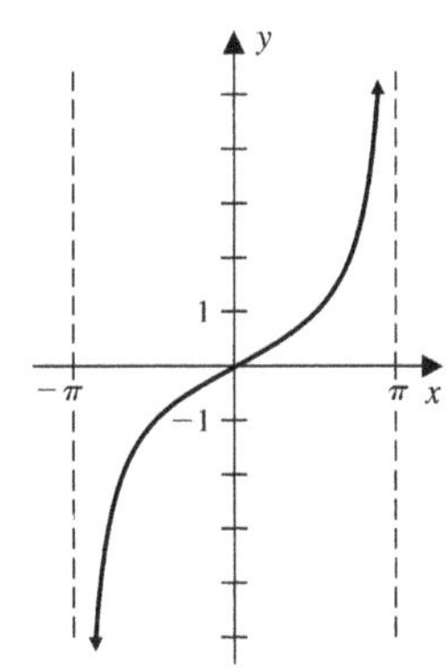

32. $y = a \tan kx$, $a < 0$ and $k > 0$

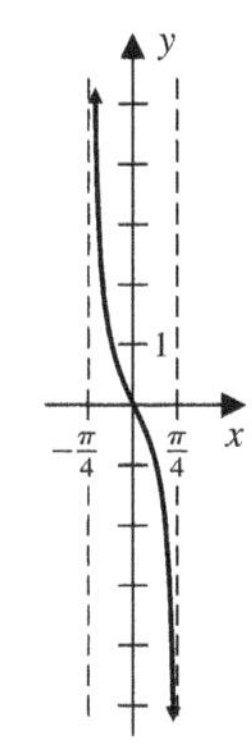

33. $y = a \csc(kx + b)$, $a > 0$, $b > 0$, $k > 0$

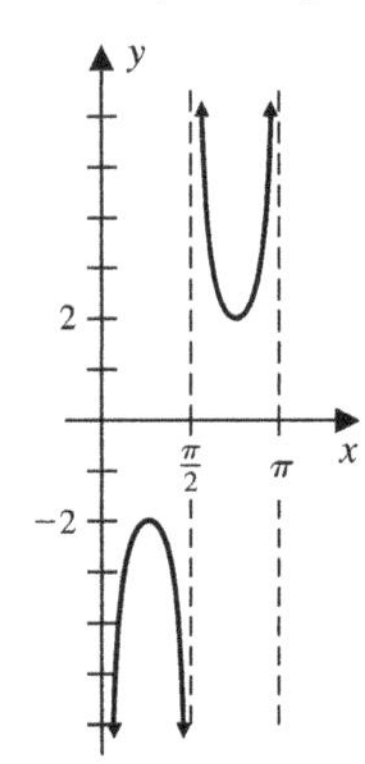

34. $y = a \sec kx$, $a > 0$ and $k > 0$

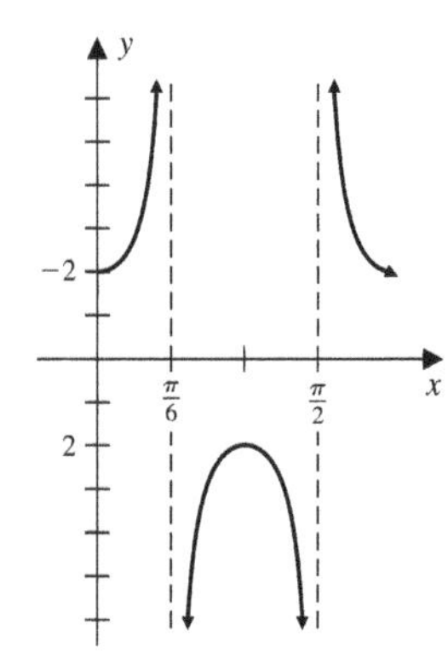

In problems 35–38, what is the distance from one vertical asymptote to the next for each function?

35. $y = \sec\left(x - \frac{\pi}{2}\right)$

36. $y = -2 \cot \frac{x}{6}$

37. $y = 2 \tan 3x$

38. $y = \csc(x + 1)$

[G] In problems 39 and 40,

(a) Use a grapher to graph the function defined by L in the interval $(0, \pi/2)$.

(b) Use the [MIN]-finding feature to approximate the location of the lowest point on the graph in part (a) to two decimal places.

39. $L = 2\sqrt{2}(\sec t + \csc t)$

40. $L = 12 \csc t + 20 \sec t$

Developing and Extending the Concepts

41. Use the graph of $y = \tan x$ to obtain the graph of

$$f(x) = |\tan x|$$

on the interval $\left(\frac{-\pi}{2}, \frac{\pi}{2}\right)$.

42. Compare the graphs of $f(x) = \sec x$ and $g(x) = \sec(-x)$. Are they the same or different? Explain.

43. Compare the graphs of $f(x) = \csc x$ and $g(x) = \csc(-x)$. Are they the same or different? Explain.

44. Sketch the graphs of

$$f(x) = \tan x$$

and

$$g(x) = \cot\left(\frac{\pi}{2} - x\right)$$

on the same coordinate system. How do the graphs compare?

45. Sketch the graphs of

$$f(x) = \sec x \quad \text{and} \quad g(x) = \csc\left(\frac{\pi}{2} - x\right)$$

on the same coordinate system. How do the graphs compare?

46. Explain why the graphs of $y = \tan x$, $f(x) = \tan(x + \pi)$, $g(x) = \tan(x - \pi)$, and $h(x) = \tan(x + 2\pi)$ are the same.

In problems 47 and 48, write an equation of the form $y = \tan kx$ or $y = \csc kx$ that satisfies the given condition.

47. (a) Tangent function of period $\frac{\pi}{6}$

(b) Cosecant function of period $\frac{5\pi}{6}$

48. (a) Tangent function of period 2

(b) Cosecant function of period 1.5

49. Show that the functions $f(x) = a \tan(kx + b)$ and $g(x) = a \cot(kx + b)$ both have period π/k and phase shift $-b/k$ for $k > 0$.

50. Show that the functions $f(x) = a \sec(kx + b)$ and $g(x) = a \csc(kx + b)$ both have period $2\pi/k$ and phase shift $-b/k$ for $k > 0$.

51. [G] Use a grapher to graph $f(x) = (\tan x)(\cos x)$ and $g(x) = \sin x$, where $-2\pi \le x \le 2\pi$. Are the graphs the same? Are the functions equal? Explain.

Interlude: Practice Problems 1.5

1. Sketch the graph of $y = \tan x + \cot x$. Find appropriate constants a, b, and k such that the graph of $y = a \sec(kx + b)$ matches the graph of $y = \tan x + \cot x$. Show clearly how you arrived at the values of the constants.

2. Sketch the graph of $y = \csc x + \cot x$. Find appropriate constants a, b, and k such that the graph of $y = a\tan(kx + b)$ matches the graph of $y = \csc x + \cot x$. Show clearly how you arrived at the values of the constants.

Objectives

1. Define Trigonometric Functions of Angles
2. Establish the Equivalency of Trigonometric Functions
3. Approximate Trigonometric Function Values of Angles
4. Solve Applied Problems

1.6 Trigonometric Functions of Angles

So far we have dealt with trigonometric functions of real numbers. Now we define the trigonometric functions of angles. We shall see how the two definitions are related.

Defining Trigonometric Functions of Angles

Here we extend the notion of trigonometric functions to include *any* angle. The angle can be measured in degrees or radians and can be positive, negative, or zero.

Definition

Trigonometric Functions of an Angle

Suppose that θ is an angle in standard position. Assume $(x, y) \neq (0, 0)$ is a point on the terminal side of θ at a distance $r = \sqrt{x^2 + y^2}$ from the origin (Figure 1).

Then the trigonometric functions of θ are defined as follows:

$$\sin\theta = \frac{y}{r} \qquad \csc\theta = \frac{r}{y},\quad y \neq 0$$

$$\cos\theta = \frac{x}{r} \qquad \sec\theta = \frac{r}{x},\quad x \neq 0$$

$$\tan\theta = \frac{y}{x},\quad x \neq 0 \qquad \cot\theta = \frac{x}{y},\quad y \neq 0$$

Figure 1

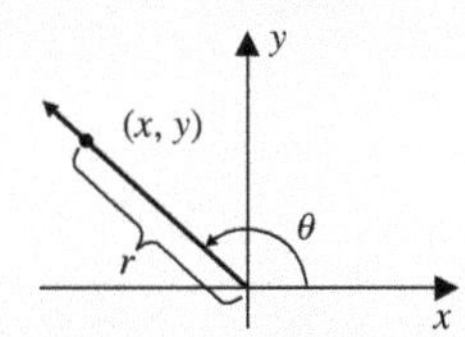

EXAMPLE 1

Evaluating Trigonometric Functions of Angles

Find the six trigonometric function values of each angle.

(a) α is an angle in standard position that contains the point $(5, -12)$ on its terminal side.

(b) $\beta = 180°$

Solution

(a) Figure 2a shows two possibilities for an angle α in standard position, one positive and one negative, whose terminal side contains the point $(5, -12)$. These two representations for α are coterminal.

Since $x = 5$ and $y = -12$, the distance r from the origin to $(5, -12)$ is given by

$$r = \sqrt{x^2 + y^2} = \sqrt{5^2 + (-12)^2} = \sqrt{25 + 144} = \sqrt{169} = 13$$

By the above definition, we have

$$\sin\alpha = \frac{y}{r} = \frac{-12}{13} \qquad \csc\alpha = \frac{r}{y} = \frac{13}{-12} = -\frac{13}{12}$$

$$\cos\alpha = \frac{x}{r} = \frac{5}{13} \qquad \sec\alpha = \frac{r}{x} = \frac{13}{5}$$

$$\tan\alpha = \frac{y}{x} = \frac{-12}{5} \qquad \cot\alpha = \frac{x}{y} = \frac{5}{-12} = -\frac{5}{12}.$$

Figure 2a

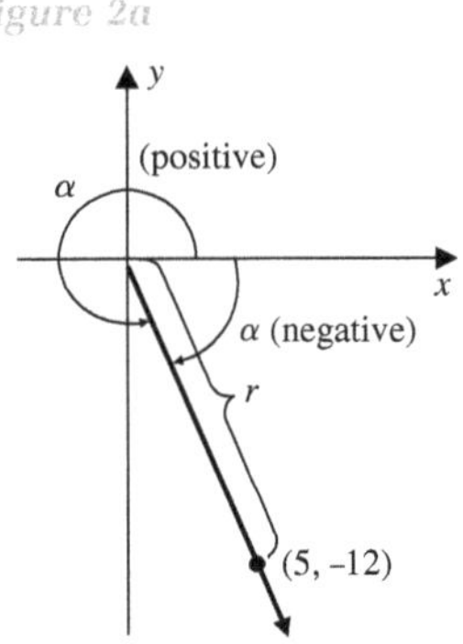

We could have chosen any point except (0, 0) on the terminal side.

(b) First, we display $\beta = 180°$ in standard position and then select a point on its terminal side. We choose the point $(-2, 0)$ (Figure 2b). For this selection,

$$x = -2 \quad \text{and} \quad y = 0,$$

and its distance from the origin is

$$r = 2.$$

By definition, we have

Figure 2b

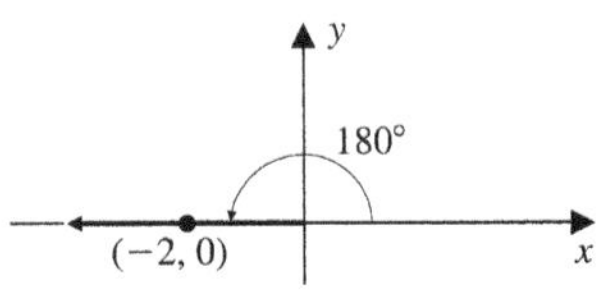

$$\sin 180^\circ = \frac{y}{r} = \frac{0}{2} = 0 \qquad \csc 180^\circ = \frac{r}{y} = \frac{2}{0}\text{ (undefined)}$$

$$\cos 180^\circ = \frac{x}{r} = \frac{-2}{2} = -1 \qquad \sec 180^\circ = \frac{r}{x} = \frac{2}{-2} = -1$$

$$\tan 180^\circ = \frac{y}{x} = \frac{0}{-2} = 0 \qquad \cot 180^\circ = \frac{x}{y} = \frac{-2}{0}\text{ (undefined).}$$

Establishing the Equivalency of Trigonometric Functions

Now we examine the connection between the trigonometric functions of real numbers or circular functions and the trigonometric functions of angles. Suppose that t is a real number and $P(t) = (x, y)$; then the trigonometric functions defined on *real numbers* (Figure 3) yields

$$\cos t = x \quad \text{and} \quad \sin t = y.$$

Next, we assume that θ is an angle in standard position so that point $P(t)$ is on the terminal side of θ (Figure 3). (The argument holds regardless of the quadrant where the terminal side of θ is located.) Since (x, y) is located on the unit circle where $r = 1$ (Figure 3), then, by using the definitions of trigonometric functions of angles, we have

Figure 3

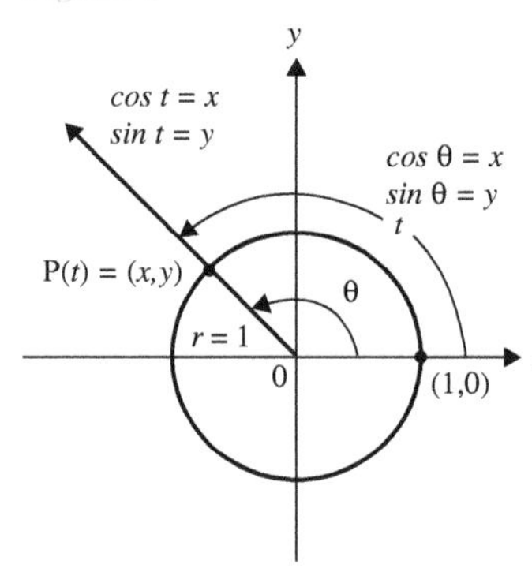

$$\cos \theta = \frac{x}{r} = \frac{x}{1} = x \quad \text{and} \quad \sin \theta = \frac{y}{r} = \frac{y}{1} = y.$$

Notice that the real number t is the same as the radian measure of θ. Thus

$$\cos t = \cos \theta \quad \text{and} \quad \sin t = \sin \theta$$

where $\cos t$ and $\sin t$ are the values of the trigonometric functions defined on real number t, and $\cos \theta$ and $\sin \theta$ are the values of the trigonometric functions defined on angle θ with radian measure t.

Similarly, if angle θ has radian measure t, we can use the definition of the trigonometric functions of *real number* t and the definition of the trigonometric functions of *angle* θ to establish

$$\tan t = \tan \theta \qquad \cot t = \cot \theta$$

$$\sec t = \sec \theta \qquad \csc t = \csc \theta.$$

Equivalency Property
Trigonometric Functions

> Consequently, we consider the trigonometric functions of angle θ to be equivalent to the trigonometric functions of a real number t, which is the radian measure of θ, and we use the terminology *trigonometric functions* regardless of whether *angles* or *real numbers* are employed.

For instance, the sine of an angle of *radian measure* $\pi/4$ is the same as the sine of the *real number* $\pi/4$. Even if an angle is given in degrees, we can associate the trigonometric functions of real numbers with the trigonometric functions of the angle by converting the degrees to radians. For example,

$$\sin 30^\circ = \sin\frac{\pi}{6} = \frac{1}{2} \quad\text{and}\quad \cos 45^\circ = \cos\frac{\pi}{4} = \frac{\sqrt{2}}{2}.$$

EXAMPLE 2

Evaluating Trigonometric Functions

Find the exact value of each expression.

(a) $\cos 120^\circ$ (b) $\sin 120^\circ$ (c) $\tan 240^\circ$

Solution By using Figure 4 and the symmetry of the unit circle, we see that

(a) $\cos 120^\circ = \cos\dfrac{2\pi}{3} = -\dfrac{1}{2}$

(b) $\sin 120^\circ = \sin\dfrac{2\pi}{3} = \dfrac{\sqrt{3}}{2}$

(c) $\tan 240^\circ = \tan\dfrac{4\pi}{3} = \sqrt{3}$

Figure 4

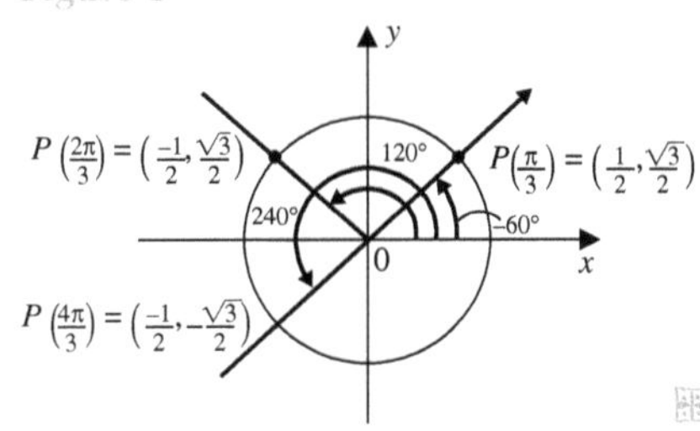

Notice that the identities and properties established earlier for trigonometric functions of real numbers are also true for trigonometric functions of angles because of the above equivalancy property. In fact, with the above property in mind, we commonly use the term *trigonometric functions* regardless of whether *angles* or *real numbers* are employed.

Note that the values of the trigonometric functions of angles in standard position depend only on the location of the terminal side of the angle but they do not depend on the choice of the point P on the terminal side of the angle. It follows that the trigonometric function values of coterminal angles are the same, no matter what particular point is selected on the terminal side of the angles.

For convenience, we use the symbol θ to represent an angle, the degree measure of the angle, or the radian measure of the angle. The context of our discussion will clarify which meaning of θ is being used.

Thus, if T represents any of the trigonometric functions we have the following result:

Coterminal Property

> $$T(\theta) = T(\theta + 2\pi n) \text{ (if radians are used)}$$
>
> or
>
> $$T(\theta) = T(\theta + 360^\circ n) \text{ (if degrees are used)}$$
>
> for any integer n.

This property reinforces the fact that the trigonometric functions have period 2π.

EXAMPLE 3

Using the Coterminal Property

Use the preceding coterminal property to find the value of each expression.

(a) $\cos(-630°)$

(b) $\sin \dfrac{25\pi}{6}$

(c) $\cot\left(-\dfrac{21\pi}{4}\right)$

(d) $\sec 1380°$

Solution

(a) Since $-630°$ is coterminal with $90°$, we have

$$\cos(-630°) = \cos[-630° + 2(360°)] = \cos 90° = \cos \frac{\pi}{2} = 0.$$

(b) Since

$$\frac{25\pi}{6} = \frac{\pi}{6} + 4\pi = \frac{\pi}{6} + 2(2\pi),$$

the angles $\dfrac{25\pi}{6}$ and $\dfrac{\pi}{6}$ are coterminal, so that

$$\sin \frac{25\pi}{6} = \sin \frac{\pi}{6} = \frac{1}{2}.$$

(c) Since

$$\frac{-21\pi}{4} = \frac{3\pi}{4} - 6\pi,$$

the angles $\dfrac{-21\pi}{4}$ and $\dfrac{3\pi}{4}$ are coterminal, so that

$$\cot\left(\frac{-21\pi}{4}\right) = \cot \frac{3\pi}{4} = -1.$$

(d) Since $1380° = 300° + 3(360°)$, the angles $1380°$ and $300°$ are coterminal, so that

$$\sec 1380° = \sec 300° = \sec \frac{5\pi}{3} = 2.$$

Approximating Trigonometric Function Values of Angles

We approximate values of trigonometric functions for angles by using a calculator and the reciprocal relationships if necessary, as we did for real numbers.

EXAMPLE 4

Using a Calculator to Approximate Values of Trigonometric Functions

Use a calculator to approximate each value to four decimal places.

(a) $\sin 162°$ (b) $\cot \dfrac{7\pi}{30}$ (c) $\csc(-4.6481)$

Solution (a) After setting the calculator in degree mode, we get

$$\sin 162° = 0.3090 \text{ (approx.)}.$$

(b) Here we set the calculator in radian mode. Then we use a reciprocal relationship to get

$$\cot \frac{7\pi}{30} = \frac{1}{\tan \frac{7\pi}{30}} = 1.1106 \text{ (approx.)}.$$

(c) We set the calculator in radian mode and we use a reciprocal relationship to get

$$\csc(-4.6481) = \frac{1}{\sin(-4.6481)} = 1.0021 \text{ (approx.)}.$$

Solving Applied Problems

The concepts of slopes of lines and angles are used in many applications and mathematical models. Highway engineers use slopes to measure the *steepness* of a roadway, referred to as the **grade** of the roadway, to indicate whether a roadway is *uphill* or *downhill*. For example, a (positive) 4% grade indicates a slope of 4/100 so that the roadway is uphill and it rises 4 feet vertically for every 100 feet of horizontal change. Figure 5 displays an uphill grade denoted by a *positive angle* and a downhill grade denoted by a *negative angle*.

Figure 5

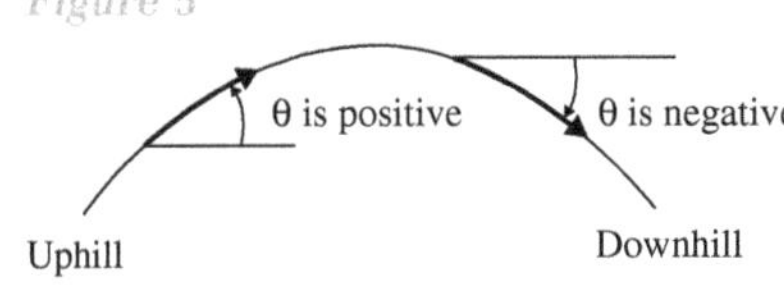

EXAMPLE 5

Modeling a Highway Ramp

Figure 6 shows an angle of $-5°$ for a downhill grade of a 200-foot ramp from an elevated service drive to an expressway.

Figure 6

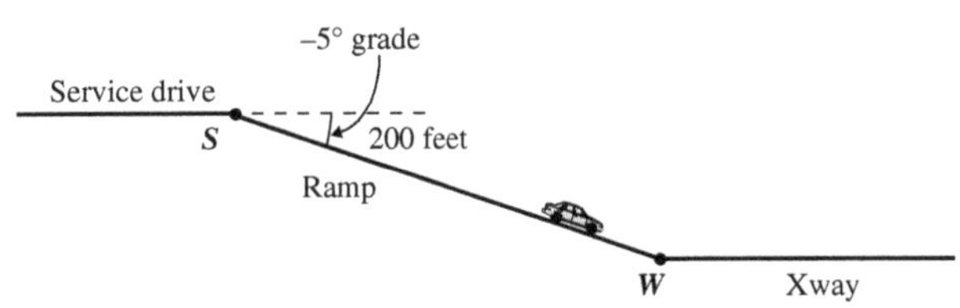

(a) Sketch a model of the situation in a Cartesian system so that the ramp entry location S is at the origin and the service drive is on the x axis.

(b) Use the model in part (a) to find the location of W where the ramp meets the expressway. Round off the answers to two decimal places. Interpret the location of W in terms of the elevation of the service drive above the expressway and the horizontal distance spanned by the ramp.

(c) What is the slope of the ramp? Round off the answer to two decimal places.

Solution (a) Figure 7 shows the representation or model in a Cartesian system. Point S, the entry location of the ramp, is at $(0, 0)$, point $W = (x, y)$ is the location where the ramp meets the expressway, and the service drive is on the x axis. Notice the grade is shown as a negative angle since the ramp is downhill.

Figure 7

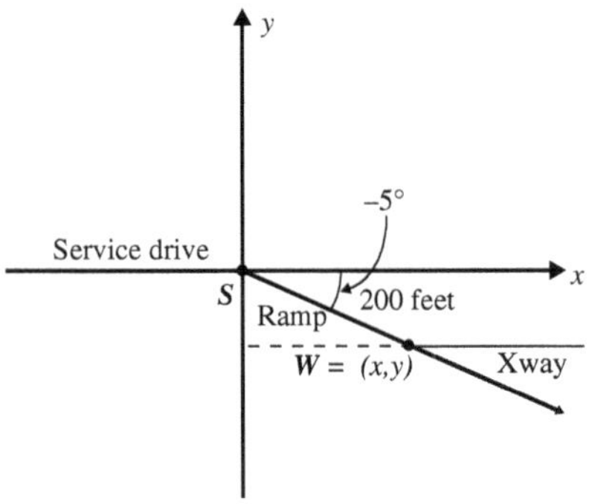

(b) We use the trigonometric functions defined for general angles to find the location of point $W = (x, y)$ as follows:

$$\cos(-5°) = \frac{x}{r} = \frac{x}{200}$$

so that

$$x = 200\cos(-5°)$$
$$= 199.24 \text{ (approx.)}$$

and

$$\sin(-5°) = \frac{y}{200}$$

so that

$$y = 200\sin(-5°)$$
$$= -17.43 \text{ (approx.)}.$$

Thus the location of W is given by

$$W = (199.24, -17.43) \text{ (approx.)}.$$

The location of W indicates that the ramp spans a horizontal distance of approximately 199.24 feet and the service drive is approximately 17.43 feet above the expressway.

(c) The line $\overline{SW}$ defined by the ramp contains points $W = (x, y)$ and the origin $S = (0, 0)$ (Figure 7). Its slope is given by

$$\text{Slope of ramp} = \frac{y - 0}{x - 0} = \frac{y}{x}$$

Slope formula: $m = \dfrac{y_1 - y_2}{x_1 - x_2}$.

To determine this ratio $\frac{y}{x}$ we can proceed in either of the following two ways:

Use the Results from Part (b)	Alternate Solution
From part (b) we found that $x = 199.24$ (approx.) and $y = -17.43$ (approx.) so that the $\text{slope} = \frac{y}{x} = \frac{-17.43}{199.24} = -0.09$ (approx.) or about −9%.	By definition, $\tan\theta = \frac{y}{x}$ so that the $\text{slope} = \tan(-5°) = -0.09$ (approx.) or about −9%.

PROBLEM SET 1.6

Mastering the Concepts

In problems 1–10, angle θ is in standard position. Find the values of the six trigonometric functions of angle θ if the terminal side of θ contains the given point (x, y) which is r units from the origin. Sketch two possible angles, one positive and one negative, that satisfy the conditions on θ. Round off the answers to three decimal places.

1. $(x, y) = (4, 3)$
2. $(x, y) = (-3, -4)$
3. $(x, y) = (-5, 12)$
4. $(x, y) = (7, -10)$
5. $(x, y) = (1.35, -2.76)$
6. $(x, y) = (-7.89, 9.39)$
7. $(x, y) = \left(a, -\frac{1}{2}\right); r = 1, a < 0$
8. $(x, y) = (8, b); r = 17, b < 0$
9. $(x, y) = (a, a); a > 0$
10. $(x, y) = (-a, a); a < 0$

In problems 11–14, find the exact values of the six trigonometric functions of each angle.

11. $90°$
12. π
13. $-\frac{\pi}{2}$
14. $-270°$

In problems 15–22, use the known locations of $P\left(\frac{\pi}{6}\right)$, $P\left(\frac{\pi}{4}\right)$, and $P\left(\frac{\pi}{3}\right)$, along with symmetry, to find the exact value of each expression.

15. (a) $\sin 150°$
    (b) $\cos 225°$
16. (a) $\cos\left(-\frac{7\pi}{6}\right)$
    (b) $\cot 330°$
17. (a) $\tan 240°$
    (b) $\cot \frac{11\pi}{6}$
18. (a) $\sec\left(-\frac{11\pi}{6}\right)$
    (b) $\csc\left(-\frac{7\pi}{6}\right)$
19. (a) $\sec 315°$
    (b) $\csc 300°$
20. (a) $\csc(-330°)$
    (b) $\cot\left(-\frac{4\pi}{3}\right)$
21. (a) $\tan\left(-\frac{\pi}{4}\right)$
    (b) $\cot 210°$
22. (a) $\sin\left(-\frac{7\pi}{4}\right)$
    (b) $\sin \frac{7\pi}{6}$

In problems 23–32, use the coterminal property to find the exact value.

23. $\sin 945°$
24. $\sin(-675°)$
25. $\tan 1020°$
26. $\cos 750°$
27. $\tan\left(-\frac{41\pi}{4}\right)$
28. $\csc\left(-\frac{8\pi}{3}\right)$
29. $\cos\left(\frac{61\pi}{6}\right)$
30. $\cot(5\pi)$
31. $\cos 900°$
32. $\sec(-7\pi)$

In problems 33–40, use a calculator to find the approximate values of the six trigonometric functions of θ. Round off the answers to four decimal places.

33. (a) $\theta = 137°$
    (b) $\theta = -95.8°$
34. (a) $\theta = 195°$
    (b) $\theta = -10°$
35. (a) $\theta = 380°$
    (b) $\theta = -\pi°$
36. (a) $\theta = -417°$
    (b) $\theta = 215°15'5''$
37. (a) $\theta = 2$
    (b) $\theta = -1.3$
38. (a) $\theta = -2$
    (b) $\theta = 1.3$
39. (a) $\theta = -5.22$
    (b) $\theta = \pi$
40. (a) $\theta = 5.22$
    (b) $\theta = \pi^{-1}$

Applying the Concepts

41. A downhill ramp of length 185 feet from a service drive to a highway has a $-4.5°$ grade.
    (a) Sketch a model of the situation in a Cartesian system so that the ramp's entry point S is at the origin and the point W where the ramp meets the highway is in quadrant IV.
    (b) Use the model in part (a) to find the location W of where the ramp meets the highway. Find the elevation of the service drive above the highway and the horizontal distance spanned by the ramp.
    (c) Find the slope of the ramp.
    Round off all answers to two decimal places.
42. Suppose a downhill ramp connecting a service drive that is 10 feet above an expressway has a $-3°$ grade.
    (a) Sketch a model of the situation in a Cartesian system so that the ramp's entry point S is at the origin and the point W where the ramp meets the expressway is in quadrant IV.
    (b) Use the model in part (a) to find the location W where the ramp meets the expressway. Find the length of the ramp and the horizontal distance spanned by the ramp.
    (c) Find the slope of the ramp.
    Round off all answers to two decimal places.

Developing and Extending the Concepts

43. A person makes the following assertion:
Assume

$$\tan\theta = \frac{2}{3}.$$

Since

$$\tan\theta = \frac{\sin\theta}{\cos\theta}$$

it follows that

$$\sin\theta = 2 \quad \text{and} \quad \cos\theta = 3.$$

Consequently, the sine and cosine of an angle can have values larger than 1. Explain why this assertion is wrong. Give an example to support your argument.

44. Suppose that (x, y) and (x_1, y_1) are two different points in quadrant I on the terminal side of θ, as shown in Figure 8.

Figure 8

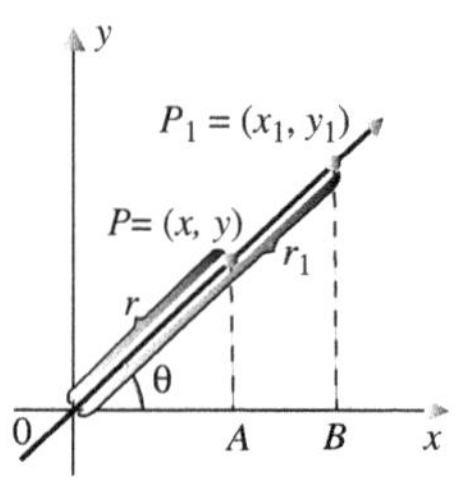

Use similar triangles to show that

$$\frac{y_1}{r_1} = \frac{y}{r} \qquad \frac{x_1}{r_1} = \frac{x}{r}$$

$$\frac{y_1}{x_1} = \frac{y}{x} \qquad \frac{r_1}{y_1} = \frac{r}{y}$$

$$\frac{r_1}{x_1} = \frac{r}{x} \qquad \frac{x_1}{y_1} = \frac{x}{y}$$

45. Show that the ratios in problem 44 also hold if angle θ is in quadrant II.

46. Show that the ratios in problem 44 also hold if angle θ is in quadrant III.

47. Show that the ratios in problem 44 also hold if angle θ is in quadrant IV.

48. A person makes the following assertion:

$$\text{If } \sin\alpha = \sin\beta, \quad \text{then} \quad \alpha = \beta.$$

Explain why this assertion is wrong. Give an example to support your argument.

49. Assume θ is an angle in standard position with terminal side lying on the line with equation $y = mx$ (Figure 9). Show that the tangent function of θ is the same as the slope of the line, that is, show that $\tan\theta = m$.

Figure 9

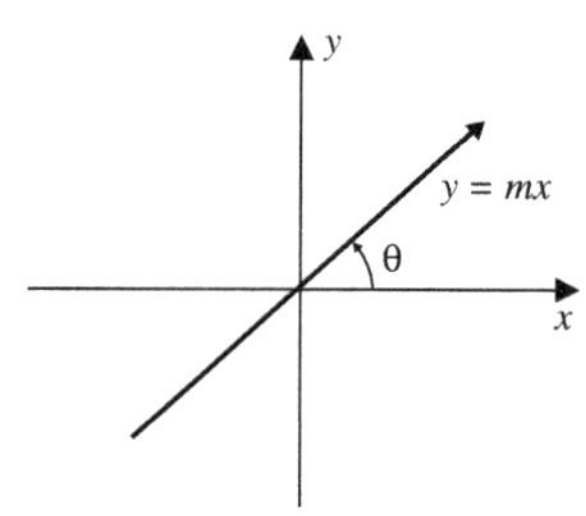

Interlude: Practice Problems 1.6

Suppose that θ is the radian measure of an angle in standard position. For each of the following, assume that the terminal side of the angle lies in the stated quadrant along the given line. Find the values of the six trigonometric functions at θ fill in the blanks).

1. Quadrant III, line with equation $y = 4x$

____

____

$\sin\theta =$ ____________________

$\cos\theta =$ ____________________

$\tan\theta =$ ____________________

$\sec\theta =$ ____________________

$\csc\theta =$ ____________________

$\cot\theta =$ ____________________

2. Quadrant IV, line with equation $5x + 3y = 0$

____

____

$\sin\theta =$ ____________________

$\cos\theta =$ ____________________

$\tan\theta =$ ____________________

$\sec\theta =$ ____________________

$\csc\theta =$ ____________________

$\cot\theta =$ ____________________

3. In aerial navigation, directions are given in degrees clockwise from north (so, for instance, east is 90°). For each of the following, give all requested results accurate to two decimal places.

(a) An airplane flies 180 miles from an airport at a direction of 164°. How far east of the airport is the plane? How far south of the airport is the plane?

(b) An airplane travels at an average speed of 180km/hr for 3 hours at a direction of 140°. Another airplane that departed an hour later travels at an average speed of 120 km/hr at a direction of 56°. How far apart are the two airplanes now?

Objectives

1. Define Trigonometric Functions of Acute Angles
2. Use Right Triangles to Evaluate Trigonometric Functions for Special Angles
3. Use Reference Triangles to Evaluate Trigonometric Functions
4. Approximate Trigonometric Values for Acute Angles
5. Solve Right Triangles
6. Solve Applied Problems

1.7 Trigonometric Functions of Acute Angles and Modeling Right Triangles

The trigonometric functions enable us to establish relationships between the acute angles and lengths of the sides of right triangles. As we shall see, this enables us to solve many applied problems that are modeled by right triangles.

Defining Trigonometric Functions of Acute Angles

Suppose we place a *right triangle* (Figure 1a) on a coordinate system with one of the acute angles α in standard position (Figure 1b), then the values of the trigonometric functions of α may be expressed in terms of the lengths of the sides of the right triangle. Using the definition of the trigonometric functions of angles on page 81 with $\theta = \alpha$, $r = c$, $x = b$, and $y = a$, we express the

Figure 1

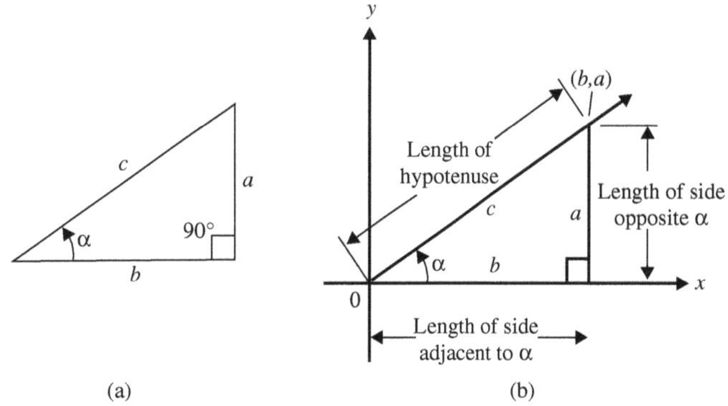

trigonometric functions of α in terms of the lengths of the sides of the right triangle as follows:

$$\sin\alpha = \frac{a}{c} = \frac{\text{length of side opposite } \alpha}{\text{length of hypotenuse}} \qquad \csc\alpha = \frac{c}{a} = \frac{\text{length of hypotenuse}}{\text{length of side opposite } \alpha}$$

$$\cos\alpha = \frac{b}{c} = \frac{\text{length of side adjacent to } \alpha}{\text{length of hypotenuse}} \qquad \sec\alpha = \frac{c}{b} = \frac{\text{length of hypotenuse}}{\text{length of side adjacent to } \alpha}$$

$$\tan\alpha = \frac{a}{b} = \frac{\text{length of side opposite } \alpha}{\text{length of side adjacent to } \alpha} \qquad \cot\alpha = \frac{b}{a} = \frac{\text{length of side adjacent to } \alpha}{\text{length of side opposite } \alpha}$$

We can abbreviate the lengths of the side **opposite** α, the side **adjacent** to α, and the **hypotenuse** as **opp**, **adj**, and **hyp**, respectively (Figure 2). Thus the relationships between the trigonometric functions of an acute angle and the ratios of the lengths of the sides of a right triangle are established in the next definition.

Definition

Trigonometric Functions of an Acute Angle in a Right Triangle

Assume α is an acute angle in a right triangle (Figure 2). Then

$$\sin\alpha = \frac{\text{opp}}{\text{hyp}} \qquad \csc\alpha = \frac{\text{hyp}}{\text{opp}}$$

$$\cos\alpha = \frac{\text{adj}}{\text{hyp}} \qquad \sec\alpha = \frac{\text{hyp}}{\text{adj}}$$

$$\tan\alpha = \frac{\text{opp}}{\text{adj}} \qquad \cot\alpha = \frac{\text{adj}}{\text{opp}}.$$

Figure 2

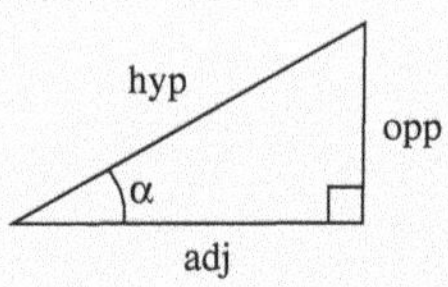

EXAMPLE 1 **Using a Right Triangle to Evaluate Trigonometric Functions**

Use the right triangle in Figure 3 to find the values of the trigonometric function of acute angle α.

Solution For angle α, we have opp = 12 centimeters and hyp = 13 centimeters, but the length of the adj isn't given. Using the Pythagorean theorem, we have

Figure 3

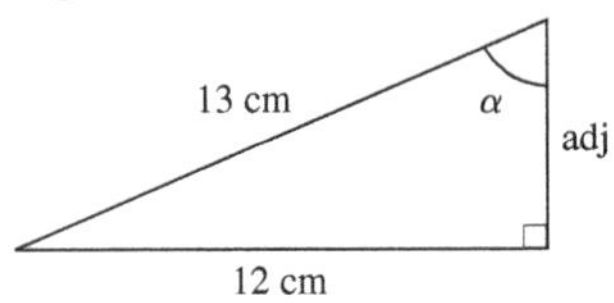

$$(\text{adj})^2 + (\text{opp})^2 = (\text{hyp})^2$$

so that

$$\begin{aligned}(\text{adj})^2 &= (\text{hyp})^2 - (\text{opp})^2\\ &= 13^2 - 12^2 = 25.\end{aligned}$$

Therefore,

$$\text{adj} = \sqrt{25} = 5 \text{ centimeters.}$$

It follows from the right triangle relationships that

$$\sin\alpha = \frac{\text{opp}}{\text{hyp}} = \frac{12}{13} \qquad \csc\alpha = \frac{\text{hyp}}{\text{opp}} = \frac{13}{12}$$

$$\cos\alpha = \frac{\text{adj}}{\text{hyp}} = \frac{5}{13} \qquad \sec\alpha = \frac{\text{hyp}}{\text{adj}} = \frac{13}{5}$$

$$\tan\alpha = \frac{\text{opp}}{\text{adj}} = \frac{12}{5} \qquad \cot\alpha = \frac{\text{adj}}{\text{opp}} = \frac{5}{12}.$$

Using Right Triangles to Evaluate Trigonometric Functions for Special Angles

Figure 4

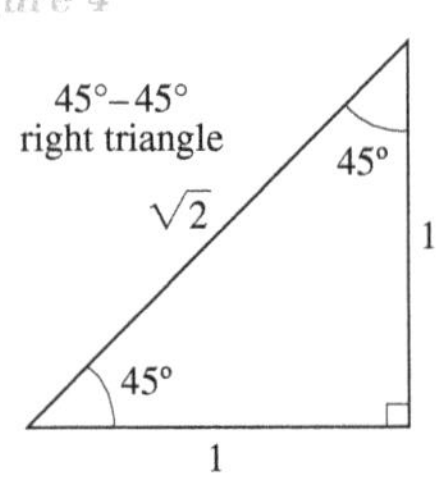

Our knowledge of right triangles provides us with another way to determine the exact values of the trigonometric functions for special angles 45°, 30°, and 60° (or equivalently $\pi/4$, $\pi/6$, and $\pi/3$). First, we construct an *isosceles right triangle* with two equal sides of length 1 unit (Figure 4). We know from plane geometry that both of the acute angles in this right triangle measure 45°. Using the Pythagorean theorem, we have

$$1^2 + 1^2 = (\text{hyp})^2$$

so that

$$\text{hyp} = \sqrt{2}.$$

From the right triangle trigonometric relationships, we obtain the familiar values:

$$\sin 45^\circ = \frac{1}{\sqrt{2}} = \frac{\sqrt{2}}{2} \qquad \csc 45^\circ = \frac{\sqrt{2}}{1} = \sqrt{2}$$

$$\cos 45^\circ = \frac{1}{\sqrt{2}} = \frac{\sqrt{2}}{2} \qquad \sec 45^\circ = \frac{\sqrt{2}}{1} = \sqrt{2}$$

$$\tan 45^\circ = \frac{1}{1} = 1 \qquad \cot 45^\circ = \frac{1}{1} = 1$$

Figure 5

A 30°–60° right triangle

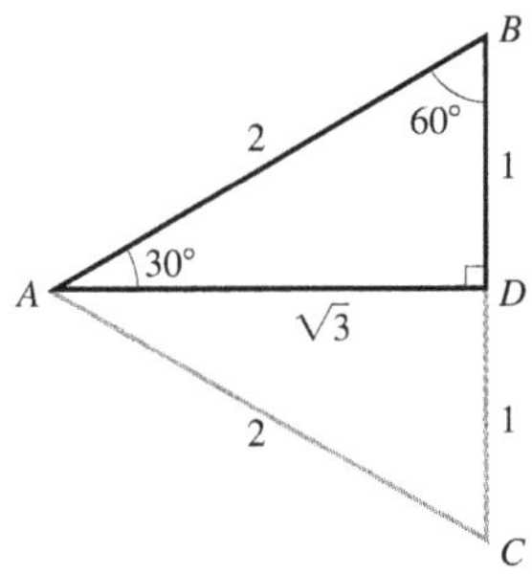

Next we consider an equilateral triangle, each of whose sides has length 2 units, such as triangle ABC in Figure 5. Recall from plane geometry that the three angles of an equilateral triangle each measure 60°. Assume that $\overline{AD}$ is the perpendicular bisector of $\overline{BC}$ (Figure 5). By using the Pythagorean theorem, we have

$$|\overline{AD}|^2 = 2^2 - 1^2 = 3 \quad \text{or} \quad |\overline{AD}| = \sqrt{3}.$$

Thus, from right triangle ADB in Figure 5, we have the known values:

$$\sin 30° = \cos 60° = \frac{1}{2} \qquad \csc 30° = \sec 60° = \frac{2}{1} = 2$$

$$\cos 30° = \sin 60° = \frac{\sqrt{3}}{2} \qquad \sec 30° = \csc 60° = \frac{2}{\sqrt{3}} = \frac{2\sqrt{3}}{3}$$

$$\tan 30° = \cot 60° = \frac{1}{\sqrt{3}} = \frac{\sqrt{3}}{3} \qquad \cot 30° = \tan 60° = \frac{\sqrt{3}}{1} = \sqrt{3}$$

Note that when we are trying to recall the values of the trigonometric functions for 45°, 30°, or 60° (or equivalently $\pi/4$, $\pi/6$, or $\pi/3$), it is useful to construct a 45°–45° right triangle (Figure 4) or a 30°–60° right triangle (triangle ADB in Figure 5) and then apply the right triangle trigonometric relationships.

Using Reference Triangles to Evaluate Trigonometric Functions

We often find it convenient to evaluate trigonometric functions of nonacute angles that are integer multiples of the special angles 30°, 45°, or 60° by making use of the concept of *reference triangles.* This procedure is described as follows:

Strategy for Reference Triangles to Evaluate Trigonometric Functions

Step 1 Sketch the given angle θ in standard position.

Step 2 Superimpose a reference triangle; that is, draw either a 45°–45° right triangle, or a 30°–60° right triangle in whichever quadrant will allow the hypotenuse to lie on the terminal side of θ and the adjacent side of the triangle to lie on the x axis.

Step 3 Use the measurements of that particular right triangle along with the quadrant location to obtain a point (x, y) on the terminal side of θ.

Step 4 Use the definitions of the trigonometric functions of angles along with the reciprocal and quotient relationships to complete the evaluation.

EXAMPLE 2 **Using a Reference Triangle to Evaluate Trigonometric Functions**

Use the above four-step procedure to evaluate the six trigonometric functions of $\theta = 135°$.

Solution Refer to Figure 6.

Step 1. We sketch $\theta = 135°$ in standard position.

Step 2. Then we draw a 45°–45° right triangle as a reference triangle in quadrant II so that the hypotenuse lies on the terminal side of θ and the adjacent leg lies on the negative x axis.

Figure 6

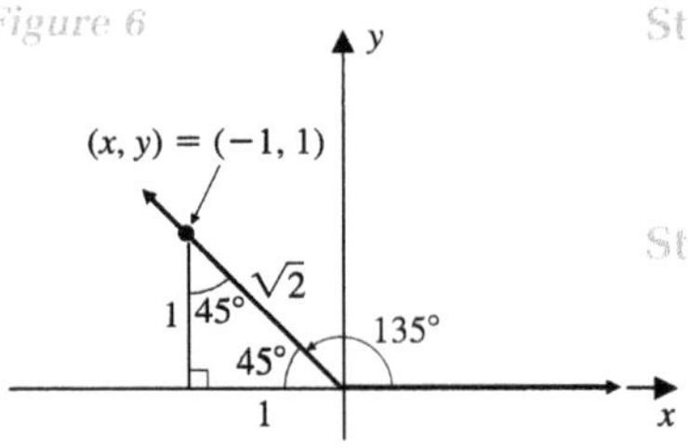

Step 3. By using the measurements of the 45°−45° right triangle, we get the point $(x, y) = (-1, 1)$ on the terminal side of θ.

It follows that $r = \sqrt{(-1)^2 + 1^2} = \sqrt{2}$.

Step 4. We get

$$\sin 135^\circ = \frac{y}{r} = \frac{1}{\sqrt{2}} \qquad \csc 135^\circ = \frac{1}{\sin 135^\circ} = \frac{1}{\left(\frac{1}{\sqrt{2}}\right)} = \sqrt{2}$$

$$\cos 135^\circ = \frac{x}{r} = \frac{-1}{\sqrt{2}} \qquad \sec 135^\circ = \frac{1}{\cos 135^\circ} = \frac{1}{\left(\frac{-1}{\sqrt{2}}\right)} = -\sqrt{2}$$

$$\tan 135^\circ = \frac{y}{x} = \frac{1}{-1} = -1 \qquad \cot 135^\circ = \frac{1}{\tan 135^\circ} = \frac{1}{-1} = -1.$$

To evaluate the trigonometric functions of a real number or of an angle given in radian measure, we can at times convert it to degree measure and then proceed as we did in Example 2. For instance, assume $\theta = (7\pi)/6$. Since $(7\pi)/6$ radians correspond to 210°, we sketch $\theta = 210^\circ$ in standard position along with its reference triangle as shown in Figure 7. After locating the point $(x, y) = (-\sqrt{3}, -1)$ on the terminal side of θ, we can find the trigonometric values of $(7\pi)/6$. For example,

Figure 7

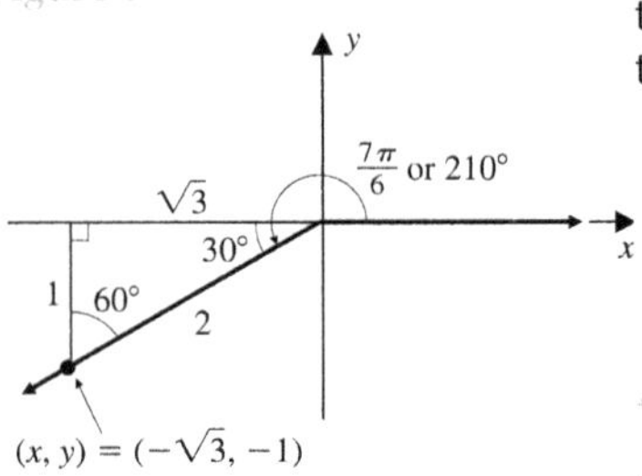

$$\sin\frac{7\pi}{6} = \sin 210^\circ = \frac{y}{r} = \frac{-1}{2} \quad \text{and} \quad \cos\frac{7\pi}{6} = \cos 210^\circ = \frac{x}{r} = \frac{-\sqrt{3}}{2}.$$

Approximating Trigonometric Values for Acute Angles

To approximate trigonometric function values for acute angles we proceed as we did for general angles.

For instance,

$$\sin 53.4^\circ = 0.8028 \text{ (approx.)};$$

and

$$\tan 75^\circ 25' 18'' = 3.8450 \text{ (approx.)}.$$

To evaluate $\cot(\pi/8)$, we set the calculator in radian mode and then we use the reciprocal identity to get

$$\cot\frac{\pi}{8} = \frac{1}{\tan(\pi/8)} = 2.4142 \text{ (approx.)}.$$

When evaluating a trigonometric function, we *input an angle* θ and *the output V is the function value* (Figure 8a).

For acute angles, this process is *reversible or invertible* in the sense that if we specify a trigonometric function value V as input, then the output is the angle θ (Figure 8b).

Later in Section 1.8 we expand the notion of inverse trigonometric functions.

Figure 8

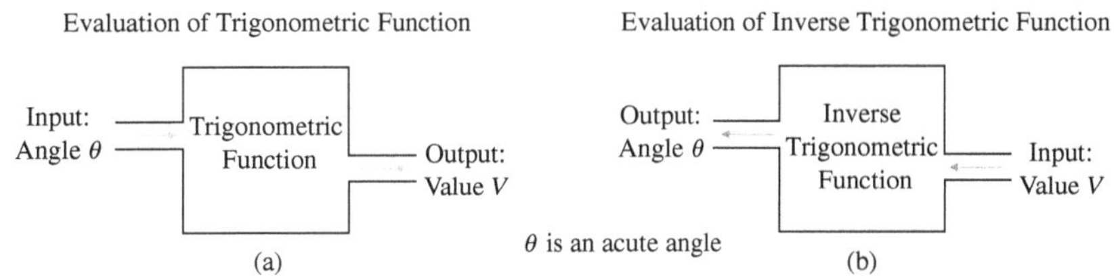

We use the superscript $^{-1}$ notation to denote this inverse function. For instance, by using the special angles, we have

We use degree measures rather than radian measures here because we are working with right triangles.

$$\cos^{-1}\left(\frac{1}{2}\right) = 60° \quad \text{because} \quad \cos 60° = \frac{1}{2}$$

$$\sin^{-1}\left(\frac{\sqrt{2}}{2}\right) = 45° \quad \text{because} \quad \sin 45° = \frac{\sqrt{2}}{2}$$

$$\tan^{-1}(1) = 45° \quad \text{because} \quad \tan 45° = 1.$$

If an inverse value is not known, then we use a calculator to approximate its value.

For instance, after setting the calculator in degree mode and using the proper key stroke entries, we obtain the approximations

Most calculators do not have an inverse secant key.

$$\sin^{-1}(0.3185) = 18.57° \quad \text{and} \quad \tan^{-1}(7.2939) = 82.19°.$$

To calculate $\sec^{-1}(1.0313)$, we proceed as follows:

$$\sec^{-1}(1.0313) = \theta \quad \text{means} \quad \sec\theta = 1.0313$$

but since

$$\sec\theta = \frac{1}{\cos\theta}$$

it follows that

$$\frac{1}{\cos\theta} = 1.0313 \quad \text{or} \quad \frac{1}{1.0313} = \cos\theta$$

so that

$$\theta = \cos^{-1}\left(\frac{1}{1.0313}\right) = 14.15° \text{ (approx.)}.$$

Solving Right Triangles

Under certain conditions, it is possible to find the measurements of unknown sides and angles of a right triangle from known sides and angles by using the trigonometric functions. This process is called *solving a right triangle.*

Consider the right triangle ACB shown in Figure 9. We'll adopt the convention of labeling the angles at the vertices A, B, and C by α, β, and γ, respectively, and the lengths of the sides opposite these angles by a, b, and c, respectively. Note that α and β are complementary angles and $\gamma = 90°$.

Figure 9

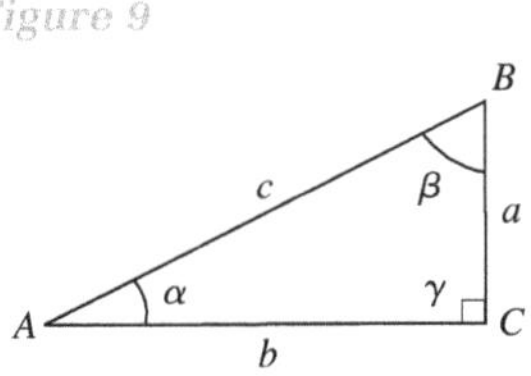

When solving a right triangle, we usually draw a sketch of the triangle approximately to scale to gain some insight into the problem.

EXAMPLE 3

Solving Right Triangles

Solve the right triangle ACB for each situation. Round off the answers to two decimal places.

(a) $a = 9$ and $\beta = 36.87°$

(b) $a = 31.42$ and $b = 26.74$

Solution

(a) Figure 10a shows a sketch of the right triangle. Here, we have to find α, b, and c. Because α and β are complementary,

$$\alpha = 90° - \beta = 90° - 36.87° = 53.13°.$$

To find b, we use the relationship $\tan \beta = \text{opp}/\text{adj}$ and a calculator to get

Figure 10a

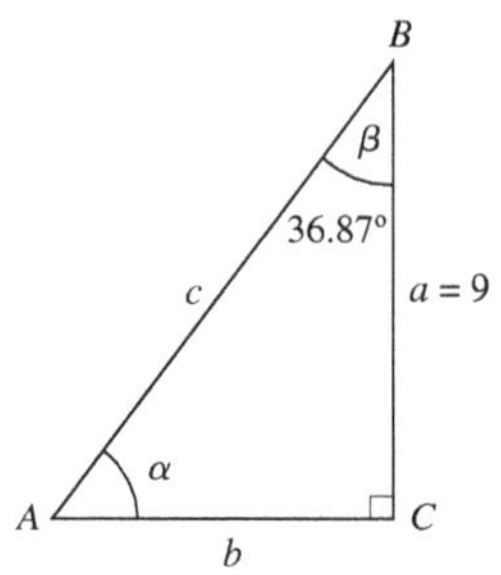

$$\tan \beta = \frac{b}{a}$$

so that

$$b = a \tan \beta = 9 \tan 36.87° = 6.75 \text{ (approx.)}.$$

To find c, we use the relationship $\cos \beta = \text{adj}/\text{hyp}$ to get

$$\cos \beta = \frac{a}{c}$$

so that

$$c = \frac{a}{\cos \beta} = \frac{9}{\cos 36.87°} = 11.25 \text{ (approx.)}.$$

Figure 10b

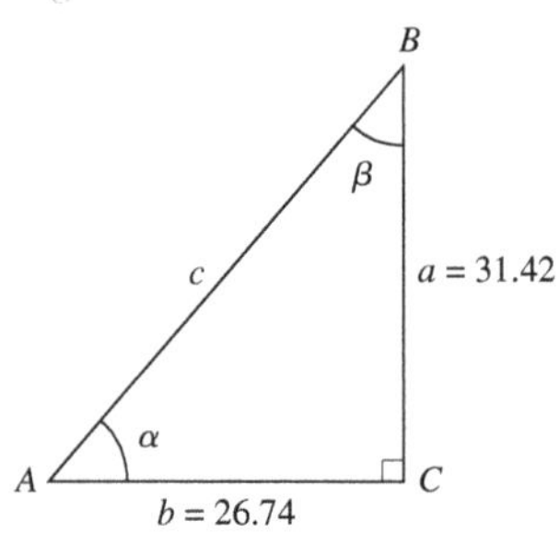

(b) Figure 10b shows a sketch of the right triangle. Here we have to find α, β, and c. We'll start by finding α. Using the definition, it follows that

$$\tan \alpha = \frac{\text{opp}}{\text{adj}} = \frac{a}{b} = \frac{31.42}{26.74}.$$

To find α means to find the acute angle whose tangent is 31.42/26.74, so we use the inverse evaluation

$$\alpha = \tan^{-1}\left(\frac{31.42}{26.74}\right) = 49.60° \text{ (approx.)}.$$

Because α and β are complementary,

$$\beta = 90° - 49.60° = 40.40° \text{ (approx.)}.$$

Finally, to find c, we use the Pythagorean theorem to get

$$c = \sqrt{a^2 + b^2} = \sqrt{31.42^2 + 26.74^2} = 41.26 \text{ (approx.)}.$$

Solving Applied Problems

One of the major reasons for the development of trigonometry was to solve applications involving right triangles. The following examples illustrate how trigonometric functions of acute angles are used to model such situations.

EXAMPLE 4

Finding the Height Reached by a Ladder

If a 24-foot extension ladder leaning against a house makes a 63° angle with the ground, then how far up the side of the house does the ladder reach? Round off the answer to two decimal places.

Figure 11

Solution The situation is shown in Figure 11. We need to determine the length $|\overline{CB}|$. By using right triangle trigonometry, we obtain

$$\sin 63^\circ = \frac{\text{opp}}{\text{hyp}} = \frac{|\overline{CB}|}{24}$$

or

$$|\overline{CB}| = 24 \sin 63^\circ = 21.38 \text{ (approx.)}.$$

Thus, the ladder reaches about 21.38 feet up the side of the house (about 21 feet and 5 inches).

At times, right triangle applications involve the measure of an acute angle formed by an observer's direct line of sight to an object and a horizontal line. If the object is above the horizontal line, this angle is called an **angle of elevation** (Figure 12a). If the object is below the horizontal line, the angle is called an **angle of depression** (Figure 12b).

A transit is an instrument used by surveyors to determine angles.

Figure 12

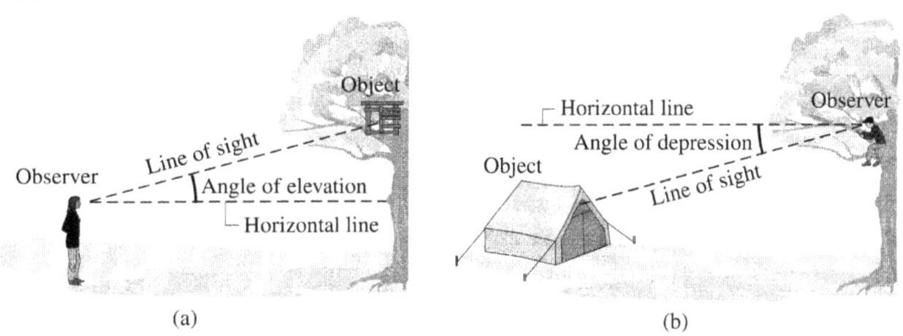

EXAMPLE 5

Using an Angle of Elevation

The Empire State Building in New York City is about 1250 feet high. A surveyor's transit is set 5 feet above ground level. The angle of elevation from the transit to a point on top of the Empire State Building is found to be 25.33°. How far is the surveyor from the building? Round off the answer to the nearest 10 feet.

Solution The situation is modeled by the right triangle in Figure 13, where d represents the desired distance. Here we have

$$\tan 25.33^\circ = \frac{\text{opp}}{\text{adj}}$$

$$= \frac{1250 - 5}{d}$$

Figure 13

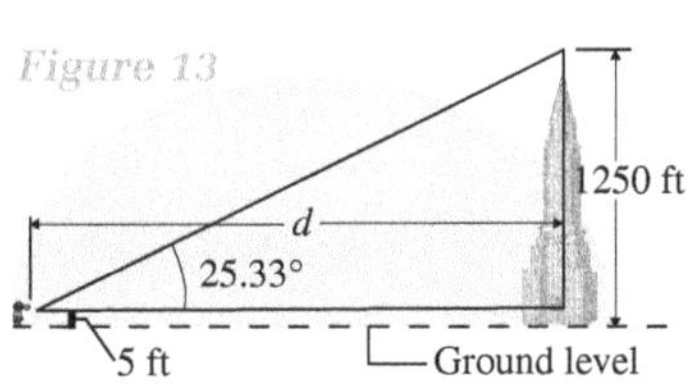

or

$$d = \frac{1245}{\tan 25.33°}$$
$$= 2630 \text{ (approx.)}.$$

Consequently, the surveyor is about 2630 feet from the building.

EXAMPLE 6

Using Angles of Depression

An air traffic controller stands in a control tower that is 150 feet high and observes two commercial jets waiting in adjacent runways for a takeoff. The angle of depression of the first jet is 15.3°, and the angle of depression of the second is 8.2°. Assuming that the tower and the two jets lie on the same plane, find the distance between the jets. Round off the answer to the nearest 10 feet.

Solution In Figure 14, A represents the location of the controller, C the location of the first jet, and D the location of the second jet. We wish to find $|\overline{CD}|$, which is equal to $|\overline{BD}| - |\overline{BC}|$.

Figure 14

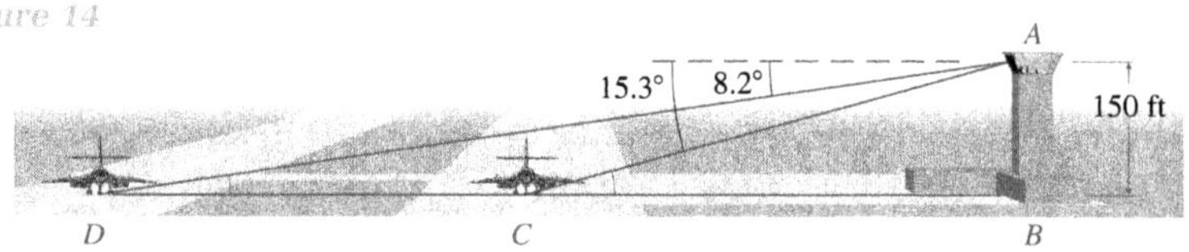

From geometry (Appendix I), we know that alternate interior angles are equal, so angle $BCA = 15.3°$ and angle $BDA = 8.2°$. In the right triangle ABC, we have

$$\tan 15.3° = \frac{|\overline{AB}|}{|\overline{BC}|} = \frac{150}{|\overline{BC}|}$$

or

$$|\overline{BC}| = \frac{150}{\tan 15.3°} = 550 \text{ (approx.)}.$$

In the right triangle ABD,

$$\tan 8.2° = \frac{|\overline{AB}|}{|\overline{BD}|} = \frac{150}{|\overline{BD}|}$$

or

$$|\overline{BD}| = \frac{150}{\tan 8.2°} = 1040 \text{ (approx.)}.$$

Hence the approximate distance between the two jets is given by

$$|\overline{CD}| = |\overline{BD}| - |\overline{BC}|$$
$$= 1040 - 550$$
$$= 490.$$

That is, the jets are approximately 490 feet apart.

PROBLEM SET 1.7

Mastering the Concepts

In problems 1–6, find the values of the six trigonometric functions of the acute angle θ for each right triangle. Round off the answers to three decimal places.

1.

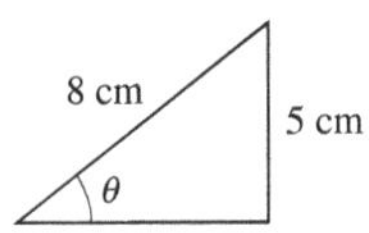

2.

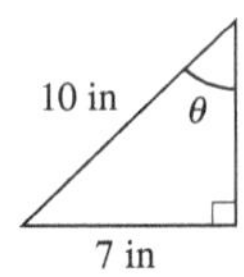

3.

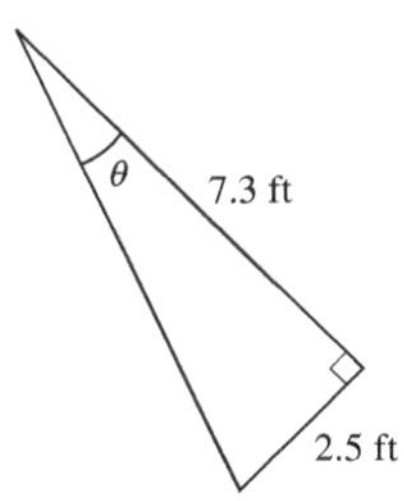

4.

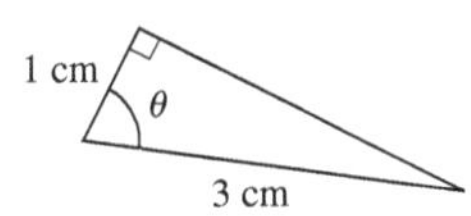

5.

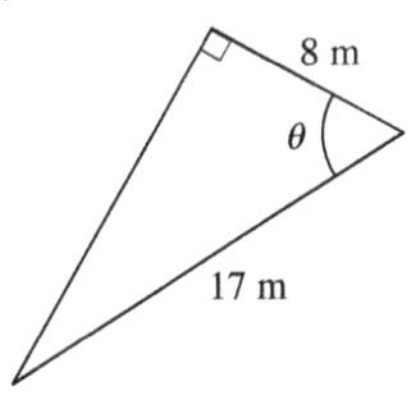

6.

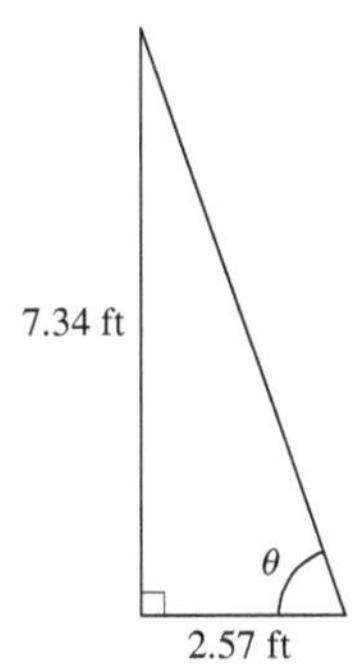

7. Let θ be an acute angle of a right triangle and

$$\sin\theta = \frac{4}{5}.$$

Draw such a triangle, and then find the exact values of the other five trigonometric functions of θ.

8. Let θ be an acute angle of a right triangle for which

$$\cos\theta = \frac{12}{13}.$$

Draw such a triangle, and then find the exact values of the other five trigonometric functions of θ.

In problems 9–16, use either a 45°–45° or a 30°–60° right triangle as a reference triangle to find the exact value of each expression.

9. (a) $\sin 150°$ (b) $\cos 225°$

10. (a) $\cos\left(-\frac{7\pi}{6}\right)$ (b) $\cot 330°$

11. (a) $\tan 240°$ (b) $\cot\frac{11\pi}{6}$

12. (a) $\sec\left(-\frac{11\pi}{6}\right)$ (b) $\csc\left(-\frac{7\pi}{6}\right)$

13. (a) $\sec 315°$ (b) $\csc 300°$

14. (a) $\csc(-330°)$ (b) $\cot\left(-\frac{4\pi}{3}\right)$

15. (a) $\tan\left(-\frac{\pi}{4}\right)$ (b) $\cot 210°$

16. (a) $\sin\left(-\frac{7\pi}{4}\right)$ (b) $\sin\frac{7\pi}{6}$

In problems 17–20, use a calculator to determine each value. Round off the answer to four decimal places.

17. (a) $\sin 43°$ (b) $\csc(61.57°)$ (c) $\cos\left(\frac{3\pi}{7}\right)$ (d) $\sec\frac{3\pi}{8}$

18. (a) $\cot 88°41'$ (b) $\sin(77°52'48'')$ (c) $\sec 1.536$ (d) $\csc 1.234$

19. (a) $\tan\frac{\pi}{5}$ (b) $\cos 71°46'32''$ (c) $\csc 53.8°$ (d) $\sin(0.453)$

20. (a) $\cos(1.233)$ (b) $\cot 37.8°$ (c) $\csc 72°12'25''$ (d) $\tan 80.3°$

In problems 21 and 22, use a calculator to find each inverse value as an acute angle in degree measure. Round off each answer to two decimal places.

21. (a) $\sin^{-1}(0.55)$ (b) $\cos^{-1}(0.98)$ (c) $\tan^{-1}(30.12)$ (d) $\cot^{-1}(1)$

22. (a) $\cos^{-1}(0.01)$ (b) $\sin^{-1}(0.10)$ (c) $\tan^{-1}(1)$ (d) $\sec^{-1}(2)$

In problems 23–32, assume that ACB is a right triangle with $\gamma = 90°$ (Figure 15). In each case, solve the triangle.

Figure 15

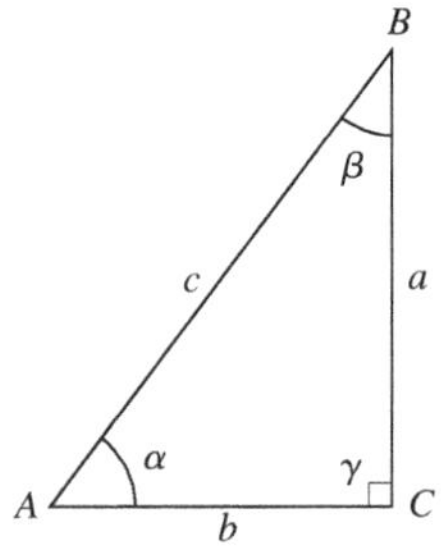

Round off angles to the nearest hundredth of a degree and side lengths to two decimal places.

23. $c = 10, \beta = 50°$
24. $c = 9, \beta = 20°$
25. $b = 4, \beta = 43°$
26. $a = 6.5, \alpha = 37.2°$
27. $b = 1500, \alpha = 31.23°$
28. $b = 567.3, \alpha = 67.41°$
29. $a = 13.2, b = 4.1$
30. $a = 31, b = 4.7$
31. $\alpha = 3\beta, a = 1$
32. $\alpha = \frac{2}{3}\beta, a = 3$

Applying the Concepts

In problems 33–54, round off the answers to two decimal places.

33. **Surveying:** The lot for a new home site is uniformly pitched from the horizontal at an angle of 0.51° (Figure 16). How many inches will the ground drop if one walks 100 feet down the slope?

Figure 16

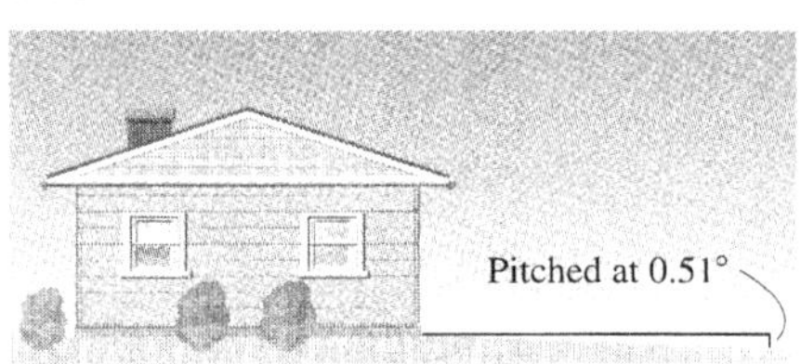

34. **Surveying:** A straight sidewalk is inclined to the horizontal at an angle of 5.3°. How far must one walk on the sidewalk to change elevation by 2 meters?

35. **Foot of a Ladder:** A ladder 24 feet long is leaning against a building. The angle formed by the ladder and the ground is 63°. How far from the building is the foot of the ladder?

36. **Length of a Shadow:** A monument is 180 meters high. What is the length of the shadow cast by the monument if the angle of elevation of the sun is 58.4°?

37. **Utility Pole:** A guy wire attached to a vertical utility pole makes an angle of 71.4° with the ground. If the end of the wire attached to the ground is 15.5 feet from the pole, how high up the pole is the other end of the wire attached?

38. **Altitude of a Shuttle:** A space shuttle rises vertically. A camera 4.3 feet above the ground is located at a point 1000 feet (on the horizontal) away from the base of the launching pad. At a certain instant, the angle of elevation of the camera, focusing on the bottom of the shuttle, is 53° (Figure 17). How high above the ground is the shuttle at that instant?

Figure 17

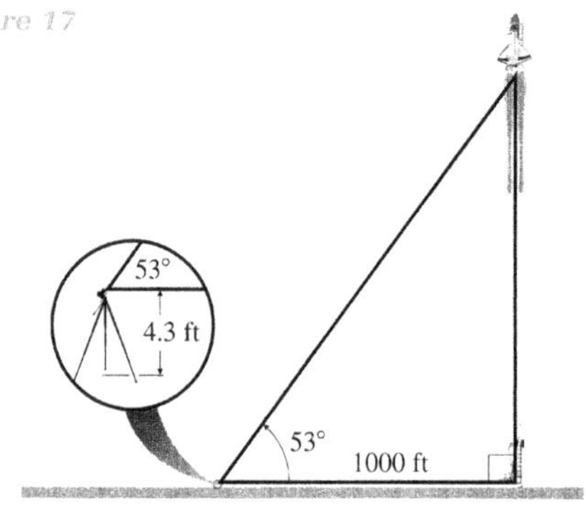

39. **Highway Patrol:** A highway patrol officer monitors traffic on a straight highway from a point A, which is 35 feet from the side of the highway (Figure 18). At a certain time, the patrol officer observed a speeding truck at point B on the highway at an angle of 25.3°. One second later, the truck was observed by the officer at point C at an angle of 10.3°. Find the speed of the truck in miles per hour.

Figure 18

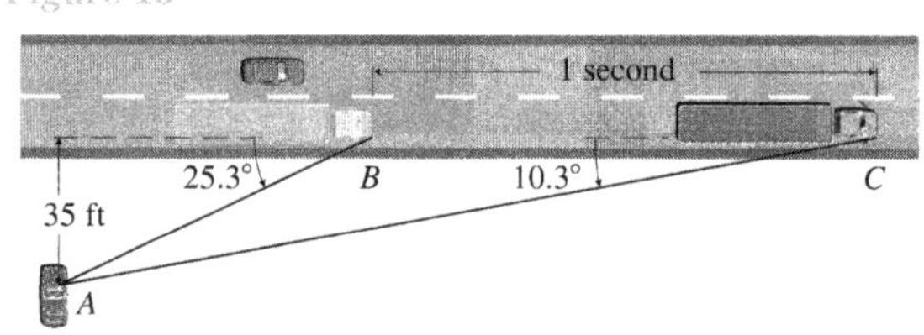

40. **Forestry:** A tree broken by the wind forms a right triangle with the ground. The broken part makes an angle of 38° with the ground, and the top of the tree is now 15 meters from the base, as shown in Figure 19. How tall was the whole tree?

Figure 19

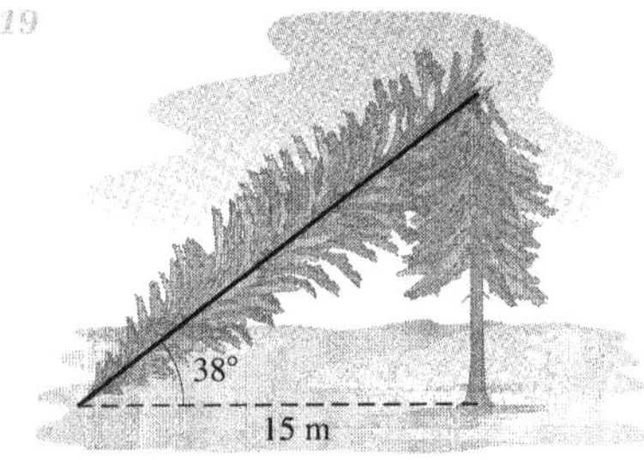

41. **Escalator:** An escalator in a department store is 50 feet long and carries customers through a vertical distance of 20 feet.
   (a) What is the angle that the escalator makes with the floor?
   (b) If it takes 25 seconds to carry a customer from the bottom of the escalator to the top, how fast (in feet per second) is the escalator moving?

42. **Surveying:** To find the length of an island, a navigator of a small plane flying at an altitude of 1540 meters determines the angles of depression of the extremities of the island to be 26.5° and 79.3° (Figure 20). What is the length of the island in kilometers?

Figure 20

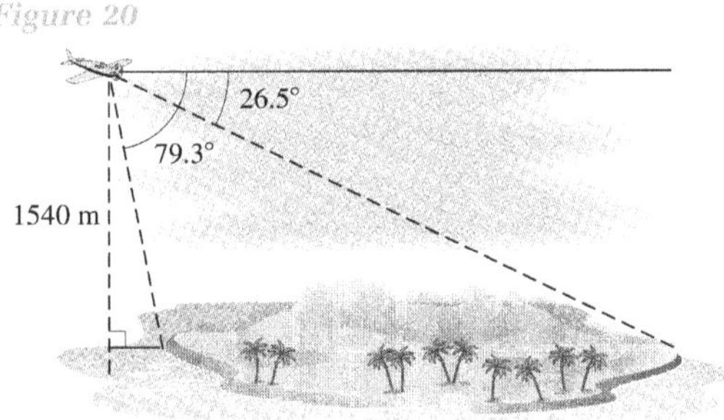

43. **Computer Monitor:** Suppose that a computer screen has a height of 7.5 inches and a width of 10 inches (Figure 21). What is the length of the diagonal d, and what angle does it make with the width?

Figure 21

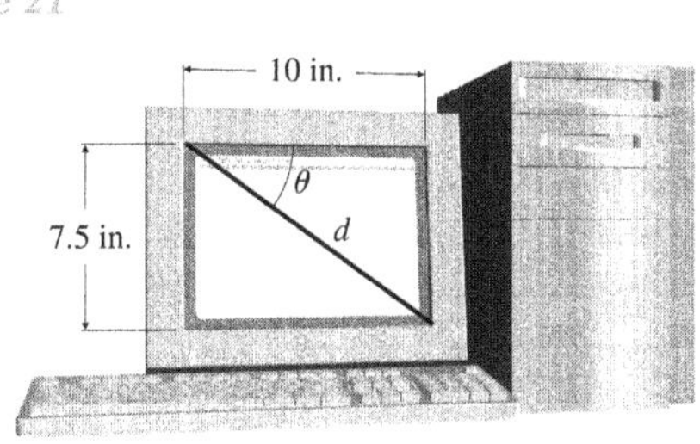

44. **The Statue of Liberty:** The torch of the Statue of Liberty is about 305 feet above water level. A sighting of the statue is taken from a ship 838 feet away. Find the angle of elevation from the ship to the top of the torch (Figure 22).

Figure 22

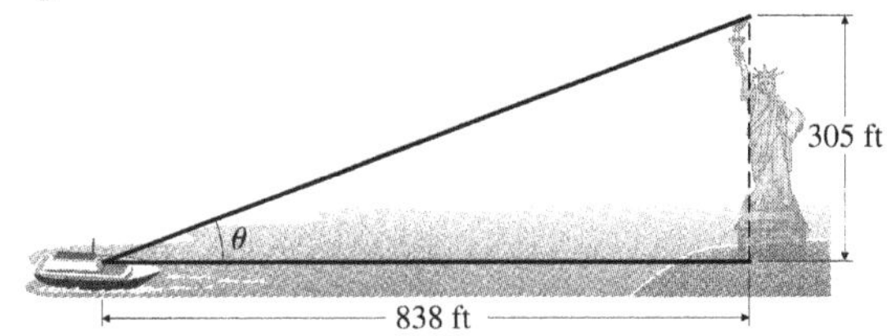

45. **Angle of Elevation of the Eiffel Tower:** The height of the Eiffel Tower (without the television mast added to the top) is approximately 300 meters (Figure 23). If the angle of elevation from a point on the ground to the top of the tower is 62.75°, how far from the center of the base of the tower is the point?

Figure 23

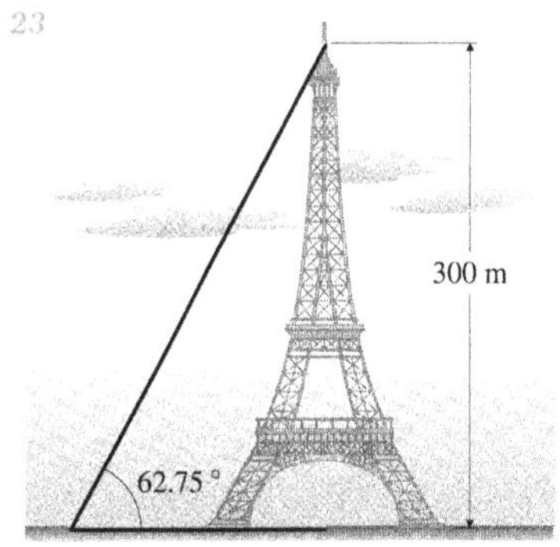

46. **Light Span of a Lamp:** A lamp is suspended 2 feet 10 inches above the center of a circular table. What is the radius of the table if the angle of depression from the lamp to the edge of the table is 56°15′, as shown in Figure 24?

Figure 24

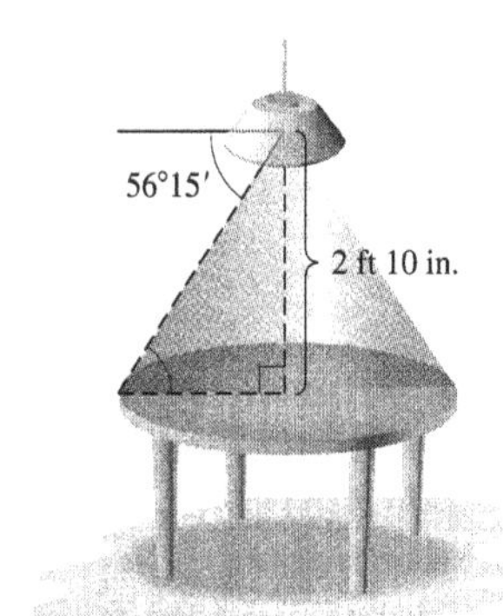

47. **Tower of Pisa:** The Leaning Tower of Pisa was designed to stand 55 meters high when vertical. However, it was built on unstable ground and is now 5.2 meters out of perpendicular (Figure 25). Find the acute angle that the tower makes with the ground.

Figure 25

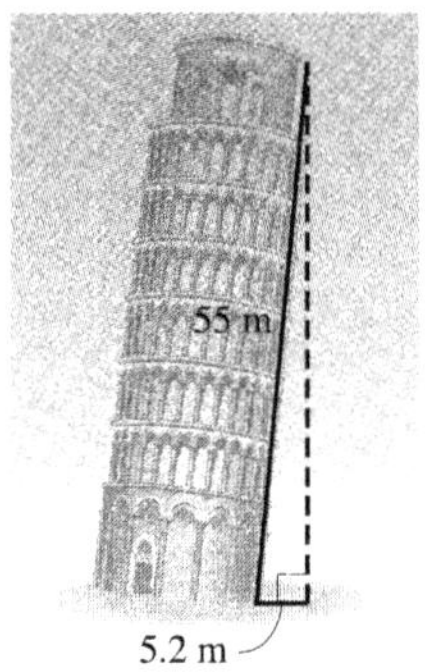

48. **Length of a Ladder:** A 6-foot fence stands 9 feet from a high wall. The shortest ladder that can reach the wall from outside the fence makes an angle of 41° with the horizontal (Figure 26). Find the length of the ladder.

Figure 26

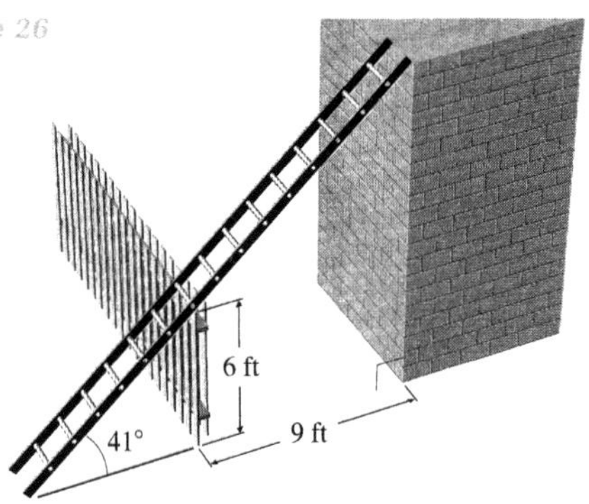

49. **Height of Al Giza Pyramid:** The distance measured along an edge from the original top of the Great Pyramid of Cheops at Al Giza, Egypt, to one of the corners of its square base is 219 meters. The angle of elevation from the corner of the base to the top is 42.06° (Figure 27). Find the original height of the Great Pyramid.

Figure 27

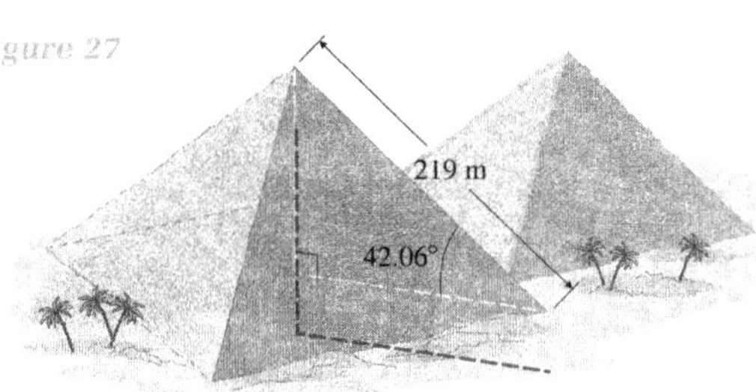

50. **Height of a Flagpole:** From the window of an apartment building 12 meters above ground, an observer determines that the angles of depression on the top and bottom of a flagpole standing on the ground (level with the base of the building) are 43° and 66°, respectively (Figure 28). Find the height of the flagpole.

Figure 28

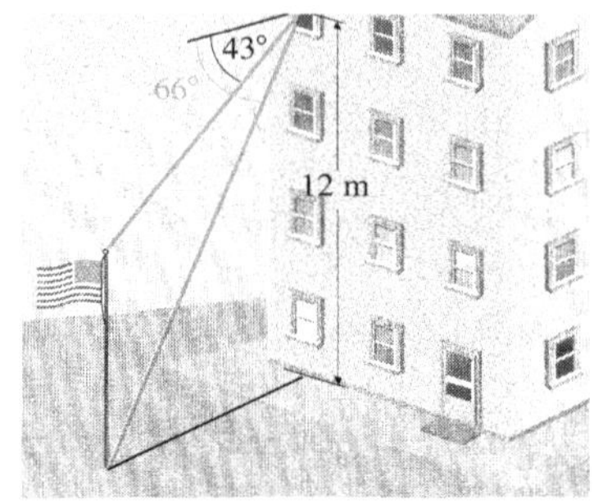

51. **Length of an Antenna:** An antenna is situated at the edge of a flat roof on top of a building that is located on level ground. From a point 100 feet from the base of the building, on the side where the antenna is placed, the angles of elevation of the top and bottom of the antenna measure 24° and 17°, respectively (Figure 29). How high is the building? How tall is the antenna?

Figure 29

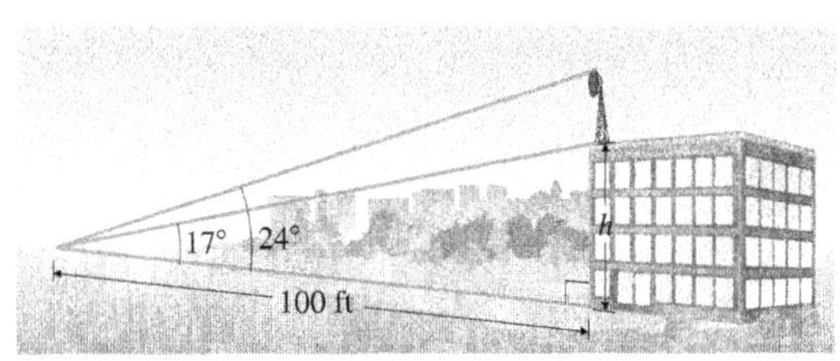

52. **Surveying:** To find the shortest distance d from a baseline l to a point T on an island offshore, a surveyor marks off a distance of 200 meters between the two points B and C on the baseline. She locates the point T on the island and then measures the angles TBC and TCB to be 68.3° and 81.5°, respectively (Figure 30). Find the distance d from the point T on the island to the nearest point on the baseline.

Figure 30

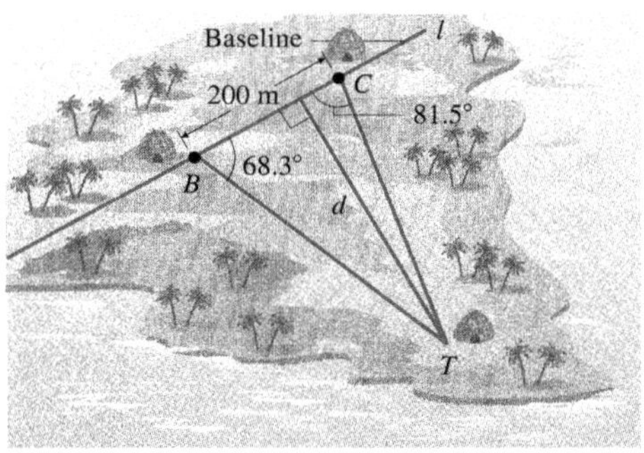

53. **Height of a Blimp:** A blimp is between two spotters who are 522 feet apart. One spotter reports that the angle of elevation of the blimp is 66°, and the other reports that it is 59°. If the blimp is directly over a line from one spotter to the other, how high is the blimp above the ground?

54. **Security:** A video security monitor is mounted at location A, 10 feet above the floor on a back wall of a drug store. The monitor scans through an angle of 51° in a vertical plane along an aisle of the store. The aisle begins at location B, 5 feet from the foot of the back wall at C and ends at D, the foot of an opposite wall (Figure 31). What is the length of the aisle; that is, what is the distance $|\overline{BD}|$?

Figure 31

Developing and Extending the Concepts

55. Use a right triangle to prove that if α and β are complementary angles, then the following equations are true. Give numerical examples to illustrate this result.

$\sin \alpha = \cos \beta \quad \tan \alpha = \cot \beta \quad \csc \alpha = \sec \beta$

56. Suppose that two straight lines EC and DB intersect at point A and right triangles ABC and ADE are formed as shown in Figure 32.

(a) Express $\tan(90° - \theta)$ in terms of a and b.

(b) Express $\sin \theta$ in terms of a and b.

Figure 32

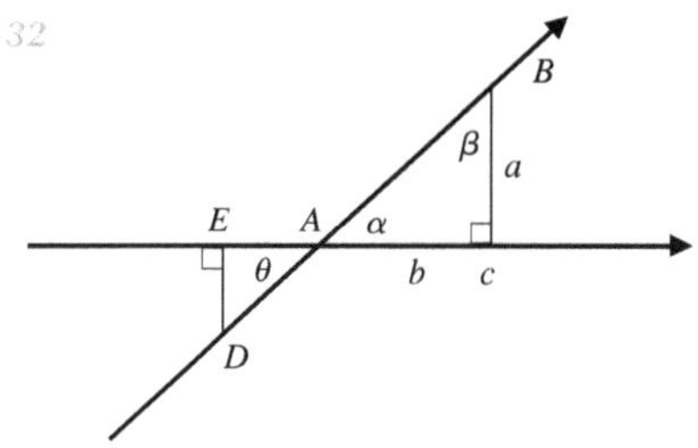

57. Two tangent lines, each of length $|T|$, are drawn from a point P and touch a circle of radius r at points M and N (Figure 33). Assume the angle MPN is θ. Express the length $|T|$ of each tangent line in terms of θ and r.

Figure 33

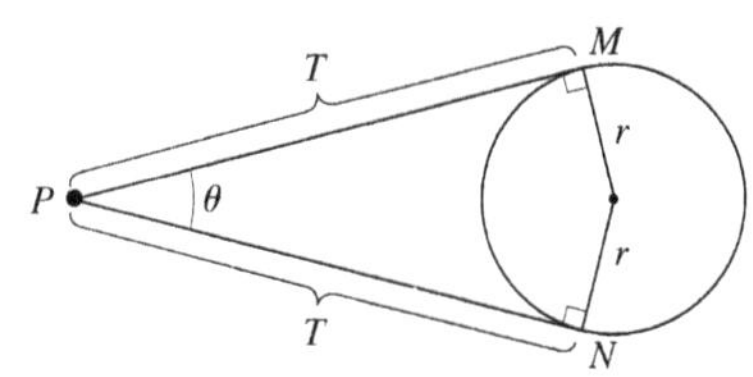

58. Figure 34 shows a regular *heptagon* (a seven-side polygon with sides of equal lengths) inscribed in a circle of radius 1. Determine the length of one side of the heptagon.

Figure 34

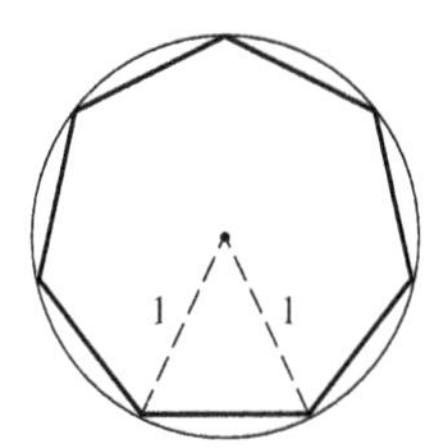

59. Explain why it is not possible to have a right triangle with a hypotenuse of 23 inches and an angle of 35° with a side opposite of length:

(a) 11 inches

(b) 14 inches

Interlude: Practice Problems 1.7, Part I

Figure 35 (not to scale)

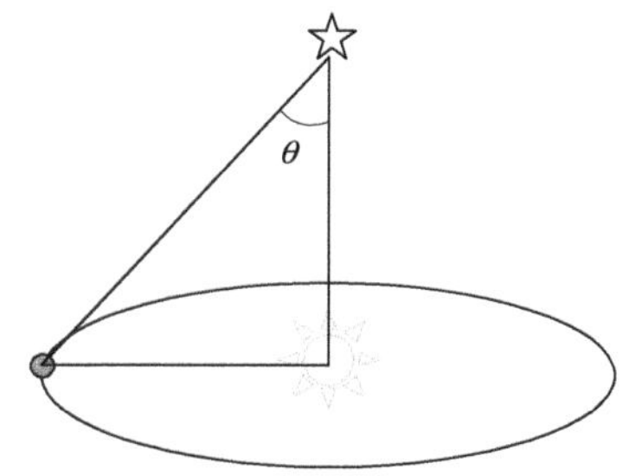

1. Consider Figure 35. The angle θ is called the *parallax*. In the late 1830s a method was developed using parallax to estimate the distance from earth to stars. Parallax angles are very small due to the enormous distances between Earth and even relatively nearby stars. One of the nearest stars is Alpha Centauri, which has a parallax of $\theta \approx 0.000212°$.

(a) Using an earth-sun distance of 93,000,000 miles, estimate the distance from Earth to Alpha Centauri, to the nearest mile.

(b) A light year is defined to be the distance light travels in one year. Light travels at about 186,282 miles per second. Determine the distance between Earth and Alpha Centauri to the nearest one tenth of a light year.

2. Two tracking stations 600 meters apart measure angles of elevation to a weather balloon to be $71.5°$ and $53.6°$, respectively.

(a) Estimate the height of the balloon, accurate to two decimal places.

(b) Estimate the respective distances between the balloon and each tracking station, accurate to two decimal places.

Interlude: Practice Problems 1.7, Part II

1. From the top of an apartment building, the angle of elevation to the top of an adjacent office building is 38.4° and the angle of depression to the base of the office building is 48.6°. Suppose that the office building is 925 feet tall.

   (a) Determine the height of the apartment building, accurate to two decimal places.

   (b) Determine how far apart the two buildings are, accurate to two decimal places.

2. A marker beacon sits at the top of a cliff at an altitude of 2960 feet. A hiker measures an angle of elevation from ground level to the beacon of 46°. After walking towards the base of the cliff for several minutes, the angle of elevation has increased to 61°. Determine how far the hiker walked during that stretch of time, accurate to two decimal places.

1.8 Inverse Trigonometric Functions

Objectives

1. Define the Inverse Sine, Cosine, and Tangent Functions
2. Simplify Expressions Involving Inverse Functions
3. Define the Other Inverse Trigonometric Functions
4. Solve Applied Problems

By examining the graphs of the six basic trigonometric functions, we observe (using the horizontal-line test) that none of them is one-to-one. So none of them has an inverse. However, by restricting domains, we can get trigonometric functions that do have inverses.

Defining the Inverse Sine, Cosine, and Tangent Functions

Table 1 lists three trigonometric functions with restricted domains—one for the sine, one for the cosine, and one for the tangent.

From the graphs in Table 1, we see that each of these three functions is one-to-one and therefore has an inverse. Let us examine each of these functions.

TABLE 1

Function	$f(x) = \sin x$	$g(x) = \cos x$	$h(x) = \tan x$
Restriction on x	$-\frac{\pi}{2} \le x \le \frac{\pi}{2}$	$0 \le x \le \pi$	$-\frac{\pi}{2} < x < \frac{\pi}{2}$
Graph	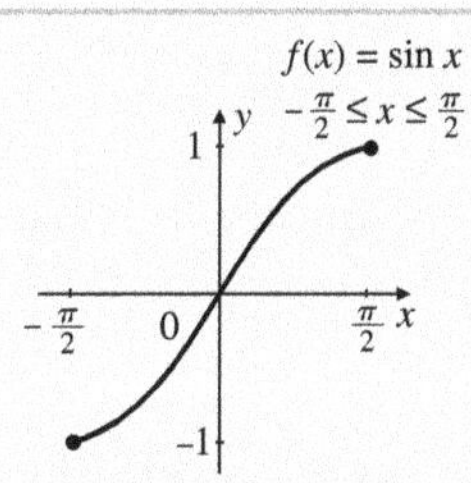	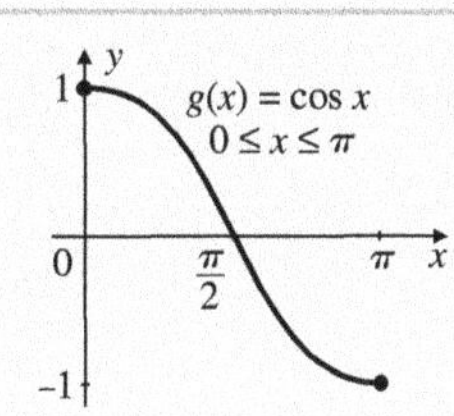	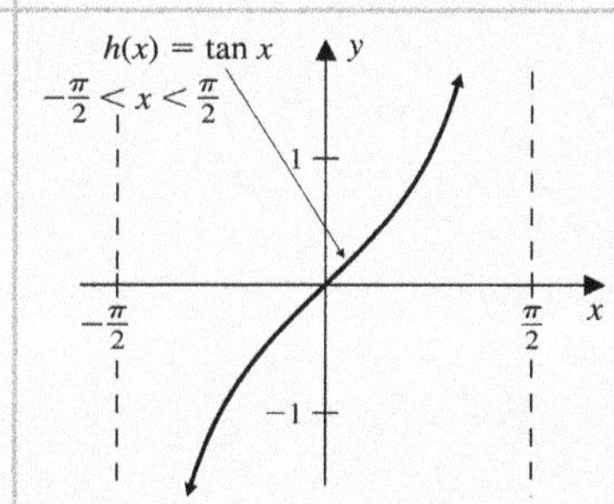

1. The Inverse Sine Function

When we refer to the *inverse sine function,* we mean the function that is the inverse of $f(x) = \sin x$, with the restriction $-\pi/2 \le x \le \pi/2$ (Table 1).

The inverse sine function is denoted by $y = \sin^{-1} x$ (read "the inverse sine of x") or by $y = \arcsin x$ (read "the arcsine of x") and is defined as follows.

Definition

The Inverse Sine Function

> The **inverse sine function,**
>
> $$y = \sin^{-1} x \quad \text{or} \quad y = \arcsin x$$
>
> is defined by
>
> $$\sin y = x, \quad \text{where} \quad -\pi/2 \le y \le \pi/2 \quad \text{and} \quad -1 \le x \le 1.$$

Note that even though we write $\sin^2 x$ to represent $(\sin x)^2$ we never use $\sin^{-1} x$ to represent $(\sin x)^{-1}$, since

$$(\sin x)^{-1} = \frac{1}{\sin x}$$

which is not the same as the inverse function of $\sin x$. This confusion arises because we use the "exponent" -1 in two different ways: to denote reciprocals and to denote inverse functions.

We can think of $\arcsin x$ or $\sin^{-1} x$ as the number (or the angle) between $-\pi/2$ and $\pi/2$ inclusive, whose sine is x. The special values learned earlier enable us to find exact inverse function values for certain numbers.

EXAMPLE 1 **Finding Exact Values of the Inverse Sine**

Find the exact value of each expression.

(a) $\sin^{-1}\frac{1}{2}$ (b) $\sin^{-1}\left(-\frac{\sqrt{2}}{2}\right)$

Solution By using the special values along with the restrictions on the inverse sine, we get the solutions.

(a) Suppose that $y = \sin^{-1}\frac{1}{2}$. This equation is equivalent to

$$\sin y = \frac{1}{2}, \text{ where } -\frac{\pi}{2} \le y \le \frac{\pi}{2}.$$

From the special values, we know that $\sin\left(\frac{\pi}{6}\right) = \frac{1}{2}$. Since $-\frac{\pi}{2} \le \frac{\pi}{6} \le \frac{\pi}{2}$, it follows that $y = \frac{\pi}{6}$. That is, we have $\sin^{-1}\left(\frac{1}{2}\right) = \frac{\pi}{6}$.

(b) Similarly, $y = \sin^{-1}\left(-\frac{\sqrt{2}}{2}\right)$ is equivalent to

$$\sin y = -\frac{\sqrt{2}}{2}, \text{ where } -\frac{\pi}{2} \le y \le \frac{\pi}{2}.$$

The value $-\frac{\pi}{4}$ satisfies the conditions for y, so $\sin^{-1}\left(-\frac{\sqrt{2}}{2}\right) = -\frac{\pi}{4}$.

The restrictions on the inverse functions are very important to keep in mind. For instance, even though

$$\sin\left(\frac{7\pi}{4}\right) = -\frac{\sqrt{2}}{2}, \quad \sin^{-1}\left(\frac{\sqrt{2}}{2}\right) \ne \frac{7\pi}{4}$$

because $\frac{7\pi}{4}$ is not in the interval $\left[-\frac{\pi}{2}, \frac{\pi}{2}\right]$.

2. The Inverse Cosine Function

We now define the inverse of $g(x) = \cos x$, with x restricted to the interval $[0, \pi]$ (Table 1).

The inverse cosine function is denoted by $y = \cos^{-1} x$ or $y = \arccos x$ and defined as follows.

Definition
The Inverse Cosine Function

The **inverse cosine function,**

$$y = \cos^{-1} x \quad \text{or} \quad y = \arccos x$$

is defined by

$$\cos y = x, \quad \text{where} \quad 0 \le y \le \pi \quad \text{and} \quad -1 \le x \le 1.$$

We can think of $\arccos x$ or $\cos^{-1} x$ as the number (or the angle) between 0 and π inclusive, whose cosine is x.

EXAMPLE 2

Finding Exact Values of the Inverse Cosine

Find the exact value of each expression.

(a) $\cos^{-1}(-1)$ (b) $\arccos\left(-\frac{\sqrt{2}}{2}\right)$

Solution

The restrictions on the inverse cosine lead us to the following results:

(a) For $y = \cos^{-1}(-1)$, we know that $\cos \pi = -1$ and $0 \le \pi \le \pi$, so $y = \pi$.

(b) For $y = \arccos\left(-\frac{\sqrt{2}}{2}\right)$, we know that $\cos\left(\frac{3\pi}{4}\right) = -\frac{\sqrt{2}}{2}$ and $0 \le \frac{3\pi}{4} \le \pi$, so $y = \frac{3\pi}{4}$.

3. The Inverse Tangent Function

The function $y = \tan^{-1} x$ or $y = \arctan x$ denotes the inverse of the function $h(x) = \tan x$, where $-\pi/2 < x < \pi/2$ (Table 1).

Definition
The Inverse Tangent Function

The **inverse tangent function**

$$y = \tan^{-1} x \quad \text{or} \quad y = \arctan x$$

is defined by

$$\tan y = x, \quad \text{where} \quad -\frac{\pi}{2} < y < \frac{\pi}{2} \text{ and } x \text{ is a real number.}$$

The values $-\pi/2$ and $\pi/2$ are excluded for the tangent function, since $\tan(-\pi/2)$ and $\tan(\pi/2)$ are not defined.

We can think of arctan x or $\tan^{-1} x$ as the number (or angle) between $-\pi/2$ and $\pi/2$ whose tangent is x. For instance,

$$\tan^{-1}(-1) = -\frac{\pi}{4}$$

since we know that

$$\tan\left(-\frac{\pi}{4}\right) = -1 \quad \text{and} \quad -\frac{\pi}{2} < -\frac{\pi}{4} < \frac{\pi}{2}.$$

When the evaluation of an inverse function cannot be done with the special values, we use a calculator to approximate the value. Normally we use the radian mode for such approximations. In right triangle trigonometry we used the degree mode to evaluate the inverse functions because we worked with triangles.

EXAMPLE 3

Approximating Values of Inverse Functions

Use a calculator to approximate each value to two decimal places.

(a) $\sin^{-1} 0.8016$ (b) $\cos^{-1}(-0.3281)$

(c) arctan 0.6235

Solution

First we set the calculator in radian mode, and then we use the appropriate inverse function key to get the following approximate values:

(a) $\sin^{-1} 0.8016 = 0.93$ (b) $\cos^{-1}(-0.3281) = 1.91$

(c) $\arctan 0.6235 = 0.56$

By reflecting each of the graphs in Table 1 on page 109 across the line $y = x$, we obtain the inverse function graphs of

$$y = \sin^{-1} x \text{ (Figure 1a)}, \; y = \cos^{-1} x \text{ (Figure 1b)}, \text{ and } y = \tan^{-1} x \text{ (Figure 1c)}.$$

Figure 1

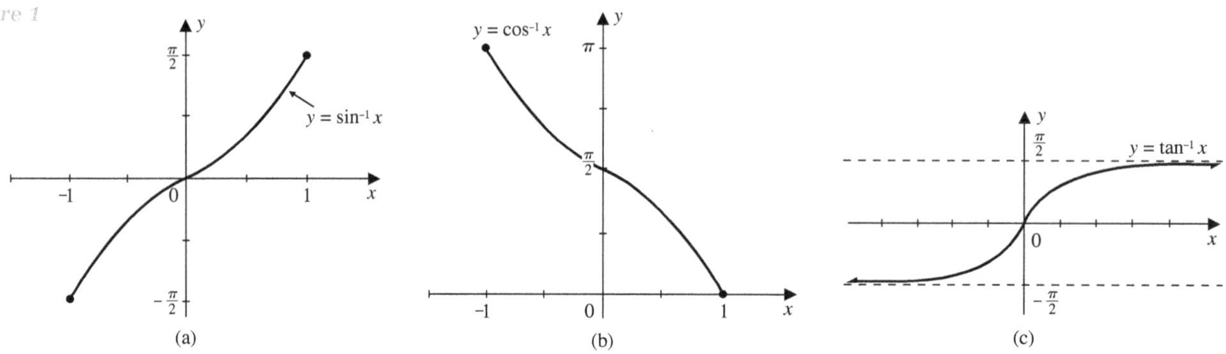

Notice that since the tangent function

$$f(x) = \tan x$$

has *vertical* asymptotes at

$$x = -\frac{\pi}{2} \quad \text{and} \quad x = \frac{\pi}{2},$$

it follows that there are *horizontal* asymptotes for the graph of

$$y = \tan^{-1} x \quad \text{at}$$

$$y = \frac{-\pi}{2} \quad \text{and} \quad y = \frac{\pi}{2} \text{ (Figure 1c).}$$

When we rewrite an equation containing an inverse trigonometric function as an equation in terms of a trigonometric function, care must be taken to keep track of the restrictions on the variables.

EXAMPLE 4

Rewriting an Inverse Function

If $y = 5 + 3\cos^{-1} 2x$, express x in terms of y. Determine the restrictions on x and y.

Solution

The given equation, $y = 5 + 3\cos^{-1} 2x$, can be written as

$$\frac{y-5}{3} = \cos^{-1} 2x.$$

This latter equation is equivalent to

$$\cos\left(\frac{y-5}{3}\right) = 2x \quad \text{or} \quad \frac{1}{2}\cos\left(\frac{y-5}{3}\right) = x$$

where

$$0 \le \frac{y-5}{3} \le \pi \quad \text{and} \quad -1 \le 2x \le 1.$$

The restriction on y is found by solving

$$0 \le \frac{y-5}{3} \le \pi$$

to get:

$$0 \le y - 5 \le 3\pi \qquad \text{Multiply by 3}$$
$$5 \le y \le 3\pi + 5 \qquad \text{Add 5}$$

The restriction on x is found by solving

$$-1 \le 2x \le 1$$

so that

$$-\frac{1}{2} \le x \le \frac{1}{2} \qquad \text{Divide by 2}$$

Thus

$$x = \frac{1}{2}\cos\left(\frac{y-5}{3}\right) \quad \text{where } 5 \le y \le 3\pi + 5 \quad \text{and} \quad -\frac{1}{2} \le x \le \frac{1}{2}.$$

Simplifying Expressions Involving Inverse Functions

Recall that

$$(f \circ f^{-1})(x) = f[f^{-1}(x)] = x \quad \text{and} \quad (f^{-1} \circ f)(x) = f^{-1}[f(x)] = x.$$

When applying this property to the inverse trigonometric functions, we must take into account the restrictions on x that are delineated in the following properties.

Property
Inverse Sine Composition

$$\sin(\sin^{-1} x) = x \text{ if } -1 \le x \le 1$$
$$\sin^{-1}(\sin x) = x \text{ if } -\frac{\pi}{2} \le x \le \frac{\pi}{2}$$

Property
Inverse Cosine Composition

$$\cos(\cos^{-1} x) = x \quad \text{if } -1 \le x \le 1$$
$$\cos^{-1}(\cos x) = x \quad \text{if } 0 \le x \le \pi$$

Property
Inverse Tangent Composition

$$\tan(\tan^{-1} x) = x \quad \text{for every real number } x$$
$$\tan^{-1}(\tan x) = x \quad \text{if } -\frac{\pi}{2} < x < \frac{\pi}{2}$$

EXAMPLE 5

Using the Inverse Trigonometric Composition Properties

Use the composition properties for the inverse trigonometric functions to find the exact value of each expression.

(a) $\sin^{-1}\left(\sin\frac{\pi}{6}\right)$ (b) $\cos\left(\cos^{-1}\frac{1}{2}\right)$ (c) $\tan^{-1}\left(\tan\frac{3\pi}{4}\right)$

Solution

(a) Since $-\frac{\pi}{2} \leq \frac{\pi}{6} \leq \frac{\pi}{2}$, then

$$\sin^{-1}\left(\sin\frac{\pi}{6}\right) = \frac{\pi}{6}.$$

As a check, we note that

$$\sin^{-1}\left(\sin\frac{\pi}{6}\right) = \sin^{-1}\frac{1}{2}$$
$$= \frac{\pi}{6}.$$

(b) Since $-1 \leq \frac{1}{2} \leq 1$, then

$$\cos\left(\cos^{-1}\frac{1}{2}\right) = \frac{1}{2}.$$

(c) We first observe that

$$\tan^{-1}\left(\tan\frac{3\pi}{4}\right) \neq \frac{3\pi}{4}$$

because $\frac{3\pi}{4}$ does not satisfy the restriction $-\frac{\pi}{2} < x < \frac{\pi}{2}$.
Thus we proceed as follows:

$$\tan^{-1}\left(\tan\frac{3\pi}{4}\right) = \tan^{-1}(-1)$$
$$= -\frac{\pi}{4}.$$

Right triangles can help us find exact values of expressions involving inverse trigonometric functions and convert trigonometric expressions into algebraic ones.

EXAMPLE 6

Simplifying Inverse Trigonometric Expressions

Use right triangles to simplify each expression.

(a) $\tan\left(\cos^{-1}\frac{3}{4}\right)$

(b) $\sin^{-1}x + \cos^{-1}x$, where $0 < x < 1$

Solution

(a) Assume that $\theta = \cos^{-1}(3/4)$.
By definition $\cos\theta = 3/4$, where $0 < \theta < \pi/2$, since the cosine is positive.

Next we construct a right triangle with one side of length 3 adjacent to angle θ and the hypotenuse of length 4 so that $\cos\theta = 3/4$ (Figure 2a).

By the Pythagorean theorem, the length of the side opposite angle θ is $\sqrt{4^2 - 3^2} = \sqrt{7}$.

Using right triangle trigonometry, we obtain

$$\tan\left(\cos^{-1}\frac{3}{4}\right) = \tan\theta$$

$$= \frac{\text{opp}}{\text{adj}} = \frac{\sqrt{7}}{3}.$$

Figure 2

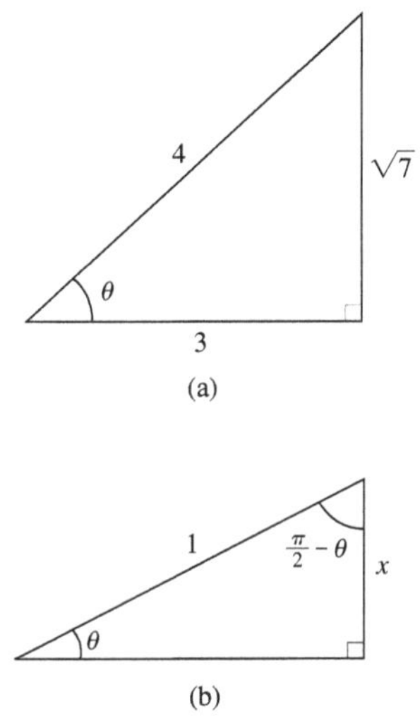

(b) Assume that $\theta = \sin^{-1} x$, so $\sin\theta = x$.
Since $0 < x < 1$, it follows that $0 < \theta < \pi/2$. We can think of θ as an acute angle in a right triangle with sides that satisfy

$$\sin\theta = \frac{\text{opp}}{\text{hyp}} = \frac{x}{1} = x \text{ (Figure 2b).}$$

Using right triangle trigonometry, we get

$$\cos\left(\frac{\pi}{2} - \theta\right) = \frac{\text{adj}}{\text{hyp}} = \frac{x}{1} = x \quad \text{where} \quad 0 < \left(\frac{\pi}{2} - \theta\right) < \frac{\pi}{2}.$$

So

$$\cos^{-1} x = \frac{\pi}{2} - \theta.$$

It follows that

$$\sin^{-1} x + \cos^{-1} x = \theta + \left(\frac{\pi}{2} - \theta\right) = \frac{\pi}{2},$$

that is, $\sin^{-1} x + \cos^{-1} x = \dfrac{\pi}{2}$.

Defining the Other Inverse Trigonometric Functions

By restricting the domains of the cosecant, secant, and cotangent functions, we get one-to-one functions that enable us to define the other three inverse trigonometric functions.

Definition

Other Inverse Trigonometric Functions

Function	Defined by
$y = \csc^{-1} x$	$\csc y = x$, where $x \le -1$ or $x \ge 1$ and $-\pi/2 \le y \le \pi/2$, $y \ne 0$
$y = \sec^{-1} x$	$\sec y = x$, where $x \le -1$ or $x \ge 1$ and $0 \le y \le \pi$, $y \ne \pi/2$
$y = \cot^{-1} x$	$\cot y = x$, where x is any real number and $0 < y < \pi$

Note that the choices of the ranges for $g(x) = \sec^{-1} x$ and $h(x) = \csc^{-1} x$ are not universally accepted and may differ in some other textbooks.

The notations arccsc x, arcsec x, and arccot x, respectively, are also used for these functions.

The graphs and properties of these inverse functions are given in Table 2.

TABLE 2 Other Inverse Trigonometric Functions

	$h(x) = \csc^{-1} x$	$g(x) = \sec^{-1} x$	$f(x) = \cot^{-1} x$
	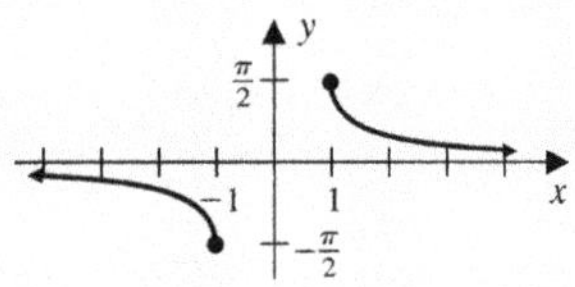	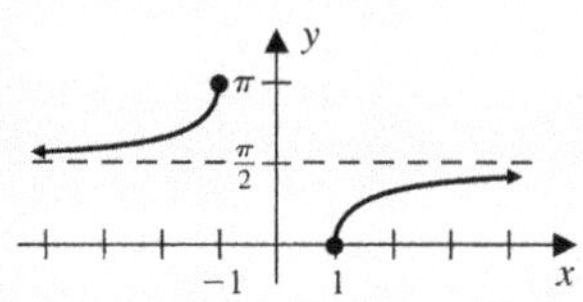	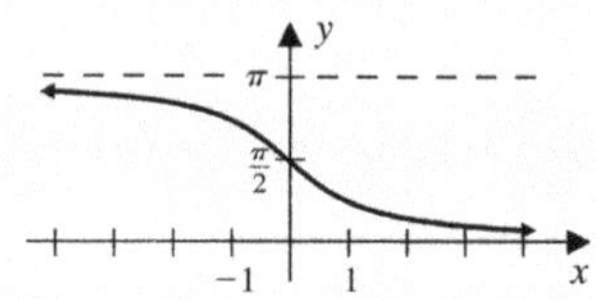
Domain	$(-\infty, -1]$ or $[1, \infty)$	$(-\infty, -1]$ or $[1, \infty)$	All real numbers
Range	$[-\pi/2, \pi/2]$, $y \neq 0$	$[0, \pi]$, $y \neq \pi/2$	$(0, \pi)$
Asymptotes	$y = 0$	$y = \pi/2$	$y = 0$ and $y = \pi$

EXAMPLE 7 **Evaluating Inverse Functions**

Find the exact value of each function.

(a) $\cot^{-1}\sqrt{3}$ (b) $\csc^{-1}(-2)$

Solution (a) $\cot^{-1}\sqrt{3}$ is the number y in the interval $(0, \pi)$ that satisfies $\cot y = \sqrt{3}$, or, equivalently, $\frac{1}{\tan y} = \sqrt{3}$ so that $\tan y = \frac{1}{\sqrt{3}}$. It follows that $y = \frac{\pi}{6}$.

(b) $\csc^{-1}(-2)$ is the number y in the interval $[-\pi/2, \pi/2]$, $y \neq 0$, that satisfies $\csc y = -2$ or, equivalently, $\frac{1}{\sin y} = -2$. That is, $\sin y = -\frac{1}{2}$ so that $y = -\frac{\pi}{6}$. ■

Most calculators do not have keys for the inverse cotangent, inverse secant, or inverse cosecant functions. The following formulas are used to find the approximate value:

1. If $|x| \geq 1$, then:

   (a) $\csc^{-1} x = \sin^{-1}\frac{1}{x}$ (b) $\sec^{-1} x = \cos^{-1}\frac{1}{x}$

2. If x is a real number, then:

$$\cot^{-1} x = \begin{cases} \tan^{-1}\frac{1}{x} & \text{if } x > 0 \\ \tan^{-1}\frac{1}{x} + \pi & \text{if } x < 0 \\ \frac{\pi}{2} & \text{if } x = 0 \end{cases}$$

EXAMPLE 8 **Approximating Values of Inverse Trigonometric Functions**

Use a calculator to approximate each value to three decimal places.

(a) $\cot^{-1} 12.3$ (b) $\sec^{-1} 1.84$ (c) $\csc^{-1}(-1.52)$

Solution First we set the calculator in radian mode. Then we use the above formulas to get the following approximate values:

(a) $\cot^{-1} 12.3 = \tan^{-1}\left(\frac{1}{12.3}\right) = 0.081$

(b) $\sec^{-1} 1.84 = \cos^{-1}\left(\frac{1}{1.84}\right) = 0.996$

(c) $\csc^{-1}(-1.52) = \sin^{-1}\left(\frac{1}{-1.52}\right) = -0.718$

Solving Applied Problems

Inverse trigonometric functions are used in mathematical modeling, as the next example illustrates.

EXAMPLE 9 **Modeling a Viewing Angle**

Figure 3

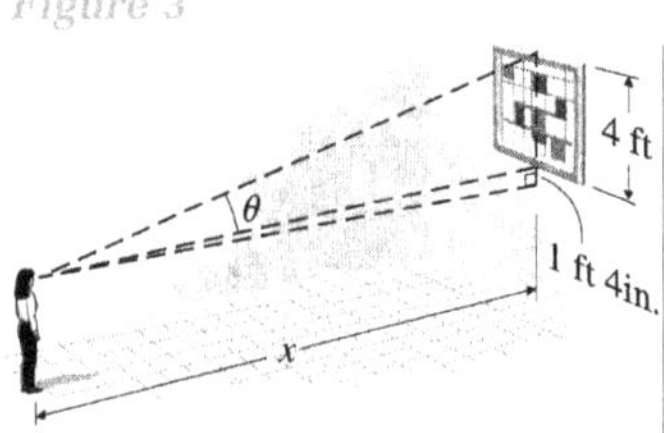

A museum plans to hang a painting for public viewing. The painting, which is 4 feet high, is mounted on a wall in such a way that its lower edge is 1 foot 4 inches above the level of the average viewer's eyes. The viewer studies the painting x feet from the wall on which the painting is mounted (Figure 3). Suppose the angle formed by the viewer's eyes and the top and bottom of the painting is θ.

(a) Use right triangle trigonometry to express the angle θ as a function of x.

(b) What is the angle θ if the viewer is standing 5 feet away from the painting? Round off the answer to two decimal places.

Solution (a) Figure 4 helps us express θ as a function of x. (Note that 1 foot 4 inches is 4/3 feet.) We see in the figure that

Figure 4

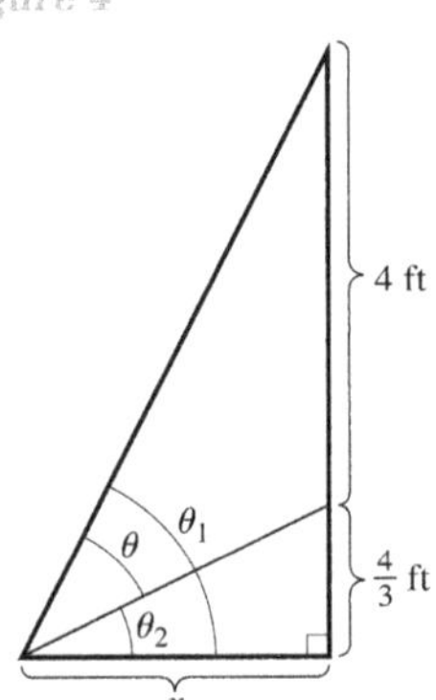

$$\theta = \theta_1 - \theta_2$$

$$\theta = \tan^{-1}\left(\frac{4 + \frac{4}{3}}{x}\right) - \tan^{-1}\left(\frac{\frac{4}{3}}{x}\right)$$

$$\theta = \tan^{-1}\frac{16}{3x} - \tan^{-1}\frac{4}{3x}.$$

This latter equation expresses θ as a function of x.

(b) Substituting 5 for x in the equation

$$\theta = \tan^{-1}\frac{16}{3x} - \tan^{-1}\frac{4}{3x}$$

we have

$$\theta = \tan^{-1}\left(\frac{16}{15}\right) - \tan^{-1}\left(\frac{4}{15}\right) = 0.56 \text{ (approx.)}.$$

Thus when the viewer stands 5 feet from the painting, the viewing angle is approximately 0.56 radian or about 31.92°.

Figure 5

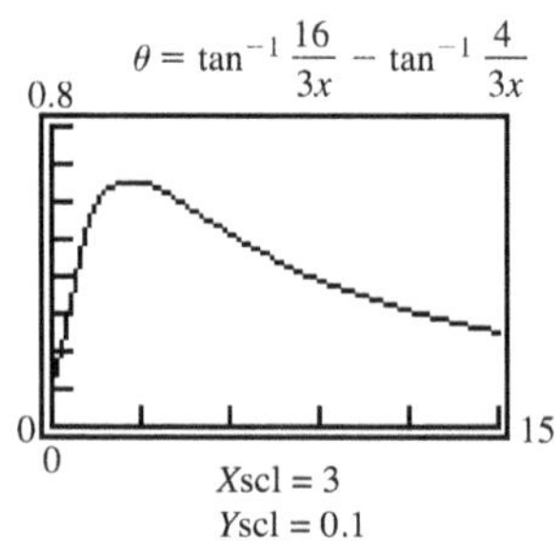

Figure 5 shows a viewing window of the graph of

$$\theta = \tan^{-1}\frac{16}{3x} - \tan^{-1}\frac{4}{3x}$$

The graph indicates that there is a high or maximum point.

By using the MAX–finding feature, we find the maximum point of the graph of θ to be approximately (2.67, 0.64), rounded to two decimal places. This means that when x is approximately 2.67, θ attains its maximum value of about 0.64 radian or about 36.67°. Thus, when the viewer stands about 2.67 feet, or 2 feet 8 inches, from the wall, the maximum angle measure of about 37° is attained.

Standing 2 feet 8 inches from a painting 4 feet high may be too close to appreciate the artistic qualities of the painting. The maximum viewing angle does not necessarily give the best location for appreciating the art.

PROBLEM SET 1.8

Mastering the Concepts

In problems 1–10, find the exact value of each expression.

1. (a) $\sin^{-1}\frac{\sqrt{2}}{2}$ (b) $\sin^{-1}\left(-\frac{1}{2}\right)$

2. (a) $\cos^{-1}\frac{\sqrt{3}}{2}$ (b) $\cos^{-1}\left(-\frac{1}{2}\right)$

3. (a) $\tan^{-1}\frac{\sqrt{3}}{3}$ (b) $\tan^{-1}\left(-\frac{\sqrt{3}}{3}\right)$

4. (a) $\arctan(-1)$ (b) $\arccos 1$

5. (a) $\arcsin\left(-\frac{\sqrt{3}}{2}\right)$ (b) $\arccos\left(-\frac{\sqrt{3}}{2}\right)$

6. (a) $\arcsin 1$ (b) $\arctan\sqrt{3}$

7. (a) $\cos^{-1}\left(\cos\frac{\pi}{6}\right)$ (b) $\tan^{-1}\left(\tan\frac{\pi}{4}\right)$

8. (a) $\sin\left(\cos^{-1}\frac{1}{2}\right)$ (b) $\sin^{-1}\left[\sin\left(-\frac{3\pi}{2}\right)\right]$

9. (a) $\sin^{-1}\left(\sin\frac{5\pi}{4}\right)$ (b) $\tan[\tan^{-1}(-1)]$

10. (a) $\sec\left(\sin^{-1}\frac{\sqrt{3}}{2}\right)$ (b) $\csc(\tan^{-1} 1)$

In problems 11–16, find the approximate values of each expression to two decimal places.

11. (a) $\sin^{-1} 0.2182$ (b) $\arcsin(-0.7771)$

12. (a) $\cos^{-1} 0.8628$ (b) $\arccos(-0.8473)$

13. (a) $\tan^{-1} 1.072$ (b) $\arctan(-41.03)$

14. (a) $\arccos 0.4037$ (b) $\arcsin(2/\pi)$

15. (a) $\arccos(-0.7112)$ (b) $\tan^{-1} 100$

16. (a) $\sin^{-1}(1 - \sqrt{2})$ (b) $\sin^{-1}(\sqrt{3} - \sqrt{2})$

In problems 17–24, express x in terms of y. Determine the restrictions on x and y.

17. $y = \sin^{-1}(x + 1)$

18. $y = \cos^{-1}(x - 2)$

19. $y = \cos^{-1}(2x - 8)$

20. $y = 2\sin^{-1}(3x + 7)$

21. $y = -3\tan^{-1}(3x - 2)$

22. $y = -2\cos^{-1}(2x - 1)$

23. $y = \frac{1}{3}\cos^{-1}(2x - 4)$

24. $y = 4 - 2\sin^{-1} 4x$

In problems 25 and 26, use the inverse trigonometric composition identities, whenever possible, to find the value of each expression. Round off the answer to four decimal places when appropriate.

25. (a) $\sin(\sin^{-1} 0.3)$
(b) $\sin^{-1}(\sin 1.3)$
(c) $\cos\left[\cos^{-1}\left(-\frac{\pi}{6}\right)\right]$
(d) $\cos^{-1}\left(\cos\frac{7\pi}{6}\right)$

26. (a) $\cos(\cos^{-1} 0.8)$
(b) $\cos^{-1}(\cos\sqrt{2})$
(c) $\tan(\tan^{-1} 90)$
(d) $\tan^{-1}\left(\tan\frac{5\pi}{4}\right)$

In problems 27–30, find the required restrictions on x in order for the equation to be true.

27. $\sin[\sin^{-1}(3x + 2)] = 3x + 2$

28. $\cos[\cos^{-1}(5 - 2x)] = 5 - 2x$

29. $\tan^{-1}\left[\tan\left(\frac{x}{3} - 1\right)\right] = \frac{x}{3} - 1$

30. $\cos^{-1}[\cos(x^2 - 4)] = x^2 - 4$

In problems 31–36, use a right triangle whenever possible to find the exact value of each expression.

31. (a) $\cos\left(\sin^{-1}\frac{4}{5}\right)$
(b) $\sin\left(\cos^{-1}\frac{5}{13}\right)$

32. (a) $\sin\left(\tan^{-1}\frac{4}{3}\right)$
(b) $\cos^{-1}\left(\cos\frac{3\pi}{2}\right)$

33. (a) $\tan\left(\sin^{-1}\frac{2\sqrt{5}}{5}\right)$
(b) $\cos\left[\cos^{-1}\left(-\frac{2\sqrt{5}}{5}\right)\right]$

34. (a) $\sin[\tan^{-1}(-2)]$
(b) $\sec\left(\sin^{-1}\frac{4}{5}\right)$

35. (a) $\csc\left(\cos^{-1}\frac{1}{4}\right)$
(b) $\sec^{-1}\left(\sec\frac{\pi}{3}\right)$

36. (a) $\cot\left[\tan^{-1}\left(-\frac{5}{12}\right)\right]$
(b) $\csc\left(\sin^{-1}\frac{12}{13}\right)$

In problems 37 and 38, use a right triangle to rewrite each expression as an algebraic expression in terms of x that does not involve an inverse trigonometric function, where $x > 0$.

37. (a) $\sin(\cos^{-1} x)$
(b) $\tan(\cos^{-1} x)$
(c) $\cot(\tan^{-1} x)$

38. (a) $\cot(\sin^{-1} x)$
(b) $\cot(\cos^{-1} x)$
(c) $\csc\left(\tan^{-1}\frac{x}{\sqrt{2}}\right)$

In problems 39–42, find the exact value of each expression.

39. (a) $\cot^{-1} 1$
(b) $\csc^{-1}(-\sqrt{2})$

40. (a) $\cot^{-1}(-1)$
(b) $\sec^{-1} 2$

41. (a) $\sec^{-1}\sqrt{2}$
(b) $\csc^{-1}\left(-\frac{2}{\sqrt{3}}\right)$

42. (a) $\sec^{-1}(-\sqrt{2})$
(b) $\csc^{-1}\left(\frac{2}{\sqrt{3}}\right)$

In problems 43–46, find the approximate value rounded to two decimal places.

43. (a) $\cot^{-1} 0.9713$
(b) $\sec^{-1} 3.4182$

44. (a) $\csc^{-1} 1.0152$
(b) $\cot^{-1} 1.0487$

45. (a) $\sec^{-1}(-2.4182)$
(b) $\csc^{-1}(-8.8952)$

46. (a) $\sec^{-1} 8.4513$
(b) $\cot^{-1}(-0.8752)$

Applying the Concepts

47. Viewing a Billboard: A billboard is to be built adjacent to a highway so that the top and bottom will be 32 feet and 24 feet, respectively, above the eye level of a passing motorist. Suppose that the motorist passes the billboard at a distance of x feet and that the angle formed by the motorist's eye level and the top and bottom of the billboard is θ (Figure 6).

(a) Express θ as a function of x.

(b) Find the angle θ if the motorist is 100 feet from the billboard. Round off the answer to two decimal places.

Figure 6

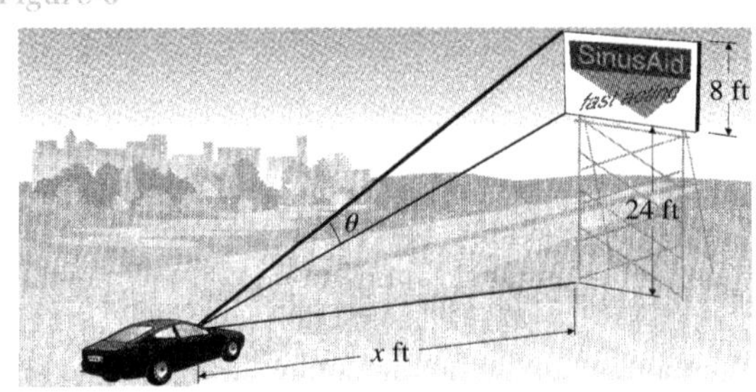

48. (Refer to problem 47.)

(a) Use a grapher to graph the function found in (a).

(b) Approximate the maximum value of θ and when it occurs. What is the degree measure for this maximum angle? Round off the answers to two decimal places.

(c) Does the maximum angle give the best view of the billboard? Explain.

49. Viewing a Window: A night watchman wants the greatest possible view of a second story window of a building as he makes his rounds. The windows of the building are 4 feet high, with the lower sill 11 feet above the ground. Assume the watchman's eyes are precisely 6 feet above the ground. Suppose that the watchman stands x feet away from the building, and that the angle formed by the watchman's eyes and the top and bottom of the window is θ (Figure 7).

(a) Express θ as a function of x.

(b) Find the angle if the watchman is 20 feet from the building. Round off the answer to two decimal places.

Figure 7

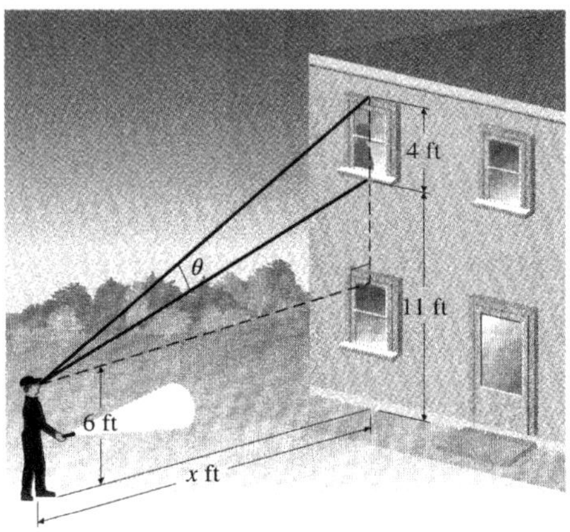

50. (Refer to problem 49.)

(a) Use a grapher to graph the function found in (a).

(b) Approximate the maximum value of θ and when it occurs. What is the degree measure for this maximum angle? Round off the answers to two decimal places.

(c) Does the maximum angle give the best view of the window? Explain.

Developing and Extending the Concepts

In problems 51–56 use transformations of the graphs in Figure 1 (see page 114) to sketch the graph of each function.

51. $y = \frac{1}{2} \sin^{-1} x$

52. $y = -3 \cos^{-1} x$

53. $y = 3 + 2 \tan^{-1} x$

54. $y = \tan^{-1} x + \frac{\pi}{2}$

55. $y = 2[\sin^{-1}(x - 1)] - \pi$

56. $y = \frac{1}{3}[\cos^{-1}(x + 1)] + \frac{\pi}{2}$

57. Show that $\sin^{-1}(-x) = -\sin^{-1} x$

58. Is it true that $\tan^{-1} x = \dfrac{\sin^{-1} x}{\cos^{-1} x}$? Justify your answer.

59. Graph

$$f(x) = (\sin x)^{-1} = \frac{1}{\sin x} = \csc x \quad \text{and} \quad g(x) = \sin^{-1} x$$

for $-1 \le x \le 1$. Are the graphs the same? Explain the use of the superscript $^{-1}$ in both equations.

60. Use a grapher to graph $y = \tan^{-1} x^2$. Then use the graph to describe the limit behavior of the function; that is, determine what happens to the values of $\tan^{-1} x^2$ as $x \to +\infty$ and as $x \to -\infty$.

61. Use a grapher to graph $y = \sin^{-1} x + \cos^{-1} x$, where $0 \le x \le 1$. Explain how the graph demonstrates the result in Example 6b.

62. Sketch the graph of the function $y = \sin(\sin^{-1} x)$. What is the domain of this function? Explain your result.

Interlude: Practice Problems 1.8, Part I

1. Write $\sec\left(\sin^{-1}\left(\frac{x}{9}\right)\right)$ as an algebraic expression in terms of x.

2. A hot-air balloon x feet above the ground is 600 feet from an observer and moving straight up. Let θ denote the angle of elevation of the balloon.

(a) Express θ as a function of x.

(b) Calculate the angle of elevation when the balloon is 185 feet above the ground. Give your answer in radians accurate to two decimal places.

3. Suppose that $t = \arcsin\left(-\frac{8}{15}\right)$. Calculate the exact values of $\csc t$, $\cos t$, and $\tan t$.

4. The graph in Figure 8 is a sketch of the graph of $y = a\cos^{-1}(bx) + k$. Determine a, b, and k.

Figure 8

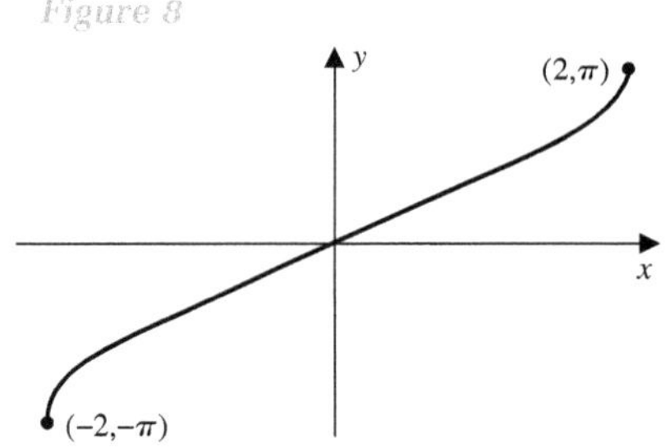

Interlude: Practice Problems 1.8, Part II

In surveying and navigation, directions are often given in terms of *bearings*. A bearing measures the acute angle along a straight path or sight line made relative to a north-south line. For instance, a bearing of S28°W indicates a path or sight line measured 28° west of due south.

1. A ship departs a port at 9 a.m. and heads due east at a speed of 25 knots (nautical miles per hour). At 11 a.m. the ship changes course to a bearing of N36°E.

(a) Determine the ship's bearing from the port as of noon, using degree measure accurate to two decimal places.

(b) Determine how far from the port the ship is at noon, accurate to two decimal places.

2. A satellite circles a planet d miles from the planet's surface. The satellite observes horizons on the planet's surface as shown in Figure 9, where r is the radius of the planet, in miles.

Figure 9

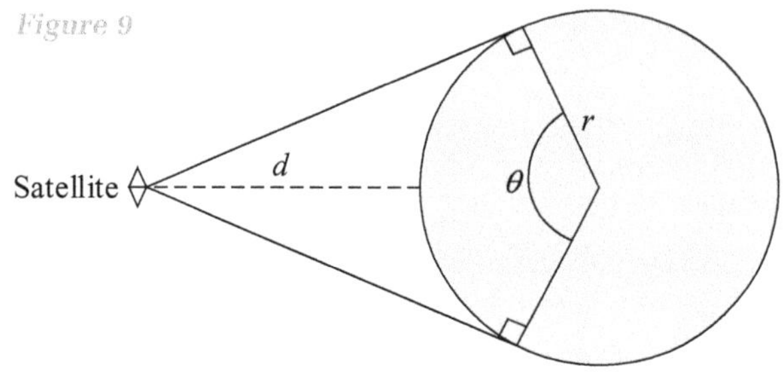

(a) Express d in terms of r and θ.

(b) Express r in terms of d and θ.

(c) Express θ in terms of r and d.

(d) Calculate θ if $r = 5200$ and $d = 310$. Give your answer in both degrees and radians, accurate to three decimal places.

Chapter 1 Readiness, Part I

Provide complete solutions in the space provided.

1. Given the diagram shown in Figure 1, calculate the values of m, n, and p, accurate to three decimal places.

Figure 1

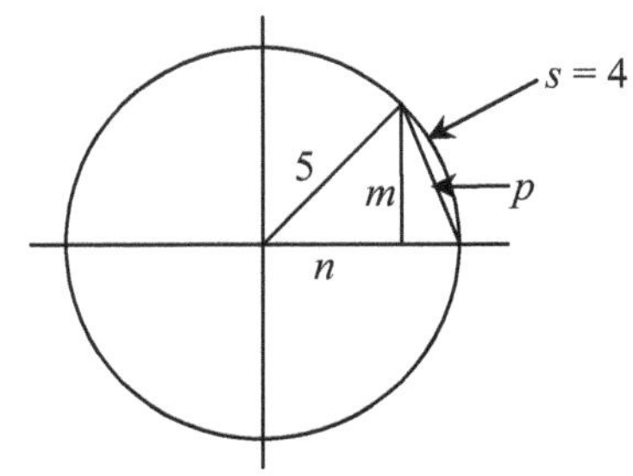

2. Calculate the exact value of $\cos\left(\arcsin\left(\frac{3}{8}\right)\right)$.

3. Calculate the degree measure, accurate to three decimal places, of the acute angle formed by the line with equation $y = -3x + 8$ and the x-axis.

4. An antenna sits atop a 50-foot pole. From a position on the ground 75 feet from the base of the pole, the antenna subtends an angle of 15°. Determine the height of the antenna, accurate to two decimal places.

Chapter 1 Readiness, Part II

Provide complete solutions in the space provided.

1. An ant is walking along a circle of radius 10 centered at the origin. The ant starts at point $P(0, 10)$ and travels counter-clockwise a distance of 24.3 units, stopping at point $Q(a, b)$. Determine a and b, accurate to three decimal places.

2. Assume that in the familiar image below, the circle is of radius 4. Calculate the total area of the six lightly-colored regions, accurate to three decimal places.

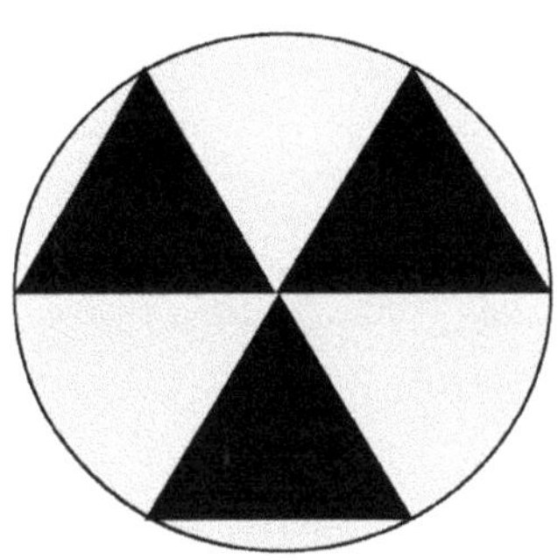

3. At a large baseball stadium, from a seat in the last row of the upper deck directly behind home plate, the angle of depression to home plate is 29.9°. Also, the angle of depression to the pitcher's mound is 24.2°. Assume that the distance between home plate and the pitcher's mound is 60.5 feet.

(a) Calculate the viewing distance from the seat to home plate, accurate to two decimal places.

(b) Calculate the viewing distance from the seat to the pitcher's mound, accurate to two decimal places.

CHAPTER 1 REVIEW PROBLEM SET

1. Express each angle measure in degrees, minutes, and seconds.
   (a) $76.25°$ (b) $-61.35°$
   (c) $143.47°$ (d) $-14.47°$
2. Express each angle as a decimal. Round off the answers to four decimal places.
   (a) $4°7'$ (b) $21°19'13''$
   (c) $-45°35'25''$ (d) $15''$
3. Convert each degree measure to radian measure. Round off the answers to four decimal places.
   (a) $15°$ (b) $-17.45°$
   (c) $45°16'51''$ (d) $36°11'25''$
4. Convert each radian measure to degree measure. Round off the answers to four decimal places.
   (a) $\frac{17\pi}{4}$ (b) $-\frac{4\pi}{7}$
   (c) 5.82 (d) -7.63
5. Suppose that angle $\theta = 36°$ is in standard position. Find the measure of each of the following angles and sketch the angles.
   (a) Complement of θ
   (b) Supplement of θ
   (c) Two angles coterminal with θ, one positive and one negative.
6. Let s denote the length of the arc intercepted on a circle of radius r by a central angle θ. In each case, find the missing measurement and the area of the sector.
   (a) $r = 7$ centimeters, $\theta = 75°$, $s = ?$
   (b) $r = 2$ inches, $s = 5$ inches, $\theta = ?$
   (c) $s = 17$ meters, $\theta = \frac{5\pi}{6}$, $r = ?$
7. Display the location of each point on the unit circle. Then find the coordinates of each point. Round off the answers to four decimal places.
   (a) $P\left(\frac{2\pi}{5}\right)$
   (b) $P(3.35)$
   (c) $P\left(-\frac{3\pi}{7}\right)$
8. Find the exact values of the six trigonometric functions of t.
   (a) $P(t) = \left(-\frac{\sqrt{3}}{2}, -\frac{1}{2}\right)$
   (b) $P(t) = \left(\frac{4}{\sqrt{17}}, \frac{1}{\sqrt{17}}\right)$
9. Graph each function. Find the period and phase shift.
   (a) $f(x) = 0.6 \tan\left(3x - \frac{\pi}{2}\right)$
   (b) $g(x) = -1.5 \sec\left(\frac{x}{4} + \pi\right)$
10. Use the known graphs of the sine and cosine functions to graph one cycle of each function. Indicate the amplitude, period, and phase shift.
    (a) $f(x) = 2 \sin\left(\frac{3x}{2} - \frac{\pi}{4}\right)$
    (b) $g(x) = -0.2 \cos\left(2x + \frac{\pi}{8}\right)$
11. Use a grapher to sketch the graph of each function for $0 \le x \le 4\pi$.
    (a) $f(x) = 2 \cos x + 3 \cos \frac{x}{2}$
    (b) $g(x) = \frac{\sin x}{x} - x$
    (c) $h(x) = \frac{1 - \cos x}{x^2}$
12. The functions for the graphs in Figure 1 have the form $y = a \cos kx$ or $y = a \sin kx$, where $k > 0$. For each graph, specify the period and amplitude, and write a specific equation for the function.

Figure 1

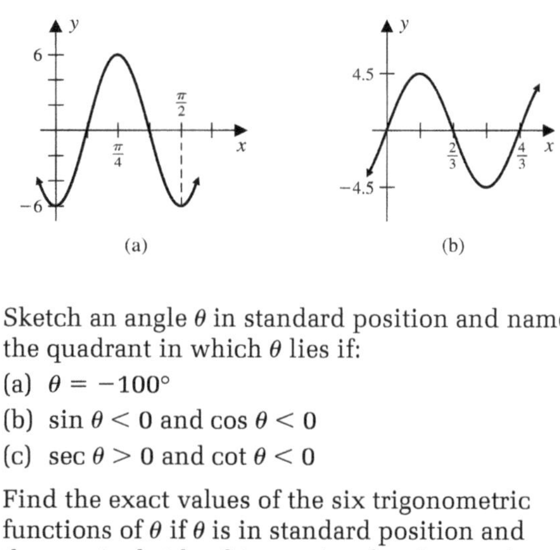

13. Sketch an angle θ in standard position and name the quadrant in which θ lies if:
    (a) $\theta = -100°$
    (b) $\sin \theta < 0$ and $\cos \theta < 0$
    (c) $\sec \theta > 0$ and $\cot \theta < 0$
14. Find the exact values of the six trigonometric functions of θ if θ is in standard position and the terminal side of θ contains the given point
    (a) $(-6, -8)$ (b) $(-3, 5)$

15. Use a calculator to find the approximate value of each expression rounded off to four decimal places.

(a) $\sin 3$ (b) $\cos(-208°)$

(c) $\tan \frac{3\pi}{5}$ (d) $\sec(-3.92)$

16. Use a reference triangle and periodicity to find the exact value of each expression.

(a) $\cos\left(-\frac{5\pi}{3}\right)$ (b) $\sec 780°$

(c) $\cot \frac{37\pi}{6}$ (d) $\tan(-600°)$

17. Find the exact values of the six trigonometric functions of the acute angle θ in Figure 2.

Figure 2

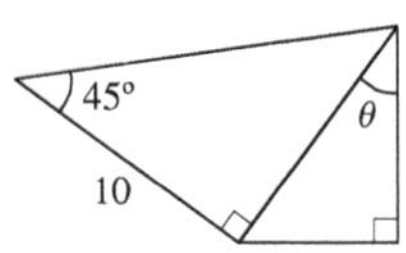

18. Assume θ is an acute angle and $\sin \theta = \frac{8}{17}$. Find

(a) $\cos \theta$ and $\tan \theta$

(b) $\sin(-\theta)$ and $\sec(-\theta)$

(c) θ, in both degree measure and radian measure. Round off the answers to two decimal places.

19. Suppose θ satisfies

$$180° < \theta < 270° \quad \text{and} \quad \tan \theta = a.$$

Express each of the following in terms of a.

(a) $\cot \theta$ (b) $\tan(\theta - 360°)$ (c) $\tan(\theta + 180°)$

20. Find the exact value of each expression.

(a) $\cos^{-1} \frac{1}{2}$

(b) $\sin^{-1}\left(-\frac{1}{2}\right)$

(c) $\tan^{-1}\left(\sin \frac{\pi}{2}\right)$

(d) $\tan\left(\sin^{-1} \frac{3}{5}\right)$

(e) $\sec\left[\tan^{-1}\left(-\frac{5}{3}\right)\right]$

21. Use a calculator to approximate the value of each expression. Give the answer in radians rounded off to two decimal places.

(a) $\sin^{-1} 0.3741$

(b) $\arccos(-0.4901)$

(c) $\sec^{-1} 9.723$

(d) $\cot^{-1} 57.29$

22. Use a right triangle to rewrite each given expression as an algebraic expression in terms of x that does not involve an inverse trigonometric function.

(a) $\sin(\tan^{-1} x)$, where $x > 0$

(b) $\tan(\sin^{-1} x)$, where $0 < x < 1$

23. In each case, solve for x in terms of y. Specify the restrictions on the values of x and y.

(a) $y = \sin^{-1} \frac{x}{2}$

(b) $y = \arctan(2x + 3)$

24. Solve the right triangle ABC if $\gamma = 90°$. Round off each answer to two decimal places.

(a) $b = 25$ inches and $\beta = 65°$

(b) $b = 7$ centimeters and $\alpha = 35°$

25. **Engineering:** A curve on a highway subtends an angle of 31° on a circle of radius 560 meters.

(a) How long is the curve subtended by this angle?

(b) How long will it take a car traveling 64 kilometers per hour to round the curve?

Round off the answers to two decimal places.

26. **Geometry:** An isosceles triangle ABC with $|\overline{AB}| = |\overline{AC}|$ and an angle $\alpha = 48°$ is inscribed in a circle. Find the radius of the circle if the side $\overline{BC}$ intercepts an arc of 6.3 centimeters on the circle. Round off the answer to two decimal places.

27. **Space Satellite:** A satellite traveling in a circular orbit of radius 6371 kilometers is known to have a linear speed of 30,720 kilometers per hour. Find the angular speed of the satellite in radians per hour. Round off the answer to two decimal places.

28. **Simple Harmonic Motion:** The position of a particle at time t is given by each equation. Find the amplitude, period, and phase shift of the path of each particle.

(a) $y = 2.7 \sin(7t - 4\pi)$

(b) $y = 4 \cos\left(3t - \frac{\pi}{3}\right)$

29. **Hours of Daylight:** Suppose the number of daylight hours N in a city at a particular time of the year is given by the model

$$N = 12 + 2.5 \sin\left[\frac{2\pi}{365}(t - 81)\right]$$

where t is the number of days, with $t = 1$ corresponding to January 1.

(a) Graph the function for $0 \le t \le 365$.

(b) Find the maximum and minimum number of hours of daylight and the corresponding values of t rounded to two decimal places.

30. **Height of a Blimp:** An observer on the ground notices that the angle of elevation of a blimp is 29°. If the observer is 2 miles away from a point directly below the blimp, what is the approximate height of the blimp? Round off the answer to two decimal places.
31. **Surveying:** A person standing on top of a building that is 120 feet high spots a car parked on top of a parking garage that is 7 feet high. The angle of depression is 57° from the person to the car. What is the approximate distance of the garage from the building? Round off the answer to two decimal places.
32. **Balloonist:** A balloonist 100 meters above ground level observed a car on the ground. If the horizontal distance from a point on the ground directly under the balloon to the car is 79 meters, find the angle of depression from the balloon to the car to the nearest tenth of a degree.

CHAPTER 1 TEST

1. (a) Convert 520° to radian measure.
   (b) Convert $3\pi/7$ to degree measure.
2. Suppose that the sides of a central angle intercept an arc of length 15 centimeters on the circumference of a circle of radius 10 centimeters.
   (a) Find the radian measure of the central angle and the area of the sector.
   (b) Express the measure of the central angle in degrees.
3. Suppose that the point $P = (3, -7)$ is on the terminal side of angle θ in standard position.
   (a) Draw two angles, one positive and one negative, coterminal with θ.
   (b) Find the exact value of each of the six trigonometric functions of θ.
4. Suppose that $P(t) = (a, 2a)$, where $a > 0$.
   (a) Determine the value of a.
   (b) Use the results from part (a) to find $\sin t$, $\cos t$, and $\tan t$.
5. Find the exact value of each expression.
   (a) $\cos(-240°)$
   (b) $\sin \frac{13\pi}{4}$
   (c) $\tan \frac{17\pi}{3}$
   (d) $\cos^{-1}\left(-\frac{\sqrt{2}}{2}\right)$
   (e) $\sin\left(\cos^{-1}\frac{3}{5}\right)$
6. Use a calculator to approximate the value of each expression rounded off to two decimal places.
   (a) $\sin 23.17°$
   (b) $\csc 2.85$
   (c) $\cos^{-1}(-0.8413)$
7. Determine the quadrant that contains the angle θ.
   (a) $\sin\theta < 0$ and $\tan\theta > 0$
   (b) $\cos\theta > 0$ and $\csc\theta < 0$
8. Suppose that $\cos\theta = 5/17$ and θ is an acute angle. Find the exact values of the other five trigonometric functions of θ.
9. Find the amplitude, period, and phase shift, and sketch the graph of one cycle of each function.
   (a) $f(x) = 3\sin(4x - \pi)$
   (b) $g(x) = -2\cos\left(\frac{\pi x}{2} - \frac{\pi}{3}\right)$
   (c) $h(x) = \tan(x + \pi)$
10. The number of hours of sunshine each month for a region is given by the model

$$f(t) = 2.04\sin\left(\frac{\pi}{6}t - 1.7\right) + 6.12,\ 1 \le t \le 12$$

where $t = 1$ represents the month of January and $t = 12$ represents the month of December.
   (a) Sketch the graph of this model.
   (b) Use the graph to determine the maximum number of hours of sunshine and in what month it occurs. Round off the answers to two decimal places.
11. A straight string of lights is to reach from the top of an 80-meter pole and make an angle of elevation of 61° with the horizontal ground. How far from the pole should the lights be fastened to the ground? Round off the answer to two decimal places.
12. One end of an 80-foot rope is attached to the top of a vertical pole standing in the center of a tent. The other end of the rope is attached to a stake 60 feet away from the base of the pole. Determine the angle of elevation of the rope with the ground.

Chapter 1: Answers to Selected Problems

Problem Set 1.1 (on page 13)

1. **(a)** $\frac{-2\pi}{3}$ **(b)** $\frac{5\pi}{2}$ **3.** **(a)** $\left(\frac{1}{2}, \frac{-\sqrt{3}}{2}\right)$ **(b)** $\left(\frac{\sqrt{2}}{2}, \frac{-\sqrt{2}}{2}\right)$ **(c)** $\left(\frac{\sqrt{3}}{2}, \frac{-1}{2}\right)$

5. **(a)** I **(b)** III **(c)** IV **(d)** I **7.** **(a)** IV **(b)** I **(c)** III **(d)** I

9. **(a)** Quadrant II; coterminal: $\frac{11\pi}{4}, -\frac{5\pi}{4}$ **(b)** Quadrant IV; coterminal: $\frac{11\pi}{6}, -\frac{13\pi}{6}$

11. **(a)** Quadrant IV; coterminal: 315°, −405° **(b)** Quadrant II; coterminal: 480°, −240°

13. **(a)** Quadrant IV; coterminal: $\frac{5\pi}{3}, -\frac{\pi}{3}$ **(b)** Quadrant I; coterminal: 20°, −340°

15. **(a)** 16°18′36″ **(b)** −87°48′36″ **(c)** −156°37′48″ **(d)** 89°44′24″ **17.** **(a)** 38.3° **(b)** 65.19° **(c)** −141.47° **(d)** 244.77° **19.** **(a)** $\frac{5\pi}{12} = 1.31$ **(b)** $-\frac{3\pi}{4} = -2.36$

21. **(a)** $-\frac{19\pi}{36} = -1.66$ **(b)** $\frac{37\pi}{15} = 7.75$ **23.** **(a)** $\frac{3\pi}{8}$, 1.18 **(b)** $\frac{339\pi}{2000}$, 0.53

25. **(a)** 120° **(b)** 1290° **27.** **(a)** 105° **(b)** −80° **29.** **(a)** 286.48° **(b)** −131.78°

31. **(a)** 4.71 in. **(b)** 16.49 sq yd **33.** **(a)** 6.82 m **(b)** 6.14 m^2 **35.** **(a)** 0.95 yd **(b)** 2.39 sq yd

37. **(a)** 1.15 mi **(b)** 6082.12 ft **39.** 20.94 ft **41.** 10.05 cm/sec

43. **(a)** 5280 rad/min **(b)** 840.34 rotations/min **45.** **(a)** 863.94 in./min **(b)** 30.56 rotations/min

47. $\omega = 0.63$ rad/sec; $\nu = 5.03$ ft/sec **49.** Radian measure depends on s as well as r. **51.** α **53.** $\frac{11\pi}{12}$

Problem Set 1.2 (on page 28)

1. $\sin t = \frac{1}{2}$
$\csc t = 2$
$\cos t = -\frac{\sqrt{3}}{2}$
$\sec t = -\frac{2\sqrt{3}}{3}$
$\tan t = -\frac{\sqrt{3}}{3}$
$\cot t = -\sqrt{3}$

3. $\sin t = -\frac{12}{13}$
$\csc t = -\frac{13}{12}$
$\cos t = \frac{5}{13}$
$\sec t = \frac{13}{5}$
$\tan t = -\frac{12}{5}$
$\cot t = -\frac{5}{12}$

5. $\sin t = \frac{2}{\sqrt{13}}$
$\csc t = \frac{\sqrt{13}}{2}$
$\cos t = -\frac{3}{\sqrt{13}}$
$\sec t = -\frac{\sqrt{13}}{3}$
$\tan t = -\frac{2}{3}$
$\cot t = -\frac{3}{2}$

7. **(a)** (0, −1); $\sin t = -1$; $\csc t = -1$; $\cos t = 0$; $\sec t$ is undefined; $\tan t$ is undefined; $\cot t = 0$
(b) Same as part (a)

9. **(a)** (1, 0); $\sin t = 0$; $\csc t$ is undefined; $\cos t = 1$; $\sec t = 1$; $\tan t = 0$; $\cot t$ is undefined
(b) (0, −1); $\sin t = -1$; $\csc t = -1$; $\cos t = 0$; $\sec t$ is undefined; $\tan t$ is undefined; $\cot t = 0$

11. $\frac{\sqrt{2}}{2}$ **13.** $-\frac{1}{2}$ **15.** $-\frac{1}{2}$ **17.** $-\frac{\sqrt{3}}{2}$ **19.** $-\frac{\sqrt{2}}{2}$ **21.** $-\frac{\sqrt{2}}{2}$ **23.** 0.1411 **25.** 0.2837

27. 1.5574 **29.** −0.6421 **31.** −2.1943 **33.** 1.1099 **35.** **(a)** (−0.9041, 0.4274)
(b) (−0.5076, 0.8616) **37.** 6.02 in.; 6.17 in.; 6.43 in.

39. $\sin t = \frac{-3\sqrt{10}}{10}$; $\cos t = \frac{-\sqrt{10}}{10}$; $\tan t = 3$; $\cot t = \frac{1}{3}$; $\sec t = -\sqrt{10}$; $\csc t = \frac{-\sqrt{10}}{3}$

41. $(\cos t)^2 \neq \cos t^2 = \cos(t^2)$; for example, $\left[\cos\left(\frac{\pi}{3}\right)\right]^2 = \left(\frac{1}{2}\right)^2 = \frac{1}{4} = 0.25$ whereas $\cos\left[\left(\frac{\pi}{3}\right)^2\right] = 0.4566$ (approx.)

43. $\sin 0.1 = 0.0998$, $g(0.1) = 0.0998$; $\sin 0.2 = 0.1987$, $g(0.2) = 0.1987$; $\sin(-0.3) = -0.2955$, $g(-0.3) = -0.2955$; $\sin(-1) = -0.8415$, $g(-1) = -0.8417$

Problem Set 1.3 (on page 39)

1. $\sec t = \frac{13}{12}$; $\csc t = \frac{13}{5}$; $\tan t = \frac{5}{12}$; $\cot t = \frac{12}{5}$

3. $\cot t = \frac{5}{12}$; $\sec t = \frac{-13}{5}$; $\sin t = \frac{-12}{13}$; $\csc t = \frac{-13}{12}$

5. $\csc t = -\sqrt{2}$; $\sec t = -\sqrt{2}$; $\tan t = 1$; $\cot t = 1$

7. $\sec t = 1$; $\tan t = 0$; $\csc t$ and $\cot t$ are undefined

9. **(a)** IV **(b)** IV **11.** **(a)** II **(b)** IV **13.** $\frac{13}{12}$ **15.** $\frac{-5}{12}$

17. **(a)** $\sin(-3.752) = 0.5732 = -\sin 3.752$ **(b)** $\cos(-5.384) = 0.6222 = \cos 5.384$

19. $\cos t = \frac{8}{17}$; $\tan t = \frac{15}{8}$; $\csc t = \frac{17}{15}$; $\sec t = \frac{17}{8}$; $\cot t = \frac{8}{15}$

21. $\sin t = \frac{3}{5}$; $\tan t = -\frac{3}{4}$; $\csc t = \frac{5}{3}$; $\sec t = -\frac{5}{4}$; $\cot t = -\frac{4}{3}$

23. $\sin t = \frac{8}{17}$; $\cos t = -\frac{15}{17}$; $\tan t = -\frac{8}{15}$; $\sec t = -\frac{17}{15}$; $\cot t = -\frac{15}{8}$

25. $\cos t = -\frac{3}{5}$; $\sin t = -\frac{4}{5}$; $\tan t = \frac{4}{3}$; $\cot t = \frac{3}{4}$; $\csc t = -\frac{5}{4}$

27. $\cos t = -0.82$; $\tan t = -0.69$; $\csc t = 1.75$; $\sec t = -1.22$; $\cot t = -1.44$

29. $\sin t = 0.84$; $\cos t = -0.55$; $\csc t = 1.19$; $\sec t = -1.83$; $\cot t = -0.65$

31. $\frac{\sqrt{3}}{2}$ **33.** $\frac{\sqrt{3}}{2}$ **35.** 1 **37.** $-\sqrt{2}$ **39.** $\frac{\sqrt{3}}{3}$

41. $2\cos t$ **43.** $3\sec t$ **45.** $2\cos t$

Problem Set 1.4 (on page 61)

1. **(a)** $[-3, 3]$ **(b)** $\left[\frac{11}{3}, \frac{13}{3}\right]$ **(c)** $\left[1, \frac{7}{3}\right]$

3. **(a)** Range: $[-3, 3]$;

Amplitude $= 3$;

Period $= 2\pi$

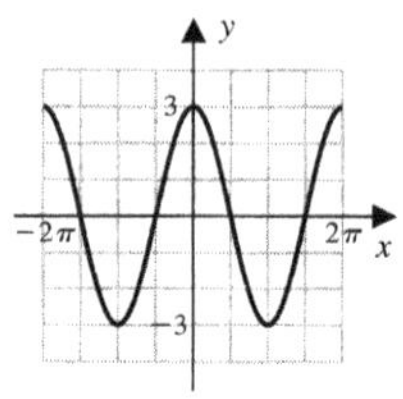

(b) Range: $[-3, 3]$;

Amplitude $= 3$;

Period $= 2\pi$

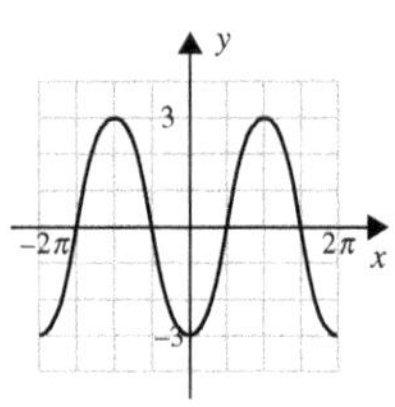

5. **(a)** Range: $\left[-\frac{2}{3}, \frac{2}{3}\right]$;

Amplitude $= \frac{2}{3}$;

Period $= 2\pi$

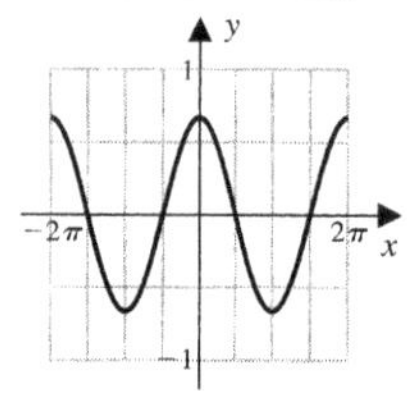

(b) Range: $\left[-\frac{1}{2}, \frac{1}{2}\right]$;

Amplitude $= \frac{1}{2}$;

Period $= 2\pi$

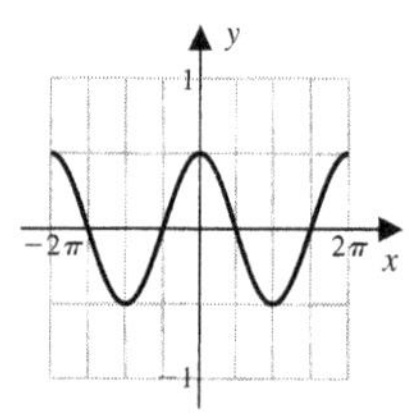

7. **(a)** Range: $[-1, 1]$;

Amplitude $= 1$;

Period $= \pi$

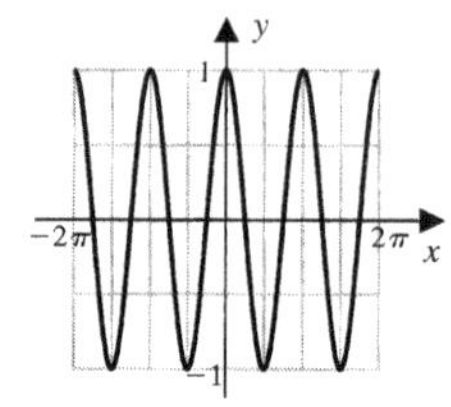

(b) Range: $[-1, 1]$;

Amplitude $= 1$;

Period $= 4\pi$

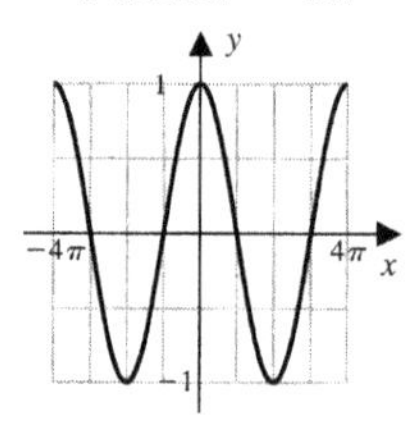

9. **(a)** Range: $[-1, 1]$;

Amplitude $= 1$;

Period $= 6$

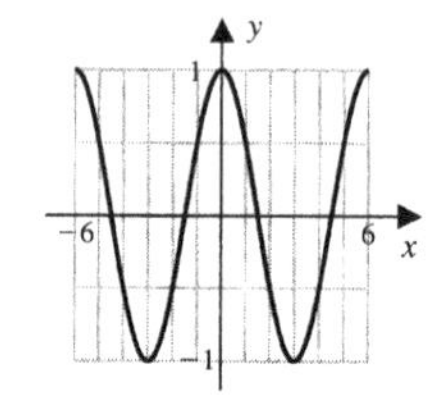

(b) Range: $[-1, 1]$;

Amplitude $= 1$;

Period $= 1$

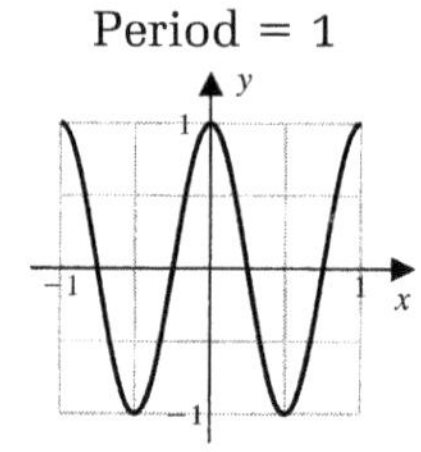

11. **(a)** Range: $\left[-\frac{5}{3}, \frac{5}{3}\right]$;

Amplitude $= \frac{5}{3}$;

Period $= \frac{\pi}{3}$

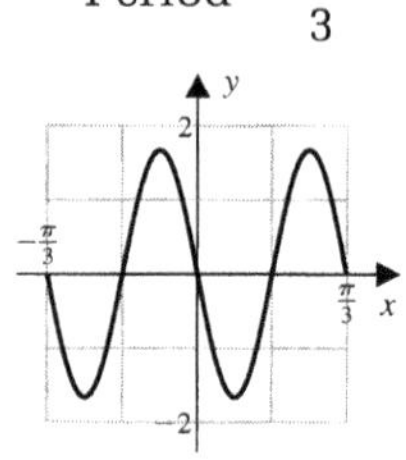

(b) Range: $[-5, 5]$;
Amplitude $= 5$;

Period $= \frac{\pi}{5}$

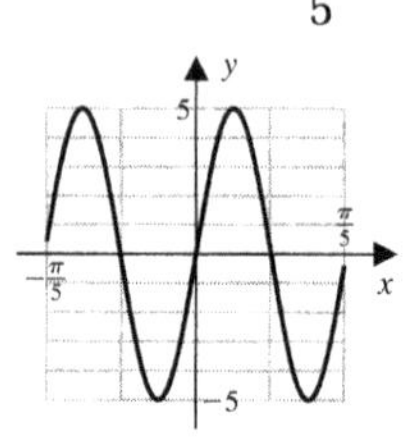

13. **(a)** Range: $[-3, 3]$;

Amplitude $= 3$;

Period $= 4$

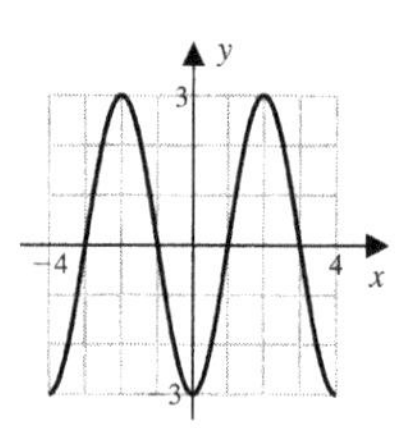

(b) Range: $[-\pi, \pi]$;
Amplitude $= \pi$;

Period $= \frac{4\pi}{5}$

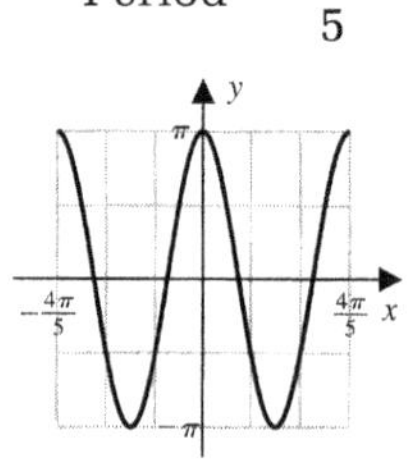

15. **(a)** Range: $[0, 2]$;

Amplitude $= 1$;

Period $= 2\pi$

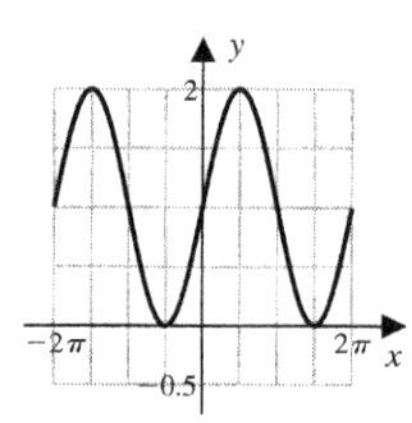

(b) Range: $[2, 4]$;
Amplitude $= 1$;

Period $= 2\pi$

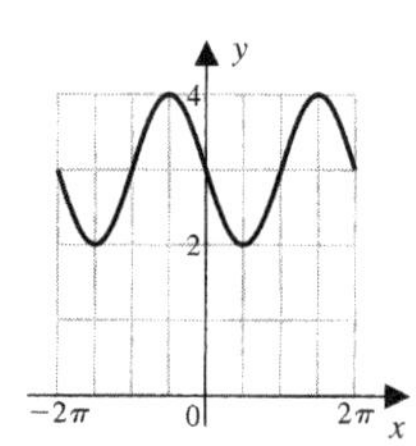

17. **(a)** Range: $[-1, 5]$;

Amplitude $= 3$;

Period $= 8$

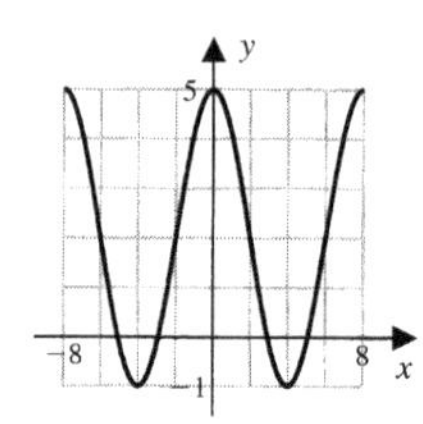

(b) Range: $[-5, 1]$;
Amplitude $= 3$;

Period $= 1$

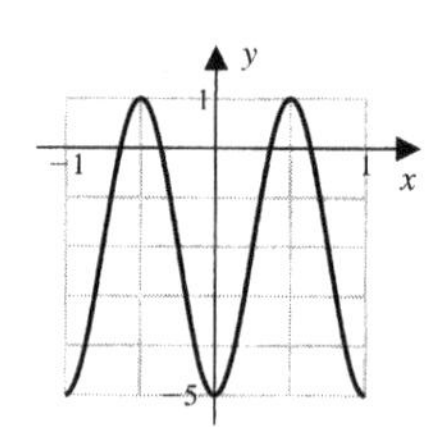

19. C; B; A

21. Amplitude = 2; Period = 2π; Phase shift = $\frac{\pi}{4}$

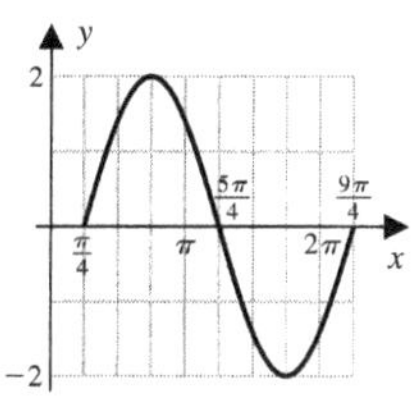

23. Amplitude = 4; Period = 2π; Phase shift = $-\frac{\pi}{6}$

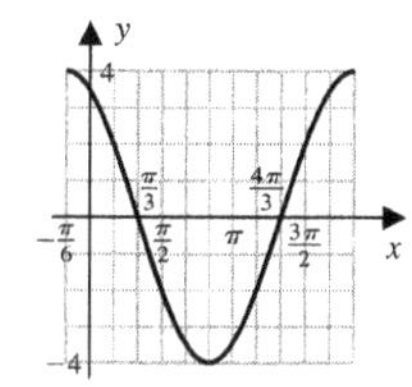

25. Amplitude = 2; Period = π; Phase shift = $\frac{\pi}{2}$

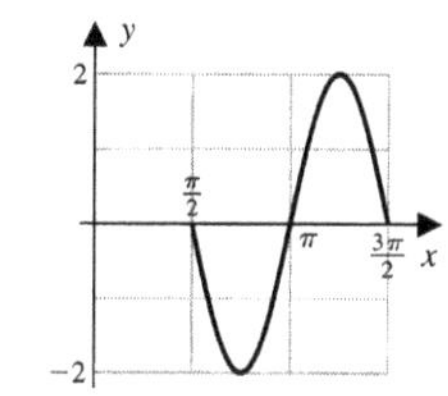

27. Amplitude = 2; Period = 2; Phase shift = 1

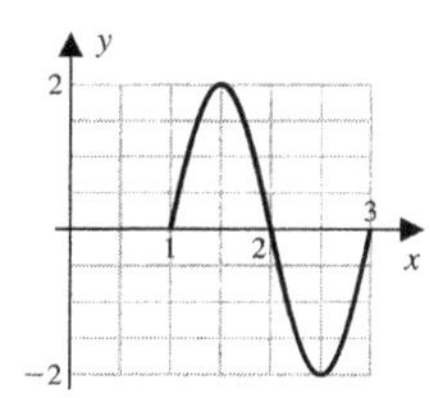

29. Amplitude = 3; Period = $\frac{2\pi}{3}$; Phase shift = $-\frac{2}{3}$

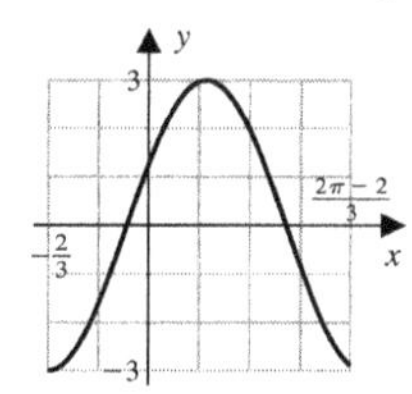

31. Amplitude = 2; Period = 4; Phase shift = $\frac{1}{2\pi}$

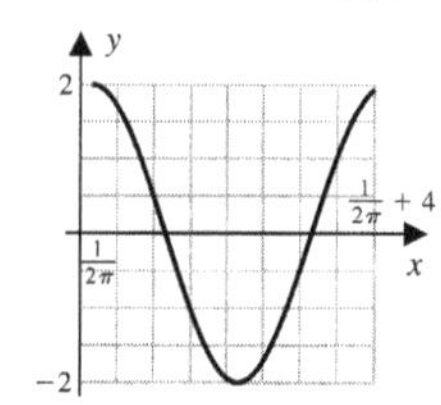

33. Amplitude = 2; Period = $\frac{2\pi}{3}$; Phase shift = $\frac{5}{3}$

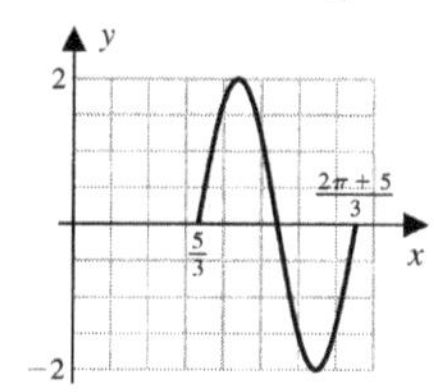

35. Amplitude = 3; Period = π; Phase shift = π

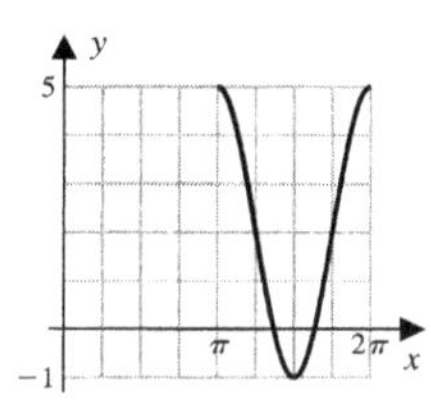

37. **(a)** $y = 2\sin\left(2x - \frac{\pi}{3}\right)$; Amplitude = 2; Period = π; Phase shift = $\frac{\pi}{6}$
(b) $y = 5\sin(4x - \pi)$; Amplitude = 5; Period = $\frac{\pi}{2}$; Phase shift = $\frac{\pi}{4}$

39. $y = 3\sin 6x$

41. **(b)** $\frac{5}{6}$ **(c)** Maximum: 140 mm Hg Minimum: 80 mm Hg

43. **(a)** 12.24 ft; −0.37 ft; 14.37 ft **(b)** 10 sec **(d)** Minimum: −1 ft; Maximum: 15 ft

45. **(b)** Amplitude: 700; Period: 12 months **(c)** \$900,000; \$900,000; \$200,000; \$1,600,000; \$200,000
(d) \$1,600,000 on July 1, 2004 and July 1, 2005

47. **(b)** Amplitude: 2.3; Period: 365 days
(c) Maximum: 14.30 hr on June 20 ($t = 171$); Minimum: 9.70 hr on December 20 ($t = 354$)

49. **(b)** $T = 16.152\sin(0.524t - 2.182) + 40.907$
(d) For March ($t = 3$), $T = 31.7$; for June ($t = 6$), $T = 54.2$; for September ($t = 9$), $T = 50.1$

51. **(a)** Amplitude = 1; Period = $\frac{1}{100}$ sec; Frequency = 100
(b) Amplitude = 2; Period = 2000 sec; Frequency = $\frac{1}{2000}$

53. **(a)** $y = -50\cos\sqrt{\frac{8}{30}}\,t$ **(c)** Amplitude = 50; Period ≈ 12.17; Frequency ≈ 0.08

55. **(a)** $\frac{\pi}{2.2}$ **(b)** $y = 0.025\sin\left(\frac{\pi}{2.2}t\right)$

57. They are the same due to the phase shift on $f(x)$.

59. They are the same.

61. **(a)** No
(b) $f(x) \to 1$ as $x \to 0$
(c) $f(x) \to 1$ as $x \to 0^-$, $f(x) \to 1$ as $x \to 0^+$

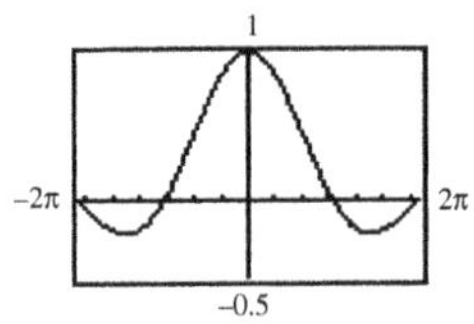

63. Period = 2π

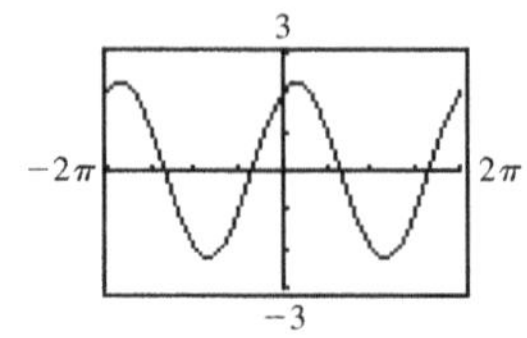

Problem Set 1.5 (on page 79)

1. Period = π;
Phase shift = 0

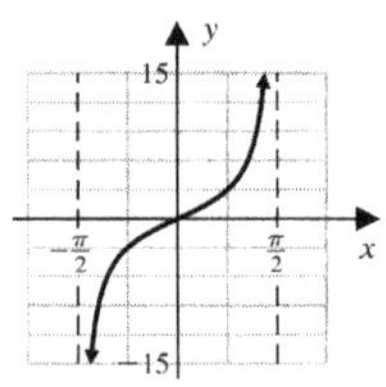

3. Period = π;
Phase shift = 0

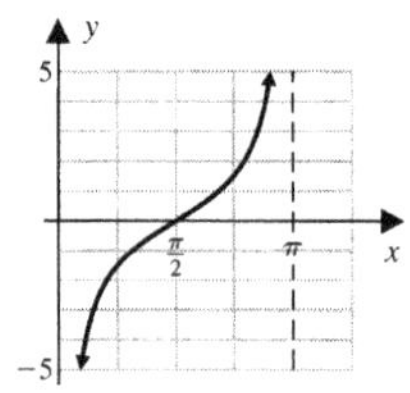

5. Period = 2π;
Phase shift = 0

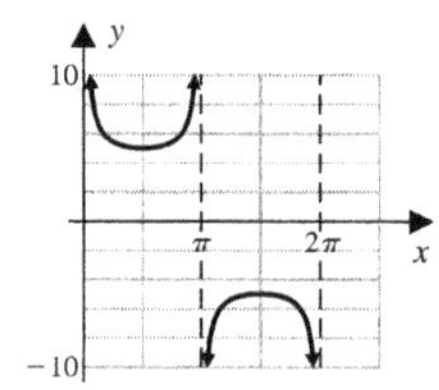

7. Period = 2π;
Phase shift = 0

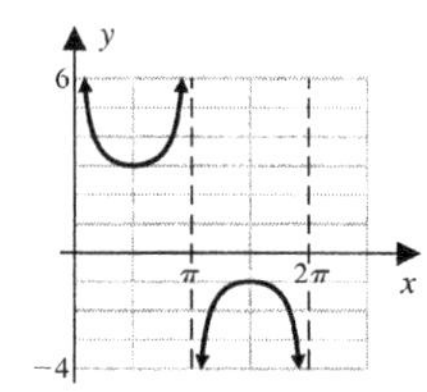

9. Period = $\frac{\pi}{4}$;
Phase shift = 0

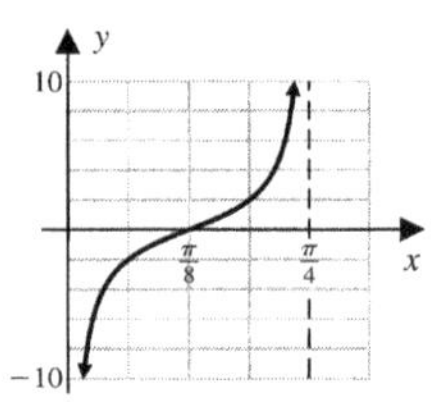

11. Period = 3π;
Phase shift = 0

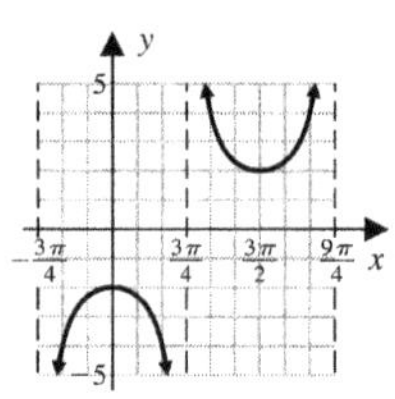

13. Period = $\frac{\pi}{2}$;
Phase shift = 0

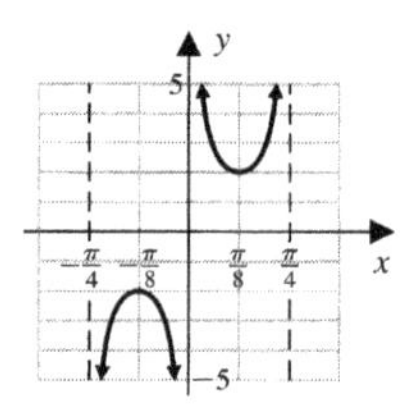

15. Period = 2π;
Phase shift = $\frac{\pi}{6}$

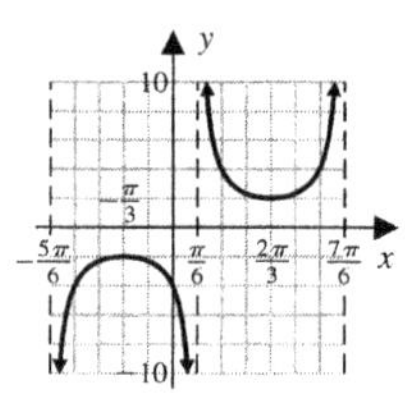

17. Period = π;
Phase shift = $\frac{\pi}{4}$

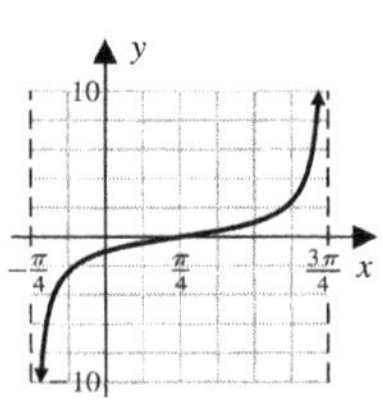

19. Period = π;
Phase shift = $\frac{\pi}{2}$

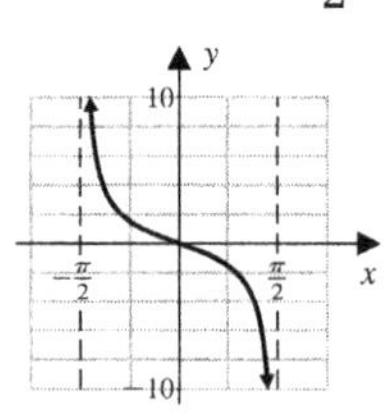

21. Period = 8π;
Phase shift = $-\pi$

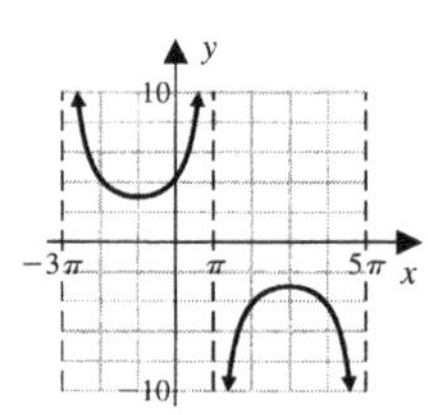

23. Period = 4;
Phase shift = 1

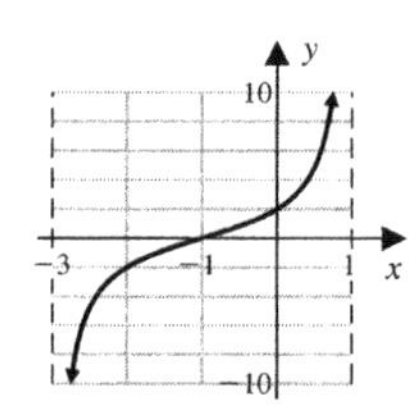

25.

	As x increases from:			
	0 to $\pi/2$	$\pi/2$ to π	π to $(3\pi)/2$	$(3\pi)/2$ to 2π
(a)	Increasing	Increasing	Increasing	Increasing
(b)	Decreasing	Decreasing	Decreasing	Decreasing
(c)	Increasing	Increasing	Decreasing	Decreasing
(d)	Decreasing	Increasing	Increasing	Decreasing

27. **(a)** 1 **(b)** $-\infty$ **29.** **(a)** ∞ **(b)** 1 **31.** $y = \tan\left(\frac{x}{2}\right)$ **33.** $y = 2\csc(2x + \pi)$

35. π **37.** $\frac{\pi}{3}$ **39.** **(b)** (0.79, 8.00)

41.

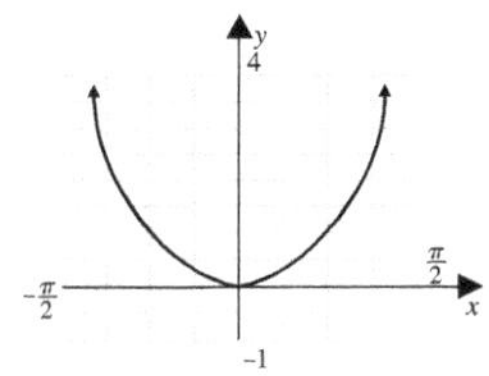

43. $f(x) \neq g(x)$; $f(x) = -g(x)$ because the cosecant is an odd function.

45. The graphs are the same.

47. **(a)** $y = \tan(6x)$

(b) $y = \csc\left(\frac{12x}{5}\right)$

51. $f(x) \neq g(x)$ since $f(x)$ is not defined when $\cos x = 0$.

Problem Set 1.6 (on page 89)

1. $\sin\theta = \frac{3}{5}$; $\csc\theta = \frac{5}{3}$; $\cos\theta = \frac{4}{5}$; $\sec\theta = \frac{5}{4}$; $\tan\theta = \frac{3}{4}$; $\cot\theta = \frac{4}{3}$

3. $\sin\theta = \frac{12}{13}$; $\csc\theta = \frac{13}{12}$; $\cos\theta = -\frac{5}{13}$; $\sec\theta = -\frac{13}{5}$; $\tan\theta = -\frac{12}{5}$; $\cot\theta = -\frac{5}{12}$

5. $\sin\theta \approx -0.898$; $\csc\theta \approx -1.113$; $\cos\theta \approx 0.439$; $\sec\theta \approx 2.276$; $\tan\theta \approx -2.044$; $\cot\theta \approx -0.489$

7. $\sin\theta = -\frac{1}{2}$; $\csc\theta = -2$; $\cos\theta = -\frac{\sqrt{3}}{2}$; $\sec\theta = -\frac{2\sqrt{3}}{2}$; $\tan\theta = \frac{\sqrt{3}}{3}$; $\cot\theta = \sqrt{3}$

9. $\sin\theta = \cos\theta = \frac{\sqrt{2}}{2}$; $\tan\theta = \cot\theta = 1$; $\csc\theta = \sec\theta = \sqrt{2}$

11. $\sin 90° = 1$; $\csc 90° = 1$; $\cos 90° = 0$; $\sec 90°$ is undefined; $\tan 90°$ is undefined; $\cot 90° = 0$

13. $\sin\left(-\frac{\pi}{2}\right) = -1$; $\csc\left(-\frac{\pi}{2}\right) = -1$; $\cos\left(-\frac{\pi}{2}\right) = 0$; $\sec\left(-\frac{\pi}{2}\right)$ is undefined; $\tan\left(-\frac{\pi}{2}\right)$ is undefined; $\cot\left(-\frac{\pi}{2}\right) = 0$

15. **(a)** $\frac{1}{2}$ **(b)** $-\frac{\sqrt{2}}{2}$ **17.** **(a)** $\sqrt{3}$ **(b)** $-\sqrt{3}$ **19.** **(a)** $\sqrt{2}$ **(b)** $\frac{-2\sqrt{3}}{3}$

21. **(a)** -1 **(b)** $\sqrt{3}$ **23.** $\frac{-\sqrt{2}}{2}$ **25.** $-\sqrt{3}$ **27.** -1 **29** $\frac{\sqrt{3}}{2}$ **31.** -1

33. **(a)** $\sin\theta \approx 0.6820$, $\csc\theta \approx 1.4663$, $\cos\theta \approx -0.7314$, $\sec\theta \approx -1.3673$, $\tan\theta \approx -0.9325$, $\cot\theta \approx -1.0724$

(b) $\sin\theta \approx -0.9949$, $\csc\theta \approx -1.0051$, $\cos\theta \approx -0.1011$, $\sec\theta \approx -9.8955$, $\tan\theta \approx 9.8448$, $\cot\theta \approx 0.1016$

35. **(a)** $\sin\theta \approx 0.3420$, $\csc\theta \approx 2.9238$, $\cos\theta \approx 0.9397$, $\sec\theta \approx 1.0642$, $\tan\theta \approx 0.3640$, $\cot\theta \approx 2.7475$

(b) $\sin\theta \approx -0.0548$, $\csc\theta \approx -18.2470$, $\cos\theta \approx 0.9985$, $\sec\theta \approx 1.0015$, $\tan\theta \approx -0.0549$, $\cot\theta \approx -18.2195$

37. **(a)** $\sin\theta \approx 0.9093$, $\csc\theta \approx 1.0998$, $\cos\theta \approx -0.4161$, $\sec\theta \approx -2.4030$, $\tan\theta \approx -2.1850$, $\cot\theta \approx -0.4577$

(b) $\sin\theta \approx -0.9636$, $\csc\theta \approx -1.0378$, $\cos\theta \approx 0.2675$, $\sec\theta \approx 3.7383$, $\tan\theta \approx -3.6021$, $\cot\theta \approx -0.2776$

39. **(a)** $\sin\theta \approx 0.8739$, $\csc\theta \approx 1.1443$, $\cos\theta \approx 0.4861$, $\sec\theta \approx 2.0572$, $\tan\theta \approx 1.7978$, $\cot\theta \approx 0.5562$

(b) $\sin\theta = 0$, $\csc\theta$ is undefined, $\cos\theta = -1$, $\sec\theta = -1$, $\tan\theta = 0$, $\cot\theta$ is undefined

41. **(a)**

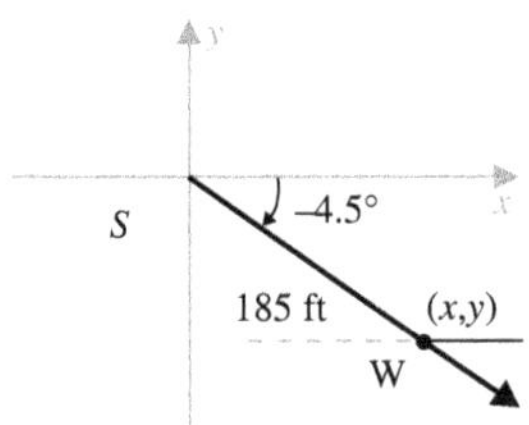

(b) (184.43, −14.51) **(c)** −0.0787

Problem Set 1.7 (on page 101)

1. $\sin\theta = 0.625$; $\cos\theta = 0.781$; $\tan\theta = 0.801$; $\csc\theta = 1.600$; $\sec\theta = 1.281$; $\cot\theta = 1.249$
3. $\sin\theta = 0.324$; $\cos\theta = 0.946$; $\tan\theta = 0.342$; $\csc\theta = 3.086$; $\sec\theta = 1.057$; $\cot\theta = 2.920$
5. $\sin\theta = 0.882$; $\cos\theta = 0.471$; $\tan\theta = 1.875$; $\csc\theta = 1.133$; $\sec\theta = 2.125$; $\cot\theta = 0.533$
7. $\cos\theta = \frac{3}{5}$; $\tan\theta = \frac{4}{3}$; $\csc\theta = \frac{5}{4}$; $\sec\theta = \frac{5}{3}$; $\cot\theta = \frac{3}{4}$ **9.** **(a)** $\frac{1}{2}$ **(b)** $\frac{-\sqrt{2}}{2}$
11. **(a)** $\sqrt{3}$ **(b)** $-\sqrt{3}$ **13.** **(a)** $\sqrt{2}$ **(b)** $\frac{-2\sqrt{3}}{3}$ **15.** **(a)** -1 **(b)** $\sqrt{3}$
17. **(a)** 0.6820 **(b)** 1.1371 **(c)** 0.2225 **(d)** 2.6131
19. **(a)** 0.7265 **(b)** 0.3127 **(c)** 1.2392 **(d)** 0.4377
21. **(a)** 33.37° **(b)** 11.48° **(c)** 88.10° **(d)** 45°
23. $\alpha = 40°$, $a = 6.43$, $b = 7.66$ **25.** $\alpha = 47°$, $a = 4.29$, $c = 5.87$ **27.** $\beta = 58.77°$, $a = 909.52$, $c = 1754.20$
29. $\alpha = 72.74°$, $\beta = 17.26°$, $c = 13.82$ **31.** $\alpha = 67.5°$, $\beta = 22.5°$, $b = 0.41$, $c = 1.08$ **33.** 10.68 in.
35. 10.90 ft **37.** 46.06 ft **39.** 80.83 mph **41.** **(a)** 23.58° **(b)** 2 ft./sec.
43. $d = 12.5$ in., $\theta = 36.87°$ **45.** 154.51 m **47.** 84.57° **49.** 146.71 m
51. Building: 30.57 ft; antenna: 13.95 ft **53.** 499.00 ft **55.** For example, let $\alpha = 30°$ and $\beta = 60°$.
57. $T = r\cot\left(\frac{\theta}{2}\right)$ **59.** **(a)–(b)** Side opposite has length 23 sin 35° or approx. 13.19 in.

Problem Set 1.8 (on page 120)

1. **(a)** $\frac{\pi}{4}$ **(b)** $-\frac{\pi}{6}$ **3.** **(a)** $\frac{\pi}{6}$ **(b)** $-\frac{\pi}{6}$ **5.** **(a)** $-\frac{\pi}{3}$ **(b)** $\frac{5\pi}{6}$
7. **(a)** $\frac{\pi}{6}$ **(b)** $\frac{\pi}{4}$ **9.** **(a)** $-\frac{\pi}{4}$ **(b)** -1 **11.** **(a)** 0.22 **(b)** −0.89
13. **(a)** 0.82 **(b)** −1.55 **15.** **(a)** 2.36 **(b)** 1.56 **17.** $x = \sin y - 1$; $-\frac{\pi}{2} \le y \le \frac{\pi}{2}$, $-2 \le x \le 0$
19. $x = \frac{1}{2}\cos y + 4$; $0 \le y \le \pi$, $\frac{7}{2} \le x \le \frac{9}{2}$ **21.** $x = \frac{1}{3}\left[2 - \tan\left(\frac{y}{3}\right)\right]$; $-\frac{3\pi}{2} < y < \frac{3\pi}{2}$, x is real
23. $x = \frac{1}{2}\cos 3y + 2$; $0 \le y \le \frac{\pi}{3}$, $\frac{3}{2} \le x \le \frac{5}{2}$
25. **(a)** 0.3 **(b)** 1.3 **(c)** −0.5236 **(d)** 2.6180 **27.** $-1 \le x \le -\frac{1}{3}$
29. $3 - \frac{3\pi}{2} < x < 3 + \frac{3\pi}{2}$ **31.** **(a)** $\frac{3}{5}$ **(b)** $\frac{12}{13}$ **33.** **(a)** 2 **(b)** $\frac{-2\sqrt{5}}{5}$
35. **(a)** $\frac{4\sqrt{15}}{15}$ **(b)** $\frac{\pi}{3}$ **37.** **(a)** $\sqrt{1 - x^2}$ **(b)** $\frac{\sqrt{1 - x^2}}{x}$ **(c)** $\frac{1}{x}$
39. **(a)** $\frac{\pi}{4}$ **(b)** $-\frac{\pi}{4}$ **41.** **(a)** $\frac{\pi}{4}$ **(b)** $-\frac{\pi}{3}$ **43.** **(a)** 0.80 **(b)** 1.27

45. **(a)** 2.00 **(b)** −0.11 **47.** **(a)** $\theta = \tan^{-1}\left(\frac{32}{x}\right) - \tan^{-1}\left(\frac{24}{x}\right)$ **(b)** 0.07 rad. (or about 4.25°)

49. **(a)** $\theta = \tan^{-1}\left(\frac{9}{x}\right) - \tan^{-1}\left(\frac{5}{x}\right)$ **(b)** 0.18 rad. (or about 10.19°)

51.

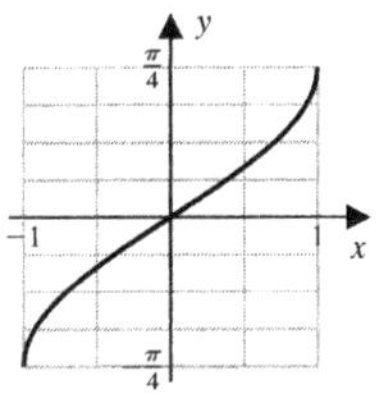

53.

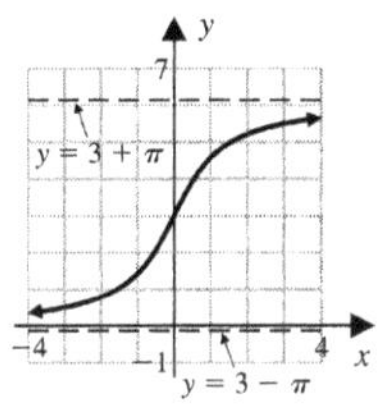

55.

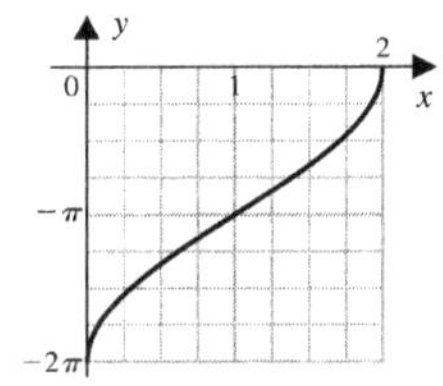

59. No **61.** For $0 \le x \le 1$, $\sin^{-1} x + \cos^{-1} x = \frac{\pi}{2}$.

Chapter 1 Review Problem Set (on page 131)

1. **(a)** 76°15′ **(b)** −61°21′ **(c)** 143°28′12″ **(d)** −14°28′12″

3. **(a)** 0.2618 **(b)** −0.3046 **(c)** 0.7903 **(d)** 0.6316 **5.** **(a)** 54° **(b)** 144° **(c)** 396°, −324°

7. **(a)** (0.3090, 0.9511) **(b)** (−0.9784, −0.2069) **(c)** (0.2225, −0.9749)

9. **(a)** Period = $\frac{\pi}{3}$; Phase shift = $\frac{\pi}{6}$ **(b)** Period = 8π; Phase shift = -4π

11. **(a)**

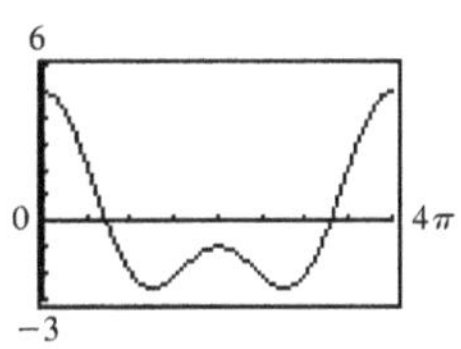

(b)

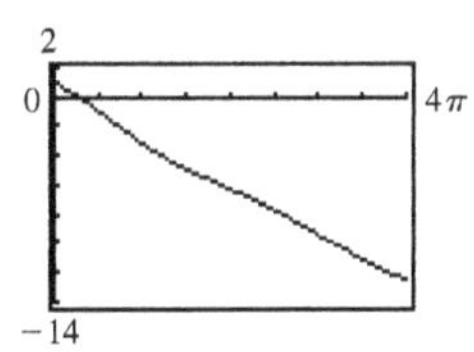

(c)

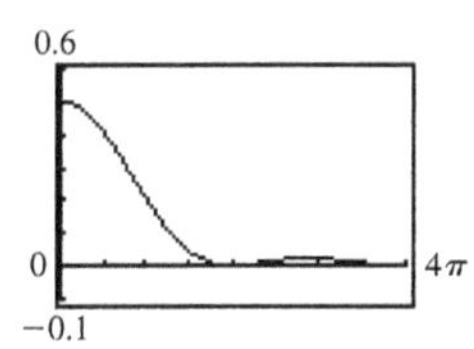

13. **(a)** III **(b)** III **(c)** IV **15.** **(a)** 0.1411 **(b)** −0.8829 **(c)** −3.0777 **(d)** −1.4044

17. $\sin\theta = \frac{3}{5}$; $\cos\theta = \frac{4}{5}$; $\tan\theta = \frac{3}{4}$; $\csc\theta = \frac{5}{3}$; $\sec\theta = \frac{5}{4}$; $\cot\theta = \frac{4}{3}$ **19.** **(a)** $\frac{1}{a}$ **(b)** a **(c)** a

21. **(a)** 0.38 **(b)** 2.08 **(c)** 1.47 **(d)** 0.02

23. **(a)** $x = 2\sin y$; $-2 \le x \le 2$, $-\frac{\pi}{2} \le y \le \frac{\pi}{2}$ **(b)** $x = \frac{1}{2}\tan y - \frac{3}{2}$; x is real; $-\frac{\pi}{2} < y < \frac{\pi}{2}$

25. **(a)** 302.99 m **(b)** 17.04 sec **27.** 4.82 rad/hr

29. **(a)**

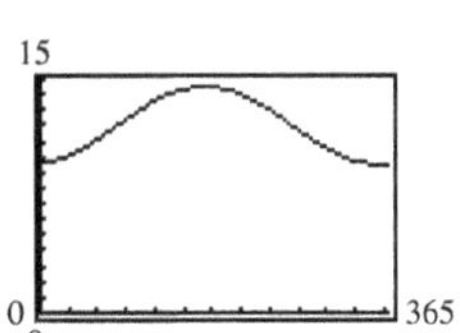

(b) Maximum: 14.5 hr for $t = 172.25$; Minimum: 9.5 hr for $t = 354.75$

31. 73.38 ft

Chapter 1 Test (on page 133)

1. **(a)** $\frac{26\pi}{9}$ or 9.08 (approx.) **(b)** $\frac{540}{7}$ or 77.14° (approx.) **2.** **(a)** 1.5; 75 cm^2 **(b)** $\frac{270}{\pi}$ or 85.94° (approx.)

3. **(a)**

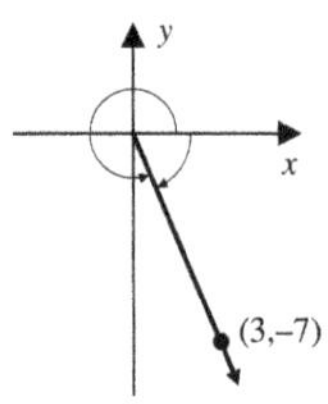

(b) $\sin\theta = -\frac{7\sqrt{58}}{58}$ $\csc\theta = -\frac{\sqrt{58}}{7}$

$\cos\theta = \frac{3\sqrt{58}}{58}$ $\sec\theta = \frac{\sqrt{58}}{3}$

$\tan\theta = -\frac{7}{3}$ $\cot\theta = -\frac{3}{7}$

4. **(a)** $\frac{\sqrt{5}}{5}$ **(b)** $\sin t = \frac{2\sqrt{5}}{5}$; $\cos t = \frac{\sqrt{5}}{5}$; $\tan t = 2$

5. **(a)** $-\frac{1}{2}$ **(b)** $-\frac{\sqrt{2}}{2}$ **(c)** $-\sqrt{3}$ **(d)** $\frac{3\pi}{4}$ **(e)** $\frac{4}{5}$ **6.** **(a)** 0.39 **(b)** 3.48 **(c)** 2.57

7. **(a)** III **(b)** IV **8.** $\sin\theta = \frac{2\sqrt{66}}{17}$; $\tan\theta = -\frac{2\sqrt{66}}{5}$; $\csc\theta = \frac{17\sqrt{66}}{132}$; $\sec\theta = -\frac{17}{5}$; $\cot\theta = -\frac{5\sqrt{66}}{132}$

9. **(a)** Amplitude = 3;
Period = $\frac{\pi}{2}$;
Phase shift = $\frac{\pi}{4}$

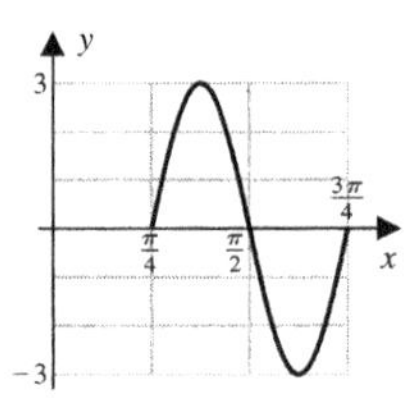

(b) Amplitude = 2;
Period = 4;
Phase shift = $\frac{2}{3}$

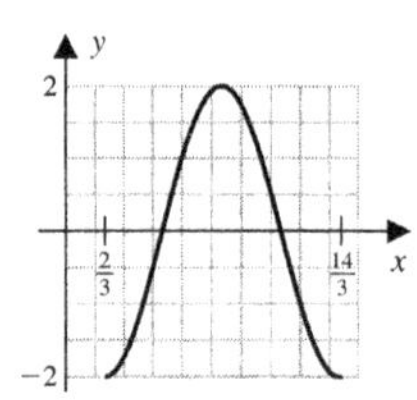

(c) Amplitude = 1;
Period = 2π;
Phase Shift = $-\pi$

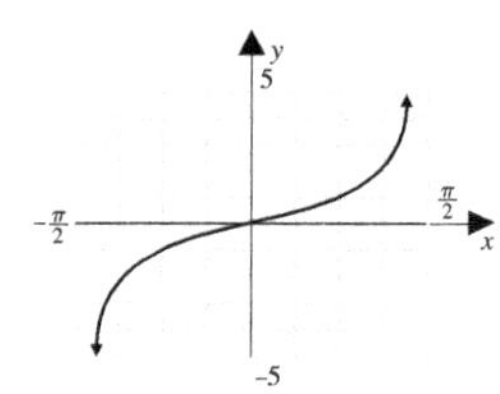

10. **(a)**

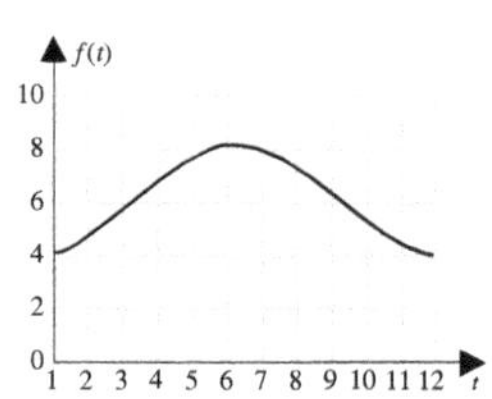

(b) max of about 8.16 hours when $t = 6.25$ (approx.) **11.** 44.34 m **12.** 41.41°

CHAPTER 2

Trigonometric Identities and Equations

Chapter Contents

Many applied problems can be modeled by trigonometric functions. For instance, suppose that an observer at point P, 111 meters away from a point Q directly beneath a hot air balloon, sights the balloon at an angle of elevation θ. At the same time, a second observer at point R, which is 74 meters closer to point Q, sights the same balloon at an angle of elevation 2θ. If the points P, R, and Q are in the same plane, find the height h of the balloon. Example 7 on page 177 gives the solution of this problem.

© JupiterImages, Inc.

In Chapter 1 we introduced special trigonometric relationships called *identities*. This chapter expands the study of trigonometric identities and examines various techniques for solving trigonometric equations.

Objectives

1. Simplify Trigonometric Expressions
2. Verify Trigonometric Identities
3. Prove a Trigonometric Equation Is Not an Identity

2.1 Fundamental Identities

So far we have four types of identities: the *reciprocal relationships,* the *quotient relationships,* the *even-odd relationships,* and the *Pythagorean identities.* They are listed below and in Appendix II.

Reciprocal Identities

1. $\csc\theta = \dfrac{1}{\sin\theta}$ 2. $\sec\theta = \dfrac{1}{\cos\theta}$ 3. $\cot\theta = \dfrac{1}{\tan\theta}$

Quotient Identities

1. $\tan\theta = \dfrac{\sin\theta}{\cos\theta}$ 2. $\cot\theta = \dfrac{\cos\theta}{\sin\theta}$

Even-Odd Identities

1. $\sin(-\theta) = -\sin\theta$ 2. $\cos(-\theta) = \cos\theta$ 3. $\tan(-\theta) = -\tan\theta$
4. $\csc(-\theta) = -\csc\theta$ 5. $\sec(-\theta) = \sec\theta$ 6. $\cot(-\theta) = -\cot\theta$

Pythagorean Identities

1. $\cos^2\theta + \sin^2\theta = 1$ 2. $1 + \tan^2\theta = \sec^2\theta$ 3. $1 + \cot^2\theta = \csc^2\theta$

Simplifying Trigonometric Expressions

At times it is possible to rewrite given trigonometric expressions in simpler forms by using known identities. The next two examples illustrate the process.

EXAMPLE 1 **Simplifying an Expression Using a Pythagorean Identity**

Use identities to simplify the expression $\cot^2 t \sin^2 t + \sin^2 t$.

Solution

$$\cot^2 t \sin^2 t + \sin^2 t = \frac{\cos^2 t}{\sin^2 t}\cdot \sin^2 t + \sin^2 t \quad \text{Quotient identity}$$
$$= \cos^2 t + \sin^2 t = 1 \quad \text{Pythagorean identity}$$

EXAMPLE 2 **Simplifying an Expression Using an Even-Odd Identity**

Use identities to simplify the expression $\dfrac{\cot\theta\cos(-\theta)}{\csc^2\theta - 1}$.

Solution

$$\frac{\cot\theta\cos(-\theta)}{\csc^2\theta - 1} = \frac{\cot\theta\cos\theta}{\csc^2\theta - 1} \quad \cos(-\theta) = \cos\theta$$
$$= \frac{\dfrac{\cos\theta}{\sin\theta}\cos\theta}{\cot^2\theta} = \frac{\left(\dfrac{\cos^2\theta}{\sin\theta}\right)}{\left(\dfrac{\cos^2\theta}{\sin^2\theta}\right)} \quad \text{Quotient and Pythagorean identities}$$
$$= \frac{\cos^2\theta}{\sin\theta}\cdot\frac{\sin^2\theta}{\cos^2\theta}$$
$$= \sin\theta \quad \text{Simplify}$$

Verifying Trigonometric Identities

The fundamental identities play an important role in solving trigonometric equations, in performing certain operations in calculus, and in deriving other identities. In these instances we are often involved in using algebra along with the identities to manipulate and convert trigonometric expressions to other forms. This process of manipulating and converting trigonometric expressions can be practiced by *verifying trigonometric identities.*

There is no general rule for proving that a trigonometric equation is an identity; however, we normally start with one side of the equation and try to convert it to the other side by means of a sequence of algebraic manipulations and substitutions utilizing known identities. Often it is helpful to start with the side containing more terms and try to convert it to the *simpler* side. Some suggestions that may help in carrying out a proof are listed below:

Suggestions for Verifying Identities

1. Combine a sum or difference of fractions into a single fraction.
2. Reduce a fraction.
3. Factor the expression.
4. Combine like terms.
5. Multiply both the numerator and denominator by the same expression.
6. Write all trigonometric expressions in terms of sines and cosines, and then simplify.

There may be more than one way of verifying that an equation is an identity, as shown in the next example.

EXAMPLE 3 **Verifying Trigonometric Identities**

Verify that the equation

$$(\cos^2 t)(1 + \tan^2 t) = 1$$

is an identity.

Solution Working with the left side of the equation, we get

$$\begin{aligned}(\cos^2 t)(1 + \tan^2 t) &= \cos^2 t \cdot \sec^2 t && \text{Pythagorean identity}\\ &= (\cos t \cdot \sec t)^2 = \left(\cos t \cdot \frac{1}{\cos t}\right)^2 && \text{Reciprocal identity}\\ &= 1^2 = 1\end{aligned}$$

Since we have transformed the left side into the right side, the equation is an identity.

Alternate Solution This strategy uses a quotient identity to express the left side in terms of $\sin t$ and $\cos t$.

$$\begin{aligned}(\cos^2 t)(1 + \tan^2 t) &= \cos^2 t\left(1 + \frac{\sin^2 t}{\cos^2 t}\right) && \text{Quotient identity}\\ &= \cos^2 t + \cos^2 t\left(\frac{\sin^2 t}{\cos^2 t}\right) && \text{Multiply and reduce}\\ &= \cos^2 t + \sin^2 t = 1 && \text{Pythagorean identity}\end{aligned}$$

At times we change the form of one side of an equation before proving that it is an identity.

EXAMPLE 4

Using Multiplication to Verify an Identity

Verify the identity $\dfrac{\cos t}{1 - \sin t} = \dfrac{1 + \sin t}{\cos t}$.

Solution

To verify this identity, it may be tempting to multiply *both sides* by either cos t or by $1 - \sin t$. But this step is incorrect because it assumes the equation is an identity before it has been verified. Instead, we work with the left side by multiplying the numerator and denominator by the expression $1 + \sin t$ to obtain

$$\begin{aligned}\frac{\cos t}{1 - \sin t} &= \frac{\cos t}{1 - \sin t} \cdot \frac{1 + \sin t}{1 + \sin t} \\ &= \frac{(\cos t)(1 + \sin t)}{1 - \sin^2 t} && \text{Multiply} \\ &= \frac{(\cos t)(1 + \sin t)}{\cos^2 t} = \frac{1 + \sin t}{\cos t} && \text{Reduce fraction}\end{aligned}$$

Since we have transformed the left side into the right side, the equation is an identity.

In the next example we verify the identity by converting each side of the equation to the same expression.

EXAMPLE 5

Verifying an Identity by Simplifying Each Side

Verify that the equation $\tan(-\theta)\sin(-\theta) = \sec\theta - \cos\theta$ is an identity.

Solution

Here we use the even-odd identities to write

$$\tan(-\theta) = -\tan\theta \qquad \text{and} \qquad \sin(-\theta) = -\sin\theta$$

so it is enough to prove that

$$\tan\theta\sin\theta = \sec\theta - \cos\theta.$$

In this situation, we verify that the equation is an identity by simplifying each side until the results become identical. We proceed as follows:

Left Side

$$\begin{aligned}\tan\theta\sin\theta &= \frac{\sin\theta}{\cos\theta} \cdot \sin\theta \\ &= \frac{\sin^2\theta}{\cos\theta}\end{aligned}$$

Right Side

$$\begin{aligned}\sec\theta - \cos\theta &= \frac{1}{\cos\theta} - \cos\theta \\ &= \frac{1 - \cos^2\theta}{\cos\theta} \\ &= \frac{\sin^2\theta}{\cos\theta}\end{aligned}$$

They are identical

Therefore, the equation is an identity.

Proving a Trigonometric Equation Is Not an Identity

In order for a trigonometric equation to be an identity, the value of each side must turn out to be the same whenever we replace the variable with any number found in the domains of both sides. One *counterexample* is sufficient to show that a proposed equation is *not* an identity.

EXAMPLE 6 **Proving That an Equation Is Not an Identity**

Show that the equation

$$\sin x \cot x = \csc x - \sin x$$

is not an identity by providing a counterexample.

Solution For the equation to be an identity, the value of each side must turn out to be the same whenever we replace x with any number found in the domains of both sides. This is not the case here. For instance, each side is defined for $x = 2$; however,

$$\sin 2 \cot 2 = -0.4161 \text{ (approx.)} \quad \text{and} \quad \csc 2 - \sin 2 = 0.1905 \text{ (approx.)}$$

are different, so we conclude that the equation is not an identity.

Viewing windows of the graphs of $y_1 = \sin x \cot x$ (Figure 1a) and $y_2 = \csc x - \sin x$ (Figure 1b) *show* that the two equations are not the same for all values in the domains of y_1 and y_2. Thus, the graphs reaffirm that the equation given in Example 6 is not an identity.

Figure 1

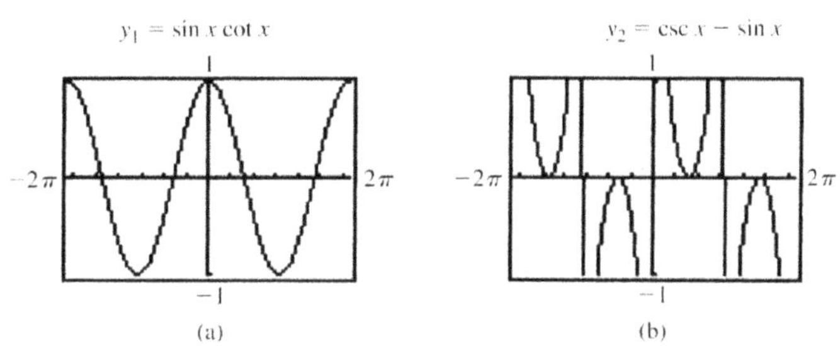

If a trigonometric expression involves radicals, then the use of absolute value is sometimes necessary, although it may be difficult to determine when to use it. In some cases we assume that the radicands are nonnegative. This means that the idenities are restricted to certain quadrants.

For example, we simplify the expression

$$\sqrt{\cos^3 x \sin x} \cdot \sqrt{\sin x}$$

for nonnegative x as follows:

$$\begin{aligned}\sqrt{\cos^3 x \sin x} \cdot \sqrt{\sin x} &= \sqrt{\cos^3 x \sin^2 x}\\ &= \sqrt{\cos^2 x \sin^2 x \cos x}\\ &= \cos x \sin x \sqrt{\cos x}\end{aligned}$$

a grapher can be used to confirm the results.

PROBLEM SET 2.1

Mastering the Concepts

In problems 1–8, use the fundamental identities to fill in the blank.

1. $\sin^2 3\theta = 1 -$ _____

2. $3\cos(-5\theta) =$ _____ $\cos(5\theta)$

3. $\sec^2 t - \tan^2 t =$ _____

4. $\cot^2 \theta - \csc^2 \theta =$ _____

5. $2\cos(-t)\sin(-t) =$ _____ $\cos t \sin t$

6. If $\sin t = \cos t$, then $\sin^2 t - \cos^2 t =$ _____

7. $\dfrac{4}{\sin^2 \theta} + \dfrac{4}{\cos^2 \theta} =$ _____ $\csc^2 \theta \sec^2 \theta$

8. $\dfrac{1 - \cos^2 t}{2\cos^2 t} =$ _____ $\tan^2 t$

In problems 9 and 10, write each expression in terms of $\sin t$.

9. (a) $\tan t \cos t$
(b) $3 + 5\cos^2 t$
(c) $\sin t \csc t - \cos^2 t$

10. (a) $(\cos t)(\tan t + \sec t)$
(b) $\sin^3 t + \sin t \cos^2 t$
(c) $\sin t \cos t \cot t$

In problems 11–16, carry out each operation and simplify.

11. (a) $\cos^4 \theta + \cos^2 \theta - \cos^2 \theta \sin^2 \theta$
(b) $\sec(-\theta)\cot(-\theta)$

12. (a) $\dfrac{4\sin^2 \theta - 1}{2\sin \theta + 1}$
(b) $\dfrac{1 - \tan^4 t}{\sec^2 t}$

13. (a) $(\csc t - \cot t)(\sec t + 1)$
(b) $(\sec \theta - \tan \theta)(\sin \theta + 1)$

14. (a) $\csc^4 t - \cot^4 t$
(b) $\cot^2 t - \csc^2 t$

15. (a) $\sin t + \dfrac{\cos^2 t}{\sin t}$
(b) $\cos \theta + \dfrac{\sin^2 \theta}{\cos \theta}$

16. (a) $\dfrac{1 - (\sin \theta - \cos \theta)^2}{\sin \theta}$
(b) $\dfrac{1 + \cos t}{\sin t} + \dfrac{\sin t}{1 + \cos t}$

In problems 17–48, verify that the equation is an identity.

17. $\sin \theta \csc \theta = 1$

18. $\tan t \cot t = 1$

19. $\sec^2 \theta(1 - \sin^2 \theta) = 1$

20. $\cot^2 \theta(\sec^2 \theta - 1) = 1$

21. $(\sin t + \cos t)^2 + 2\sin(-t)\cos(-t) = 1$

22. $\cos(-t)\tan(-t) + \sin t = 0$

23. $(\cos t - \sin t)^2 + (\cos t + \sin t)^2 = 2$

24. $\sec^7 \theta \tan \theta - \tan \theta \sec^5 \theta = \sec^5 \theta \tan^3 \theta$

25. $\dfrac{\sin \theta}{\tan \theta} + \dfrac{\cos \theta}{\cot \theta} = \sin \theta + \cos \theta$

26. $\tan^2 \theta - \sin^2 \theta = \dfrac{\sin^4 \theta}{\cos^2 \theta}$

27. $\dfrac{\sin \theta}{\cot \theta + \csc \theta} - \dfrac{\sin \theta}{\cot \theta - \csc \theta} = 2$

28. $\dfrac{\cos \theta \cot \theta}{\cot \theta - \cos \theta} = \dfrac{\cot \theta + \cos \theta}{\cos \theta \cot \theta}$

29. $\dfrac{\sin t}{1 + \cos t} + \dfrac{1 + \cos t}{\sin t} = 2\csc t$

30. $\dfrac{\tan \theta + \cot \theta}{\tan \theta - \cot \theta} = \dfrac{\sec^2 \theta}{\tan^2 \theta - 1}$

31. $\dfrac{\cos \theta}{1 - \sin \theta} + \dfrac{\cos \theta}{1 + \sin \theta} = 2\sec \theta$

32. $\dfrac{\sin \theta}{\csc \theta - \cot \theta} = 1 + \cos \theta$

33. $\dfrac{1}{1 + \tan \theta} - \dfrac{\cot \theta}{1 + \cot \theta} = 0$

34. $(\sec \theta - \tan \theta)^2 = \dfrac{1 - \sin \theta}{1 + \sin \theta}$

35. $\dfrac{\sec \theta + 1}{\tan \theta} + \dfrac{\tan \theta}{\sec \theta + 1} = 2\csc \theta$

36. $\dfrac{1}{1 + \sin t} + \dfrac{1}{1 - \sin t} = 2\sec^2 t$

37. $\dfrac{1 - \sin \theta}{1 - \sec \theta} + \dfrac{1 + \sin \theta}{1 + \sec \theta} = 2\cot \theta(1 - \cot \theta)$

38. $\dfrac{\cos \theta + \sin^2(-\theta)\sec \theta}{\csc \theta} = \tan \theta$

39. $\dfrac{\sin t}{1 + \cos t} + \dfrac{\sin t}{1 - \cos t} = 2\csc t$

40. $\dfrac{\tan t}{\sin t(1 + \tan^2 t)} = \cos t$

41. $\dfrac{1}{\cos^2 \theta} + 1 + \dfrac{\sin^2 \theta}{\cos^2 \theta} = 2\sec^2 \theta$

42. $1 - \dfrac{\cos^2 t}{1 + \sin t} = \sin t$

43. $\dfrac{1 + \tan t}{\sin t} - \sec t = \csc t$

44. $\cot t + \dfrac{\sin t}{1 + \cos t} = \csc t$

45. $\dfrac{\tan t + \cot t}{\tan t \cot t} = \sec t \csc t$

46. $\dfrac{\sec t - \csc t}{\sec t + \csc t} = \dfrac{\tan t - 1}{\tan t + 1}$

47. $\dfrac{\cot t - \tan t}{\cot t + \tan t} = \cos^2 t - \sin^2 t$

48. $\dfrac{\sin t}{1 - \cos t} = \csc t + \cot t$

In problems 49–52, show that the given equation is *not* an identity by giving a value of t for which the given equation is false.

49. $\sin t(1 + \cot t) = 2 \sin t + \cos t$

50. $\cos t(\tan t + \cot t) = -\csc t$

51. $\sec t - \cos t = 3 \tan t \cos t$

52. $\sin t \cot t + \cos t \tan t = \sin t - \cos t$

Developing and Extending the Concepts

In problems 53–56, each equation is an identity in certain quadrants associated with t. Determine the quadrants in each case and verify the identity.

53. $\dfrac{\sin t}{\sqrt{1 - \sin^2 t}} = -\tan t$

54. $\dfrac{\cos t}{\sqrt{1 - \cos^2 t}} = \cot t$

55. $\dfrac{1}{\cos t} - \dfrac{1}{\cot t} = \sqrt{\dfrac{1 - \sin t}{1 + \sin t}}$

56. $\dfrac{\sec t - 1}{\tan t} = \sqrt{\dfrac{1 - \cos t}{1 + \cos t}}$

In problems 57–60, verify that each equation is an identity.

57. $-2 \csc(-\theta) - \dfrac{\sin \theta}{1 + \cos(-\theta)} = \dfrac{1 + \cos(-\theta)}{-\sin(-\theta)}$

58. $\dfrac{\sec(-t)}{\csc(-t)[\tan(-t) + \cot(-t)]} = \sin^2 t$

59. $\ln|\sec t + \tan t| = -\ln|\sec t - \tan t|$

60. $\dfrac{1 - \sin t \cos t}{\cos t(\sec t - \csc t)} \cdot \dfrac{\sin^2 t - \cos^2 t}{\sin^3 t + \cos^3 t} = \sin t$

61. (a) Use a grapher to get two separate graphs for

$$y_1 = \cos^4 t - \sin^4 t$$

and

$$y_2 = \cos^2 t - \sin^2 t$$

for the same window settings.

(b) Compare the graphs.

(c) Is there an identity that relates y_1 and y_2? Explain.

62. Use a grapher to help determine whether or not the equation

$$\ln(1 - \cos t) - 2 \ln|\sin t| = -\ln(1 + \cos t)$$

is an identity. If the equation appears to be an identity, then prove it.

Interlude: Practice Problems 2.1, Part I

1. Determine in which quadrants $\sec \theta = \sqrt{1 + \tan^2 \theta}$ is an identity.

2. Verify that $\dfrac{\cos x}{1 + \sin x} + \tan x = \sec x$ is an identity by transforming the left side of the equation into the right side.

3. Verify that $\dfrac{\cot^2 x}{1 + \csc x} = \dfrac{1 - \sin x}{\sin x}$ is an identity by transforming the left side of the equation into the right side.

4. Completely simplify the expression $\dfrac{2\sin^2 t + \sin t - 3}{\csc t - \cot t - 1}$

Interlude: Practice Problems 2.1, Part II

1. Verify that $\dfrac{1 + \csc(-t)}{\cos(-t) + \cot(-t)} = \sec t$ is an identity by transforming the left side of the equation into the right side.

2. Verify that $\dfrac{\sin \alpha}{1 - \cos \alpha} - \dfrac{1 + \cos \alpha}{\sin \alpha} = 0$ is an identity by transforming the left side of the equation into the right side.

3. Verify that $\dfrac{1}{\cot x + 1} + \dfrac{1}{\tan x + 1} = 1$ is an identity by transforming the left side of the equation into the right side.

4. Verify that $\dfrac{1 + \csc(-\beta)}{\cos(-\beta) + \cot(-\beta)} = \sec \beta$ is an identity by transforming the left side of the equation into the right side.

2.2 Addition and Subtraction Identities

Objectives

1. Use the Addition and Subtraction Cosine Identities
2. Use the Addition and Subtraction Sine Identities
3. Simplify Trigonometric Expressions and Verify Identities
4. Solve Applied Problems

In Section 2.1 we worked with identities that express relationships among trigonometric functions of a *single variable*. In this section we develop trigonometric identities involving the sum or difference of *two variables* for

$$\cos(u - v),\quad \cos(u + v),\quad \sin(u - v),\quad \sin(u + v),\quad \text{and}\quad \tan(u - v),\quad \tan(u + v)$$

These identities are referred to as the *sum and difference identities.*

Using the Addition and Subtraction Cosine Identities

Suppose we want to express the value of $\cos(u + v)$ in terms of values of trigonometric functions of u and v. We might be *tempted* to say that

$$\cos(u + v) \text{ is the same as } \cos u + \cos v.$$

To find out whether this is true, we compare the values of the expressions

$$\cos(30^\circ + 60^\circ)$$

and

$$\cos 30^\circ + \cos 60^\circ.$$

Clearly, the results are different (see Table 1).

TABLE 1 $\cos(30^\circ + 60^\circ) \neq \cos 30^\circ + \cos 60^\circ$

$\cos(30^\circ + 60^\circ)$	$\cos 30^\circ + \cos 60^\circ$
$= \cos 90^\circ$	$= \frac{\sqrt{3}}{2} + \frac{1}{2}$
$= 0$	$= \frac{\sqrt{3} + 1}{2}$

Thus, in general

$$\cos(u + v) \neq \cos u + \cos v.$$

The next result shows how it is possible to express $\cos(u + v)$ and $\cos(u - v)$ in terms of trigonometric functions of u and v.

Addition and Subtraction Identities for the Cosine

(i) $\cos(u - v) = \cos u \cos v + \sin u \sin v$
(ii) $\cos(u + v) = \cos u \cos v - \sin u \sin v$

The results are true for all possible values of u *and* v.

In words, the first identity states that

> *the cosine of the difference of two angles (or real numbers) equals the cosine of the first times the cosine of the second, plus the sine of the first times the sine of the second.*

The second identity can be worded in a similar way.

Proofs

(i) To prove identity (i), let u and v represent angles in standard position whose terminal sides intersect the unit circle at the points $P_1 = (x_1, y_1)$ and $P_2 = (x_2, y_2)$, respectively. Figure 1a illustrates one possible situation.

Next we place the angle $u - v$ in standard position and label the point of intersection of its terminal side and the unit circle as $P_3 = (x_3, y_3)$ (Figure 1b).

Figure 1

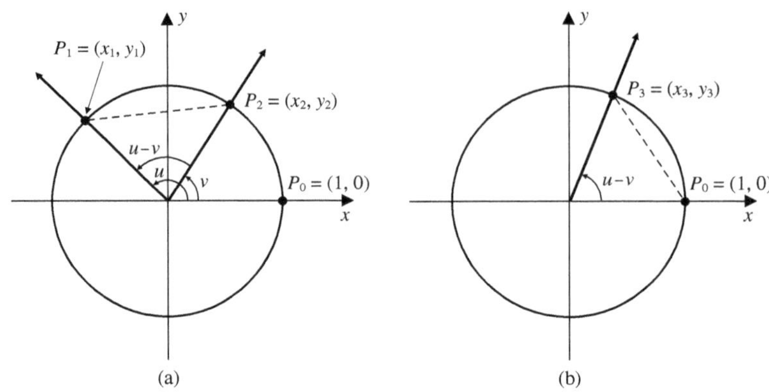

If $P_0 = (1, 0)$, then the length of arcs $\overparen{P_1P_2}$ and $\overparen{P_0P_3}$ are equal. It follows from geometry that the corresponding chord lengths are equal. Thus,

$$|\overline{P_0P_3}| = |\overline{P_1P_2}|$$

$$|\overline{P_0P_3}|^2 = |\overline{P_1P_2}|^2 \quad \text{Square each side}$$

$$(x_3 - 1)^2 + y_3^2 = (x_1 - x_2)^2 + (y_1 - y_2)^2 \quad \text{Distance formula}$$

$$(x_3^2 + y_3^2) + 1 - 2x_3 = (x_1^2 + y_1^2) + (x_2^2 + y_2^2) - 2x_1x_2 - 2y_1y_2 \quad \text{Multiply and rearrange}$$

$$1 + 1 - 2x_3 = 1 + 1 - 2x_1x_2 - 2y_1y_2 \quad \text{Points on unit circle satisfy } x^2 + y^2 = 1$$

$$-2x_3 = -2x_1x_2 - 2y_1y_2 \quad \text{Subtract 2 from each side}$$

[1] $$x_3 = x_1x_2 + y_1y_2 \quad \text{Divide each side by } -2$$

However, we know from the definition of the trigonometric functions that

$$\cos u = x_1, \quad \cos v = x_2, \quad \cos(u - v) = x_3,$$
$$\sin u = y_1 \quad \text{and} \quad \sin v = y_2$$

Consequently, after substituting these latter expressions into the above equation labeled [1], we get

$$\cos(u - v) = \cos u \cos v + \sin u \sin v$$

(ii) To prove identity (ii), we proceed as follows:

$$\cos(u + v) = \cos[u - (-v)]$$
$$= \cos u \cos(-v) + \sin u \sin(-v) \quad \text{Identity (i)}$$
$$= \cos u \cos v - \sin u \sin v \quad \text{Cosine is even; sine is odd}$$

Using the Addition and Subtraction Cosine Identities

In Examples 1 and 2 find the exact value of each expression by using the addition or subtraction cosine identities. Compare the result to the calculator value, rounded off to four decimal places.

EXAMPLE 1 $\cos 15°$

Solution Using $15° = 45° - 30°$ and the identity for $\cos(u - v)$, we get

$$\begin{aligned}\cos 15° &= \cos(45° - 30°)\\ &= \cos 45° \cos 30° + \sin 45° \sin 30°\\ &= \frac{\sqrt{2}}{2} \cdot \frac{\sqrt{3}}{2} + \frac{\sqrt{2}}{2} \cdot \frac{1}{2}\\ &= \frac{\sqrt{6}}{4} + \frac{\sqrt{2}}{4}\\ &= \frac{\sqrt{6} + \sqrt{2}}{4}\end{aligned}$$

Rounding off $(\sqrt{6} + \sqrt{2})/4$ to four decimal places gives 0.9659. Using a calculator to evaluate $\cos 15°$ and rounding off to four decimal places, we again get 0.9659.

EXAMPLE 2 $\cos \frac{7\pi}{12}$

Solution Applying the identity for $\cos(u + v)$ and using

$$\frac{7\pi}{12} = \frac{\pi}{3} + \frac{\pi}{4},$$

we get

$$\begin{aligned}\cos \frac{7\pi}{12} &= \cos\left(\frac{\pi}{3} + \frac{\pi}{4}\right)\\ &= \cos \frac{\pi}{3} \cos \frac{\pi}{4} - \sin \frac{\pi}{3} \sin \frac{\pi}{4}\\ &= \frac{1}{2} \cdot \frac{\sqrt{2}}{2} - \frac{\sqrt{3}}{2} \cdot \frac{\sqrt{2}}{2} = \frac{\sqrt{2}}{4} - \frac{\sqrt{6}}{4} = \frac{\sqrt{2} - \sqrt{6}}{4}\end{aligned}$$

After rounding off $(\sqrt{2} - \sqrt{6})/4$ to four decimal places, we get -0.2588. Using a calculator to evaluate $\cos[7\pi/12]$ and rounding off to four decimal places, we get the same approximate value, -0.2588.

We can use the identity for $\cos(u - v)$ to establish some identities that relate the trigonometric functions to their corresponding *cofunctions*.

Cofunction Identities

If v is a real number or the radian measure of an angle, then:

(i) $\cos\left(\frac{\pi}{2} - v\right) = \sin v$ (ii) $\sin\left(\frac{\pi}{2} - v\right) = \cos v$

(iii) $\tan\left(\frac{\pi}{2} - v\right) = \cot v$ (iv) $\cot\left(\frac{\pi}{2} - v\right) = \tan v$

(v) $\sec\left(\frac{\pi}{2} - v\right) = \csc v$ (vi) $\csc\left(\frac{\pi}{2} - v\right) = \sec v$

Proof

(i) Using the subtraction identity for the cosine, we have

$$\cos\left(\frac{\pi}{2} - v\right) = \cos\frac{\pi}{2}\cos v + \sin\frac{\pi}{2}\sin v$$
$$= 0 + \sin v = \sin v.$$

This gives us the first identity.

(ii) If we replace v by $\pi/2 - v$ in the first identity, we get

$$\cos\left[\frac{\pi}{2} - \left(\frac{\pi}{2} - v\right)\right] = \sin\left(\frac{\pi}{2} - v\right)$$
$$\cos v = \sin\left(\frac{\pi}{2} - v\right)$$

that is,
$$\sin\left(\frac{\pi}{2} - v\right) = \cos v.$$

(iii) We use the first two identities to obtain the cofunction identity for the tangent as follows:

$$\tan\left(\frac{\pi}{2} - v\right) = \frac{\sin\left(\frac{\pi}{2} - v\right)}{\cos\left(\frac{\pi}{2} - v\right)} = \frac{\cos v}{\sin v} = \cot v$$

The proofs of the remaining identities are similar (problem 45).

Naturally, if we measure an angle θ in degrees instead of radians, the cofunction identities still hold. Thus

(i) $\cos(90° - \theta) = \sin\theta$
(ii) $\sin(90° - \theta) = \cos\theta$
(iii) $\tan(90° - \theta) = \cot\theta$
(iv) $\cot(90° - \theta) = \tan\theta$
(v) $\sec(90° - \theta) = \csc\theta$
(vi) $\csc(90° - \theta) = \sec\theta$

Using the Addition and Subtraction Sine Identities

Using the cofunction identities and the addition and subtraction identities for the cosine, we can derive identities for the sine.

Addition and Subtraction Identities for the Sine

(i) $\sin(u - v) = \sin u \cos v - \cos u \sin v$
(ii) $\sin(u + v) = \sin u \cos v + \cos u \sin v$

Proofs

(i) To prove the identity for $\sin(u - v)$, we use the cofunction identity

$$\cos\left(\frac{\pi}{2} - v\right) = \sin v.$$

Reading from right to left and replacing v with $u - v$, we get

$$\sin(u - v) = \cos\left[\frac{\pi}{2} - (u - v)\right]$$

$$= \cos\left(\frac{\pi}{2} - u + v\right)$$

$$= \cos\left[\left(\frac{\pi}{2} - u\right) + v\right].$$

So,

$$\sin(u - v) = \cos\left(\frac{\pi}{2} - u\right)\cos v - \sin\left(\frac{\pi}{2} - u\right)\sin v \qquad \left\{\text{Addition identity for the cosine}\right.$$

$$= \sin u \cos v - \cos u \sin v \qquad \left\{\text{Cofunction identities (i) and (ii)}\right.$$

(ii) After replacing v by $-v$ in part (i), we obtain

$$\sin(u + v) = \sin[u - (-v)]$$

$$= \sin u \cos(-v) - \cos u \sin(-v) \qquad \text{Identity (i)}$$

$$= \sin u \cos v + \cos u \sin v \qquad \left\{\text{Cosine is even; sine is odd}\right.$$

Note that by using the addition identity for the sine, we have

$$\sin\left(x + \frac{\pi}{2}\right) = \sin x \cos\frac{\pi}{2} + \cos x \sin\frac{\pi}{2}$$

$$= \sin x \cdot 0 + \cos x \cdot 1$$

$$= \cos x$$

that is, $\cos x = \sin\left(x + \frac{\pi}{2}\right)$.

Since $\cos x = \sin\left(x + \frac{\pi}{2}\right) = \sin\left[x - \left(-\frac{\pi}{2}\right)\right]$, it follows that the graph of $y = \cos x$ can be obtained by shifting the graph of $f(x) = \sin x$ horizontally $\pi/2$ unit to the left.

EXAMPLE 3 **Evaluating an Expression by Recognizing the Identity**

Find the exact value of the expression

$$\sin 81° \cos 21° - \cos 81° \sin 21°$$

by recognizing the applicable sine identity.

Solution By recognizing the subtraction sine identity, from right to left, we have

$$\sin 81° \cos 21° - \cos 81° \sin 21° = \sin(81° - 21°)$$

$$= \sin 60°$$

$$= \frac{\sqrt{3}}{2}$$

EXAMPLE 4

Finding the Exact Value of the Sine of a Difference

Suppose that α and β are angles in standard position, where α is in quadrant I, $\cos\alpha = 4/5$, β is in quadrant II, and $\sin\beta = 3/5$.

Use identities to determine the exact value of $\sin(\alpha - \beta)$.

Solution

Before we can apply the subtraction identity, we need to determine both $\sin\alpha$ and $\cos\beta$.

Since the terminal side of α is in quadrant I, $\sin\alpha$ is positive, so

$$\sin\alpha = \sqrt{1 - \cos^2\alpha} = \sqrt{1 - (16/25)} = \frac{3}{5}.$$

Also, since the terminal side of β is in quadrant II, $\cos\beta$ is negative, so

$$\cos\beta = -\sqrt{1 - \sin^2\beta} = -\sqrt{1 - 9/25} = -\frac{4}{5}.$$

It follows that

$$\sin(\alpha - \beta) = \sin\alpha\cos\beta - \cos\alpha\sin\beta = \left(\frac{3}{5}\right)\left(-\frac{4}{5}\right) - \left(\frac{4}{5}\right)\left(\frac{3}{5}\right) = -\frac{24}{25}.$$

EXAMPLE 5

Finding a Value Involving Inverse Functions

Find the exact value of $\sin[\cos^{-1}(5/13) + \sin^{-1}(3/5)]$ by using the addition identity for the sine function and right triangle trigonometry.

Solution

First we let $\alpha = \cos^{-1}(5/13)$ and $\beta = \sin^{-1}(3/5)$, so $\cos\alpha = 5/13$, $\sin\beta = 3/5$, and α and β are acute angles.

Using right triangles, we see in Figure 2a that $\sin\alpha = 12/13$ and in Figure 2b that $\cos\beta = 4/5$.

Thus

$$\sin\left(\cos^{-1}\frac{5}{13} + \sin^{-1}\frac{3}{5}\right) = \sin(\alpha + \beta) = \sin\alpha\cos\beta + \cos\alpha\sin\beta = \left(\frac{12}{13}\right)\left(\frac{4}{5}\right) + \left(\frac{5}{13}\right)\left(\frac{3}{5}\right) = \frac{48}{65} + \frac{15}{65} = \frac{63}{65}$$

Figure 2

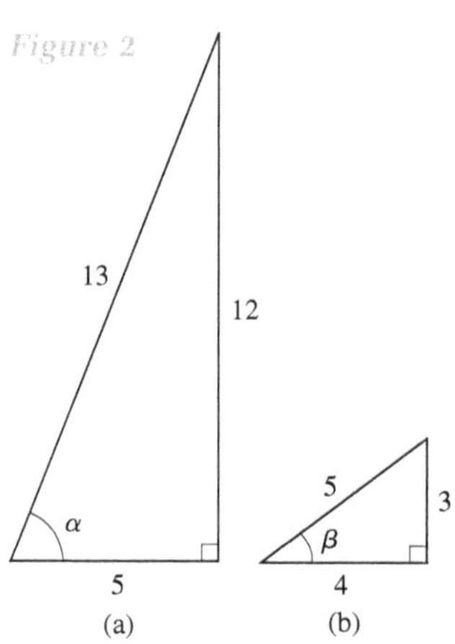

Using the addition and subtraction identities for the sine and cosine, we can derive addition and subtraction identities for the tangent function.

Addition and Subtraction Identities for the Tangent

(i) $\tan(u - v) = \dfrac{\tan u - \tan v}{1 + \tan u \tan v}$ (ii) $\tan(u + v) = \dfrac{\tan u + \tan v}{1 - \tan u \tan v}$

Proof (i) We prove the first identity as follows:

$$\tan(u - v) = \frac{\sin(u - v)}{\cos(u - v)}$$

$$= \frac{\sin u \cos v - \cos u \sin v}{\cos u \cos v + \sin u \sin v} \quad \text{Sine and cosine subtraction identities}$$

$$= \frac{\dfrac{\sin u \cos v}{\cos u \cos v} - \dfrac{\cos u \sin v}{\cos u \cos v}}{\dfrac{\cos u \cos v}{\cos u \cos v} + \dfrac{\sin u \sin v}{\cos u \cos v}} \quad \text{Divide numerator and denominator by } \cos u \cos v$$

$$= \frac{\tan u - \tan v}{1 + \tan u \tan v} \quad \text{Reduce fractions and use } \tan t = \frac{\sin t}{\cos t}$$

The proof of part (ii) is left as an exercise (problem 46).

Care must be taken when applying the trigonometric identities. For example, if we replace u by $\pi/2$ in the subtraction identity for $\tan(u - v)$, we get

$$\tan\left(\frac{\pi}{2} - v\right) = \frac{\tan\dfrac{\pi}{2} - \tan v}{1 + \tan\dfrac{\pi}{2}\tan v}.$$

Since $\tan(\pi/2)$ is not defined, we might erroneously conclude that $\tan[(\pi/2) - v]$ is not defined for any value of v. However, we know from the cofunction identity that $\tan[(\pi/2) - v]$ can be written as

$$\tan\left(\frac{\pi}{2} - v\right) = \cot v.$$

This illustration emphasizes the important fact that the identities derived in trigonometry are applicable only when the values of the variables are in the domains of all the functions contained in the identities.

Simplifying Trigonometric Expressions and Verifying Identities

We can use the addition and subtraction identities to simplify certain trigonometric expressions and to verify other identities.

EXAMPLE 6 **Simplifying a Trigonometric Expression**

Write the expression $\sin 5t \cos 2t - \sin 2t \cos 5t$ in terms of the sine function.

Solution We recognize that the expression fits the identity for $\sin(u - v)$, with $u = 5t$ and $v = 2t$. So

$$\sin 5t \cos 2t - \sin 2t \cos 5t = \sin(5t - 2t)$$
$$= \sin 3t.$$

In Section 1.4 we showed that the tangent function has a period π. In the next example, we confirm this fact by using an identity.

EXAMPLE 7

Proving That the Tangent Function Has Period π

Prove that $f(x) = \tan x$ has period π.

Solution Using the addition identity for the tangent, we get

$$\begin{aligned}\tan(x + \pi) &= \frac{\tan x + \tan \pi}{1 - \tan x \tan \pi}\\ &= \frac{\tan x + 0}{1 - 0}\\ &= \tan x.\end{aligned}$$

Thus the tangent function has period π.

In some applications it is necessary to transform expressions from the form $P \cos u + Q \sin u$ to the form $A \cos(u - v)$.

EXAMPLE 8

Writing $P \cos u + Q \sin u$ in the Form $A \cos(u - v)$

Let P and Q be constants, and let v be an angle in standard position with the point (P, Q) on the terminal side of v.

(a) Show that if $A = \sqrt{P^2 + Q^2}$, then

$$P = A \cos v \quad \text{and} \quad Q = A \sin v.$$

(b) Use part (a) to show that

$$P \cos u + Q \sin u = A \cos(u - v)$$

for any angle u.

Solution (a) Figure 3 shows angle v in standard position with the point (P, Q) on the terminal side. Notice that the distance A from the point $(0, 0)$ to (P, Q) is given by $A = \sqrt{P^2 + Q^2}$.

Using the cosine and sine definitions, we have

$$\cos v = \frac{P}{A} \quad \text{and} \quad \sin v = \frac{Q}{A}.$$

So

$$P = A \cos v \quad \text{and} \quad Q = A \sin v.$$

Figure 3

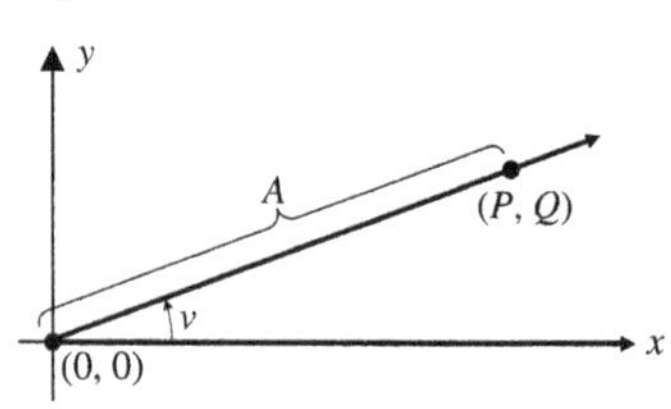

(b) Using the results from part (a), we get

$$\begin{aligned}P \cos u + Q \sin u &= (A \cos v)(\cos u) + (A \sin v)(\sin u)\\ &= A(\cos v \cos u + \sin v \sin u)\\ &= A \cos(u - v)\end{aligned}$$

for any angle u.

Solving Applied Problems

Applied problems from different fields such as physics, engineering, and meteorology are modeled by functions of the form

$$y = P\cos u + Q\sin u.$$

EXAMPLE 9

Modeling a Weather Pattern

The U.S. Weather Bureau conducted a study on the temperature fluctuation for a desert region. Using data collected over several years, it was determined that the average temperature T over one year (in degrees Fahrenheit) is approximated by the model

$$T = 80 - 12\cos\frac{\pi t}{6} + 12\sqrt{3}\sin\frac{\pi t}{6}$$

where t is the number of elapsed months, and $t = 0$ represents the month of March in each year.

(a) Write the equation representing this model in the form

$$T = 80 + A\cos(u - v).$$

(b) Find the maximum average temperature and in what month it occurs over a 12-month period starting in March.
(c) Find the minimum average temperature and in what month it occurs.

Solution (a) We begin by rewriting the given function as

$$T = 80 + y \quad \text{where} \quad y = -12\cos\frac{\pi t}{6} + 12\sqrt{3}\sin\frac{\pi t}{6}.$$

Figure 4

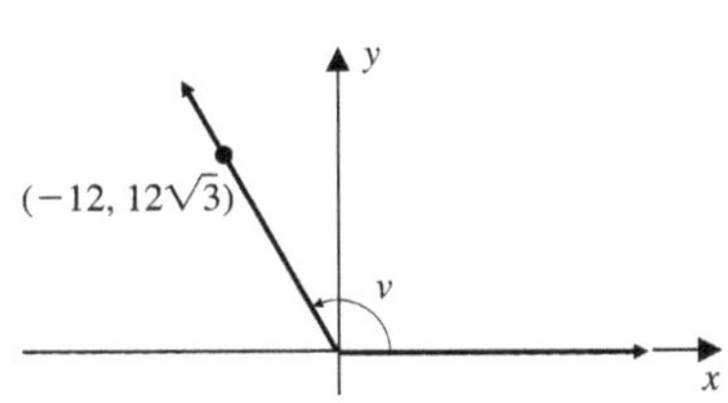

The expression for y is of the form

$$P\cos u + Q\sin u,$$
$$\text{where} \quad P = -12, \quad Q = 12\sqrt{3}, \quad \text{and} \quad u = (\pi t)/6.$$

Proceeding as we did in Example 8, we locate the point $(P, Q) = (-12, 12\sqrt{3})$ on the terminal side of an angle v in standard position (Figure 4).

Since

$$\tan v = \frac{Q}{P} = \frac{12\sqrt{3}}{-12} = -\sqrt{3}$$

and the point is in quadrant II, it follows that one possibility for angle v is $v = (2\pi)/3$.

Also

$$\begin{aligned} A &= \sqrt{P^2 + Q^2} \\ &= \sqrt{(-12)^2 + (12\sqrt{3})^2} \\ &= 24 \end{aligned}$$

So

$$y = A\cos(u - v)$$
$$= 24\cos\left(\frac{\pi t}{6} - \frac{2\pi}{3}\right)$$

Therefore

$$T = 80 + y$$
$$= 80 + 24\cos\left(\frac{\pi t}{6} - \frac{2\pi}{3}\right)$$

(b) The maximum value of T occurs when the cosine value attains its maximum value of 1. So the maximum value of T is given by

$$T = 80 + 24(1) = 104°\text{F}$$

and it occurs when

$$\cos\left(\frac{\pi t}{6} - \frac{2\pi}{3}\right) = 1$$

that is, for the first time, when

$$\frac{\pi t}{6} - \frac{2\pi}{3} = 0$$

or

$$t = 4.$$

Thus the maximum average temperature occurs in July (when $t = 4$).

(c) The minimum value of T occurs when the cosine value attains its minimum value of -1. This occurs when

$$\cos\left(\frac{\pi t}{6} - \frac{2\pi}{3}\right) = -1$$

that is, for the first time, when

$$\frac{\pi t}{6} - \frac{2\pi}{3} = \pi \quad \text{or} \quad \frac{\pi t}{6} = \frac{5\pi}{3}, \quad t = 10.$$

Thus the minimum value of T is given by

$$T = 80 + 24(-1)$$
$$= 56°\text{F}$$

and it occurs in the month of January (when $t = 10$).

Graphers are useful in performing *trend analysis*. For instance, in Example 9 we derived the result that the average temperature T (in degrees Fahrenheit) for a certain desert region is approximated by the model

$$T = 80 + 24\cos\left(\frac{\pi t}{6} - \frac{2\pi}{3}\right)$$

Figure 5

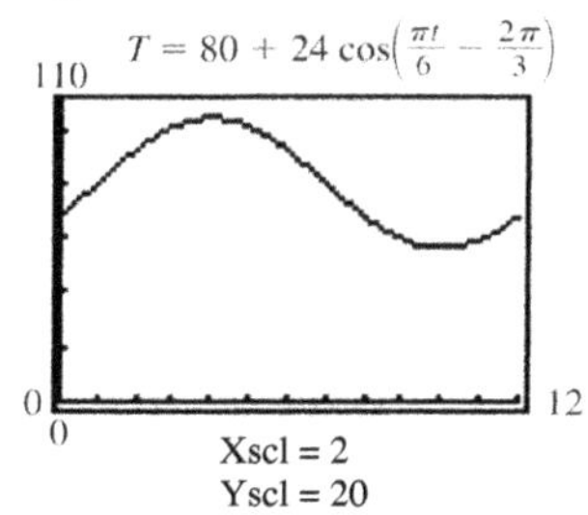

where t is the number of elapsed months, and $t = 0$ represents the month of March.

A viewing window of the graph of T is shown in Figure 5 for $0 \le t \le 12$. By using the TRACE -feature, we find that starting in March ($t = 0$) the average temperature first exceeds 100°F when t is approximately 2.9 and then falls below 100°F when t is approximately 5.1. This means that the average temperature is above 100°F from about the end of May ($t = 2$) through the beginning of August ($t = 5$).

PROBLEM SET 2.2

Mastering the Concepts

In problems 1 and 2, use an appropriate addition or subtraction identity to find the exact value of each expression. In each case compare this value to the calculator value (rounding off to four decimal places).

1. (a) $\sin\left(\frac{\pi}{6} + \frac{3\pi}{4}\right)$
(b) $\cos(330° + 45°)$

2. (a) $\cos\left(\frac{5\pi}{4} - \frac{\pi}{3}\right)$
(b) $\tan(60° - 45°)$

In problems 3 and 4, find the exact value of each expression by using an appropriate addition or subtraction identity and $(5\pi)/12 = (\pi/6) + (\pi/4)$ or $165° = 210° - 45°$. In each case compare this value to the calculator value (rounding off to four decimal places).

3. (a) $\sin\frac{5\pi}{12}$
(b) $\cos 165°$
(c) $\tan\frac{5\pi}{12}$

4. (a) $\cos\frac{5\pi}{12}$
(b) $\sin 165°$
(c) $\tan 165°$

In problems 5 and 6, use an appropriate addition or subtraction identity to find the exact value of each expression. In each case compare this value to the calculator value (rounding off to four decimal places).

5. (a) $\cos 105°$
(b) $\sin\frac{19\pi}{12}$

6. (a) $\tan\frac{13\pi}{12}$
(b) $\cos 285°$

In problems 7 and 8, find the exact value of each expression by recognizing the applicable identity. Also, use a calculator to evaluate each expression to four decimal places. Compare the results in both approaches.

7. (a) $\sin 33° \cos 27° + \sin 27° \cos 33°$
(b) $\dfrac{\tan\frac{4\pi}{5} - \tan\frac{3\pi}{10}}{1 + \tan\frac{4\pi}{5}\tan\frac{3\pi}{10}}$

8. (a) $\cos\frac{5\pi}{7}\sin\frac{2\pi}{7} + \cos\frac{2\pi}{7}\sin\frac{5\pi}{7}$
(b) $\dfrac{\tan 17° + \tan 43°}{1 - \tan 17° \tan 43°}$

In problems 9–12, suppose that u and v are angles in standard position. Use the given information to find the value of each expression. Round off the answers in problems 11 and 12 to three decimal places.

(a) $\sin(u + v)$
(b) $\cos(u + v)$
(c) $\sin(u - v)$
(d) $\cos(u - v)$
(e) $\tan(u + v)$
(f) $\tan(u - v)$

9. $\sin u = \frac{12}{13}$, u is in quadrant II, $\cos v = \frac{-4}{5}$, and v is in quadrant II

10. $\sin u = \frac{3}{5}$, u is in quadrant II, $\cos v = \frac{3}{5}$, and v is in quadrant IV

11. $\cos u = 0.47$, u is in quadrant IV, $\sin v = -0.96$, and v is in quadrant III

12. $\tan u = 0.42$, u is in quadrant I, $\tan v = 0.59$, and v is in quadrant III

In problems 13–16, find the exact value of each expression by using the addition and subtraction identities and right triangle trigonometry.

13. $\cos\left(\sin^{-1}\frac{3}{5} - \cos^{-1}\frac{4}{5}\right)$

14. $\sin\left[\sin^{-1}\left(-\frac{3}{5}\right) - \cos^{-1}\frac{4}{5}\right]$

15. $\sin\left[\cos^{-1}\frac{5}{13} + \sin^{-1}\left(-\frac{12}{13}\right)\right]$

16. $\cos\left(\sin^{-1}\frac{8}{17} - \cos^{-1}\frac{5}{17}\right)$

In problems 17–26, rewrite each expression in terms of the sine, cosine, or tangent by using the addition and subtraction identities. Simplify each result.

17. (a) $\sin\left(\frac{5\pi}{2} - t\right)$
(b) $\cos(270° + t)$

18. (a) $\cos\left(\frac{\pi}{6} - t\right)$
(b) $\sin(\theta - 45°)$

19. (a) $\tan\left(t + \frac{\pi}{4}\right)$
(b) $\sin(45° - \theta)$

20. (a) $\sin(\pi - t)$
(b) $\tan(\theta + 60°)$

21. (a) $\csc(90° + \theta)$
(b) $\sin\left(\frac{3\pi}{2} - t\right)$

22. (a) $\cot(10\pi - t)$
(b) $\tan(1080° - \theta)$

23. (a) $\cos 7t \cos t - \sin 7t \sin t$
(b) $\sin 7t \cos 3t - \cos 7t \sin 3t$

24. (a) $\sin(-t) \cos 4t - \cos(-t) \sin 4t$
(b) $\cos 3\theta \cos 2\theta + \sin 3\theta \sin 2\theta$

25. (a) $\cos \frac{2x}{3} \cos \frac{x}{3} - \sin \frac{2x}{3} \sin \frac{x}{3}$
(b) $\frac{\tan 4t + \tan 3t}{1 - \tan 4t \tan 3t}$

26. (a) $\sin(-5t) \cos 2t - \cos(-5t) \sin 2t$
(b) $\frac{\tan 5t - \tan 2t}{1 + \tan 5t \tan 2t}$

In problems 27–44, verify that the equation is an identity.

27. $\sin(t - 3\pi) = -\sin t$

28. $\cos(t - 450°) = \sin t$

29. $\sin(30° + t) = \frac{1}{2}\cos t + \frac{\sqrt{3}}{2}\sin t$

30. $\sin(60° + t) - \cos(30° + t) = \sin t$

31. $\tan(45° + t) = \frac{1 + \tan t}{1 - \tan t}$

32. $\sin(\pi - t) - \cot t \sin\left(t - \frac{\pi}{2}\right) = \csc t$

33. $\sin\left(\frac{\pi}{4} + t\right) - \sin\left(\frac{\pi}{4} - t\right) = \sqrt{2}\sin t$

34. $\cos\left(\frac{\pi}{6} + t\right)\cos\left(\frac{\pi}{6} - t\right) - \sin\left(\frac{\pi}{6} + t\right)\sin\left(\frac{\pi}{6} - t\right) = \frac{1}{2}$

35. $\sin(t + s) \sin(t - s) = \sin^2 t - \sin^2 s$

36. $\sin t \cos s - \sin\left(t + \frac{\pi}{2}\right)\sin(-s) = \sin(s + t)$

37. $\cos(s + t) \cos(s - t) = \cos^2 s + \cos^2 t - 1$

38. $\tan s - \tan t = \frac{\sin(s - t)}{\cos s \cos t}$

39. $\cot s - \tan t = \frac{\cos(s + t)}{\sin s \cos t}$

40. $\frac{1}{\tan s + \tan t} = \frac{\csc(s + t)}{\sec s \sec t}$

41. $\frac{\cos(u + v)}{\cos(u - v)} = \frac{1 - \tan u \tan v}{1 + \tan u \tan v}$

42. $\sin(\pi - s - t) = \sin s \cos t + \cos s \sin t$

43. $\frac{\sin(s + t)}{\sin(s - t)} = \frac{\tan s + \tan t}{\tan s - \tan t}$

44. $\cot(u + v) = \frac{\cot u \cot v - 1}{\cot u + \cot v}$

45. Verify that each equation is an identity.
(a) $\cot\left(\frac{\pi}{2} - v\right) = \tan v$
(b) $\sec\left(\frac{\pi}{2} - v\right) = \csc v$
(c) $\csc\left(\frac{\pi}{2} - v\right) = \sec v$

46. Use the identity for

$$\tan(u - v)$$

to derive the identity for

$$\tan(u + v).$$

In problems 47–50, express each function in the form

$$y = A\cos(u - v)$$

where v is the smallest positive angle. Show that the function obtained is equal to the given function.

47. $y = \sin t + \cos t$

48. $y = \sqrt{3}\cos \pi t + \sin \pi t$

49. $y = \cos 2\pi t - \sqrt{3}\sin 2\pi t$

50. $y = \cos 4t + \sqrt{3}\sin 4t$

Applying the Concepts

51. Height of a Tide: In a certain harbor, the mathematical model

$$h(t) = 0.3\cos\frac{\pi t}{6} + 0.4\sin\frac{\pi t}{6}$$

is used to predict the height h (in meters) of the tide above or below mean sea level t hours after midnight, over a 12-hour period.

(a) Write the equation of this model in the form

$$h(t) = A\cos(u - v).$$

Round off v to three decimal places.

(b) Find the maximum height and when it occurs. Round off the answer to one decimal place.

(c) Find the minimum height and indicate when it occurs. Round off the answer to one decimal place.

52. (Refer to problem 51.)

(a) Use a grapher to graph the function found in problems 51(a) for $0 \le t \le 12$ and interpret it.

(b) Use a grapher to find when the height of the tide is more than 0.3 meter above mean sea level. Round off the answer to two decimal places.

53. **Physics:** An object attached to a spring vibrates vertically according to the mathematical model

$$d(t) = 4\cos 4t + 3\sin 4t$$

where $d(t)$ is the distance of the object from its rest position measured in centimeters, t seconds after the start of the motion.

(a) Write the equation of this model in the form

$$d(t) = A\cos(u - v).$$

Round off v to three decimal places.

(b) Determine the amplitude of the vibration of the object.

(c) Determine the period of the vibration of the object.

54. (Refer to Problem 53.)

(a) Use a grapher to graph the function found in problem 52(a) and interpret it for $0 \le t \le 10$.

(b) Use a grapher to find when the object is more than 2 centimeters from its rest position. Round off the answer to two decimal places.

Developing and Extending the Concepts

55. Consider the triangle ABC in Figure 6.

(a) Express $\sin(\alpha + \beta)$ in terms of a trigonometric function of γ.

(b) Express $\cos(\alpha + \beta)$ in terms of a trigonometric function of γ.

Figure 6

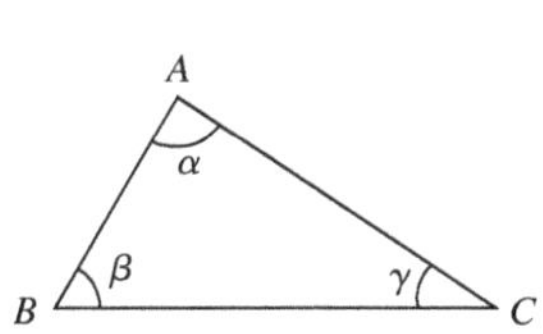

56. Figure 7 shows intersecting lines L_1 and L_2 with slopes m_1 and m_2, respectively.

(a) Show that $\tan\theta_1 = m_1$ and

$$\tan\theta_2 = m_2.$$

(b) Show that the angle $\theta = \theta_2 - \theta_1$ between L_1 and L_2 is given by

$$\tan\theta = \frac{m_2 - m_1}{1 + m_1 m_2}.$$

(c) Use part (b) to show that if L_1 and L_2 are perpendicular, then $m_1 m_2 = -1$.

Figure 7

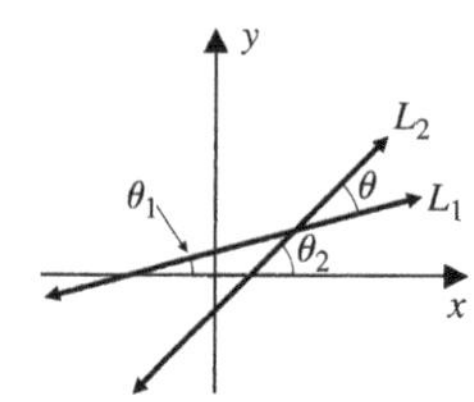

57. (a) Show that

$$\cos\left(\frac{\pi}{2} - x\right) = \cos\left(x - \frac{\pi}{2}\right).$$

(b) Explain why

$$\sin x = \cos\left(x - \frac{\pi}{2}\right).$$

(c) In view of the identity in part (b), compare the graphs of

$$f(x) = \sin x \quad \text{and} \quad g(x) = \cos\left(x - \frac{\pi}{2}\right).$$

How can the graph of $y = \cos x$ be used to obtain the graph of $f(x) = \sin x$?

58. (a) Find the value of the expression

$$\sin 1^\circ + \sin 2^\circ + \sin 3^\circ + \cdots + \sin 357^\circ + \sin 358^\circ + \sin 359^\circ$$

by using the identity

$$\sin\theta + \sin(360^\circ - \theta) = \sin\theta - \sin\theta = 0.$$

(b) Use the identity

$$\cos\theta + \cos(360^\circ - \theta) = 2\cos\theta$$

to evaluate the expression

$$\cos 1^\circ + \cos 2^\circ + \cos 3^\circ + \cdots + \cos 357^\circ + \cos 358^\circ + \cos 359^\circ.$$

59. Let $f(x) = \sin x$.

(a) Show that

$$\frac{f(t+h) - f(t)}{h} = \sin t\left(\frac{\cos h - 1}{h}\right) + \cos t\left(\frac{\sin h}{h}\right).$$

(b) Use a grapher to approximate the value of the expressions

$$\frac{\cos h - 1}{h} \quad \text{as} \quad h \to 0$$

and

$$\frac{\sin h}{h}$$

as $h \to 0$

(c) Explain why

$$\frac{f(t+h) - f(t)}{h} \to \cos t \quad \text{as} \quad h \to 0$$

60. Let $g(x) = \cos x$.

(a) Show that

$$\frac{g(t+h) - g(t)}{h} = \cos t\left(\frac{\cos h - 1}{h}\right) - \sin t\left(\frac{\sin h}{h}\right)$$

(b) Use part (b) of problem 59 to show that

$$\frac{g(t+h) - g(t)}{h} \to -\sin t \quad \text{as} \quad h \to 0$$

Interlude: Practice Problems 2.2, Part I

1. Fill in the blanks.

(a) If $\sin t = \frac{1}{3}$ and $0 < t < \frac{\pi}{2}$, then $\sin\left(\frac{\pi}{6} + t\right) =$ __________

(b) If $\sec t = -\frac{8}{5}$ and $\frac{3\pi}{2} < t < 2\pi$, then $\cos\left(\frac{\pi}{4} - t\right) =$ __________

(c) If $\cot t = \frac{4}{3}$ and $\pi < t < \frac{3\pi}{2}$, then $\cos\left(\frac{2\pi}{3} + t\right) =$ __________

(d) If $\sin t = \frac{1}{6}$ and $\frac{\pi}{2} < t < \pi$, then $\csc\left(\frac{3\pi}{4} + t\right) =$ __________

2. Verify that $\cot 2x = \dfrac{\cot^2 x - 1}{2 \cot x}$ is an identity by transforming the left side of the equation into the right side.

3. Suppose that u and v are real numbers with $\dfrac{\pi}{2} < u < \pi$ and $\pi < v < \dfrac{3\pi}{2}$. If $\sin u = \dfrac{1}{3}$ and $\cos v = -\dfrac{3}{4}$, then calculate the exact values of $\cos(u + v)$, $\sin(u + v)$, and $\tan(u + v)$. In addition, determine if $\pi < u + v < \dfrac{3\pi}{2}$ or $\dfrac{3\pi}{2} < u + v < 2\pi$.

Interlude: Practice Problems 2.2, Part II

1. An antenna sits atop a pole. The pole is 40 feet tall and from a point 50 feet from the base of the pole, the antenna subtends an angle of 15°. Determine the length of the antenna (accurate to two decimal places) **without** using an inverse trigonometric function.

2. Assume that $0 < x < \frac{\pi}{2}$. Completely simplify $\sin(\arcsin x + \arccos x)$.

3. Given the right triangle shown in Figure 8, write the following expressions in terms of θ (fill in the blanks).

Figure 8

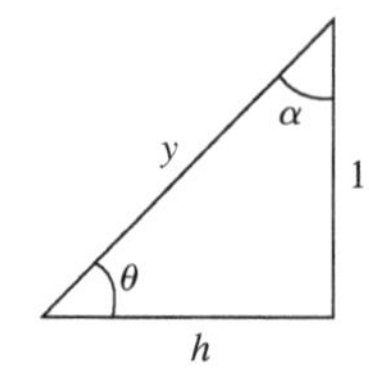

____

____

____

(a) $y =$ ____________________

(b) $\cos\alpha =$ ____________________

(c) $h =$ ____________________

(d) $\tan\alpha =$ ____________________

(e) $\alpha =$ ____________________

(f) $\sin(\pi - \alpha) =$ ____________________

(g) $\sec(\alpha + \theta) =$ ____________________

(h) $\cot(\alpha - \theta) =$ ____________________

Objectives

1. Use Double-Angle Identities
2. Use Half-Angle Identities
3. Solve Applied Problems

2.3 Double-Angle and Half-Angle Identities

In this section we'll study identities that express trigonometric functions of twice an angle and half an angle in terms of trigonometric functions of the angle. These identities follow from the addition identities. They are used in calculus.

Using Double-Angle Identities

Suppose that we want to express the value of $\sin 2\theta$ in terms of trigonometric functions of θ. We might be tempted to say that $\sin 2\theta = 2 \sin \theta$. However, this is *not* correct.

For example, assume $\theta = 30°$,
then

$$\sin 2\theta = \sin[2(30°)] = \sin 60° = \frac{\sqrt{3}}{2}$$

whereas

$$2 \sin \theta = 2 \sin 30° = 2 \cdot \frac{1}{2} = 1.$$

So, in general

$$\sin 2\theta \neq 2 \sin \theta.$$

The identities that express the trigonometric functions of 2θ in terms of functions of θ are called *double-angle identities.*

Double-Angle Identities

(i) $\sin 2\theta = 2 \sin \theta \cos \theta$

(ii) $\cos 2\theta = \cos^2 \theta - \sin^2 \theta$
$= 2 \cos^2 \theta - 1$
$= 1 - 2 \sin^2 \theta$

(iii) $\tan 2\theta = \dfrac{2 \tan \theta}{1 - \tan^2 \theta}$

Proofs

(i) To prove the first identity, we start with the addition identity for $\sin(u + v)$ and replace both u and v by θ throughout to get

$$\sin(\theta + \theta) = \sin \theta \cos \theta + \cos \theta \sin \theta$$
$$\sin 2\theta = 2 \sin \theta \cos \theta.$$

(ii) To prove the second identity, we use the identity for $\cos(u + v)$ and replace both u and v by θ to obtain

$$\cos(\theta + \theta) = \cos \theta \cos \theta - \sin \theta \sin \theta$$
$$\cos 2\theta = \cos^2 \theta - \sin^2 \theta$$

To obtain the second form for $\cos 2\theta$, we substitute the Pythagorean identity

$$\sin^2 \theta = 1 - \cos^2 \theta$$

into $\cos 2\theta = \cos^2\theta - \sin^2\theta$ to get

$$\begin{aligned}\cos 2\theta &= \cos^2\theta - (1 - \cos^2\theta)\\ &= \cos^2\theta - 1 + \cos^2\theta\\ &= 2\cos^2\theta - 1.\end{aligned}$$

Using the alternate form of the Pythagorean identity, $\cos^2\theta = 1 - \sin^2\theta$, we have

$$\begin{aligned}\cos 2\theta &= \cos^2\theta - \sin^2\theta\\ &= (1 - \sin^2\theta) - \sin^2\theta\\ &= 1 - 2\sin^2\theta\end{aligned}$$

which is the third form of the identity.

(iii) The identity for $\tan 2\theta$ is obtained by replacing both u and v by θ in the addition identity for $\tan(u + v)$ to get

$$\tan(\theta + \theta) = \frac{\tan\theta + \tan\theta}{1 - \tan\theta\tan\theta}$$

$$\tan 2\theta = \frac{2\tan\theta}{1 - \tan^2\theta}$$

EXAMPLE 1 **Using Double-Angle Identities to Evaluate**

If $\sin\theta = 12/13$ and θ is an angle in quadrant I, find the exact value of each expression.

(a) $\sin 2\theta$ (b) $\cos 2\theta$ (c) $\tan 2\theta$

Solution

First we need to find $\cos\theta$. Because θ is in quadrant I, we know that $\cos\theta$ is positive, so

$$\cos\theta = \sqrt{1 - \sin^2\theta} = \sqrt{1 - (12/13)^2} = \sqrt{25/169} = 5/13.$$

Thus we get the following results:

(a) $\sin 2\theta = 2\sin\theta\cos\theta = 2\left(\frac{12}{13}\right)\left(\frac{5}{13}\right) = \frac{120}{169}$

(b) $\cos 2\theta = \cos^2\theta - \sin^2\theta = \left(\frac{5}{13}\right)^2 - \left(\frac{12}{13}\right)^2 = \frac{25}{169} - \frac{144}{169} = -\frac{119}{169}$

(c) Since

$$\tan\theta = \frac{\sin\theta}{\cos\theta} = \frac{\left(\frac{12}{13}\right)}{\left(\frac{5}{13}\right)} = \frac{12}{5}$$

it follows that

$$\tan 2\theta = \frac{2\tan\theta}{1-\tan^2\theta} = \frac{2\left(\frac{12}{5}\right)}{1-\left(\frac{12}{5}\right)^2} = \frac{\left(\frac{24}{5}\right)}{1-\left(\frac{144}{25}\right)} = \frac{120}{25-144} = -\frac{120}{119}.$$

A more efficient way of finding tan 2θ after finding sin 2θ and cos 2θ is to proceed as follows:

$$\tan 2\theta = \frac{\sin 2\theta}{\cos 2\theta} = \frac{\left(\frac{120}{169}\right)}{\left(-\frac{119}{169}\right)} = -\frac{120}{119}$$

EXAMPLE 2

Rewriting a Trigonometric Expression

Use a double-angle identity to rewrite the expression $\cos^2 3t - \sin^2 3t$ as a single trigonometric function.

Solution We use the double-angle identity $\cos 2\theta = \cos^2\theta - \sin^2\theta$ by replacing θ with $3t$ to get

$$\cos^2 3t - \sin^2 3t = \cos[2(3t)] = \cos 6t.$$

EXAMPLE 3

Finding a Value Involving Inverse Functions

Find the exact value of $\sin[2\cos^{-1}(8/17)]$. Compare this value to the calculator value, rounded to four decimal places.

Solution First we let $\theta = \cos^{-1}(8/17)$, so $\cos\theta = 8/17$.

Next we draw a right triangle with an acute angle θ whose cosine is 8/17 (Figure 1). By the pythogorean theorem, we get $b = 15$.

From right triangle trigonometry, it follows that

Figure 1

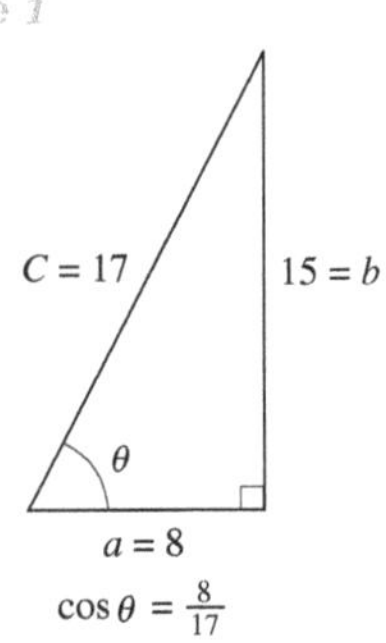

$$\sin\theta = 15/17.$$

Thus

$$\begin{aligned}\sin\left(2\cos^{-1}\frac{8}{17}\right) &= \sin 2\theta \\ &= 2\sin\theta\cos\theta \\ &= 2\left(\frac{15}{17}\right)\left(\frac{8}{17}\right) \\ &= \frac{240}{289}\text{ (approx. 0.8304).}\end{aligned}$$

The calculator value of the expression $\sin[2\cos^{-1}(8/17)]$ also equals 0.8304, rounded to four decimal places.

EXAMPLE 4 **Verifying an Identity Using Double-Angle Identities**

Verify the identity: $\dfrac{2\cos 2t}{\sin 2t - 2\sin^2 t} - 1 = \cot t.$

Solution To verify that the equation is an identity, we proceed as follows:

$$\begin{aligned}
\frac{2\cos 2t}{\sin 2t - 2\sin^2 t} - 1 &= \frac{2(\cos^2 t - \sin^2 t)}{2\sin t\cos t - 2\sin^2 t} - 1 && \left\{\begin{matrix}\cos 2t \text{ and}\\ \sin 2t \text{ identities}\end{matrix}\right. \\
&= \frac{2(\cos t - \sin t)(\cos t + \sin t)}{2(\sin t)(\cos t - \sin t)} - 1 && \text{Factor} \\
&= \frac{\cos t + \sin t}{\sin t} - 1 && \text{Reduce} \\
&= \frac{\cos t}{\sin t} + \frac{\sin t}{\sin t} - 1 && \frac{A+B}{C} = \frac{A}{C} + \frac{B}{C} \\
&= \cot t + 1 - 1 = \cot t
\end{aligned}$$

Therefore, the equation is an identity.

Using Half-Angle Identities

Consider the double-angle identities

$$\cos 2t = 2\cos^2 t - 1 \quad \text{and} \quad \cos 2t = 1 - 2\sin^2 t.$$

If we solve the first for $\cos^2 t$ and the second for $\sin^2 t$ in terms of $\cos 2t$, we get the following identities:

Square Identities

$$\text{(i)}\ \cos^2 t = \frac{1 + \cos 2t}{2} \qquad \text{(ii)}\ \sin^2 t = \frac{1 - \cos 2t}{2}$$

For example, if we replace t with $3x$ in identity (i), we get

$$\cos^2 3x = \frac{1 + \cos 6x}{2}.$$

Similarly, if we replace t with $5x$ in identity (ii), we obtain

$$\sin^2 5x = \frac{1 - \cos 10x}{2}.$$

By replacing t with $\theta/2$ in the square identities, we obtain the equivalent forms

$$\text{(i)}\ \cos^2 \frac{\theta}{2} = \frac{1 + \cos\theta}{2} \quad \text{and} \quad \text{(ii)}\ \sin^2 \frac{\theta}{2} = \frac{1 - \cos\theta}{2}.$$

Next we take the square roots of both sides of the latter identities to get the following results:

Half-Angle Identities

$$\text{(i)}\ \cos\frac{\theta}{2} = \pm\sqrt{\frac{1 + \cos\theta}{2}} \quad \text{and} \quad \text{(ii)}\ \sin\frac{\theta}{2} = \pm\sqrt{\frac{1 - \cos\theta}{2}}.$$

The sign depends on the quadrant in which the terminal side of the angle (in standard position) with measure $\theta/2$ lies.

These two identities can be used to derive the identity

$$\text{(iii)} \quad \tan\frac{\theta}{2} = \frac{1 - \cos\theta}{\sin\theta}$$

EXAMPLE 5

Using a Half-Angle Identity to Evaluate

Find the exact value of $\cos(\pi/12)$ by using the half-angle cosine identity. Compare this value to the calculator value, rounded to four decimal places.

Solution

Using half-angle identity (i) for the cosine and the fact that $\cos(\pi/12)$ is positive, we get

$$\cos\frac{\pi}{12} = \cos\left[\frac{1}{2}\left(\frac{\pi}{6}\right)\right] = \sqrt{\frac{1 + \cos(\pi/6)}{2}} = \sqrt{\frac{1 + (\sqrt{3}/2)}{2}}$$

$$= \sqrt{\frac{2 + \sqrt{3}}{4}} = \frac{\sqrt{2 + \sqrt{3}}}{2} \text{ (approx. 0.9659).}$$

The calculator value of $\cos(\pi/12)$ also equals 0.9659, to four decimal places.

EXAMPLE 6

Evaluating Functions Involving Half-Angle Identities

Suppose that θ is a positive obtuse angle in standard position, measured in degrees, and $\sin\theta = 3/5$. Find the exact value of each expression.

(a) $\sin\frac{\theta}{2}$ (b) $\cos\frac{\theta}{2}$ (c) $\tan\frac{\theta}{2}$

Solution

Since θ is obtuse, it is in quadrant II, where $\cos\theta$ is negative. So

$$\cos\theta = -\sqrt{1 - \sin^2\theta} = -\sqrt{1 - \left(\frac{3}{5}\right)^2} = -\sqrt{\frac{16}{25}} = -\frac{4}{5}$$

Also, because $90° < \theta < 180°$, we have $45° < \theta/2 < 90°$, so $\theta/2$ is in quadrant I. Thus $\sin(\theta/2)$ and $\cos(\theta/2)$ are both positive, and we have:

$$\text{(a)} \quad \sin\frac{\theta}{2} = \sqrt{\frac{1 - \cos\theta}{2}} = \sqrt{\frac{1 - (-\frac{4}{5})}{2}} = \sqrt{\frac{9}{10}} = \frac{3}{\sqrt{10}} = \frac{3\sqrt{10}}{10}$$

$$\text{(b)} \quad \cos\frac{\theta}{2} = \sqrt{\frac{1 + \cos\theta}{2}} = \sqrt{\frac{1 + (-\frac{4}{5})}{2}} = \sqrt{\frac{1}{10}} = \frac{1}{\sqrt{10}} = \frac{\sqrt{10}}{10}$$

$$\text{(c)} \quad \tan\frac{\theta}{2} = \frac{\sin(\theta/2)}{\cos(\theta/2)} = \frac{(3\sqrt{10})/10}{\sqrt{10}/10} = 3$$

Solving Applied Problems

Some applied problems can be modeled by trigonometric functions of multiple angles as the next example shows.

EXAMPLE 7

Modeling a Balloon Sighting

Suppose that an observer at point P, 111 meters away from point Q, directly beneath a hot air balloon at T, sights the balloon at an angle of elevation θ (Figure 2). At the same time, a second observer at point R, which is 74 meters closer to point Q, sights the same balloon at an angle of elevation 2θ. If the points P, R, and Q are in the same plane, find the height h of the balloon. Round off the answer to one decimal place.

Solution Using the right triangles QPT and QRT in Figure 2, we have

Figure 2

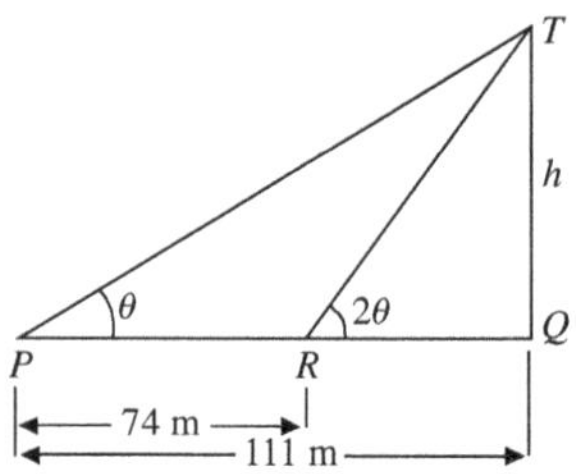

$$\tan 2\theta = \frac{h}{37} \quad \text{and} \quad \tan\theta = \frac{h}{111}.$$

Using $\tan 2\theta = \dfrac{2\tan\theta}{1-\tan^2\theta}$, we have $\dfrac{h}{37} = \dfrac{\left(\frac{2h}{111}\right)}{1-\frac{h^2}{(111)^2}}$ Substitute for tangents

so that $1 - \dfrac{h^2}{(111)^2} = \dfrac{\left(\frac{2h}{111}\right)}{\left(\frac{h}{37}\right)} = \dfrac{2}{3}\dfrac{h^2}{(111)^2} = 1 - \dfrac{2}{3} = \dfrac{1}{3}$

$$3h^2 = (111)^2 \quad h^2 = \frac{\overset{37}{(111)}(111)}{3} \quad h = \sqrt{(37)(111)} = 37\sqrt{3}$$

Therefore, the height h, rounded off to one decimal place, is 64.1 meters.

EXAMPLE 8

Maximizing the Distance of a Football

A place kicker determines that if he kicks the ball at a specified angle θ he can achieve the maximum distance. If the ball is kicked at an initial velocity V_0, then the distance R (in feet) (Figure 3) is given by the model

$$R = \frac{V^2}{32}\sin 2\theta, \quad 0° \le \theta \le 45°.$$

(a) Find the distance R if $\theta = 37°$ and $V = 70$ feet per second. Round off the answer to two decimal places.

(b) If the football is kicked with an initial velocity of $V = 77$ feet per second, determine the value of θ that maximizes R and find the maximum value of R. Round off the answers to two decimal places.

Figure 3

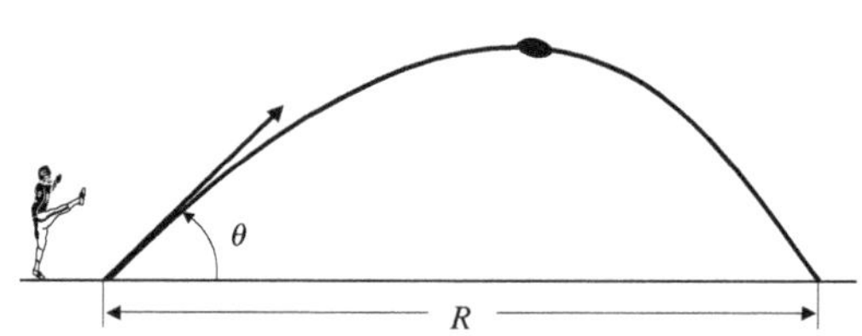

Solution (a) Replacing V by 70 and θ by 37° in the above formula we have

$$R = \frac{(70)^2}{32}\sin 74°$$
$$= 147.19.$$

So the distance R when the ball is kicked at 37° with an initial velocity of 70 feet per second is approximately 147.19 feet or about 49.06 yards.

(b) Replacing $V = 77$ in the above equation and rounding off to two decimal places we have

$$R = 185.28 \sin 2\theta.$$

Since the amplitude of $y = \sin 2\theta$ is 1, it follows that the maximum value of

$\sin 2\theta$ is 1.

Consequently, the maximum value of R is attained when $\sin 2\theta = 1$, that is, when

$$2\theta = 90°$$

or

$$\theta = 45°.$$

At $\theta = 45°$, $R = 185.28 \cdot \sin 90° = 185.28$.

So the maximum value of R, about 185.28 feet or 61.76 yards, is attained if the ball is kicked at 45°.

PROBLEM SET 2.3

Mastering the Concepts

1. Evaluate $\cos 2t$ and $2 \cos t$ for $t = \pi/4$. What conclusion can be made about $\cos 2t$ and $2 \cos t$?
2. Evaluate $\tan 2t$ and $2 \tan t$ for $t = \pi/4$. What conclusion can be made about $\tan 2t$ and $2 \tan t$?

In problems 3–10, use the given information to find the exact value of each expression.

(a) $\sin 2t$
(b) $\cos 2t$
(c) $\tan 2t$

3. $\sin t = \frac{4}{5}$; t in quadrant I
4. $\cos t = -\frac{12}{13}$; t in quadrant III
5. $\cos t = -\frac{7}{25}$; t in quadrant III
6. $\sin t = -\frac{5}{13}$; t in quadrant IV
7. $\tan t = -\frac{5}{12}$; t in quadrant II
8. $\sec t = -\frac{5}{2}$; t in quadrant III
9. $\cot t = -\frac{7}{25}$; t in quadrant IV
10. $\tan t = \frac{18}{5}$; t in quadrant I

In problems 11–18, use the double-angle identities to rewrite each expression in terms of a single trigonometric function.

11. $2 \sin 3t \cos 3t$
12. $1 - 2 \sin^2 3\pi x$
13. $2 \cos^2 8t - 1$
14. $4 \sin^2 7\theta \cos^2 7\theta$
15. $\cos^2 5t - \sin^2 5t$
16. $2 - 4 \sin^2 \frac{t}{2}$
17. $\dfrac{2 \tan 4\theta}{1 - \tan^2 4\theta}$
18. $\dfrac{\tan(t/2)}{\frac{1}{2} - \frac{1}{2}\tan^2(t/2)}$

In problems 19 and 20, use an identity and right triangle trigonometry to find the exact value of each expression. Compare each value to the calculator value, rounded to four decimal places.

19. (a) $\sin\left(2\sin^{-1}\frac{1}{2}\right)$

(b) $\sin\left(2\cos^{-1}\frac{\sqrt{3}}{2}\right)$

(c) $\tan\left(2\tan^{-1}\frac{4}{3}\right)$

20. (a) $\cos(2\sin^{-1} 1)$

(b) $\cos\left(2\sin^{-1}\frac{\sqrt{2}}{2}\right)$

(c) $\tan\left[2\cos^{-1}\left(-\frac{24}{25}\right)\right]$

In problems 21–26, use the half-angle identities to find the exact value of each expression. Compare this value to the calculator value, rounded to four decimal places.

21. (a) $\cos 22.5°$ (b) $\sin\frac{5\pi}{12}$

22. (a) $\sin 75°$ (b) $\cos\frac{\pi}{8}$

23. (a) $\cos\frac{5\pi}{12}$ (b) $\tan\frac{3\pi}{8}$

24. (a) $\sin\left(-\frac{7\pi}{12}\right)$ (b) $\tan\frac{5\pi}{8}$

25. (a) $\cos 112.5°$ (b) $\sin 67.5°$

26. (a) $\tan\frac{7\pi}{12}$ (b) $\cos 67.5°$

In problems 27–34, use the given information to find the exact values of each expression if $0 < t < 2\pi$.

(a) $\sin\frac{t}{2}$

(b) $\cos\frac{t}{2}$

(c) $\tan\frac{t}{2}$

27. $\sin t = \frac{24}{25}$; t in quadrant I

28. $\cos t = \frac{7}{25}$; t in quadrant IV

29. $\cos t = -\frac{4}{5}$; t in quadrant II

30. $\sin t = -\frac{5}{13}$; t in quadrant III

31. $\cot t = \frac{8}{15}$; t in quadrant III

32. $\sec t = \frac{13}{5}$; t in quadrant I

33. $\tan t = -\frac{8}{15}$; t in quadrant IV

34. $\cot t = \frac{3}{4}$; t in quadrant III

In problems 35 and 36, use the half-angle identities to rewrite each expression as a single trigonometric function.

35. (a) $\sqrt{\frac{1+\cos 4t}{2}}$ (b) $\sqrt{\frac{1-\cos 6t}{2}}$

36. (a) $\sqrt{\frac{1-\cos 2\pi t}{2}}$ (b) $-\sqrt{\frac{1-\cos 2t}{1+\cos 2t}}$

In problems 37–48, verify that the equation is an identity.

37. (a) $\sin^2 t + \cos 2t = \cos^2 t$

(b) $\csc t \sec t = 2\csc 2t$

38. (a) $\frac{\sin 2t}{2\sin t} = \cos t$

(b) $\cos^2\theta(1 - \tan^2\theta) = \cos 2\theta$

39. (a) $(\sin 2t + \cos 2t)^2 - \sin 4t = 1$

(b) $(\sin t - \cos t)^2 + 2\sin 2t = 1 + \sin 2t$

40. (a) $\frac{1-\cos 2t}{\sin 2t} = \tan t$

(b) $\frac{\sin 4\theta}{1-\cos 4\theta} = \cot 2\theta$

41. (a) $\frac{\sin^2 2\theta}{\sin^2\theta} = 4\cos^2\theta$

(b) $\sin\theta\tan\frac{\theta}{2}\csc^2\frac{\theta}{2} = 2$

42. (a) $2\tan\frac{\theta}{2}\csc\theta = \sec^2\frac{\theta}{2}$

(b) $2\cos\frac{\theta}{2} = (1+\cos\theta)\sec\frac{\theta}{2}$

43. $\frac{\sin 3\theta}{\sin\theta} - \frac{\cos 3\theta}{\cos\theta} = 2$

44. $\frac{\sin 5t}{\sin t} - \frac{\cos 5t}{\cos t} = 4\cos 2t$

45. $\frac{\cos^3 t - \sin^3 t}{\cos t - \sin t} = \frac{2+\sin 2t}{2}$

46. $\frac{\sec^2 t}{2-\sec^2 t} = \sec 2t$

47. $\sin^2 t\cos^2 t = \frac{1}{8}(1-\cos 4t)$

48. $\sin^4\theta = \frac{3}{8} - \frac{1}{2}\cos 2\theta + \frac{\cos 4\theta}{8}$

Applying the Concepts

49. **Area:** Suppose that triangle PBC is an isosceles triangle where each of the equal sides $\overline{PB}$ and $\overline{PC}$ has a length of 4 centimeters and θ is the angle BPC, as shown in Figure 4.

Figure 4

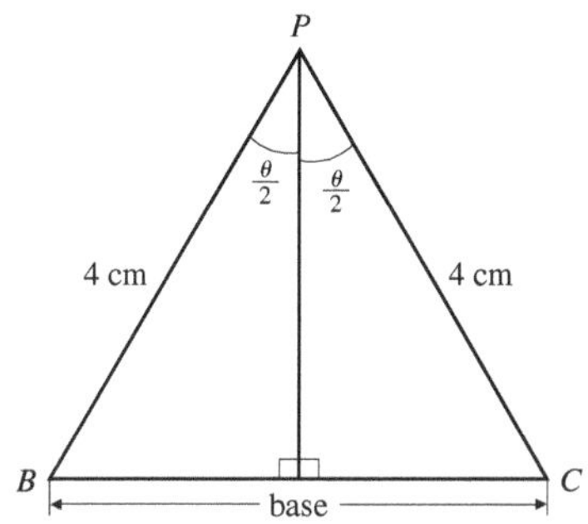

(a) Show that the area A of the triangle is given by $A = 8 \sin \theta$.

(b) Find the area A of the triangle when

$$\theta = 36°.$$

Round off the answer to two decimal places.

(c) Find the domain of the function in part (a).

(d) Find the maximum area and the corresponding value of θ.

50. (Refer to problem 49.)

(a) Graph the function in part (a) with the restrictions found in part (c). Interpret the graph.

(b) Find θ when the area is 4.

51. **Area:** Consider an isosceles trapezoid with the dimensions shown in Figure 5.

Figure 5

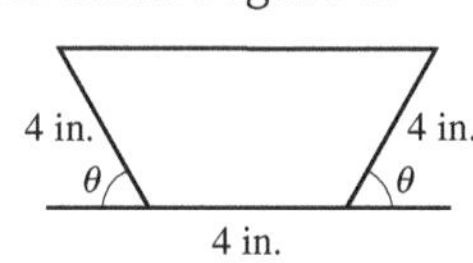

(a) Show that the area A of the trapezoid is given by

$$A = 64 \sin \frac{\theta}{2} \cos^3 \frac{\theta}{2}.$$

(b) Find the area of A of the trapezoid when

$$\theta = 36°.$$

Round off the answer to two decimal places.

(c) Find the domain of the function in part (a).

52. (Refer to problem 51.)

(a) Use a grapher to graph the function in part (a) with the restrictions found in part (c). Interpret the graph.

(b) Use the graph in part (a) to find the maximum area and the value of θ when it occurs. Round off to two decimal places.

53. **Household Electrical Circuit:** Suppose that the voltage V (in volts) and the amperage I (in amperes) in a household electrical circuit at t seconds are given by the models

$$V = 280 \sin(120\pi t) \quad \text{and} \quad I = 7 \sin(120\pi t).$$

(a) Write the wattage $W = VI$ in the form

$$W = a \cos(k\pi t) + d$$

where a, k, and d are constants.

(b) Find the period of each of the three functions V, I, and W, and compare these periods.

(c) What is the maximum wattage?

54. **Electrical Household Appliance:** If a household appliance is plugged into an outlet, the wattage W used is given by the model

$$W = I^2R$$

where $I = 12 \sin(120\pi t)$ is the amperage and R is a constant that measures the resistance (in ohms) and t is in seconds.

(a) If $R = 12$ ohms, write an expression for the wattage in the form

$$W = a \cos(240\pi t) + d$$

where a and d are constants.

(b) Find the period of the amperage I and wattage W of the appliance and compare these periods.

(c) What is the maximum wattage?

Developing and Extending the Concepts

55. Some techniques in calculus make use of the substitution $z = \tan(\theta/2)$. Use this substitution to show that:

(a) $\cos \theta = \dfrac{1 - z^2}{1 + z^2}$

(b) $\sin \theta = \dfrac{2z}{1 + z^2}$

56. Use the results from problem 55 to rewrite each expression in terms of z and simplify.

(a) $\dfrac{3}{2 \sin \theta + 3 \cos \theta}$

(b) $\dfrac{1}{3 + 2 \sin \theta}$

57. Suppose that P is a point on a unit circle, and

$$P(t) = \left(\frac{3}{5}, \frac{-4}{5}\right).$$

Find:

(a) $P(2t)$
(b) $P(t/2)$

58. (a) Suppose that $u + v = \pi/2$. Verify that $\sin(u - v) = -\cos 2u$.

(b) Suppose that $u + v = \pi$. Verify that $\sin(u - v) = -\sin 2u$.

59. (a) Express $\sec 2t$ in terms of $\sec t$.

(b) Find $\sec 2t$ if $\tan t = 3$ and $\sec t < 0$.

60. Suppose that α and β are acute angles in a right triangle. Show that

$$\sin 2\alpha = \sin 2\beta.$$

61. G (a) Verify that $\cos 3t = 4\cos^3 t - 3\cos t$.

(b) Use a grapher to demonstrate the validity of the identity in part (a) by comparing the graphs of

$$y_1 = \cos 3t \text{ and } y_2 = 4\cos^3 t - 3\cos t.$$

62. G (a) Use a grapher to graph the equation

$$y = \frac{6\sin^2 2x}{\sin 4x}.$$

(b) Find an equation of the form $y = a \tan kx$ that has the same graph as the equation in part (a).

Interlude: Practice Problems 2.3, Part I

1. Verify that $\cos 3\theta = 4\cos^3\theta - 3\cos\theta$ is an identity by transforming the left side of the equation into the right side.

2. Verify that $\cot \frac{\alpha}{2} = \frac{1 + \cos \alpha}{\sin \alpha}$ is an identity by transforming the left side of the equation into the right side.

3. Write $\sin 4\theta$ in terms of $\sin \theta$ and constants.

4. Verify that $\sin^4 t = \frac{1}{8}(3 - 4\cos 2t + \cos 4t)$ is an identity by transforming the left side of the equation into the right side.

Interlude: Practice Problems 2.3, Part II

1. Suppose that $\pi/2 < t < \pi$ and that $\cos t = -4/7$. Calculate each of the following (fill in the blanks).

____

____

____

(a) $\sin 2t =$ ____________

(b) $\cos 2t =$ ____________

(c) $\cot \frac{t}{2} =$ ____________

(d) $\sec \frac{t}{2} =$ ____________

2. Verify that $\cos^2 \frac{x}{2} = \frac{\sin x + \tan x}{2 \tan x}$ is an identity by transforming one side of the equation into the other side.

3. Verify that $\tan \phi = \csc 2\phi - \cot 2\phi$ is an identity by transforming one side of the equation into the other side.

Objectives

1. Use Reference Angles
2. Solve Trigonometric Equations
3. Solve Applied Problems

2.4 Trigonometric Equations

We know that trigonometric identities are equations that are *true for all values* for which both sides of the equation are defined. By contrast, *conditional trigonometric equations* are equations that are *true for certain values* but false for others. Examples of such equations are

$$\sin t = 1, \quad 2\cos\theta - 1 = 0, \quad \sin 2t = 2\cos t \quad \text{and} \quad \tan^2\theta + \sec\theta = 1$$

In this section, we learn how to solve conditional trigonometric equations.

Let us begin by considering the equation

$$\sin\theta = \frac{1}{2}, \text{ where } \theta \text{ represents degree measure.}$$

Suppose we want to solve this equation for θ under each of the following three conditions:

(i) $0° < \theta < 90°$ (ii) $0° < \theta < 360°$ (iii) No restriction on θ

We proceed as follows:

(i) To solve

$$\sin\theta = \frac{1}{2}, \quad 0° < \theta < 90°$$

means to find an acute angle, say θ_R, such that

$$\sin\theta_R = \frac{1}{2}.$$

Figure 1

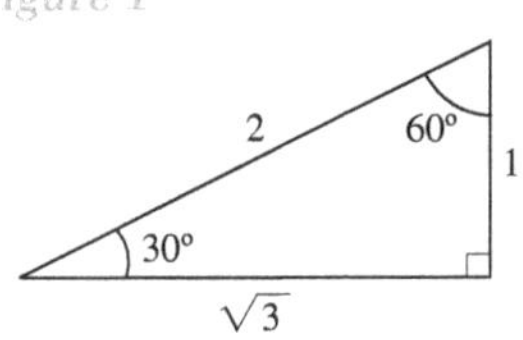

From right triangle trigonometry, we use a 30°–60° triangle (Figure 1) to observe that

$$\sin 30° = \frac{1}{2}.$$

So $\theta_R = 30°$ and the solution of the equation is 30° for the given restriction.

(ii) To solve

$$\sin\theta = \frac{1}{2}, \quad 0° < \theta < 360°$$

Figure 2

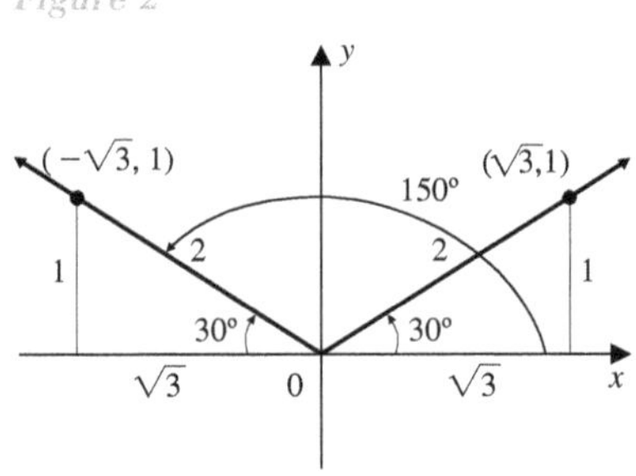

we first observe that, when placed in standard position, θ is in either quadrants I or II because the sine value is positive. To determine the exact locations of θ, we use the result in part (i), $\theta_R = 30°$, to construct reference triangles in quadrants I and II as shown in Figure 2.

Upon examining this figure, we see that both

$$\sin 30° = \frac{1}{2} \quad \text{and} \quad \sin 150° = \frac{1}{2}.$$

So, under the given restrictions, the solution of the equation includes both 30° and 150°.

(iii) Next we solve

$$\sin\theta = \frac{1}{2} \quad (\theta \text{ is unrestricted}).$$

From part (ii) we notice that there are solutions in both quadrants I and II, 30° and 150°. However, since θ is not restricted, any other angles coterminal with 30° and 150° are also solutions. Thus we conclude that the solution of this equation includes all angles of the form

$$\theta = 30^\circ + k \cdot 360^\circ$$

$$\text{or} \quad \theta = 150^\circ + k \cdot 360^\circ, \quad \text{where } k \text{ is an integer.}$$

It is important to notice the key role that the angle $\theta_R = 30^\circ$ had in solving the equation in all three situations. In this context, θ_R is called a *reference angle.*

Using Reference Angles

Before attempting to solve other trigonometric equations, it is important to understand the general concept of reference angles.

Reference angles are defined as follows:

Definition
Reference Angle

If θ is an angle in standard position whose terminal side does not lie on either coordinate axis, then the **reference angle** θ_R for θ is defined to be the *positive acute angle* formed by the terminal side of θ and the x axis.

For instance,
the reference angle for $\theta = 115^\circ$ is

$$\theta_R = 180^\circ - 115^\circ = 65^\circ \text{ (Figure 3a);}$$

the reference angle for $\theta = 3.78$ is given by

$$\theta_R = 3.78 - \pi \quad \text{or approximately } 0.64 \text{ (Figure 3b).}$$

Figure 3

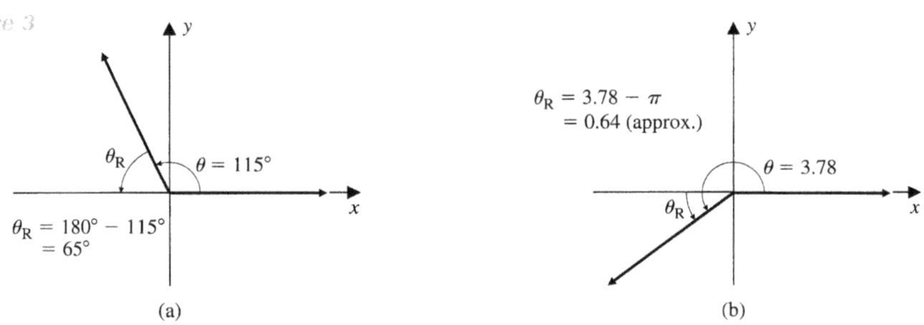

It is possible to express trigonometric function values of an angle in terms of values of its reference angle because of symmetry.
For instance, if $\theta = 115^\circ$, its reference angle is $\theta_R = 65^\circ$.
As shown in Figure 4, we see that

$$\sin 115^\circ = \frac{y}{r} = \sin 65^\circ \qquad \text{and} \qquad \cos 115^\circ = \frac{-x}{r} = -\frac{x}{r} = -\cos 65^\circ$$

Figure 4

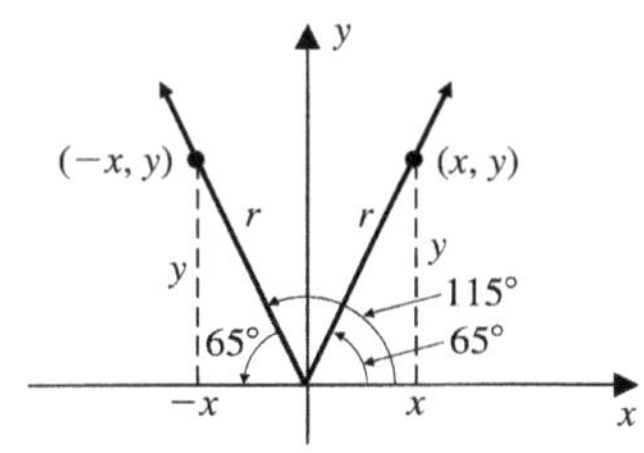

This observation leads us to the following general result

Evaluating Trigonometric Functions of Reference Angles

> If $T(\theta)$ represents any of the six trigonometric functions of θ, and if θ_R is the reference angle of θ, then either
>
> $$T(\theta) = T(\theta_R) \quad \text{or} \quad T(\theta) = -T(\theta_R)$$
>
> depending on the function and the quadrant in which the terminal side of θ lies, when θ is in standard position.

EXAMPLE 1

Relating Trigonometric Function Evaluations to Reference Angles

For each expression, draw the angle in standard position, and then determine and display the reference angle. Rewrite the given expression in terms of the reference angle. Use a calculator to confirm the result (round off to four decimal places).

(a) $\cos 163°$ (b) $\tan 5.1416$

Solution

Figure 5

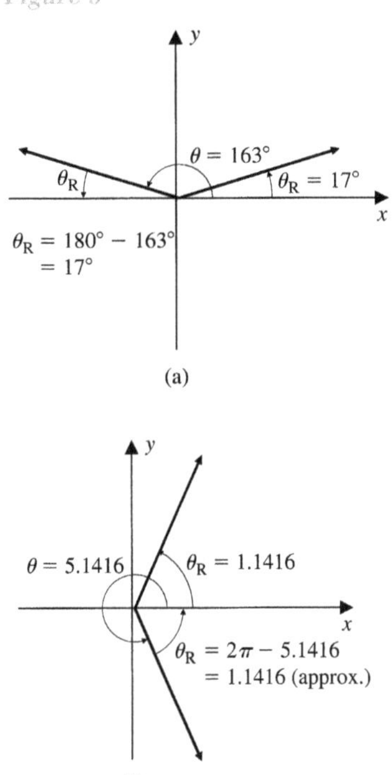

(a) Figure 5a, shows that the reference angle for

$$\theta = 163° \quad \text{is} \quad \theta_R = 17°.$$

Since the value of the cosine is negative in quadrant II, it follows that

$$\cos 163° = -\cos 17°.$$

By using a calculator, we find that both $\cos 163°$ and $-\cos 17°$ are approximately equal to -0.9563.

(b) Figure 5b shows that the reference angle for

$$\theta = 5.1416 \quad \text{is approximately} \quad \theta_R = 1.1416.$$

Since the tangent is negative in quadrant IV, it follows that

$$\tan 5.1416 = -\tan 1.1416.$$

Both $\tan 5.1416$ and $-\tan 1.1416$ are approximately equal to -2.1850. ∎

Solving Trigonometric Equations

Analytic techniques for solving trigonometric equations such as

$$\sin t = 0.8931$$
$$2\cos^2 t - 7\cos t = -3$$
$$\tan^4(2\theta) = 9$$
$$3\sin\theta = 2\cos^2\theta$$
$$\sin(2\theta) = \sin\theta$$

and

$$\tan\theta + 1 = \sec(3\theta)$$

involve the use of properties of the trigonometric functions such as periodicity, reference angles, signs of the values, and identities.

Commonly, trigonometric equations have restrictions that include only angles that fall within one positive rotation between 0° and 360° if degrees are used or between 0 and 2π if radians or real numbers are used.

EXAMPLE 2

Using a Reference Angle to Solve a Trigonometric Equation

Solve the equation $\tan t = -1.2137$, for $0 \le t < 2\pi$. Round off the answers to four decimal places.

Solution

Figure 6

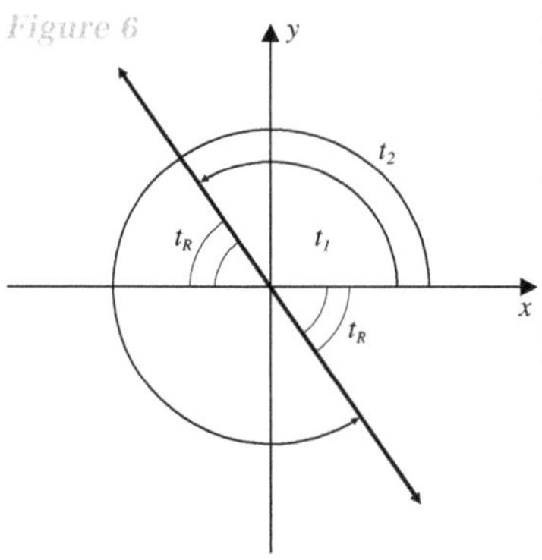

Since the value of $\tan t$ is negative, it follows that there are two possible solutions for $0 \le t < 2\pi$, one in quadrant II, labeled t_1 and one in quadrant IV labeled t_2 as shown in Figure 6. The reference angle t_R also displayed in Figure 6 satisfies $\tan t_R = 1.2137$.

A calculator in radian mode gives us

$$t_R = \tan^{-1}(1.2137) = 0.8816 \text{ (approx.)}.$$

It follows that the two solutions are given by

$$t_1 = \pi - t_R = 2.2600 \text{ (approx.)}$$
$$\text{and } t_2 = 2\pi - t_R = 5.4016 \text{ (approx.)}.$$

EXAMPLE 3

Using a Known Solution to Solve a Trigonometric Equation

Use the solution of $\sin\theta = 1/2$ to solve

$$\sin(\theta - 60°) = 1/2, \quad \text{for } 0° \le \theta < 360°$$

Solution

Earlier (see page 188) we determined that if θ is not restricted the solution of $\sin\theta = 1/2$ includes all values of the form

$$\theta = 30° + k \cdot 360° \quad \text{or} \quad \theta = 150° + k \cdot 360°$$

where k is an integer.

This means that the solutions of

$$\sin(\theta - 60°) = 1/2$$

must satisfy

$$\theta - 60° = \ldots -690°, -330°, 30°, 390°, \ldots$$
$$\text{or} \quad \theta - 60° = \ldots -210°, 150°, 510°, \ldots$$

Solving the latter equations for θ by adding 60 to each side of each equation, we get

$$\theta = \ldots -630°, -270°, 90°, 450°, \ldots$$
$$\text{or} \quad \theta = \ldots -150°, 210°, 570°, \ldots$$

However, because of the restriction $0° \leq \theta < 360°$, we conclude that the solution includes $\theta = 90°$ and $\theta = 210°$ only.

EXAMPLE 4 **Solving a Trigonometric Equation by Factoring**

Solve the equation

$$\sin\theta\tan\theta - \sin\theta = 0, \quad \text{for} \quad 0° \leq \theta < 360°.$$

Solution First we factor the given equation

$$\sin\theta\tan\theta - \sin\theta = 0$$

to get

[1] $$\sin\theta(\tan\theta - 1) = 0.$$

Setting each factor equal to 0, we have two equations

$$\sin\theta = 0 \quad \text{and} \quad \tan\theta - 1 = 0.$$

For

$$\sin\theta = 0,$$

we get the solutions

$$\theta = 0° \quad \text{or} \quad \theta = 180°.$$

For $\tan\theta - 1 = 0$, we solve the equivalent equation

$$\tan\theta = 1$$

as follows:

Figure 7

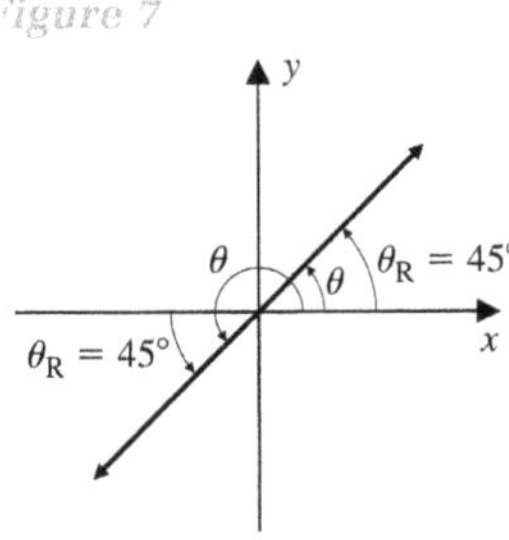

Since the value of the tangent is a positive number, it follows that the solutions of

$$\tan\theta = 1$$

are in quadrants I and III. Also, the reference angle for $\tan\theta_R = 1$ is $\theta_R = 45°$ (Figure 7).

So the solution of $\tan\theta = 1$ in quadrant I is $\theta = 45°$, and in quadrant III is $\theta = 180° + 45° = 225°$.

Combining the results, we conclude that the solution of the original equation includes 0°, 45°, 180°, and 225°.

Note that we do not divide both sides of an equation by an expression containing a variable. For instance, in Example 4, if we had divided both sides of the above equation denoted by [1] by $\sin\theta$, we would have obtained the equation $\tan\theta - 1 = 0$ and two of the solutions, 0° and 180°, would have been lost.

Trigonometric equations involving multiple angles require special treatment, as illustrated in the next example.

EXAMPLE 5

Solving a Trigonometric Equation Involving a Multiple Angle

Solve the equation

$$\cos 3\theta = \sqrt{3}/2, \text{ for } 0° \le \theta < 360°.$$

Solution To solve the equation, we let $\alpha = 3\theta$ in the given equation to get

$$\cos 3\theta = \cos\alpha = \frac{\sqrt{3}}{2}.$$

Since

$$0° \le \theta < 360°, \quad \text{then} \quad 0° \le 3\theta < 1080°;$$

so that

$$\cos\alpha = \frac{\sqrt{3}}{2} \quad \text{for } 0° \le \alpha < 1080°.$$

The reference angle α_R for the latter equation is given by

$$\alpha_R = \cos^{-1}\frac{\sqrt{3}}{2} = 30°.$$

Since the cosine is positive in quadrants I and IV, it follows that

$$\alpha = 30° \quad \text{and} \quad \alpha = 330°$$

Figure 8

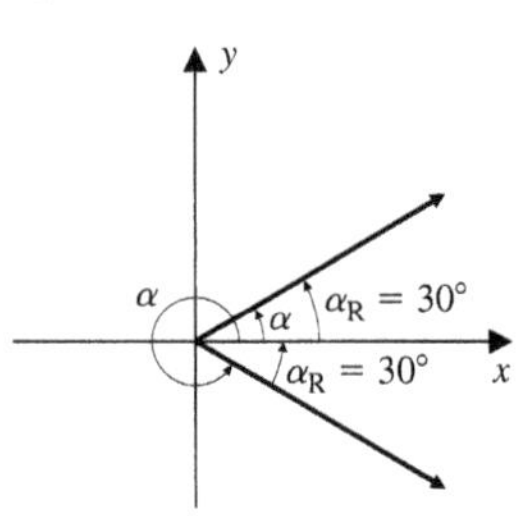

are two solutions (Figure 8).

Next we use periodicity by adding 360° to these two angle measures to get two more solutions:

$$\alpha = 30° + 360° = 390° \quad \text{and} \quad \alpha = 330° + 360° = 690°$$

both of which satisfy the condition $0° \le \alpha < 1080°$.

Finally, adding 360° again, we obtain two additional solutions:

We stop at 750° and 1050° because adding 360° again would result in angle measures exceeding the 1080° restriction.

$$\alpha = 390° + 360° = 750° \quad \text{and} \quad \alpha = 690° + 360° = 1050°.$$

Thus for $0° \le \alpha < 1080°$, we have

$$\alpha = 3\theta = 30°, 330°, 390°, 690°, 750°, \text{ or } 1050°.$$

The values of θ, the original unknown, are found by dividing each value of $\alpha = 3\theta$ by 3 to obtain

$$\theta = 10°, 110°, 130°, 230°, 250°, \text{ or } 350°,$$

which are the solutions of the original equation.

The next example employs an identity to solve a trigonometric equation.

EXAMPLE 6

Solving an Equation by Using an Identity

Solve the equation $\cos 2t = \cos t$ for t in the interval $[0, 2\pi)$.

Solution We begin by using the double-angle identity

$$\cos 2t = 2\cos^2 t - 1$$

to rewrite the given equation

$$\cos 2t = \cos t \quad \text{as} \quad 2\cos^2 t - 1 = \cos t.$$

The latter equation is a quadratic equation in $\cos t$, which we can solve as follows:

$$2\cos^2 t - 1 = \cos t$$

$$2\cos^2 t - \cos t - 1 = 0 \qquad \text{Subtract } \cos t \text{ from each side}$$

$$(2\cos t + 1)(\cos t - 1) = 0 \qquad \text{Factor}$$

$2\cos t + 1 = 0$	or	$\cos t - 1 = 0$
$2\cos t = -1$		
$\cos t = -\frac{1}{2}$		$\cos t = 1$
So $t = \frac{2\pi}{3}$ or $\frac{4\pi}{3}$		$t = 0$

Set each factor equal to zero, and solve the resulting equations with the restriction $0 \le t < 2\pi$

Therefore, the solutions are 0, $(2\pi)/3$ (approx. 2.0944) and $(4\pi)/3$ (approx. 4.1888).

Graphers enable us to solve equations containing trigonometric expressions that cannot be solved by the analytic methods that we have studied so far in algebra and trigonometry.

EXAMPLE 7

G Solving an Equation Graphically

(a) Solve the equation $\cos x = x$ graphically. Round off the result to four decimal places.

(b) Use the result from part (a) to solve the inequality $\cos x < x$.

Solution (a) The equation $\cos x = x$ is equivalent to $\cos x - x = 0$.
The viewing window in Figure 9 shows that the graph of $y = \cos x - x$ has one x intercept, so there is only one solution.

By using the ZERO -finding feature of a grapher, we find that the solution is approximately 0.7391.

Figure 9

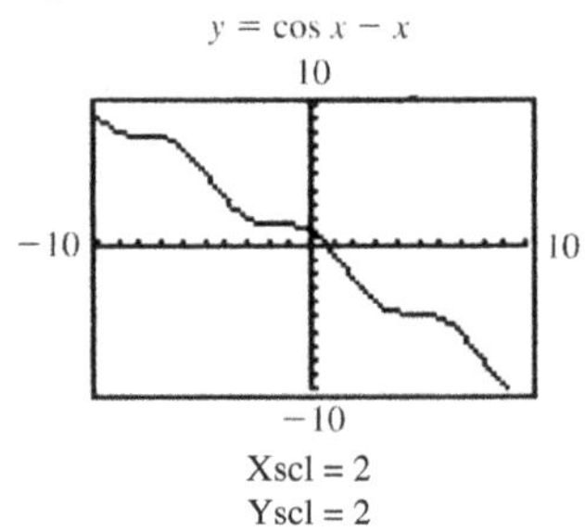

(b) The given inequality is equivalent to $\cos x - x < 0$.
The solution of this inequality incudes all values of x for which the graph of $y = \cos x - x$ is below the x axis. Since the x intercept is approximately 0.7391, it follows that the solution includes all numbers (approximately) in the interval $(0.7391, \infty)$, that is, $x > 0.7391$.

Solving Applied Problems

At times, mathematical models of real-world situations involve solving trigonometric equations.

EXAMPLE 8

Modeling the Design of a Window

Figure 10

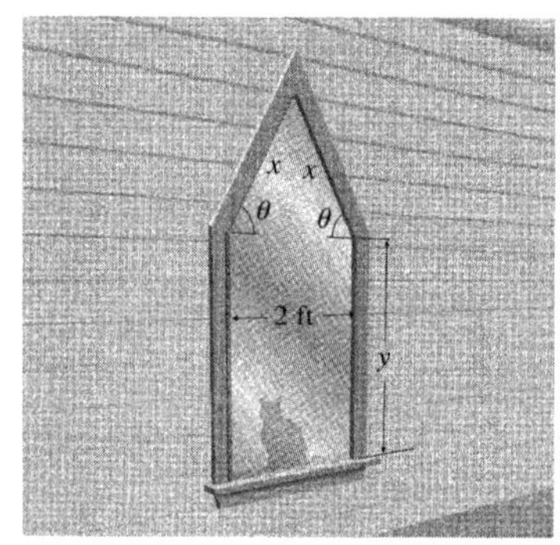

(Refer to Figure 10.) A custom window is to be designed in the shape of a rectangle 2 feet wide, topped by an isosceles triangle whose equal sides make an angle θ with the width. Let y (in feet) be the height of the rectangular part of the window; and let x (in feet) be the length of each of the equal sides of the isosceles triangle. Assume the perimeter of the entire window is to be 12 feet.

(a) Express x and y, in terms of θ. Then write A, the total area (in square feet) of the window as a function of θ.

(b) What are the restrictions on θ?

(c) Find θ if y is 3 feet. Then use the value of θ, along with the results in part (a), to find x and A. Round off the answers to two decimal places.

Solution

(a) Figure 11 shows the isosceles triangle that forms the top of the window. It follows from right triangle trigonometry that

Figure 11

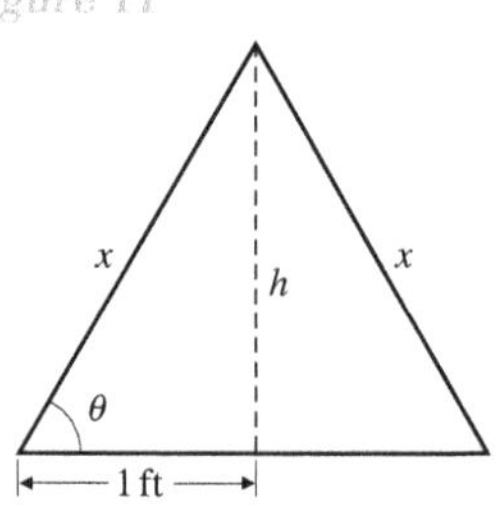

$$\sec \theta = \frac{x}{1} \quad \text{or} \quad x = \sec \theta.$$

Since the perimeter of the entire window is 12 feet, we have

$$2y + 2x + 2 = 12.$$

Substituting for x in the latter equation, we get

$$\begin{aligned} 2y + 2x + 2 &= 12 \\ 2y + 2\sec \theta + 2 &= 12 \\ 2y + 2\sec \theta &= 10. \end{aligned}$$

Solving for y, we have

$$y = 5 - \sec \theta.$$

Note that the height h of the triangle in Figure 11 is given by

$$\tan \theta = h/1 = h.$$

Thus the area A of the window can be rewritten as follows:

$$\begin{aligned} A &= (\text{area of rectangle}) + (\text{area of triangle}) \\ &= (\text{width})(\text{height of rectangle}) + \frac{1}{2}(\text{base})(\text{height of triangle}) \\ &= 2y + \frac{1}{2}(2)\,h && \text{Substitute} \\ &= 2y + h && \text{Simplify} \\ &= 2(5 - \sec\theta) + \tan\theta && \text{Substitute} \\ &= 10 - 2\sec\theta + \tan\theta && \text{Multiply} \end{aligned}$$

So we have

$$x = \sec\theta, \quad y = 5 - \sec\theta, \quad \text{and} \quad A = 10 - 2\sec\theta + \tan\theta.$$

(b) Since the sum of the angles of a triangle is 180°, it follows that

$$0^\circ < 2\theta < 180^\circ \quad \text{or} \quad 0^\circ < \theta < 90^\circ$$

that is, θ is an acute angle.

(c) If $y = 3$, then, after substituting this value into the equation $y = 5 - \sec\theta$ which was derived in part (a), we get

$$\begin{aligned} 3 &= 5 - \sec\theta \\ -2 &= -\sec\theta \\ 2 &= \sec\theta, \quad \text{where } \theta \text{ is an acute angle.} \end{aligned}$$

The solution of this equation is $\theta = 60^\circ$.

It follows that

$$x = \sec\theta = \sec 60^\circ = 2$$

and

$$\begin{aligned} A &= 10 - 2\sec\theta + \tan\theta \\ &= 10 - 2\sec 60^\circ + \tan 60^\circ \\ &= 10 - 2(2) + \sqrt{3} \\ &= 6 + \sqrt{3} = 7.73 \text{ (approx.).} \end{aligned}$$

So, if $y = 3$, then x, the length of the two sides of the triangular top is 2 feet, and A, the total surface area, is approximately 7.73 square feet.

In the Cartesian coordinate system, we graph trigonometric functions with real number domains (corresponding to radian measure). However, graphers also enable us to *display* the horizontal axis in terms of degrees rather than just real numbers. For instance, in Example 8, we established that the area A of the window is given by

$$A = 10 - 2\sec\theta + \tan\theta, \quad \text{where } 0^\circ < \theta < 90^\circ.$$

To find the angle θ when the area $A = 7.5$, we need to solve the equation

$$7.5 = 10 - 2\sec\theta + \tan\theta$$

Figure 12

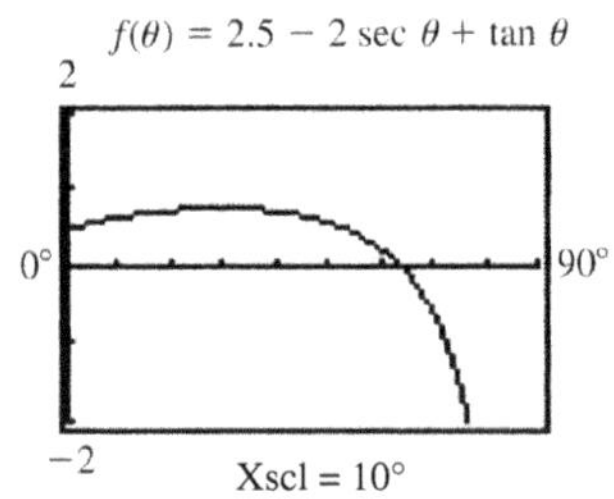

or, equivalently,

$$0 = 2.5 - 2 \sec \theta + \tan \theta.$$

Figure 12 shows a viewing window of the graph of

$$f(\theta) = 2.5 - 2 \sec \theta + \tan \theta \quad \text{for } 0° < \theta < 90°$$

(*set in degree mode*). By using the ZERO -finding feature, we find that

$$\theta \text{ is approximately } 63.83°.$$

Using the results in Example 8, we have

x (the length of the equal sides of the triangular top)

$$= \sec \theta = \sec 63.83° = 2.27 \text{ feet (approx.)}$$

and

y (the height of the rectangle) $= 5 - \sec \theta = 5 - \sec 63.83°$

$$= 2.73 \text{ feet (approx.)}.$$

PROBLEM SET 2.4

Mastering the Concepts

In problems 1 and 2, verify that the given values are solutions of the indicated equation.

1. (a) $2 \sin \theta - 1 = 0$; $30°$, $150°$

(b) $\tan\left(3t - \frac{\pi}{4}\right) = 1$; $-\frac{\pi}{6}, \frac{\pi}{6}$

2. (a) $2 \sin 2t - \sqrt{3} = 0$; $\frac{7\pi}{6}, \frac{4\pi}{3}$

(b) $2 \sin(\theta + 30°) = -1$; $-180°$, $300°$

3. Solve the equation $\sin \theta = -\sqrt{2}/2$ if:

(a) $0° \le \theta < 90°$

(b) $0° \le \theta < 180°$

(c) $0° \le \theta < 360°$

(d) θ has no restriction

4. Solve the equation $\cos t = -\sqrt{2}/2$ if:

(a) $0 \le t < \pi/2$

(b) $0 \le t < \pi$

(c) $0 \le t < 2\pi$

(d) t has no restriction

In problems 5 and 6, draw the angle in standard position. Then determine and display its reference angle. Rewrite the given expression in terms of the reference angle. Use a calculator to confirm the result (rounding off to four decimal places).

5. (a) $\sin 190°$ (b) $\tan(-113°)$

(c) $\sec 372°$ (d) $\cos(-1.9)$

(e) $\cot 7.2$

6. (a) $\cos 219°$ (b) $\cot 114°$

(c) $\cos(-295°)$ (d) $\sin \frac{11\pi}{8}$

(e) $\sec 5.9$

In problems 7–14, assume that $0 \le t < 2\pi$ and $0° \le \theta < 360°$. Find the exact solution of each equation.

7. (a) $\cos t = -\frac{1}{2}$

(b) $\sec t = -\sqrt{2}$

8. (a) $\sin t = \frac{\sqrt{2}}{2}$

(b) $3 \csc \theta = -2\sqrt{3}$

9. (a) $\tan \theta = \sqrt{3}$

(b) $3 \cot \theta = -\sqrt{3}$

10. (a) $\csc t = \sqrt{2}$

(b) $\cot t = -1$

11. (a) $2 \sin t - \sqrt{3} = 0$

(b) $2 \sin \theta + \sqrt{2} = 0$

12. (a) $\tan \theta - 1 = 0$

(b) $\cot t + \sqrt{3} = 0$

13. (a) $\sqrt{2} \cos t = -1$

(b) $\sqrt{2} \cos\left(t - \frac{\pi}{4}\right) = -1$

14. (a) $\tan \theta = -1$

(b) $\tan(\theta + 60°) = -1$

In problems 15–18, assume $0 \le t < 2\pi$. Solve each equation. Round off the answers to four decimal places.

15. (a) $\sin t = 0.8134$
(b) $\cot t = -6.6173$

16. (a) $\cos t = -0.4176$
(b) $\tan t = 0.6696$

17. (a) $\sec t = -\frac{6}{5}$
(b) $\tan t = 10$

18. (a) $\csc t = \pi$
(b) $\cot t = -10$

In problems 19–36, assume that $0 \le t < 2\pi$ and $0° \le \theta < 360°$. Solve each equation. Round off the answers to four decimal places, when applicable.

19. $\sin t \cos t - \sin t = 0$

20. $2 \sin \theta \cos \theta - \sqrt{2} \cos \theta = 0$

21. $2 \sin \theta \cos \theta + \sqrt{3} \cos \theta = 0$

22. $\sin t \cos t - \cos t = 0$

23. $\tan^2 t - \sqrt{3} \tan t = 0$

24. $2 \cos^2 \theta - \cos \theta = 0$

25. $\sec^2 \theta + \sec \theta = 0$

26. $\sqrt{3} \csc^2 \theta + 2 \csc \theta = 0$

27. $\tan^2 t - 3 = 0$

28. $\cot^2 t - 1 = 0$

29. $4 \sin^2 \theta - 3 = 0$

30. $\csc^2 \theta - 4 = 0$

31. $3 - 2 \cos^2 t = 3 - \cos t$

32. $1 - 2 \sin^2 t = 1 - \sin t$

33. (a) $2 \cos 3\theta = -\sqrt{2}$
(b) $2 \cos(3\theta - 30°) = -\sqrt{2}$

34. (a) $2 \sin 2t = 1$
(b) $2 \sin\left(2t - \frac{\pi}{4}\right) = 1$

35. (a) $\sec \frac{t}{2} = \sqrt{2}$
(b) $\sec\left(\frac{t}{2} + \frac{\pi}{5}\right) = \sqrt{2}$

36. (a) $\tan \frac{\theta}{2} + \sqrt{3} = 0$
(b) $\tan\left(\frac{\theta}{2} - 40°\right) + \sqrt{3} = 0$

In problems 37–52, assume that

$$0 \le t < 2\pi \quad \text{and} \quad 0° \le \theta < 360°.$$

Solve each equation by using identities. Round off the answers to four decimal places, when applicable.

37. $\sin^2 t = \cos^2 t$

38. $2 \cos^2 t - 1 = 2 \sin^2 t$

39. $\cot^2 \theta = \cot \theta$

40. $2 \tan \theta - \sec^2 \theta = 0$

41. $2 - \sin t = 2 \cos^2 t$

42. $\cos^2 t - \sin^2 t - \sin t - 1 = 0$

43. $\cos \theta - \sin \theta = 1$

44. $\tan \theta + \sec \theta = 1$

45. $\sin 2t = \sqrt{2} \cos t$

46. $2 - \cos^2 t - 4 \sin^2 \frac{t}{2} = 0$

47. $\sin \frac{t}{2} + \cos t = 1$

48. $\cos 2\theta = 2 \sin^2 \theta$

49. $\cos 2\theta - 3 \sin \theta = 2$

50. $\cos 2\theta + \sin^2 \theta = 1$

51. $\cos 2t + 2 \cos^2\left(\frac{t}{2}\right) - 2 = 0$

52. $\cos 2\theta - \cos \theta = 0$

Applying the Concepts

In problems 53 and 54, round off each answer to two decimal places.

53. **Physics:** In physics, Snell's law deals with the change in direction of a ray of light as it passes through a medium, as shown in Figure 13, where the dashed line is perpendicular to the surface of the medium:

Figure 13

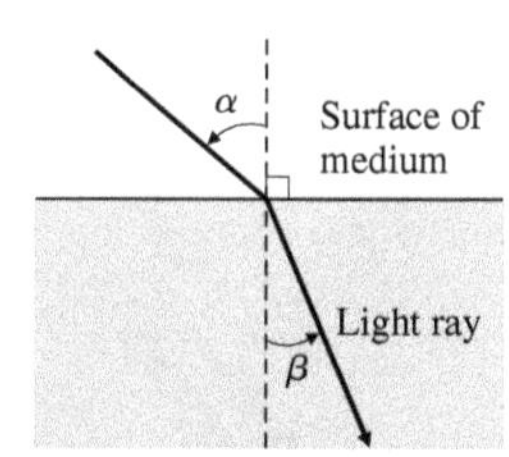

The angle α is called the **angle of incidence,** and the angle β is called the **angle of refraction. Snell's law** states that for a given medium,

$$\frac{\sin \alpha}{\sin \beta} = C$$

where C is a constant. The constant C is called the **index of refraction** of the medium.

(a) Find the degree measure of the angle of refraction of a light ray traveling through ice if

the angle of incidence is 45°,

and

the index of refraction is 1.31.

(b) Find the degree measure of the angle of incidence of a light ray striking a rock-salt crystal if

the angle of refraction of the ray is 35°,

and

index of refraction of rock-salt is 1.54.

54. **Animal Population:** Suppose that the number of rodents N in a certain population after t years is given by the mathematical model

$$N(t) = 3200 + 800 \sin \frac{3t}{2} \quad \text{where } 0 \le t < 5.$$

(a) How many rodents are initially in the population?
(b) Find how many years later it takes for the population to drop to 3000.
(c) Find the maximum number of rodents and when it occurs.

55. **Length of a Ladder:** A 5-foot fence stands 4 feet away from a high wall. A ladder on the ground leaning against the wall touches the top of the fence and makes an angle θ with the ground (Figure 14).

(a) Show that the length L of the ladder is given by

$$L = 4 \sec \theta + 5 \csc \theta.$$

(b) What are the restrictions on θ?

Figure 14

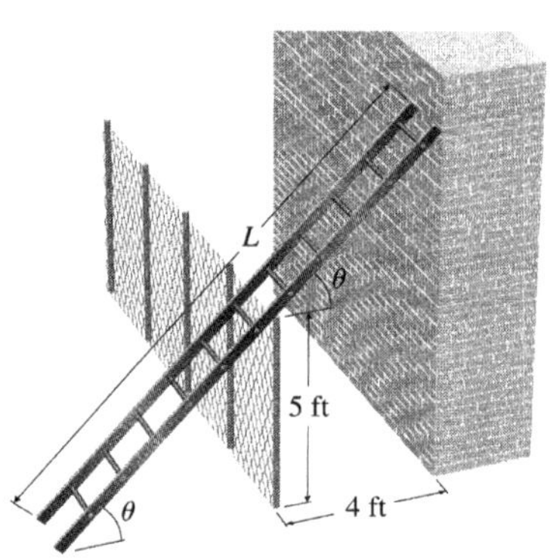

56. (Refer to problem 55.) Use a grapher to find the values in degrees of θ when the length of the ladder is 13 feet. Round off to two decimal places.

Developing and Extending the Concepts

57. (a) Let θ be an angle in quadrant II in standard position and let θ_R be its reference angle. Show that the value of any trigonometric function of θ is the same as the value for the function of θ_R, except possibly for a change of algebraic sign.
(b) Repeat part (a) for θ in quadrant III.
(c) Repeat part (a) for θ in quadrant IV.

58. When asked to find a number in the interval

$$\left[\frac{-\pi}{2}, \frac{\pi}{2}\right]$$

that solves the equation $\frac{\sin x}{x} = 1$ a student responds as follows:

First multiply both sides of the given equation by x to obtain

$$x \cdot \frac{\sin x}{x} = x \cdot 1$$

$$\sin x = x.$$

Since $\sin 0 = 0$, it follows that a solution of the given equation is $x = 0$.

Do you agree with this answer? Explain.

Can you find a solution for

$$\frac{\sin x}{x} = 1$$

in the interval

$$\left[-\frac{\pi}{2}, \frac{\pi}{2}\right]?$$

In problems 59 and 60, assume $0 \le t < 2\pi$. For what values of k does the equation have:

(a) No solution?
(b) One solution?
(c) Two solutions?
(d) More than two solutions?

59. $\cos t = k$

60. $\sec t = k$

61. Find the smallest positive angle θ in radians such that

$$\frac{3.43 \times 10^6}{1 - 0.21 \cos \theta} = 3.08 \times 10^6$$

Round off the answer to two decimal places.

62. Suppose that

$$x = a \cos \theta - b \sin \theta$$
$$y = a \sin \theta + b \cos \theta$$

(a) Find the product xy in terms of a, b, and θ.
(b) Find the smallest positive angle θ so that the coefficient of ab in the result from part (a) is zero.

In problems 63–68:

(a) Solve each equation graphically if $0 \le t < 2\pi$. Round off the answers to three decimal places.

(b) Use the results from part (a) to solve the corresponding inequality.

63. $15 \sin^2 t - 8 \sin t + 1 = 0$; $15 \sin^2 t - 8 \sin t + 1 > 0$

64. $\tan^2 t - 3 \tan t + 2 = 0$; $\tan^2 t - 3 \tan t + 2 < 0$

65. $t + \sin t = 0.32$; $t + \sin t \le 0.32$

66. $\sin t + \ln t = 2t$; $\sin t + \ln t > 2t$

67. $e^{-t} - \cos t = 0$; $e^{-t} - \cos t \ge 0$

68. $\tan t = t^2 + 1$; $\tan t \le t^2 + 1$

Interlude: Practice Problems 2.4, Part I

1. Consider the equation $\sin t + \cos t = 1$.

(a) Using algebraic methods, find all solutions to the equation in the interval $[0, 2\pi)$. Give both exact solutions and approximate solutions accurate to three decimal places.

(b) Verify your result above using graphical methods. Show the graph(s) you use.

2. Consider the equation $\sec 2t - \tan 2t = 1$.

(a) Using algebraic methods, find all solutions to the equation in the interval $[0, 2\pi)$. Give both exact solutions and approximate solutions accurate to three decimal places.

(b) Verify your result above using graphical methods. Show the graph(s) you use.

Interlude: Practice Problems 2.4, Part II

1. Consider the equation $3\cos^2\theta + \sin\theta = 3$.

(a) Using algebraic methods, find all solutions to the equation in the interval $[0, 2\pi)$. Give both exact solutions and approximate solutions accurate to three decimal places.

(b) Verify your result above using graphical methods. Show the graph(s) you use.

2. Consider the equation $2\sin^2\theta + \sin 2\theta + \cos 2\theta = 0$.

(a) Using algebraic methods, find all solutions to the equation in the interval $[0, 2\pi)$. Give both exact solutions and approximate solutions accurate to three decimal places.

(b) Verify your result above using graphical methods. Show the graph(s) you use.

Objectives

1. Use Product-to-Sum Identities
2. Use Sum-to-Product Identities
3. Solve Equations by Using the Sum-to-Product Identities
4. Solve Applied Problems

2.5 Other Identities

In calculus it is necessary at times to rewrite a product of sine and cosine functions as a sum or difference of such functions, or vice versa. This can be done by using the identities in this section.

Using Product-to-Sum Identities

The addition and subtraction identities can be used to derive the following identities that express the products of sines and cosines in terms of sums or differences.

Product-to-Sum Identities

$$\text{(i)}\quad \sin u \cos v = \frac{1}{2}[\sin(u + v) + \sin(u - v)]$$

$$\text{(ii)}\quad \cos u \sin v = \frac{1}{2}[\sin(u + v) - \sin(u - v)]$$

$$\text{(iii)}\quad \cos u \cos v = \frac{1}{2}[\cos(u - v) + \cos(u + v)]$$

$$\text{(iv)}\quad \sin u \sin v = \frac{1}{2}[\cos(u - v) - \cos(u + v)]$$

Proofs

(i) We use the addition and subtraction identities for the sine to prove the first identity as follows:

$$\begin{aligned}\frac{1}{2}[\sin(u + v) + \sin(u - v)] &= \frac{1}{2}\,[(\sin u \cos v + \cos u \sin v)\\ &\quad + (\sin u \cos v - \cos u \sin v)]\\ &= \frac{1}{2}[2 \sin u \cos v]\\ &= \sin u \cos v\end{aligned}$$

The proofs of parts (ii)–(iv) are left as exercises (problems 41–43).

EXAMPLE 1

Using a Product-to-Sum Identity to Evaluate

(a) Find the exact value of

$$\cos \frac{\pi}{12} \sin \frac{7\pi}{12}$$

by using a product-to-sum identity.

(b) Compare the result in part (a) to the calculator value of the product, rounded to four decimal places.

Solution (a) We proceed as follows:

$$\begin{aligned}
\cos\frac{\pi}{12}\sin\frac{7\pi}{12} &= \frac{1}{2}\left[\sin\left(\frac{\pi}{12}+\frac{7\pi}{12}\right)-\sin\left(\frac{\pi}{12}-\frac{7\pi}{12}\right)\right] && \text{Identity (ii)}\\
&= \frac{1}{2}\left[\sin\frac{8\pi}{12}-\sin\left(-\frac{6\pi}{12}\right)\right] && \text{Add/Subtract}\\
&= \frac{1}{2}\left[\sin\frac{2\pi}{3}-\sin\left(-\frac{\pi}{2}\right)\right] && \text{Reduce fractions}\\
&= \frac{1}{2}\left[\sin\frac{2\pi}{3}+\sin\frac{\pi}{2}\right] && \text{Sine is an odd function}\\
&= \frac{1}{2}\left[\frac{\sqrt{3}}{2}+1\right] && \text{Evaluate}\\
&= \frac{\sqrt{3}}{4}+\frac{1}{2} && \text{Multiply}
\end{aligned}$$

(b) The approximate value of $\frac{\sqrt{3}}{4}+\frac{1}{2}$

to four decimal places is 0.9330.

Using a calculator to evaluate the product

$$\cos\frac{\pi}{12}\sin\frac{7\pi}{12},$$

and rounding off to four decimal places, we again get 0.9330.

EXAMPLE 2 **Rewriting a Product as a Sum**

Express cos $3t$ cos $4t$ as a sum involving only cosines.

Solution Using identity (iii), with $u = 3t$ and $v = 4t$, we get

$$\begin{aligned}
\cos 3t\cos 4t &= \frac{1}{2}[\cos(3t-4t)+\cos(3t+4t)]\\
&= \frac{1}{2}[\cos(-t)+\cos 7t]\\
&= \frac{1}{2}\cos t+\frac{1}{2}\cos 7t. && \text{Cosine is an even function}
\end{aligned}$$

Using Sum-to-Product Identities

The sum-to-product identities enable us to rewrite the sums of sines and cosines as products.

Sum-to-Product Identities

$$\text{(i)}\ \sin \omega + \sin t = 2 \sin\left(\frac{\omega + t}{2}\right) \cos\left(\frac{\omega - t}{2}\right)$$

$$\text{(ii)}\ \sin \omega - \sin t = 2 \cos\left(\frac{\omega + t}{2}\right) \sin\left(\frac{\omega - t}{2}\right)$$

$$\text{(iii)}\ \cos \omega + \cos t = 2 \cos\left(\frac{\omega + t}{2}\right) \cos\left(\frac{\omega - t}{2}\right)$$

$$\text{(iv)}\ \cos \omega - \cos t = -2 \sin\left(\frac{\omega + t}{2}\right) \sin\left(\frac{\omega - t}{2}\right)$$

Proofs

(i) To prove the first identity, we let $\omega = u + v$ and $t = u - v$. Then

$$\omega + t = (u + v) + (u - v) = 2u.$$

So $$u = \frac{\omega + t}{2}.$$

Also,

$$\omega - t = (u + v) - (u - v) = 2v.$$

So $$v = \frac{\omega - t}{2}.$$

By substituting $(\omega + t)/2$ for u and $(\omega - t)/2$ for v in product-to-sum identity (i), we have

$$\sin\left(\frac{\omega + t}{2}\right) \cos\left(\frac{\omega - t}{2}\right) = \frac{1}{2}(\sin \omega + \sin t)$$

or

$$\sin \omega + \sin t = 2 \sin\left(\frac{\omega + t}{2}\right) \cos\left(\frac{\omega - t}{2}\right).$$

The proofs of parts (ii)–(iv) are left as exercises (problems 44–46).

EXAMPLE 3 **Using a Sum-to-Product Identity to Evaluate**

(a) Use a sum-to-product identity to find the exact value of

$$\sin 75° + \sin 15°.$$

(b) Compare the result from part (a) to the calculator value, rounded to four decimal places.

Solution (a) We use sum-to-product identity (i) as follows:

$$\sin 75° + \sin 15° = 2 \sin\left(\frac{75° + 15°}{2}\right) \cos\left(\frac{75° - 15°}{2}\right)$$
$$= 2 \sin 45° \cos 30°$$
$$= 2\left(\frac{\sqrt{2}}{2}\right)\left(\frac{\sqrt{3}}{2}\right) = \frac{\sqrt{6}}{2}$$

(b) Approximating $\sqrt{6}/2$ to four decimal places gives 1.2247. The calculator value of $\sin 75° + \sin 15°$, when rounded to four decimal places, also gives 1.2247.

EXAMPLE 4 **Rewriting a Difference as a Product**

Rewrite $\cos 8\theta - \cos 3\theta$ as a product.

Solution Using sum-to-product identity (iv) with $\omega = 8\theta$ and $t = 3\theta$, we get

$$\cos 8\theta - \cos 3\theta = -2 \sin\left(\frac{8\theta + 3\theta}{2}\right) \sin\left(\frac{8\theta - 3\theta}{2}\right)$$
$$= -2 \sin \frac{11\theta}{2} \sin \frac{5\theta}{2}.$$

EXAMPLE 5 **Verifying an Identity**

Verify the identity

$$\frac{\cos 3t - \cos 5t}{\sin 3t + \sin 5t} = \tan t.$$

Solution Using the sum-to-product identities for the expressions in the numerator and the denominator, we have:

$$\frac{\cos 3t - \cos 5t}{\sin 3t + \sin 5t} = \frac{-2 \sin\left(\frac{3t + 5t}{2}\right) \sin\left(\frac{3t - 5t}{2}\right)}{2 \sin\left(\frac{3t + 5t}{2}\right) \cos\left(\frac{3t - 5t}{2}\right)} \quad \begin{matrix}\text{Identity (iv)}\\ \text{Identity (i)}\end{matrix}$$
$$= \frac{-2 \sin 4t \sin(-t)}{2 \sin 4t \cos(-t)} \quad \text{Simplify}$$
$$= \frac{2 \sin 4t \sin t}{2 \sin 4t \cos t} \quad \text{Sine is odd and cosine is even}$$
$$= \frac{\sin t}{\cos t} \quad \text{Reduce}$$
$$= \tan t$$

Solving Equations by Using the Sum-to-Product Identities

We can employ sum-to-product identities to solve equations such as

$$\sin 3t + \sin 2t = 0$$
$$\cos 4t - \cos 3t = 0,$$

and

$$\cos 4t + \cos 2t = \cos t$$

EXAMPLE 6

Solving a Trigonometric Equation

Solve the equation $\sin 3t + \sin t = 0, \quad 0 \leq t < 2\pi$.

Solution Using sum-to-product identity (i), we rewrite the left side as

$$\sin 3t + \sin t = 2\sin\left(\frac{3t + t}{2}\right)\cos\left(\frac{3t - t}{2}\right) = 2\sin 2t \cos t$$

Next we solve the resulting equation as follows:

$$2\sin 2t \cos t = 0 \qquad \text{Given}$$

$$\sin 2t \cos t = 0 \qquad \text{Divide each side by 2}$$

Thus there are two possibilities:

$\sin 2t = 0$	$\cos t = 0$
$2t = 0, \pi, 2\pi, 3\pi$	$t = \frac{\pi}{2}, \frac{3\pi}{2}$
$t = 0, \frac{\pi}{2}, \pi, \frac{3\pi}{2}$	

Therefore, the solution includes $0, \frac{\pi}{2}, \pi,$ and $\frac{3\pi}{2}$.

Solving Applied Problems

If two tones from different musical instruments are sounded simultaneously, they will interact in a phenomenon called *beats* or **heterodyning.** When the two instruments are in tune, the beats will disappear. The human ear hears beats because the sound pressure slowly rises and falls as a result of slight variations in the frequency. Consider two tones with frequencies ν_1 and ν_2 (in hertz or cycles per second) and pressures P_1 and P_2 (in pounds per square foot) given by

$$P_1 = a_1 \cos(2\pi\nu_1 t) \text{ and } P_2 = a_2 \cos(2\pi\nu_2 t)$$

where t is the time (in seconds).

The superposition

$$P = P_1 + P_2$$

can be regarded as a total pressure on the eardrum of frequency $(\nu_1 + \nu_2)/2$. If the two tones have the same intensity, then $a_1 = a_2$.

EXAMPLE 7

Modeling Musical Beats

Consider the two tones of sound waves with the same intensity and with frequencies 110 and 116 hertz. The pressure on an eardrum is given by the models

$$P_1 = 0.035\cos(220\pi t) \quad \text{and} \quad P_2 = 0.035\cos(232\pi t)$$

where P_1 and P_2 are in pounds per square foot and t is the time in seconds.

(a) Write a model for

$$P = P_1 + P_2$$

and determine the frequency of P.

(b) Use a sum-to-product identity to write

$$P = P_1 + P_2$$

as a product of trigonometric expressions.

Solution (a) A model for $P = P_1 + P_2$ is given by

$$P = 0.035\cos(220\pi t) + 0.035\cos(232\pi t)$$

The frequencies ν_1 and ν_2 are 110 and 116 cycles per second and so the frequency of P is $(110 + 116)/2 = 113$ cycles per second.

(b) Using the result from part (a), we have

$$\begin{aligned} P &= 0.035\cos(220\pi t) + 0.035\cos(232\pi t) \\ &= 0.035\left[2\cos\left(\frac{220\pi t + 232\pi t}{2}\right)\cos\left(\frac{220\pi t - 232\pi t}{2}\right)\right] && \text{Sum-to-product identity (iii)} \\ &= 0.07\cos 226\pi t\cos(-6\pi t) && \text{Simplify} \\ &= 0.07\cos 226\pi t\cos 6\pi t && \text{Cosine is even} \end{aligned}$$

PROBLEM SET 2.5

Mastering the Concepts

In problems 1–8:

(a) Use a product-to-sum identity to find the exact value.

(b) Compare the result from part (a) to the calculator value of the product, rounded to four decimal places.

1. $\sin\frac{5\pi}{12}\cos\frac{7\pi}{12}$

2. $\sin\frac{\pi}{12}\cos\frac{5\pi}{12}$

3. $\cos 195° \sin 75°$

4. $\cos 112.5° \sin 67.5°$

5. $\cos\frac{17\pi}{12}\cos\frac{7\pi}{12}$

6. $\cos\frac{11\pi}{12}\cos\frac{5\pi}{12}$

7. $\sin 67.5° \sin 22.5°$

8. $\sin 165° \sin 75°$

In problems 9–16, rewrite each product as a sum or difference using the product-to-sum identities.

9. $\sin 4w \sin w$ **10.** $\cos 3\theta \sin 9\theta$

11. $\cos 8\nu \cos 4\nu$ **12.** $\cos 7t \cos 3t$

13. $\cos 3t \sin 4t$ **14.** $\cos 5t \cos 2t$

15. $\sin 4t \sin 5t$ **16.** $\cos 3t \sin t$

In problems 17–24:

(a) Use a sum-to-product identity to find the exact value.

(b) Compare the result from part (a) to the calculator value of the expression, rounded to four decimal places.

17. $\sin \frac{7\pi}{18} + \sin \frac{\pi}{18}$ **18.** $\sin \frac{13\pi}{12} + \sin \frac{7\pi}{12}$

19. $\sin 75° - \sin 165°$ **20.** $\sin 105° - \sin 75°$

21. $\cos \frac{7\pi}{12} + \cos \frac{\pi}{12}$ **22.** $\cos \frac{19\pi}{12} + \cos \frac{13\pi}{12}$

23. $\cos 75° - \cos(-15°)$

24. $\cos(-75°) - \cos 16.5°$

In problems 25–32, rewrite each sum as a product using the sum-to-product identities.

25. $\sin 3\theta + \sin \theta$ **26.** $\sin 5x + \sin 4x$

27. $\sin 5t - \sin 7t$ **28.** $\sin 3v - \sin v$

29. $\cos 2t + \cos 3t$ **30.** $\cos 3v + \cos v$

31. $\cos 5x - \cos 6x$ **32.** $\cos 2x - \cos x$

In problems 33 and 34, verify that the equation is an identity.

33. $\frac{\sin 2t + \sin 4t}{\cos 2t - \cos 4t} = \cot t$

34. $\frac{\sin t + \sin 3t}{\cos t + \cos 3t} = \tan 2t$

In problems 35–38, apply the sum-to-product identities to solve each equation for $0 \le t < 2\pi$.

35. $\sin 2t - \sin 5t = 0$

37. $\cos t - \cos 3t = 0$

36. $\cos 3t + \cos 6t = 0$

38. $\sin 7t - \sin 3t = 0$

Solving Applied Problems

39. Musical Tones and Beats: Suppose that two tones of sound waves with the same intensity and with frequencies 64 and 72 hertz are emitted simultaneously. The pressures on an eardrum, P_1 and P_2 (in pounds per square feet), and in t seconds are given, respectively, by the models

$$P_1 = 0.5 \cos 128\pi t \quad \text{and} \quad P_2 = -0.05 \cos 144\pi t.$$

(a) Write a model for $P = P_1 + P_2$ and determine the frequency of P.

(b) Use a sum-to-product identity to write $P_1 + P_2$ as a product of two sine functions.

40. Musical Tones and Beats: Suppose that two tones of sound waves with the same intensity and with frequencies 128 and 144 hertz are emitted simultaneously. The pressures exerted on an eardrum, P_1 and P_2 (in pounds per square feet), in t seconds are given, respectively, by the models

$$P_1 = 0.25 \sin 256\pi t \quad \text{and} \quad P_2 = 0.25 \sin 288\pi t.$$

(a) Write a model for $P = P_1 + P_2$ and determine the frequency of P.

(b) Use a sum-to-product identity to write P as a product of trigonometric functions.

Developing and Extending the Concepts

In problems 41–46, prove each identity.

41. $\cos u \sin v = \frac{1}{2}[\sin(u + v) - \sin(u - v)]$

42. $\cos u \cos v = \frac{1}{2}[\cos(u - v) + \cos(u + v)]$

43. $\sin u \sin v = \frac{1}{2}[\cos(u - v) - \cos(u + v)]$

44. $\sin \omega - \sin t = 2 \cos\left(\frac{\omega + t}{2}\right) \sin\left(\frac{\omega - t}{2}\right)$

45. $\cos \omega + \cos t = 2 \cos\left(\frac{\omega + t}{2}\right) \cos\left(\frac{\omega - t}{2}\right)$

46. $\cos \omega - \cos t = -2 \sin\left(\frac{\omega + t}{2}\right) \sin\left(\frac{\omega - t}{2}\right)$

G In problems 47–50, use a grapher to graph each given expression and the corresponding resulting expression found in problems 25–28 separately, using the same viewing window settings. Compare the results.

47. See problem 25.

48. See problem 26.

49. See problem 27.

50. See problem 28.

Interlude: Practice Problems 2.5, Part I

1. Verify that $\dfrac{\cos 4x - \cos 2x}{2 \sin 3x} = -\sin x$ is an identity by transforming the left side of the equation into the right side.

2. Calculate the exact value of $\cos 195° + \cos 105°$.

3. Verify that $\tan 2t = \dfrac{\sin t + \sin 3t}{\cos t + \cos 3t}$ is an identity by transforming the left side of the equation into the right side.

4. Write $2\cos(9x)\cos(15x)$ as a sum involving only the cosine function and no products of trigonometric functions.

Interlude: Practice Problems 2.5, Part II

1. Consider the equation $\sin 2t - \sin 10t = 0$. Using algebraic methods, find all solutions to the equation in the interval $[0, 2\pi)$. Give both exact solutions and approximate solutions accurate to three decimal places.

2. Write $\sin x \sin y + \sin x \cos y + \cos x \cos y$ as an equivalent expression that does not involve a product of trigonometric functions.

Chapter 2 Readiness, Part I

Provide complete solutions in the space provided.

1. Calculate the exact value of each of the following by using appropriate identities (fill in the blanks).

____

____

____

(a) $\cos \frac{\pi}{8}$ ______________

(b) $\sin \frac{\pi}{8}$ ______________

(c) $\tan \frac{\pi}{12}$ ______________

(d) $\cos 160°$ ______________

(e) $\sec \frac{7\pi}{12}$ ______________

2. Consider the expression $\sqrt{4 - u^2}$. Assume that $0 < \theta < \pi$. Using the substitution $u = 2\cos\theta$, rewrite the given expression as a trigonometric expression containing no radical.

3. Determine if $\dfrac{\sin(-t)}{\tan(-t)\cos(-t)} = 1$ is an identity. If so, verify the identity by transforming the left side of the equation into the right side. If not, give a value of t that shows the equation is not an identity.

4. Consider the graph shown in Figure 1. Calculate the radian measure of θ, accurate to three decimal places.

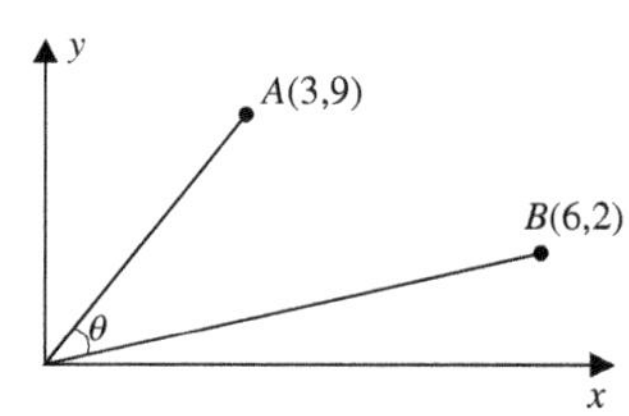

Figure 1

Chapter 2 Readiness, Part II

Provide complete solutions in the space provided.

1. The voltage in a 220-volt electrical circuit is modeled by $V = 310 \sin 120\pi t$ and the amperage flow through a heater is given by $I = 7 \sin 120\pi t$. The wattage consumed by the heater is $W = VI$. Find appropriate constants a, b, and k so that $W = a + b \cos kt$.

2. Consider the equation $\cos(3x)\cos(2x) + \sin(3x)\sin(2x) = 1$. Using algebraic methods, find all solutions to the equation in the interval $[0, 2\pi)$. Give both exact solutions and approximate solutions accurate to three decimal places.

3. Consider the equation $\frac{\cos 2x}{\cos x + \sin x} = 0$. Using algebraic methods, find all solutions to the equation in the interval $[0, 2\pi)$. Give both exact solutions and approximate solutions accurate to three decimal places.

4. Verify that $\frac{\cos(u - v)}{\sin(u - v)} = \frac{\cot u + \tan v}{1 + \cot u \tan v}$ is an identity by transforming the one side of the equation into the other side.

CHAPTER 2 REVIEW PROBLEM SET

In problems 1–4, prove that each equation is an identity.

1. (a) $\cos t \sin t \csc t \sec t = 1$
(b) $\cot\theta \cos\theta + \sin\theta = \csc\theta$

2. (a) $\sec t - \cos t = \sin t \tan t$
(b) $(\csc t + \cot t)(1 - \cos t) = \sin t$

3. (a) $\dfrac{\sin^2 t \cos t + \cos^3 t}{\cot t} = \sin t$
(b) $\cos^2 t - \sin^2 t = \dfrac{1 - \tan^2 t}{1 + \tan^2 t}$

4. (a) $\dfrac{\sin^4 t - \cos^4 t}{\sin^2 t - \cos^2 t} = 1$
(b) $\dfrac{\sec^2 t + 2\tan t}{2 - \sec^2 t} = \dfrac{1 + \tan t}{1 - \tan t}$

In problems 5–10, simplify each expression by using identities.

5. (a) $\sin 7t \cos 2t - \cos 7t \sin 2t$
(b) $\sin(t + s)\cos s - \cos(t + s)\sin s$

6. (a) $\cos 51° \cos 39° - \sin 51° \sin 39°$
(b) $\cos 33° \cos 27° - \sin 33° \sin 27°$

7. (a) $\sin(3\pi - t)$
(b) $2\sin^2 4\theta - 1$

8. (a) $\csc(90° - \theta)\cos\theta + \cot^2\theta$
(b) $\cos(270° + \theta)$

9. (a) $\cos\left(\dfrac{\pi}{2} - t\right)\tan\left(\dfrac{\pi}{2} - t\right)$
(b) $\cos\left(\dfrac{\pi}{3} - t\right) - \cos\left(\dfrac{\pi}{3} + t\right)$

10. (a) $\dfrac{\tan 5t + \tan t}{1 - \tan 5t \tan t}$
(b) $\dfrac{\tan 7t - \tan 2t}{1 + \tan 7t \tan 2t}$

In problems 11–16, use an identity to find the exact value of each expression.

11. (a) $\cos\left(\dfrac{\pi}{4} + \dfrac{\pi}{6}\right)$
(b) $\cos\dfrac{\pi}{4} + \cos\dfrac{\pi}{6}$

12. (a) $\sin\left(\dfrac{3\pi}{4} - \dfrac{\pi}{3}\right)$
(b) $\sin\dfrac{3\pi}{4} - \sin\dfrac{\pi}{3}$

13. (a) $\sin 165°$; $165° = 120° + 45°$
(b) $\cos 195°$; $195° = 150° + 45°$

14. (a) $\cos 285°$; $285° = 240° + 45°$
(b) $\cos 255°$; $255° = 210° + 45°$

15. (a) $\sin 22.5°$; $22.5° = \dfrac{1}{2}(45°)$
(b) $\cos\dfrac{11\pi}{12}$; $\dfrac{11\pi}{12} = \dfrac{1}{2}\left(\dfrac{11\pi}{6}\right)$

16. (a) $\cos 75°$; $75° = \dfrac{1}{2}(150°)$
(b) $\sin\dfrac{7\pi}{12}$; $\dfrac{7\pi}{12} = \dfrac{1}{2}\left(\dfrac{7\pi}{6}\right)$

17. Use the fundamental identities to find the exact values of the other five trigonometric functions under the given conditions.
(a) $\sin\theta = \dfrac{3}{5}$; θ in quadrant II
(b) $\cos\theta = \dfrac{3}{8}$; θ in quadrant IV

18. Use the fundamental identities to simplify each expression.
(a) $\csc^2 t \tan^2 t - \tan^2 t$
(b) $\sec t - \sin t \tan t$
(c) $\sin(-\theta)\sec(-\theta)$

19. Suppose that $\sin t = 1/4$, where t is in quadrant I, and $\cos s = -5/17$, where s is in quadrant II. Find:
(a) $\sin(t - s)$
(b) $\cos(t - s)$
(c) $\sin 2t$
(d) $\cos\left(\dfrac{s}{2}\right)$
(e) $\tan 2t$

20. Suppose that $\tan t = -4/3$, where t is in quadrant II, and $\tan s = -5/12$, where s is in quadrant IV. Find:
(a) $\tan(t + s)$
(b) $\sin(t + s)$
(c) $\cos(t + s)$
(d) $\cos 2s$
(e) $\sin\left(\dfrac{t}{2}\right)$

In problems 21–28, prove that each equation is an identity.

21. $2\csc 2\theta \cot\theta = 1 + \cot^2\theta$

22. $\csc\theta \sin 2\theta = 2\cos\theta$

23. $\tan\left(t + \dfrac{3\pi}{4}\right) = \dfrac{\tan t - 1}{\tan t + 1}$

24. $\dfrac{\cos\left(t - \dfrac{\pi}{4}\right)}{\cos t \sin\dfrac{\pi}{4}} = \tan t + 1$

25. $\dfrac{1 - \cos 2\theta}{\sin 2\theta} = \tan \theta$

26. $\dfrac{\cos 2\theta}{1 - \sin^2 \theta} = 2 - \sec^2 \theta$

27. $\sin \dfrac{t}{2} \cos \dfrac{t}{2} = \dfrac{\sin t}{2}$

28. $\dfrac{1 - \tan^2 \theta}{1 + \tan^2 \theta} = \cos 2\theta$

In problems 29–32, solve each equation, where $0 \le t < 2\pi$ or $0° \le \theta < 360°$. Round off part (b) of each problem to two decimal places.

29. (a) $\sin \theta = -1$
(b) $\cos \theta = 0.3217$

30. (a) $\tan t = \sqrt{3}$
(b) $\sec t = 1.4931$

31. (a) $5 \sin^2 t - 4 \sin t - 1 = 0$
(b) $(2 \sin t - 1)\left(\sin t - \dfrac{1}{3}\right) = 0$

32. (a) $3 \tan^2 \theta - \sqrt{3} \tan \theta = 0$
(b) $\tan^2 t - \tan t - 2 = 0$

In problems 33 and 34, find the exact value of each expression without using a calculator.

33. (a) $\cos 15° \cos 105°$
(b) $\cos \dfrac{11\pi}{12} \sin \dfrac{7\pi}{12}$

34. (a) $\sin 195° \cos 15°$
(b) $\sin \dfrac{\pi}{12} \cos \dfrac{7\pi}{12}$

In problems 35 and 36, write each expression as a sum or difference of sine or cosine functions.

35. (a) $2 \sin t \cos 2t$
(b) $2 \cos 3t \cos 5t$

36. (a) $\sin \dfrac{5t}{2} \cos \dfrac{t}{2}$
(b) $\sin 7t \sin 3t$

In problems 37 and 38, write each expression as a product of sine or cosine functions.

37. (a) $\sin 4t - \sin t$
(b) $\cos 7t + \cos 3t$

38. (a) $\sin 10t + \sin 2t$
(b) $\cos 6t - \cos 2t$

In problems 39 and 40, use the features on a grapher to approximate the solution of each equation to two decimal places for $0 \le t < 2\pi$.

39. $\cos t = 3t$

40. $2 \sin t = 2t - 1$

In problems 41–44, write each equation in the form $y = A \cos(u - v)$.

41. $y = \sqrt{3} \sin x + \cos x$

42. $y = -\sin x - \sqrt{3} \cos x$

43. $y = -\sin x + \cos x$

44. $y = -\sin x - \cos x$

45. Electrical Voltage: Suppose the voltage V (in volts) in an electrical circuit is given by the model

$$V = 80 \cos\left(120\pi t - \frac{\pi}{4}\right)$$

where t is in seconds.
(a) Graph one cycle of the function.
(b) Determine the smallest positive value of t when the voltage is 60 volts. Round off to three decimal places.

46. **Business Profit:** A company's profit during a 12-month period is modeled by the function

$$p(t) = 100{,}000(t - 2 \cos t)$$

where $p(t)$ (in dollars) is the profit after t months.
(a) Use a grapher to graph p for $0 \le t < 12$.
(b) Use the grapher to approximate the value of t to three decimal places, when $p(t) = \$650{,}000$.

47. Billiard Ball Motion: Figure 1 shows a billiard ball (at position A) that travels a path to hit a ball at position B. Assume the billiard ball strikes the rail of the table and rebounds so that the angles labeled θ are equal.

Figure 1

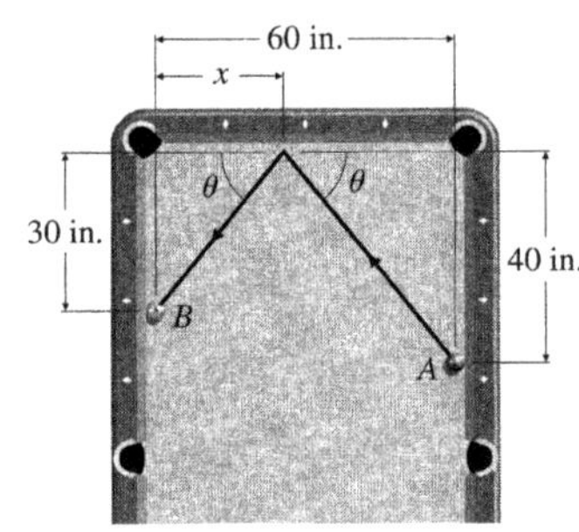

(a) Find the distance x denoted in the figure.
(b) Find $\tan \theta$, and then solve the equation for θ. Round off the answer to the nearest degree.

48. Hours of Daylight: Suppose the number of hours N of daylight on a given day of the year for a certain area is given by the model

$$N = 12 + 3 \sin\left[\frac{2\pi}{365}(t - 85)\right]$$

where t is the day of the year, and $t = 0$ is January 1.
(a) Graph the function.
(b) Determine which days of the year have 14 hours of daylight. Round off the answers to two decimal places.

CHAPTER 2 TEST

1. Use

$$255° = 225° + 30°$$

to find the exact value of each expression.

(a) $\sin 255°$ (b) $\cos 255°$ (c) $\tan 255°$

2. Suppose that $\sin s = -3/5$, where s is in quadrant IV, and $\cos t = 4/5$, where t is in quadrant IV. Find the exact value of each expression.

(a) $\sin(s + t)$ (b) $\cos(s + t)$ (c) $\tan(s + t)$
(d) $\sin 2t$ (e) $\cos 2t$ (f) $\sin 4t$
(g) $\sin^2\left(\frac{t}{2}\right)$ (h) $\sin(s + 2t)$

3. Use $67.5° = (1/2)(135°)$ to find the exact value of each expression.

(a) $\sin 67.5°$ (b) $\cos 67.5°$ (c) $\tan 67.5°$

4. Simplify each trigonometric expression by using identities.

(a) $1 - 2\cos^2 47°$

(b) $\sin\frac{5\pi}{7}\cos\frac{5\pi}{7}$

(c) $\sin 25° \cos 35° + \sin 35° \cos 25°$

(d) $\cos\frac{5\pi}{7}\cos\frac{2\pi}{7} - \sin\frac{5\pi}{7}\sin\frac{2\pi}{7}$

(e) $\cos\frac{3t}{4}\cos\frac{t}{4} + \sin\frac{3t}{4}\sin\frac{t}{4}$

5. Solve each equation for $0° \le \theta < 360°$ or $0 \le t < 2\pi$. Round off the answers to two decimal places.

(a) $4\cos^2 t - 3 = 0$

(b) $\sin 2\theta = \frac{1}{2}$

(c) $(\tan t - 1)(2\tan t + 3) = 0$

6. Prove that each equation is an identity.

(a) $(\csc^2 t - 1)(1 + \tan^2 t) = \csc^2 t$

(b) $\sin 2t \sec t = 2\sin t$

(c) $\sin\left(t + \frac{\pi}{6}\right)\sin\left(t - \frac{\pi}{6}\right) = \sin^2 t - \frac{1}{4}$

(d) $\frac{1 - \cos 2t}{\sin t} = 2\sin t$

7. Use a grapher to solve the equation $3x + \cos x = 0$. Round off the answer to two decimal places.

8. An observer in a balloon finds that the angle of depression from a balloon to a car is θ (measured in degrees). The balloon is x feet high and the distance horizontal from a point on the ground directly under it to the car is 240 feet.

(a) Express $\cos 2\theta$ as a function of x.

(b) If $x = 296$ feet, what is the measure of θ? Round off the answer to two decimal places.

Chapter 2: Answers to Selected Problems

Problem Set 2.1 (on page 148)

1. $\cos^2 3\theta$ **3.** 1 **5.** -2 **7.** 4 **9.** **(a)** $\sin t$ **(b)** $8 - 5\sin^2 t$ **(c)** $\sin^2 t$
11. **(a)** $2\cos^4\theta$ **(b)** $-\csc\theta$ **13.** **(a)** $\tan t$ **(b)** $\cos\theta$ **15.** **(a)** $\csc t$ **(b)** $\sec\theta$
49. False for $t = \frac{\pi}{2}$ **51.** False for $t = \frac{\pi}{4}$ **53.** QII and III **55.** Q_I and IV
61. **(b)** Graphs are the same. **(c)** Yes.

Problem Set 2.2 (on page 165)

1. **(a)** $\frac{\sqrt{6}-\sqrt{2}}{4} = 0.2588$ **(b)** $\frac{\sqrt{6}+\sqrt{2}}{4} = 0.9659$
3. **(a)** $\frac{\sqrt{6}+\sqrt{2}}{4} = 0.9659$ **(b)** $\frac{-\sqrt{6}-\sqrt{2}}{4} = -0.9659$ **(c)** $2+\sqrt{3} = 3.7321$
5. **(a)** $\frac{\sqrt{2}-\sqrt{6}}{4} = -0.2588$ **(b)** $\frac{-\sqrt{6}-\sqrt{2}}{4} = -0.9659$ **7.** **(a)** $\frac{\sqrt{3}}{2} = 0.8660$ **(b)** $\tan\frac{\pi}{2}$; undefined
9. **(a)** $-\frac{63}{65}$ **(b)** $-\frac{16}{65}$ **(c)** $-\frac{33}{65}$ **(d)** $\frac{56}{65}$ **(e)** $\frac{63}{16}$ **(f)** $-\frac{33}{56}$
11. **(a)** -0.204 **(b)** -0.979 **(c)** 0.698 **(d)** 0.716 **(e)** 0.208 **(f)** 0.976 **13.** 1 **15.** 0
17. **(a)** $\cos t$ **(b)** $\sin t$ **19.** **(a)** $\frac{\tan t + 1}{1 - \tan t}$ **(b)** $\frac{\sqrt{2}}{2}(\cos\theta - \sin\theta)$
21. **(a)** $\frac{1}{\cos\theta}$ **(b)** $-\cos t$ **23.** **(a)** $\cos 8t$ **(b)** $\sin 4t$ **25.** **(a)** $\cos x$ **(b)** $\tan 7t$
47. $y = \sqrt{2}\cos\left(t - \frac{\pi}{4}\right)$ **49.** $y = 2\cos\left(2\pi t - \frac{5\pi}{3}\right)$
51. **(a)** $h(t) = 0.5\cos\left(\frac{\pi t}{6} - 0.927\right)$ **(b)** Maximum height is 0.5 m when t is about 1.8.
(c) Minimum height is $-$ 0.5 m when t is about 7.8. **53.** **(a)** $d(t) = 5\cos(4t - 0.644)$ **(b)** 5 **(c)** $\frac{\pi}{2}$
55. **(a)** $\sin\gamma$ **(b)** $-\cos\gamma$ **57.** **(c)** Same; shift the graph of $y = \cos x$ to the right $\frac{\pi}{2}$ units.
59. **(b)** 0 and 1

Problem Set 2.3 (on page 179)

1. $\cos 2t \neq 2\cos t$ **3.** **(a)** $\frac{24}{25}$ **(b)** $-\frac{7}{25}$ **(c)** $-\frac{24}{7}$ **5.** **(a)** $\frac{336}{625}$ **(b)** $-\frac{527}{625}$ **(c)** $-\frac{336}{527}$
7. **(a)** $-\frac{120}{169}$ **(b)** $\frac{119}{169}$ **(c)** $-\frac{120}{119}$ **9.** **(a)** $-\frac{175}{337}$ **(b)** $-\frac{288}{337}$ **(c)** $\frac{175}{288}$ **11.** $\sin 6t$ **13.** $\cos 16t$
15. $\cos 10t$ **17.** $\tan 8\theta$ **19.** **(a)** $\frac{\sqrt{3}}{2} = 0.8660$ **(b)** $\frac{\sqrt{3}}{2} = 0.8660$ **(c)** $-\frac{24}{7} = -3.4286$
21. **(a)** $\frac{\sqrt{2+\sqrt{2}}}{2} = 0.9239$ **(b)** $\frac{\sqrt{2+\sqrt{3}}}{2} = 0.9659$
23. **(a)** $\frac{\sqrt{2-\sqrt{3}}}{2} = 0.2588$ **(b)** $\sqrt{3+2\sqrt{2}} = 2.4142$
25. **(a)** $-\frac{\sqrt{2-\sqrt{2}}}{2} = -0.3827$ **(b)** $\frac{\sqrt{2+\sqrt{2}}}{2} = 0.9239$

27. **(a)** $\frac{t}{2}$ in quadrant I: $\frac{3}{5}$ **(b)** $\frac{t}{2}$ in quadrant I: $\frac{4}{5}$ **(c)** $\frac{t}{2}$ in quadrant I: $\frac{3}{4}$

29. **(a)** $\frac{t}{2}$ in quadrant I: $\frac{3}{\sqrt{10}}$ **(b)** $\frac{t}{2}$ in quadrant I: $\frac{1}{\sqrt{10}}$ **(c)** $\frac{t}{2}$ in quadrant I: 3

31. **(a)** $\frac{t}{2}$ in quadrant II: $\frac{5}{\sqrt{34}}$ **(b)** $\frac{t}{2}$ in quadrant II: $-\frac{3}{\sqrt{34}}$ **(c)** $\frac{t}{2}$ in quadrant II: $-\frac{5}{3}$

33. **(a)** $\frac{t}{2}$ in quadrant II: $\frac{1}{\sqrt{17}}$ **(b)** $\frac{t}{2}$ in quadrant II: $-\frac{4}{\sqrt{17}}$ **(c)** $\frac{t}{2}$ in quadrant II: $-\frac{1}{4}$

35. **(a)** $\cos 2t$ **(b)** $\sin 3t$ **49.** **(b)** 4.70 cm^2 **(c)** $0 < \theta < 180°$ **(d)** Maximum area is 8 cm^2 when $\theta = 90°$

51. **(b)** 17.09 in^2 **(c)** $0° < \theta < 90°$

53. **(a)** $W = 980 - 980\cos(240\pi t)$ **(b)** V, Period $= \frac{1}{60}$; I, Period $= \frac{1}{60}$; W, Period $= \frac{1}{120}$ **(c)** 1960

57. **(a)** $\left(-\frac{7}{25}, -\frac{24}{25}\right)$ **(b)** $\left(-\frac{2}{\sqrt{5}}, \frac{1}{\sqrt{5}}\right)$ if t is positive **59.** **(a)** $\frac{\sec^2 t}{2 - \sec^2 t}$ **(b)** $-\frac{5}{4}$

Problem Set 2.4 (on page 196)

3. **(a)** No solution **(b)** No solution **(c)** 225°, 315°
(d) $225° + (360°)k$ or $315° + (360°)k$, where k is an integer

5. **(a)** $\sin 190° = -\sin 10° = -0.1736$ **(b)** $\tan(-113°) = \tan 67° = 2.3559$
(c) $\sec 372° = \sec 12° = 1.0223$ **(d)** $\cos(-1.9) = -0.3233$; $-\cos 1.2416 = -0.3233$
(e) $\cot 7.2 = 0.7665$; $\cot 0.9168 = 0.7665$

7. **(a)** 2.0944 or $\frac{2\pi}{3}$; 4.1888 or $\frac{4\pi}{3}$ **(b)** 2.3562 or $\frac{3\pi}{4}$; 3.9270 or $\frac{5\pi}{4}$

9. **(a)** 60° or 240° **(b)** 120° or 300°

11. **(a)** 1.0472 or $\frac{\pi}{3}$; 2.0944 or $\frac{2\pi}{3}$ **(b)** 225° or 315°

13. **(a)** 2.3562 or $\frac{3\pi}{4}$; 3.9270 or $\frac{5\pi}{4}$ **(b)** 3.1416 or π; 4.7124 or $\frac{3\pi}{2}$

15. **(a)** 0.9500 or 2.1916 **(b)** 2.9916 or 6.1332 **17.** **(a)** 2.5559 or 3.7273 **(b)** 1.4711 or 4.6127

19. 0, 3.1416 or π **21.** 90°, 240°, 270°, 300° **23.** 0, 1.0472 or $\frac{\pi}{3}$, 3.1416 or π, 4.1888 or $\frac{4\pi}{3}$

25. 180° **27.** 1.0472 or $\frac{\pi}{3}$, 2.0944 or $\frac{2\pi}{3}$, 4.1888 or $\frac{4\pi}{3}$; 5.2360 or $\frac{5\pi}{3}$

29. 60°, 120°, 240°, 300° **31.** 1.0472 or $\frac{\pi}{3}$, 1.5708 or $\frac{\pi}{2}$, 4.7124 or $\frac{3\pi}{2}$, 5.2360 or $\frac{5\pi}{3}$

33. **(a)** 45°, 75°, 165°, 195°, 285°, 315° **(b)** 55°, 85°, 175°, 205°, 295°, 325°

35. **(a)** 1.5708 or $\frac{\pi}{2}$ **(b)** 0.3142 or $\frac{\pi}{10}$

37. 0.7854 or $\frac{\pi}{4}$, 2.3562 or $\frac{3\pi}{4}$, 3.9270 or $\frac{5\pi}{4}$, 5.4978 or $\frac{7\pi}{4}$ **39.** 45°, 90°, 225°, 270°

41. 0, 0.5234 or $\frac{\pi}{6}$, 2.6180 or $\frac{5\pi}{6}$, 3.1416 or π **43.** 0°, 270°

45. 0.7854 or $\frac{\pi}{4}$, 1.5708 or $\frac{\pi}{2}$, 2.3562 or $\frac{3\pi}{4}$, 4.7124 or $\frac{3\pi}{2}$ **47.** 0, 1.0472 or $\frac{\pi}{3}$, 5.2360 or $\frac{5\pi}{3}$

49. 210°, 270°, 330° **51.** 0.6749 or 5.6083 **53.** **(a)** 32.67° **(b)** 62.04° **55.** **(b)** $0° < \theta < 90°$

59. **(a)** $k < -1$ or $k > 1$ **(b)** $k = -1$ or $k = 1$ **(c)** $-1 < k < 1$ **(d)** No values of k

61. 2.14 radians **63.** **(a)** 0.201, 0.340, 2.802, 2.940 **(b)** (0, 0.201) or (0.340, 2.802) or (2.940, 6.283)

65. **(a)** 0.160 **(b)** [0, 0.160] **67.** **(a)** 0, 1.293, 4.721; $t = 0$ **(b)** [1.293, 4.721]; $t = 0$

Problem Set 2.5 (on page 210)

1. **(a)** $-\frac{1}{4}$ **(b)** -0.2500 **3.** **(a)** $\frac{(-2-\sqrt{3})}{4}$ **(b)** -0.9330 **5.** **(a)** $\frac{(2-\sqrt{3})}{4}$ **(b)** 0.0670

7. **(a)** $\frac{\sqrt{2}}{4}$ **(b)** 0.3536 **9.** $\frac{1}{2}(\cos 3w - \cos 5w)$ **11.** $\frac{1}{2}(\cos 4v + \cos 12v)$ **13.** $\frac{1}{2}(\sin 7t + \sin t)$

15. $\frac{1}{2}(\cos t - \cos 9t)$ **17.** **(a)** $\sqrt{3}\sin\frac{2\pi}{9}$ **(b)** 1.1133 **19.** **(a)** $\frac{\sqrt{2}}{2}$ **(b)** 0.7071

21. **(a)** $\frac{\sqrt{2}}{2}$ **(b)** 0.7071 **23.** **(a)** $\frac{-\sqrt{2}}{2}$ **(b)** -0.7071 **25.** $2\sin 2\theta\cos\theta$ **27.** $-2\cos 6t\sin t$

29. $2\cos\frac{5t}{2}\cos\frac{t}{2}$ **31.** $2\sin\frac{11x}{2}\sin\frac{x}{2}$ **35.** $0, \frac{\pi}{7}, \frac{3\pi}{7}, \frac{2\pi}{3}, \frac{4\pi}{3}, 2\pi$ **37.** $0, \frac{\pi}{2}, \pi, \frac{3\pi}{2}, 2\pi, 3\pi$

39. **(a)** $P = 0.5\cos(128\pi t) - 0.5\cos(144\pi t)$; frequency of $P = 68$ hertz. **(b)** $\sin(136\pi t)\sin(8\pi t)$

Chapter 2 Review Problem Set (on page 221)

5. **(a)** $\sin 5t$ **(b)** $\sin t$ **7.** **(a)** $\sin t$ **(b)** $-\cos 8\theta$ **9.** **(a)** $\cos t$ **(b)** $\sqrt{3}\sin t$

11. **(a)** $\frac{\sqrt{6}-\sqrt{2}}{4}$ **(b)** $\frac{\sqrt{2}+\sqrt{3}}{2}$ **13.** **(a)** $\frac{\sqrt{6}-\sqrt{2}}{4}$ **(b)** $\frac{-\sqrt{6}-\sqrt{2}}{4}$

15. **(a)** $\frac{\sqrt{2-\sqrt{2}}}{2}$ **(b)** $\frac{-\sqrt{2+\sqrt{3}}}{2}$

17. **(a)** $\cos\theta = -\frac{4}{5}$; $\tan\theta = -\frac{3}{4}$; $\csc\theta = \frac{5}{3}$; $\sec\theta = -\frac{5}{4}$; $\cot\theta = -\frac{4}{3}$.

(b) $\sin\theta = -\frac{\sqrt{55}}{8}$; $\tan\theta = -\frac{\sqrt{55}}{3}$; $\csc\theta = -\frac{8}{\sqrt{55}}$; $\sec\theta = \frac{8}{3}$; $\cot\theta = -\frac{3}{\sqrt{55}}$.

19. **(a)** $\frac{-5-6\sqrt{110}}{68}$ **(b)** $\frac{2\sqrt{66}-5\sqrt{15}}{68}$ **(c)** $\frac{\sqrt{15}}{8}$ **(d)** $\frac{\sqrt{102}}{17}$ **(e)** $\frac{\sqrt{15}}{7}$

29. **(a)** $270°$ **(b)** $71.23°, 288.77°$

31. **(a)** $\frac{\pi}{2}$, 3.34, 6.08 **(b)** $\frac{\pi}{6}, \sin^{-1}\frac{1}{3}, \pi - \sin^{-1}\frac{1}{3}, \frac{5\pi}{6}$ or $t = 0.34, 0.52, 2.62, 2.80$

33. **(a)** $-\frac{1}{4}$ **(b)** $\frac{(-2-\sqrt{3})}{4}$ **35.** **(a)** $\sin 3t - \sin t$ **(b)** $\cos 2t + \cos 8t$

37. **(a)** $2\cos\frac{5t}{2}\sin\frac{3t}{2}$ **(b)** $2\cos 5t\cos 2t$ **39.** 0.32 **41.** $y = 2\cos\left(x - \frac{\pi}{3}\right)$

43. $y = \sqrt{2}\cos\left(x - \frac{7\pi}{4}\right)$ or $y = \cos\left(x + \frac{\pi}{4}\right)$ **45.** **(b)** 0.004 sec

47. **(a)** 25.71 in. **(b)** $\tan\theta = \frac{7}{6}$; $\theta = 49°$

Chapter 2 Test (on page 223)

1. **(a)** $\frac{(-\sqrt{6}-\sqrt{2})}{4}$ **(b)** $\frac{(\sqrt{2}-\sqrt{6})}{4}$ **(c)** $2+\sqrt{3}$

2. **(a)** $-\frac{24}{25}$ **(b)** $\frac{7}{25}$ **(c)** $-\frac{24}{7}$ **(d)** $-\frac{24}{25}$ **(e)** $\frac{7}{25}$ **(f)** $-\frac{336}{625}$ **(g)** $\frac{1}{10}$ **(h)** $-\frac{117}{125}$

3. **(a)** $\dfrac{(\sqrt{2+\sqrt{2}})}{2}$ **(b)** $\dfrac{(\sqrt{2-\sqrt{2}})}{2}$ **(c)** $\sqrt{2}+1$

4. **(a)** $-\cos 94^\circ$ **(b)** $\dfrac{1}{2}\sin\dfrac{10\pi}{7}$ **(c)** $\sin 60^\circ$ **(d)** $\cos\pi$ **(e)** $\cos\dfrac{t}{2}$

5. **(a)** 0.52 or $\dfrac{\pi}{6}$, 2.62 or $\dfrac{5\pi}{6}$, 3.67 or $\dfrac{7\pi}{6}$, 5.76 or $\dfrac{11\pi}{6}$

(b) 15°, 75°, 195°, 255° **(c)** $\dfrac{\pi}{4}, \dfrac{5\pi}{4}$, 2.16, 5.30; or 0.79, 2.16, 3.93, 5.30

7. −0.32 **8.** **(a)** $\dfrac{57{,}600 - x^2}{57{,}600 + x^2}$ **(b)** 50.96°

CHAPTER 3

Additional Applications of Trigonometry

Chapter Contents

So far we have used trigonometry to solve right triangles. However, it can also be used to solve other types of triangles. For instance, a gear designer wishes to mount three circular gears with centers at *A*, *B*, and *C* at the vertices of a triangle *ABC*. The gear with center at *A* has a 2-inch radius, the gear with center at *B* has a 3-inch radius, and the gear with center *C* has a 5-inch radius. What are the measures of the angles of the triangle *ABC* to make this mounting possible? The solution to this problem is given in Example 5 on page 237.

In fields such as surveying, navigation, and engineering, it is often necessary to solve triangles that do not necessarily contain a right angle. In this chapter, we will study ways of solving such triangles. We will also use trigonometry to explore the polar coordinate system, to study vectors, and to extend our understanding of the complex number system.

Objectives

1. Use the Law of Cosines to Solve Triangles
2. Use the Law of Sines to Solve Triangles
3. Solve the Ambiguous Case
4. Solve Applied Problems

3.1 The Law of Cosines and the Law of Sines

In this section, we derive two formulas—the *law of cosines* and the *law of sines*—that enable us to find the missing parts of a triangle under certain conditions.

For convenience, we shall adopt standard notation to label the sides and angles of a triangle. Thus in triangle ABC (Figure 1) the angle at vertex A is denoted by α, the angle at B is denoted by β, and the angle at C is denoted by γ. The length of the side opposite α is a, the length of the side opposite β is b, and the length of the side opposite γ is c.

Figure 1

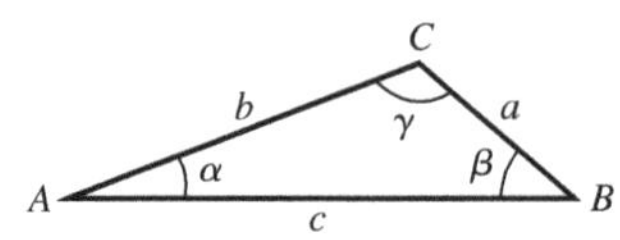

Using the Law of Cosines to Solve Triangles

We begin by establishing the *law of cosines* which is a generalization of the Pythagorean theorem.

The Law of Cosines

In any triangle ABC:

$$\text{(i)}\quad c^2 = a^2 + b^2 - 2ab\cos\gamma$$
$$\text{(ii)}\quad b^2 = a^2 + c^2 - 2ac\cos\beta$$
$$\text{(iii)}\quad a^2 = b^2 + c^2 - 2bc\cos\alpha$$

Proofs of Formulas

Figure 2

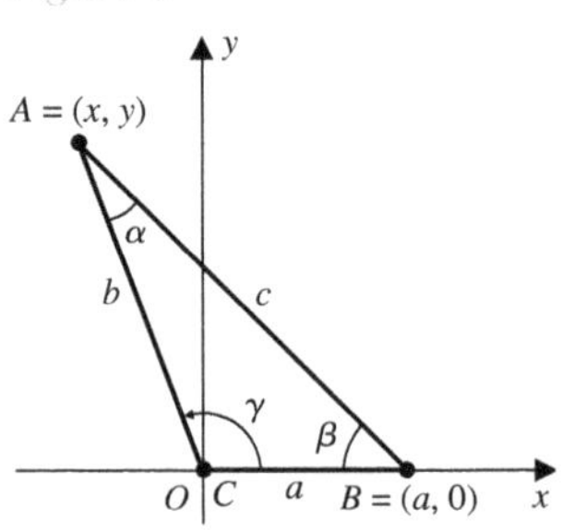

(i) To derive the law of cosines, consider triangle ABC in a Cartesian plane with γ in standard position and assume that vertex A has coordinates (x, y) (Figure 2).

By the distance formula, we have

$$
\begin{aligned}
c^2 &= (x-a)^2 + y^2 \\
&= a^2 + (x^2 + y^2) - 2ax && \text{Multiply and rearrange terms} \\
&= a^2 + b^2 - 2ax && x^2 + y^2 = b^2 \\
&= a^2 + b^2 - 2ab\cos\gamma && \cos\gamma = \frac{x}{b},\ \text{so } b\cos\gamma = x
\end{aligned}
$$

Formulas (ii) and (iii) are obtained by placing the other two angles in standard position and repeating the steps in the above proof. ∎

The three formulas (i), (ii), and (iii) can be consolidated into one word form for the law of cosines as follows:

The square of the length of any side of a triangle is equal to the sum of the squares of the lengths of the other two sides, minus twice the product of these two sides and the cosine of their included angle.

The law of cosines is used to solve a triangle when given either

(i) two sides and the angle between them (SAS),
or
(ii) three sides (SSS).

EXAMPLE 1

Using the Law of Cosines

Use the law of cosines to find the indicated parts. Round off the answers to two decimal places.

(a) In triangle ABC (Figure 3), find c if $a = 8$, $b = 6$, and $\gamma = 60°$.

(b) Use the result in part (a) to find the measures of α and β.

Figure 3

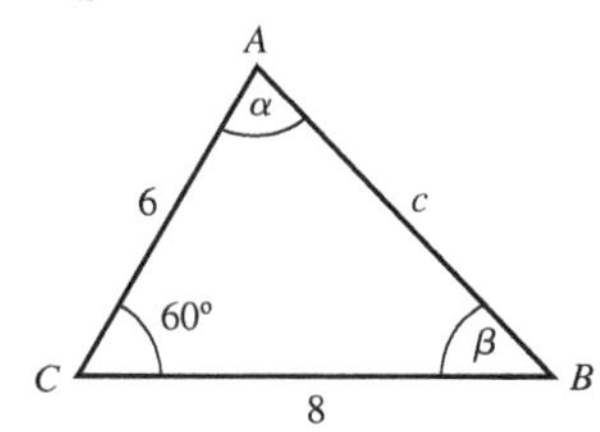

Solution

(a) We observe in Figure 3 that two sides and the included angle of the triangle are given. This corresponds to the SAS case, so the law of cosines is applicable. To find c, we write:

$$c^2 = a^2 + b^2 - 2ab\cos\gamma \qquad \text{Formula}$$

$$= 64 + 36 - 2(8)(6)\left(\frac{1}{2}\right) = 100 - 48 = 52 \qquad \text{Substitute}$$

So $c = \sqrt{52} = 2\sqrt{13}$ or 7.21 (approx.) Compute

(b) We now know the lengths of all sides of triangle ABC:

$$a = 8,\ b = 6, \quad \text{and} \quad c = 2\sqrt{13}$$

This corresponds to the SSS situation, so the law of cosines can be used to find either α or β.

To find α we proceed as follows:

$$a^2 = b^2 + c^2 - 2bc\cos\alpha \qquad \text{Law of cosines}$$

$$8^2 = 6^2 + (2\sqrt{13})^2 - 2(6)(2\sqrt{13})\cos\alpha \qquad \text{Substitute}$$

$$\cos\alpha = \frac{36 + 52 - 64}{24\sqrt{13}} = \frac{1}{\sqrt{13}} \qquad \text{Simplify and solve for } \cos\alpha$$

Therefore

We could have solved for β first, and then found α second.

$$\alpha = \cos^{-1}\left(\frac{1}{\sqrt{13}}\right) = 73.90° \text{ (approx.)}.$$

The other angle β can be found by using the fact that the sum of the interior angles of a triangle is 180°.

So,

$$\beta = 180° - 60° - 73.90° = 46.10° \text{ (approx.)}.$$

Using the Law of Sines to Solve Triangles

Next we derive another formula used to solve triangles the *law of sines.*

The Law of Sines

In any triangle *ABC:*

$$\frac{\sin\alpha}{a} = \frac{\sin\beta}{b} = \frac{\sin\gamma}{c}$$

Proof To derive the law of sines,

let h_1 be the height of triangle ABC from vertex C to side $\overline{AB}$

and

let h_2 be the height from vertex B to side $\overline{AC}$ (Figure 4).

Figure 4

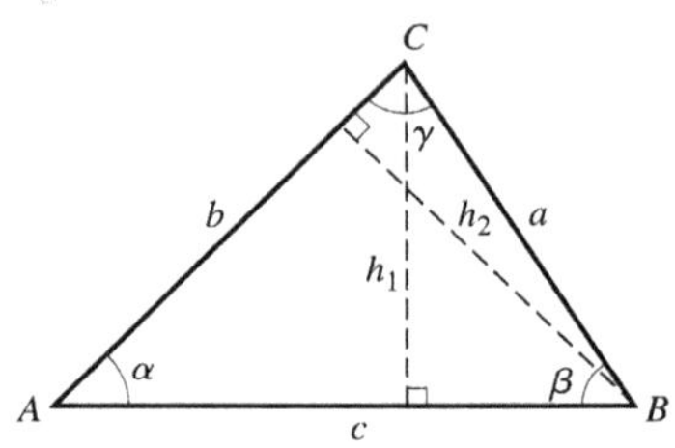

By using right triangle trigonometry, we have

$$\sin\alpha = \frac{h_1}{b} \quad \text{and} \quad \sin\beta = \frac{h_1}{a}$$

or

$$h_1 = b\sin\alpha \quad \text{and} \quad h_1 = a\sin\beta.$$

So

$$b\sin\alpha = a\sin\beta;$$

that is,

$$\frac{\sin\alpha}{a} = \frac{\sin\beta}{b}.$$

Similarly, we have

$$\sin\alpha = \frac{h_2}{c} \quad \text{and} \quad \sin\gamma = \frac{h_2}{a}.$$

So

$$h_2 = c\sin\alpha = a\sin\gamma,$$

or

$$\frac{\sin\alpha}{a} = \frac{\sin\gamma}{c}.$$

Thus

$$\frac{\sin\alpha}{a} = \frac{\sin\beta}{b} = \frac{\sin\gamma}{c}.$$

Note that the law of sines consists of the following three formulas

(i) $\dfrac{\sin\alpha}{a} = \dfrac{\sin\beta}{b}$ (ii) $\dfrac{\sin\alpha}{a} = \dfrac{\sin\gamma}{c}$ (iii) $\dfrac{\sin\beta}{b} = \dfrac{\sin\gamma}{c}$

and that it can also be written in the equivalent reciprocal form

$$\frac{a}{\sin\alpha} = \frac{b}{\sin\beta} = \frac{c}{\sin\gamma}.$$

The law of sines is used to solve a triangle when given

The AAS case can always be converted to the ASA case by first solving for the third angle.

(i) Two angles and the side between them (ASA)

or

(ii) Two angles and a side (not the included one) (AAS)

or

(iii) Two sides and an angle opposite one of them (SSA)

EXAMPLE 2

Using the Law of Sines (ASA)

In triangle ABC (Figure 5), $a = 20$, $\gamma = 51°$, and $\beta = 42°$. Solve the triangle for α, b, and c. Round off the answers to two decimal places.

Figure 5

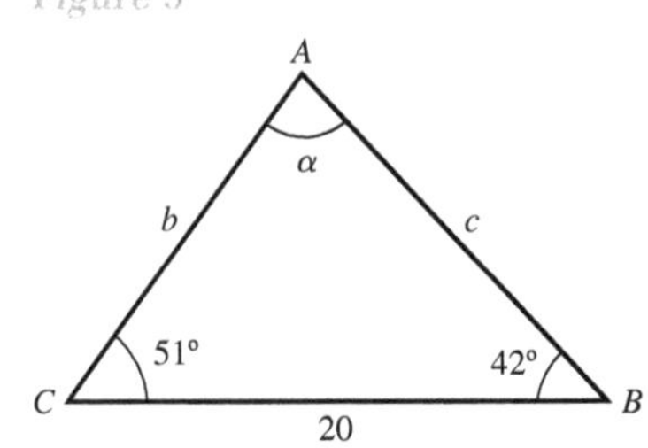

Solution Because $\alpha + \beta + \gamma = 180°$,

$$\alpha = 180° - (\beta + \gamma) = 180° - 93° = 87°.$$

Since we have an ASA situation, the law of sines is applicable. Thus

$$\frac{b}{\sin \beta} = \frac{a}{\sin \alpha}$$

$$\frac{b}{\sin 42°} = \frac{20}{\sin 87°}.$$

So
$$b = \frac{20 \sin 42°}{\sin 87°} = 13.40 \text{ (approx.)}$$

To find c, we again use the law of sines to get

$$\frac{c}{\sin \gamma} = \frac{a}{\sin \alpha}$$

$$\frac{c}{\sin 51°} = \frac{20}{\sin 87°}.$$

So
$$c = \frac{20 \sin 51°}{\sin 87°} = 15.56 \text{ (approx.)}.$$

Solving the Ambiguous Case

Data such as those given in Example 2 lead to exactly one triangle. To decide whether or not we have enough information to determine a triangle, it is often helpful to make a sketch. For instance, if we are given two angles and the included side as shown in Figure 5, then only one triangle can be formed. However, when we are given two sides and an angle opposite one of them (SSA), a unique triangle is not always determined. In this case, there are *three* possible outcomes: no triangle; only one triangle; two different triangles. These three possible outcomes are illustrated in Table 1. For this reason, we sometimes refer to the SSA situation as the **ambiguous case** of the law of sines.

TABLE 1 Ambiguous Case (SSA) Possibilities (Given a, b, and α)

$h = b \sin \alpha$	Number of Possible Triangles	Illustration
$0 < a < h$	0	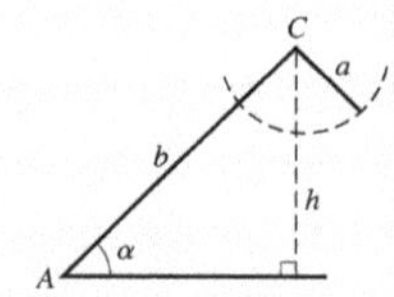
$h < a < b$	2: ΔAB_1C and ΔAB_2C	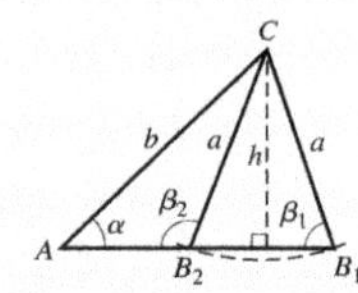
$a \geq b$	1: ΔABC	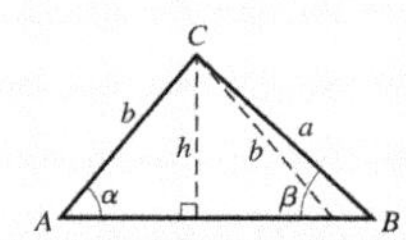
$a = h$	1	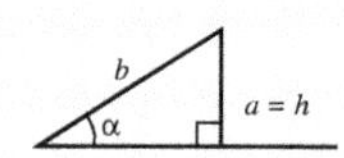

Memorizing special rules to handle each of these possibilities is unnecessary. Rather, we apply the law of sines and reason our way through the problem, as illustrated in the next example.

EXAMPLE 3

Using the Law of Sines (SSA)

Solve each triangle (if possible). Determine all angles to the nearest hundredth of a degree and all lengths to two decimal places.

(a) $a = 5$, $b = 20$, $\alpha = 30°$ (b) $a = 5$, $b = 9$, $\alpha = 33°$

(c) $a = 20$, $b = 15$, $\alpha = 29°$

Solution (a) Using the law of sines with $a = 5$, $b = 20$, and $\alpha = 30°$, we have

Figure 6

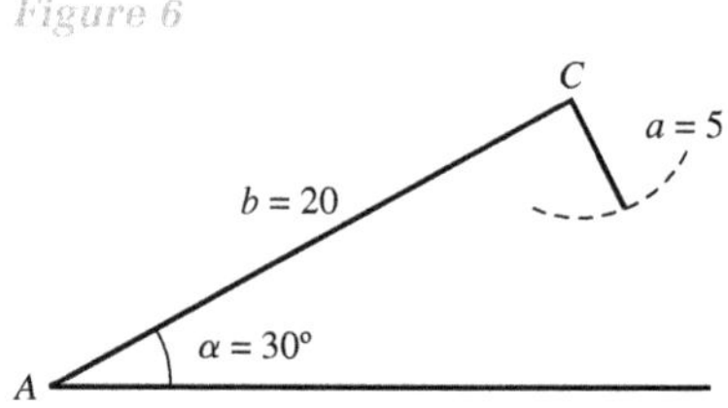

$$\frac{\sin \beta}{20} = \frac{\sin 30°}{5}$$

or
$$\sin \beta = \frac{20 \sin 30°}{5} = 2 > 1.$$

Since the sine cannot be greater than 1, there is *no possible triangle* satisfying the given conditions. Figure 6 illustrates this situation.

(b) Using the law of sines with $a = 5$, $b = 9$, and $\alpha = 33°$, we have

$$\frac{\sin \beta}{9} = \frac{\sin 33°}{5}$$

or
$$\sin \beta = \frac{9 \sin 33°}{5}.$$

One possible angle that satisfies this equation is the acute angle β_1 given by

$$\beta_1 = \sin^{-1}\left(\frac{9 \sin 33^\circ}{5}\right) = 78.62^\circ \text{ (approx.)}.$$

However, using 78.62° as a reference angle, we find an obtuse angle $\beta_2 = 180^\circ - 78.62^\circ = 101.38^\circ$ (approx.) that also satisfies the equation (Figure 7a).

Figure 7

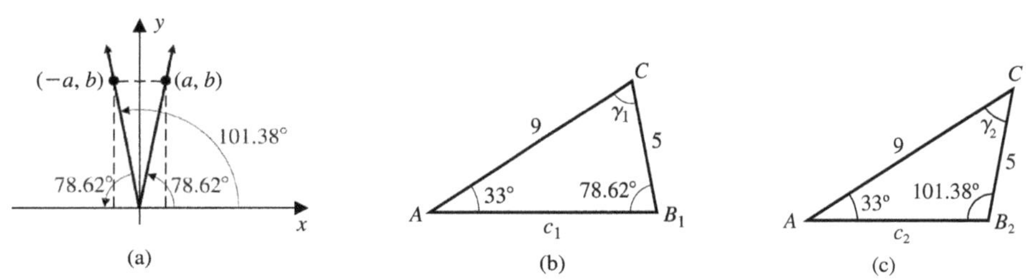

The solution $\beta_1 = 78.62^\circ$ leads to the triangle AB_1C shown in Figure 7b. Here,

$$\gamma_1 = 180^\circ - (33^\circ + 78.62^\circ) = 68.38^\circ \text{ (approx.)}.$$

So, by the laws of sines

$$\frac{c_1}{\sin \gamma_1} = \frac{a}{\sin \alpha}$$

or

$$c_1 = \frac{a \sin \gamma_1}{\sin \alpha} = \frac{5 \sin 68.38^\circ}{\sin 33^\circ} = \frac{5 \sin 68.38^\circ}{\sin 33^\circ} = 8.53 \text{ (approx.)}.$$

The solution $\beta_2 = 101.38^\circ$ leads to a second possibility, triangle AB_2C, shown in Figure 7c. For this triangle,

$$\gamma_2 = 180^\circ - (33^\circ + 101.38^\circ) = 45.62^\circ$$

So

$$\frac{c_2}{\sin \gamma_2} = \frac{a}{\sin \alpha}$$

or

$$c_2 = \frac{a \sin \gamma_2}{\sin \alpha} = \frac{5 \sin 45.62^\circ}{\sin 33^\circ} = 6.56 \text{ (approx.)}.$$

Thus in this situation we have *two possible triangles* with the given parts.

(c) Applying the law of sines with $a = 20$, $b = 15$, and $\alpha = 29°$, we have

$$\frac{\sin \beta}{b} = \frac{\sin \alpha}{a} \quad \text{or} \quad \sin \beta = \frac{b \sin \alpha}{a} = \frac{15 \sin 29°}{20}.$$

We find that an acute angle given by

$$\beta_1 = \sin^{-1}\left(\frac{15 \sin 29°}{20}\right) = 21.32° \text{ (approx.)}$$

Figure 8

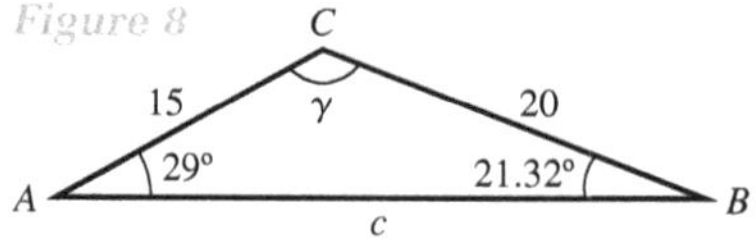

and an obtuse angle given by

$$\beta_2 = 180° - 21.32° = 158.68° \text{ (approx.)}.$$

both β_1 and β_2 give the same value of $\sin \beta$.

However, it is impossible to have angle $\beta_2 = 158.68°$ in a triangle that is known to have an angle of 29°, since

$$\alpha + \beta_2 = 29° + 158.68° = 187.68° > 180°.$$

So the β_2 solution must be rejected. There is *only one possible triangle* (Figure 8) with the given parts.
The solution $\beta_1 = 21.32°$ leads to

$$\gamma = 180° - (29° + 21.32°) = 129.68° \text{ (approx.)}$$

and
$$\frac{c}{\sin 129.68°} = \frac{20}{\sin 29°}$$

or
$$c = \frac{20 \sin 129.68°}{\sin 29°} = 31.75 \text{ (approx.)}.$$

Solving Applied Problems

The following examples illustrate applications of the law of cosines and the law of sines in mathematical modeling involving triangles.

EXAMPLE 4

Measuring Distance Indirectly

A straight tunnel with end points A and B is to be blasted through a mountain. The distance between A and B cannot be measured directly. So, from a third point C, a surveyor finds the distance from C to A to be 573 meters and from C to B to be 819 meters (Figure 9). If the measure of angle ACB is 67°, find the length of the tunnel; that is, find $|\overline{AB}|$. Round off the answer to two decimal places.

Figure 9

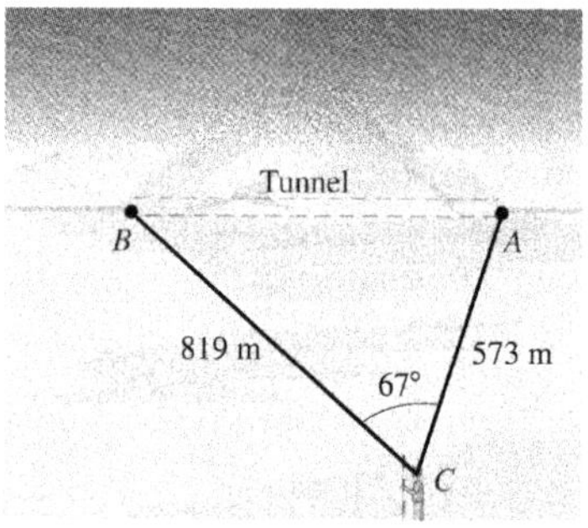

Solution Here we have an SAS case, so we apply the law of the cosines to triangle ACB to get

$$\begin{aligned} |\overline{AB}|^2 &= |\overline{CB}|^2 + |\overline{CA}|^2 - 2|\overline{CB}||\overline{CA}| \cos \gamma \\ &= 819^2 + 573^2 - 2(819)(573) \cos 67°. \end{aligned}$$

Hence

$$\begin{aligned} |\overline{AB}| &= \sqrt{819^2 + 573^2 - 2(819)(573) \cos 67°} \\ &= 795.21 \text{ (approx.)} \end{aligned}$$

so the length of the tunnel is about 795.21 meters.

EXAMPLE 5

Gear Designing

A gear designer wishes to mount three circular gears with centers at A, B, and C as the vertices of a triangle ABC. The gear with center at A has a 2-inch radius, the gear with center at B has a 3-inch radius, and the gear with center at C has a 5-inch radius. What are the measures of the angles of the triangle ABC to make this mounting possible? Round off the answers to the nearest hundredth of a degree.

Solution

This is an SSS situation with $a = 8$, $b = 7$, and $c = 5$ (Figure 10), and so we apply the law of cosines.

Figure 10

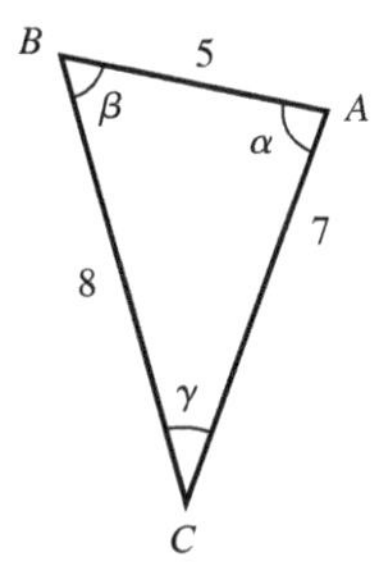

To find angle α, we write

$$a^2 = b^2 + c^2 - 2bc\cos\alpha \qquad \text{Law of sines}$$

$$64 = 49 + 25 - 2(7)(5)\cos\alpha \qquad \text{Substitute}$$

$$\cos\alpha = \frac{1}{7} \qquad \text{Solve for } \cos\alpha$$

So $$\alpha = \cos^{-1}\left(\frac{1}{7}\right) = 81.79° \qquad \text{Approximate}$$

To find angle β, we write

$$b^2 = a^2 + c^2 - 2ac\cos\beta \qquad \text{Law of cosines}$$

$$49 = 64 + 25 - 2(8)(5)\cos\beta \qquad \text{Substitute}$$

$$\cos\beta = \frac{1}{2} \qquad \text{Solve for } \cos\beta$$

$$\beta = 60° \qquad \cos^{-1}\left(\frac{1}{2}\right) = 60°$$

To find angle γ, we use

$$\gamma = 180° - (81.79° + 60°)$$
$$= 38.21°$$

Therefore, the angles α, β, and γ are approximately, 81.79°, 60°, and 38.21°, respectively.

In applications of trigonometry to surveying and navigation, the **direction** or **bearing** of a point Q as viewed from a point P is defined as the positive acute angle θ between the ray from P through Q and the north–south line through P (Figure 11). Bearings are usually measured in degrees east of north, west of north, east of south, or west of south. For instance, the symbol N25°E is read as 25° east of north.

Figure 11

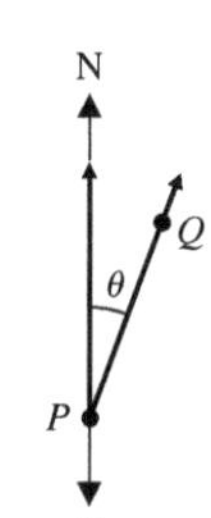

Figure 12 shows four bearings.

Figure 12

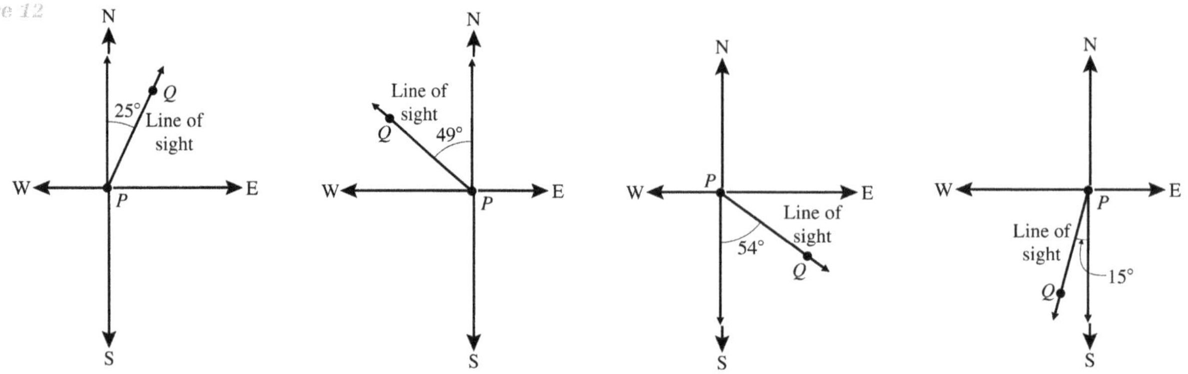

(a) N25°E (b) N49°W (c) S54°E (d) S15°W

EXAMPLE 6

Sighting a Fire from Observation Posts

A forest ranger at observation post A sights a fire at location C in the direction N66.5°E. Another ranger at observation post B, 8 kilometers directly east of A, sights the same fire at N32.2°W. Figure 13 illustrates the situation. How far is the fire from observation post A? Round off the answer to two decimal places.

Figure 13

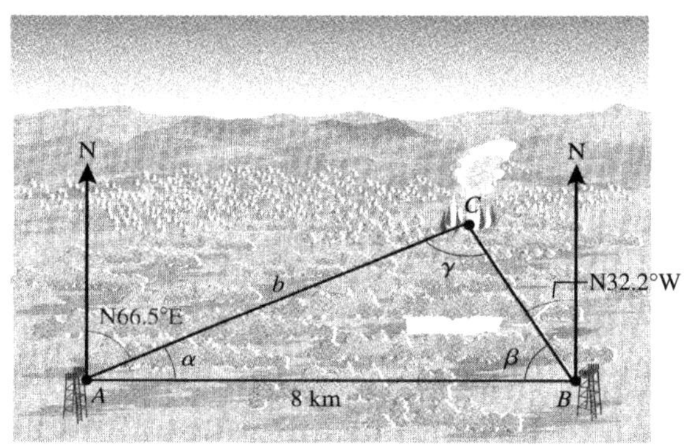

Solution Because the interior angles α and β at vertices A and B are complementary to the known angles, we find that they measure 23.5° and 57.8°, respectively. So the angle γ at C is given by

$$\gamma = 180° - (23.5° + 57.8°) = 98.7°.$$

This ASA case can now be solved by using the law of sines to get

$$\frac{b}{\sin 57.8°} = \frac{8}{\sin 98.7°}$$

or

$$b = \frac{8 \sin 57.8°}{\sin 98.7°}$$

$$= 6.85 \text{ (approx.)}.$$

Therefore, the fire is about 6.85 kilometers from the observation post A.

EXAMPLE 7

Modeling the Distance between Two Ships

Two ships leave a port at the same time, traveling along straight-line paths at 13 nautical miles per hour and 20 nautical miles per hour, respectively. Assume that the angle between their directions of travel is θ, where $0° < \theta \leq 180°$ and both ships maintain constant speeds.

(a) Express the distance d between them 1 hour later as a function of θ.

(b) Use the result in part (a) to find the distance between the ships if θ is 45°, 60°, or 75°. Round off the answers to two decimal places.

(c) Find angle θ if the distance between the ships is 10 nautical miles. Round off the answer to the nearest degree.

Solution

Figure 14

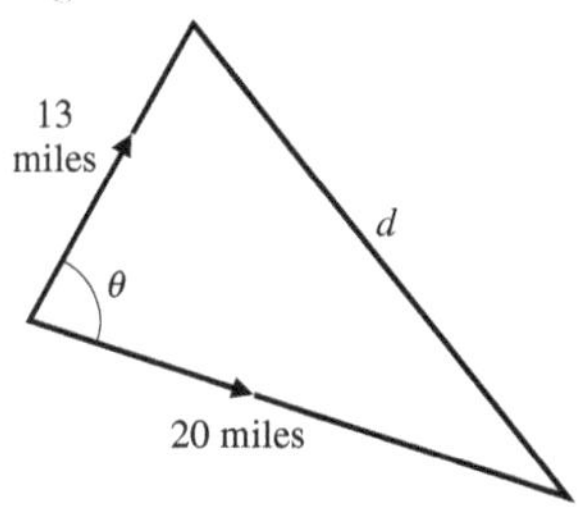

(a) After 1 hour, the ship traveling 20 nautical miles per hour will have traveled 20 nautical miles, and the ship traveling 13 nautical miles per hour will have traveled 13 nautical miles (Figure 14).

Using the law of cosines, we have

$$d^2 = 13^2 + 20^2 - 2(13)(20)\cos\theta$$

so that $$d = \sqrt{169 + 400 - 520\cos\theta}$$

or $$d = \sqrt{569 - 520\cos\theta}$$

which expresses d as a function of θ.

(b) To find the distance between the ships for the given values of θ, we substitute into the function

$$d = \sqrt{569 - 520\cos\theta}$$

to get the following approximate values:

for $\theta = 45°$, $d = \sqrt{569 - 520\cos 45°} = 14.19$ nautical miles;

for $\theta = 60°$, $d = \sqrt{569 - 520\cos 60°} = 17.58$ nautical miles;

for $\theta = 75°$, $d = \sqrt{569 - 520\cos 75°} = 20.84$ nautical miles.

(c) To find the angle θ if the distance between the ships is 10 nautical miles, we substitute 10 for d in the model to obtain the equation

$$10 = \sqrt{569 - 520\cos\theta}.$$

Thus, $$100 = 569 - 520\cos\theta$$ Square both sides

$$\cos\theta = \frac{469}{520}$$ Solve for $\cos\theta$

So, $$\theta = \cos^{-1}\left(\frac{469}{520}\right) = 26° \text{ (approx.)}.$$

PROBLEM SET 3.1

Mastering the Concepts

In problems 1–12, indicate whether the case is SAS or SSS, and then use the law of cosines to find the specified unknown part of triangle *ABC*. Round off angles to the nearest hundredth of a degree and side lengths to two decimal places.

1. a; if $b = 3$, $c = 4$, $\alpha = 30°$
2. c; if $a = 6$, $b = 4$, $\gamma = 59°$
3. b; if $a = 3$, $c = 4$, $\beta = 105°$
4. a; if $c = 2.4$, $b = 3.2$, $\alpha = 117°$
5. c; if $a = 5.8$, $b = 4.8$, $\gamma = 35.25°$
6. b; if $a = 6.1$, $c = 5.2$, $\beta = 5.67°$
7. γ; if $a = 2$, $b = 3$, $c = 4$
8. β; if $a = 14.3$, $b = 10.6$, $c = 8.4$
9. β; if $a = 23.7$, $b = 31.4$, $c = 40.6$
10. α; if $a = 0.6$, $b = 0.3$, $c = 0.5$
11. α; if $a = 81$, $b = 193$, $c = 253$
12. γ; if $a = 6$, $b = 5$, $c = 5.5$

In problems 13–22, indicate whether the case is

ASA or AAS,

and then use the law of sines to solve each triangle *ABC*. Round off angles to the nearest hundredth of a degree and side lengths to two decimal places.

13. $a = 20$, $\alpha = 29°$, $\beta = 136°$
14. $c = 32$, $\beta = 57°$, $\alpha = 38°$
15. $b = 30$, $\alpha = 80°$, $\gamma = 41°$
16. $a = 10.5$, $\alpha = 41°$, $\gamma = 77°$
17. $b = 19.7$, $\alpha = 42.17°$, $\gamma = 61.33°$
18. $b = 20$, $\alpha = 30°$, $\gamma = 52°$
19. $c = 5.4$, $\beta = 50.83°$, $\gamma = 70.5°$
20. $a = 13.1$, $\gamma = 100°$, $\alpha = 12.67°$
21. $b = 19$, $\alpha = 40°$, $\gamma = 62°$
22. $b = 38.8$, $\alpha = 103.45°$, $\gamma = 27.19°$

In problems 23–32, if possible, solve the ambiguous case

SSA

by using the law of sines. Round off angles to the nearest hundredth of a degree and side lengths to two decimal places. Be sure to find all possible triangles that satisfy the given conditions.

23. $b = 12$, $c = 11$, $\gamma = 81°$
24. $a = 140$, $c = 115$, $\gamma = 53.54°$
25. $a = 27$, $b = 52$, $\alpha = 70°$
26. $a = 1$, $b = 1.8$, $\alpha = 26°$
27. $a = 12.4$, $b = 8.7$, $\beta = 36.67°$
28. $a = 12.41$, $b = 81.69$, $\beta = 36.67°$
29. $a = 21.3$, $b = 18.9$, $\alpha = 65.18°$
30. $a = 263.6$, $c = 574.3$, $\alpha = 32.32°$
31. $a = 30$, $b = 44.5$, $\alpha = 33.33°$
32. $a = 10.1$, $b = 15.2$, $\alpha = 67.67°$

In problems 33–40, identify the case

SSS, SAS, ASA, AAS, or SSA

and then use the law of sines or the law of cosines to solve triangle *ABC*. Round off all angles to the nearest hundredth of a degree and side lengths to two decimal places. Be sure to find all possible triangles that satisfy the given conditions.

33. $a = 10$, $b = 50$, $\alpha = 22°$
34. $a = 14$, $\alpha = 21°$, $\beta = 35°$
35. $a = 4$, $b = 4$, $c = 6$
36. $a = 5$, $b = 10$, $c = 10$
37. $a = 14$, $\alpha = 12°$, $\beta = 97°$
38. $a = 10$, $b = 12$, $\gamma = 108°$
39. $b = 3.7$, $\alpha = 100.30°$, $\beta = 5.50°$
40. $b = 98$, $c = 39$, $\alpha = 17°$

Applying the Concepts

In problems 41–58, round off angles to the nearest hundredth of a degree and lengths to two decimal places.

41. **Surveying:** A vertical tower that is 70 feet high stands on a hill that makes an angle of 14° with the horizontal. A surveyor locates a point C downhill, 110 feet from the base of the tower at A (Figure 15). Find the distance from C to the top of the tower, B; that is, find $|\overline{CB}|$.

Figure 15

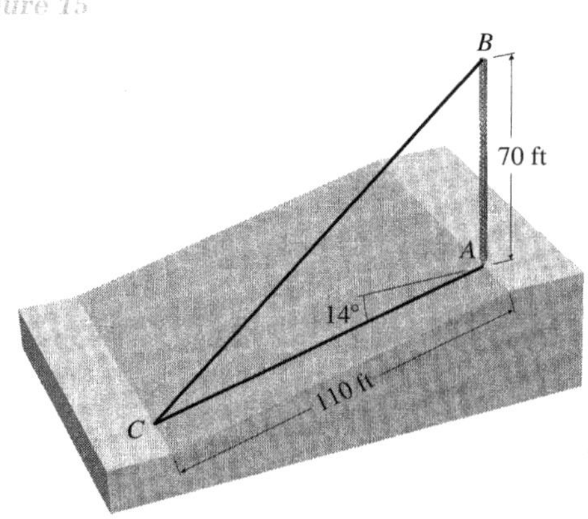

42. **Distance between Jets:** Two jets leave an air base at the same time and fly at the same speed along straight courses forming an angle of 112.45° with each other (Figure 16). After the jets have each flown 504 kilometers, how far apart are they?

Figure 16

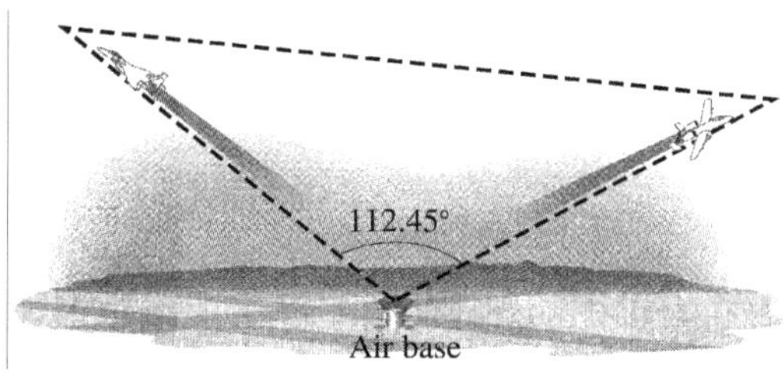

43. **Surveying:** Engineers are planning to construct a straight tunnel from point A on one side of a hill to point B on the other side. Since its length cannot be measured directly, point C is chosen by a surveyor who is 473 meters from A and 367 meters from B (Figure 17). If the measure of angle ACB is 46.2°, find the length of the tunnel.

Figure 17

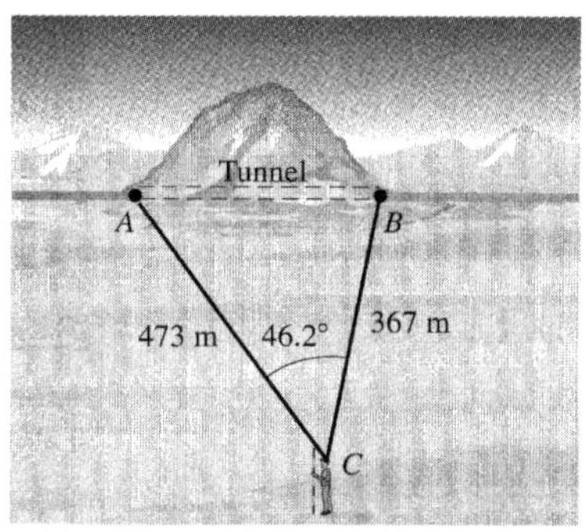

44. **Surveying:** To measure the length of a lake between points A and B, a point C on land is chosen. The distance between C and A is found to be 137 meters, and the distance between C and B is 405 meters (Figure 18). If angle ACB is 42.5°, how long is the lake?

Figure 18

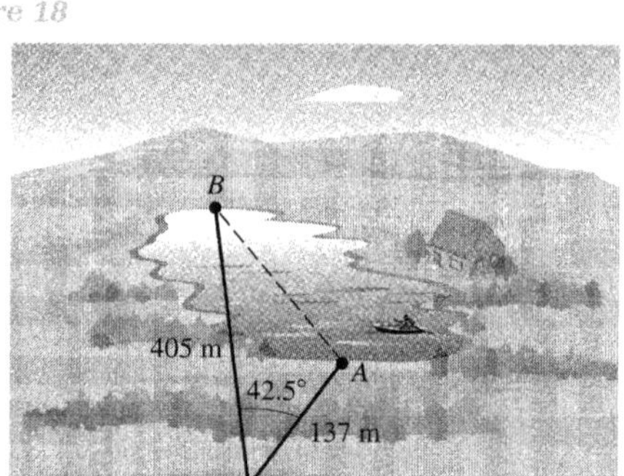

45. **Joggers' Courses:** Two straight roads intersect at 65° (Figure 19). A jogger on one road is 5 miles from the intersection and moving away from it at a constant rate of 7 miles per hour. At the same instant, another jogger on the other road is 4 miles from the intersection and moving away from it at a constant rate of 5 miles per hour. If the two joggers maintain constant speeds, what is the distance between them 30 minutes later?

Figure 19

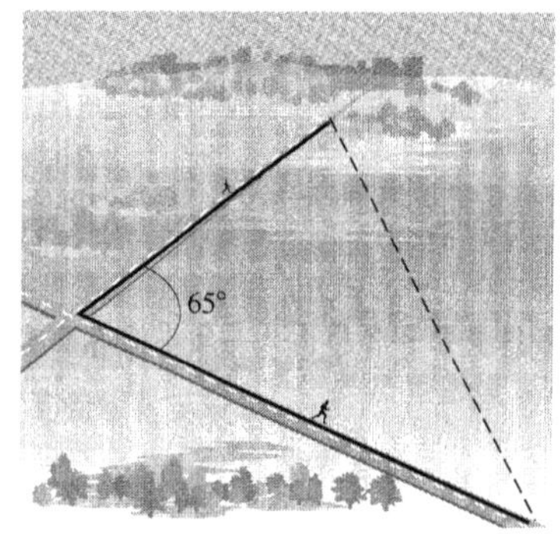

46. **Navigation:** A navigator of a ship plots a straight course from point A to point B. Because of an error, the ship proceeds from point A along a slightly different straight-line course. After traveling 3 hours on the wrong course, the error is discovered and the ship is turned through an angle of 22.5° so that it can reach point B (Figure 20). After 5 more hours, the ship reaches point B. Assuming that the ship was proceeding at a constant speed of 10 nautical miles per hour, find the time lost (to the nearest minute) because of the error.

Figure 20

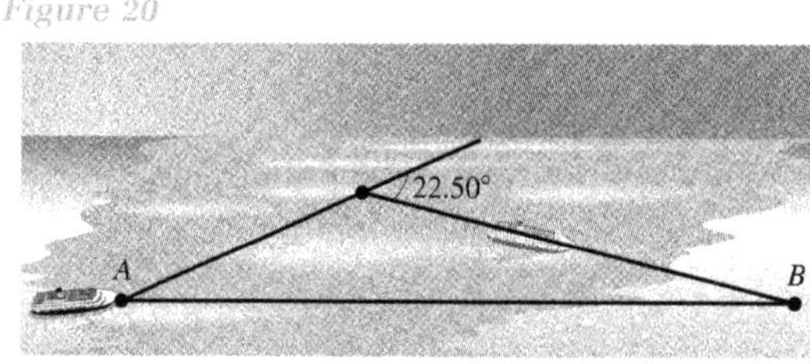

47. **Lighthouse Lookout:** As shown in Figure 21, two lighthouses along a straight shoreline of a lake are located 2.43 kilometers apart at points A and B. The keepers of the lighthouses at A and B sight an overturned boat at point C at angles 70.3° and 48.2°, respectively. A Coast Guard rescue boat at D, located 1.2 kilometers downshore from B, is to rescue the boat. How far is the Coast Guard boat from the overturned boat?

Figure 21

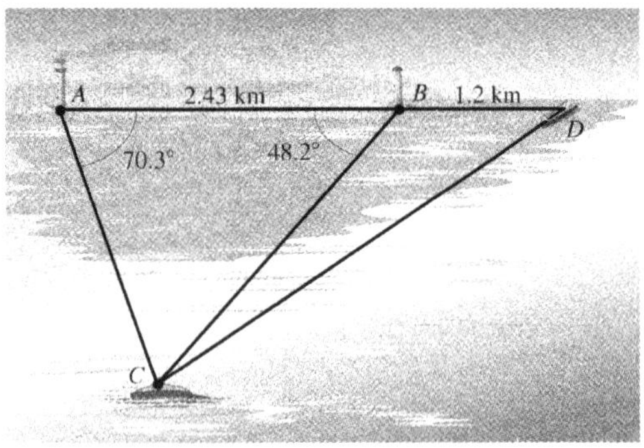

48. Baseball: Figure 22 shows a baseball diamond. It is in the form of a square that is 90 feet long on each side. The pitcher's mound at P is 60.5 feet from home plate, H, along the diagonal from home plate to second base. How far is the pitcher's mound P from first base, located at F?

Figure 22

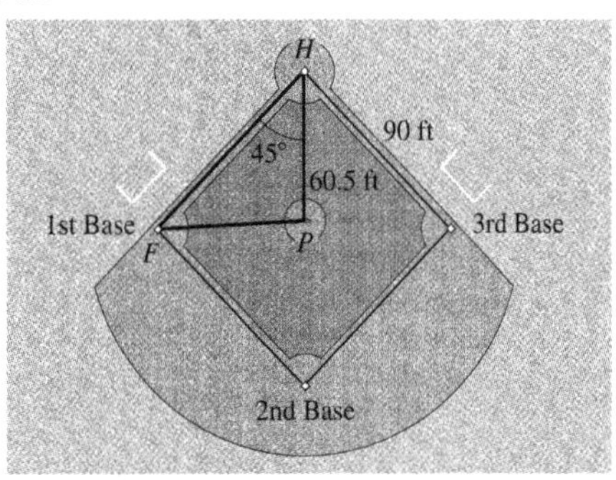

49. Fire Lookout: Two forest rangers are 1.2 kilometers apart. One ranger sights a fire at an angle of 41.43° from the line between the two observation points. The other ranger sights the same fire at an angle of 61.4° from the same line between the two observation points (Figure 23). How far is the fire from each observation point?

Figure 23

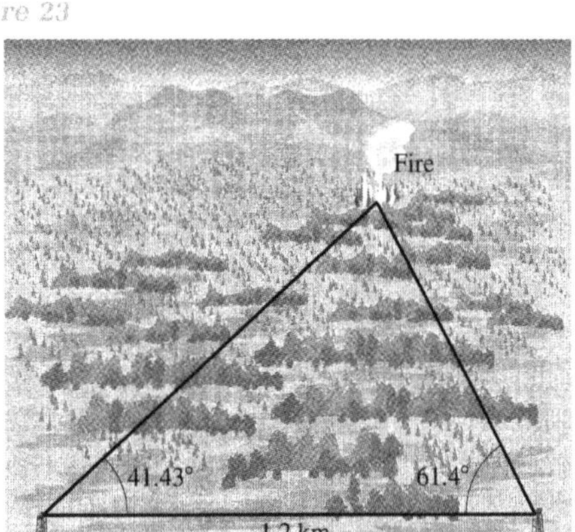

50. Distance to Top of a Tree: A biologist standing on the ground on the same horizontal plane as the base of a tree determines that the angle of elevation from a point A to the top of the tree is 26.2°. After walking 150 feet closer to the tree to a point B, she finds the angle of elevation from point B to the top of the tree to be 37.1°. Find the distance from point B to the top of the tree.

51. Distance from a Parachutist: A sports parachutist is sighted simultaneously by two observers 3 miles apart on opposite sides of the parachutist. The observed angles of elevation are 25.5° and 17.8°. Assuming that the parachutist and the two observers lie in the same vertical plane, find the distance of the parachutist from the farther observer.

52. Engineering: The crankshaft $\overline{OA}$ of an engine is 5.8 centimeters long, and the connecting rod $\overline{AP}$ is 20.3 centimeters long (Figure 24). Find angle AOP at the instant when angle APO is 12°.

Figure 24

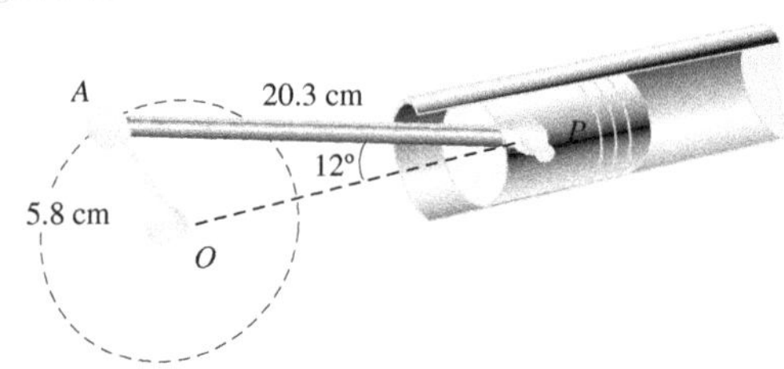

53. Streetlight Design: A streetlight is to be mounted on a brick wall. The length of brace $\overline{BC}$ is 1.3 meters, angle BCA is 24.8°, and angle CAB is 37.8° (Figure 25). Find the length of the supporting brace $\overline{AC}$.

Figure 25

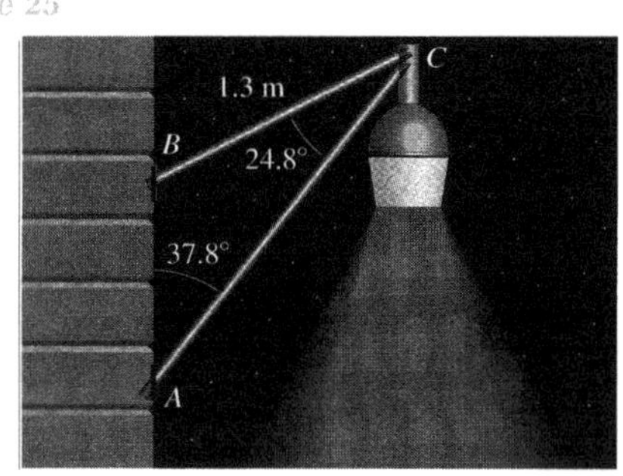

54. Communications Satellite: A communications satellite traveling in a circular orbit 1600 kilometers above the surface of the earth is located at position A by a tracking station B on the earth at a certain time (Figure 26). If the tracking antenna is aimed 35° above the horizon and if the radius of the earth is 6400 kilometers, what is the distance from the antenna to the satellite?

Figure 26

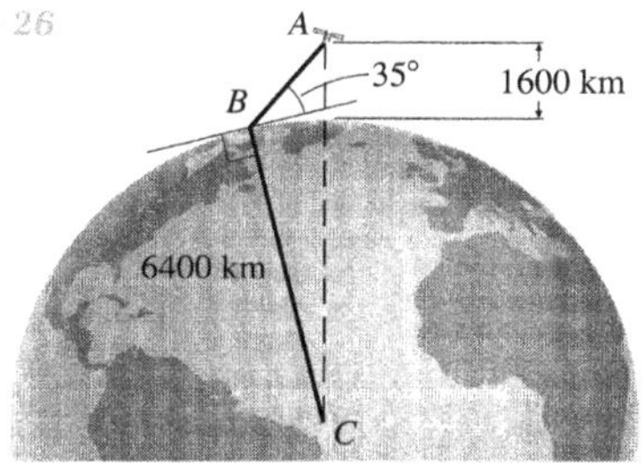

55. **Navigation:** A ship sails 19 nautical miles in the direction S29.3°W, and then turns onto a course S51.7°W and sails 24 nautical miles. How far is the ship from the starting point?

56. **Navigation:** A ship leaves a harbor at noon and sails S50°W at 18 nautical miles per hour until 2:30 PM. At that time, it changes course and sails N20°W at a reduced speed of 13 nautical miles per hour until 4:00 PM. Find:
   (a) The distance of the ship from the harbor at 4:00 PM
   (b) The ship's bearing from the harbor at 4:00 PM

57. **Distance Model:** Two straight roads intersect at an angle of 70°. A bus on one of the roads is 8 kilometers from the intersection and moving away from it at a rate of 90 kilometers per hour. At the same instant, a truck on the other road is 16 kilometers from the intersection and moving away from it at 100 kilometers per hour. Assume the bus and the truck maintain constant speeds.
   (a) Express the distance d between them t hours later as a function of t.
   (b) Find the distance between the two vehicles after 30 minutes, after 60 minutes, and after 75 minutes.
   (c) Approximate how long it takes for the distance between the vehicles to be 50 kilometers, 100 kilometers, and 150 kilometers. Round off the answers to two decimal places.

58. **Navigation—Distance Model:** Two ships leave a port at the same time, traveling along straight-line paths at 12 nautical miles per hour and 16 nautical miles per hour, respectively. Assume that the angle between their directions of travel is θ, where $0° < \theta \leq 180°$, and both ships maintain constant speeds.
   (a) Express the distance d between them 1 hour later as a function of θ.
   (b) Use the result in part (a) to find the distance between the ships if θ is 10°, 90°, or 120°.
   (c) Find angle θ if the distance between the ships is 5 nautical miles, 20 nautical miles, or 25 nautical miles. Round off the answers to the nearest degree.

Developing and Extending the Concepts

59. **Geometry:** Figure 27 shows three circles with centers at P, Q, and R, and with radii 5, 8, and 8.4 centimeters, respectively. If the circles are tangent to one another, find the three angles of triangle PQR formed by the line segments connecting their centers.

Figure 27

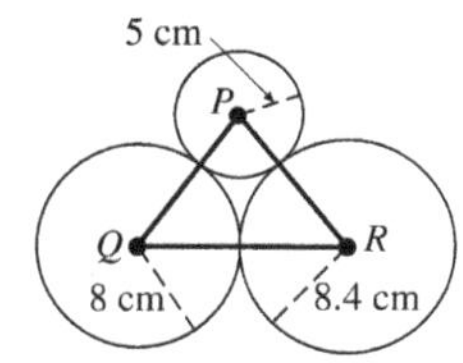

60. **Geometry:** Assume that a diagonal of a parallelogram is 80 inches long and that one end of it forms angles of 35° and 27°, respectively, with the two sides. Find the lengths of the sides of the parallelogram.

61. Find the lengths of the diagonals of parallelogram $ABCD$ with side $\overline{AB}$ of length 10 inches and side $\overline{BC}$ of length 15 inches, and with angle ABC of 110°.

62. Rewrite formula (i) of the law of cosines if $\gamma = 90°$. Explain the result.

63. Explain why it is not possible to have a triangle with side lengths of 7, 8, and 16. What happens if we attempt to find an angle by using the law of cosines under the SSS case?

64. Assume that triangle ABC has height $h = b \sin \alpha$ (Figure 28).

Figure 28

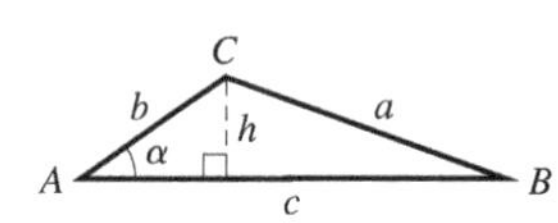

Using the law of cosines, we have

$$a^2 = b^2 + c^2 - 2bc \cos \alpha$$

or $\quad b^2 + c^2 - a^2 - 2bc \cos \alpha = 0.$

So $\quad c^2 - (2b \cos \alpha)c + (b^2 - a^2) = 0.$

   (a) Use the quadratic formula with c as the unknown to show that
   $$c = b \cos \alpha \pm \sqrt{a^2 - b^2 \sin^2 \alpha}.$$
   (b) Under what conditions on a and h are there no real solutions for c? Relate the result to the ambiguous case when no triangles exist for given values of a, b, and α.
   (c) If $a^2 - b^2 \sin^2 \alpha > 0$, how many possible triangles exist? Explain.

Interlude: Practice Problems 3.1

1. The course for a boat race starts at a point S and proceeds N44°W to point A, then N38°E to a point B that is exactly 15 miles north of point S, and then proceeds back to point S. Calculate the total distance of the race course, accurate to three decimal places.

2. A plane flies 830 km with a bearing of S24°E from point A to point B. The plane then turns and flies 102 km northeast to a point C that is due east of point A. Find the bearing of the flight from B to C, accurate to two decimal places.

3. The leading edge of each wing of the B-2 Stealth Bomber measures 105.6 feet in length. The angle between these leading edges is 109.05°. Calculate the wing span of the B-2, accurate to 3 decimal places.

4. A ship travels 36 miles with a bearing of N78°E and then adjusts to a bearing of N138°E for an additional 48 miles. Determine both the distance and bearing of the boat from its starting point, accurate to two decimal places.

Objectives

1. Plot Points in the Polar Coordinate System
2. Convert Points from Polar to Cartesian Form and Vice Versa
3. Convert Equations from Cartesian to Polar Form and Vice Versa
4. Graph Polar Equations

3.2 Introduction to Polar Coordinates

So far, we have used the Cartesian (rectangular) coordinate system to represent points in the plane. There is another system, called the *polar coordinate system,* where we use angles and radial distances to represent points. For certain procedures, it is easier to work with equations written in terms of polar coordinates rather than Cartesian coordinates. In this section, we'll study the polar coordinate system and use trigonometry to relate it to the Cartesian coordinate system.

Plotting Points in the Polar Coordinate System

The frame of reference for the polar coordinate system consists of a fixed point O, called the **pole,** and a fixed ray with end point O, called the **polar axis** (Figure 1a). The position of a point P in this system is given by coordinates (r, θ), called **polar coordinates** of P (Figure 1b).

Figure 1

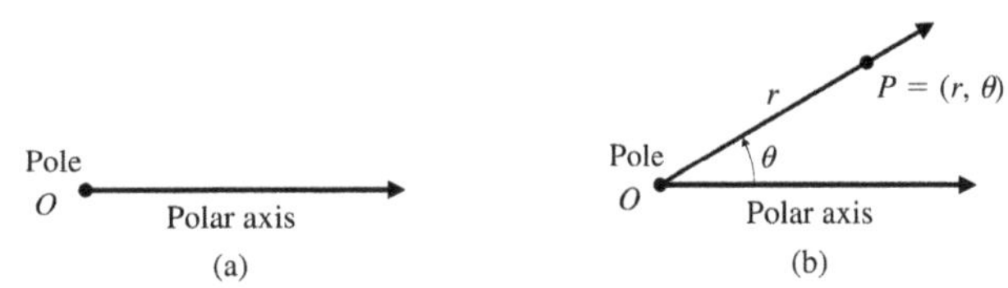

θ is *any* angle having the polar axis as its initial side and the ray from the pole through point P as its terminal side; r is the directed distance along the terminal side of θ from the pole O to P.

Figure 2 shows a sample grid for the polar coordinate system. The angle θ can be measured either in degrees or radians. It may be positive or negative, depending on whether it is generated by a counterclockwise or clockwise rotation.

Figure 2

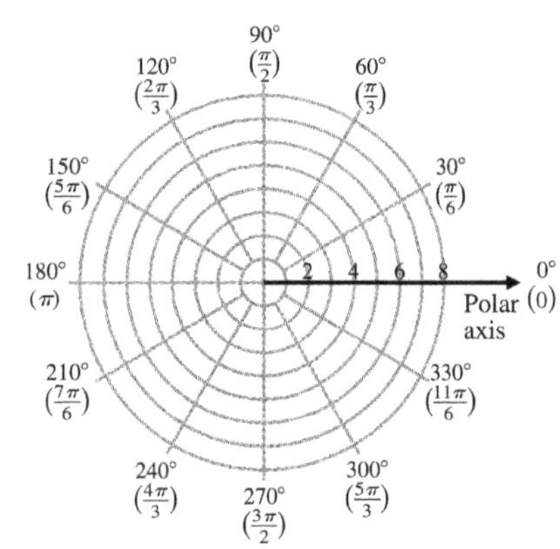

EXAMPLE 1

Plotting Points in the Polar Coordinate System

Plot the points with the given polar coordinates.

(a) $(3, 60°)$ (b) $(4, -30°)$ (c) $\left(4, \dfrac{3\pi}{4}\right)$

Solution

(a) To plot the point $(3, 60°)$, we first form an angle of $60°$ in such a way that the polar axis is the initial side of the angle. Then we move 3 units from the pole on the terminal side of this angle to locate the point (Figure 3).

(b)–(c) The other points in parts (b) and (c) are plotted in a similar way (Figure 3).

Figure 3

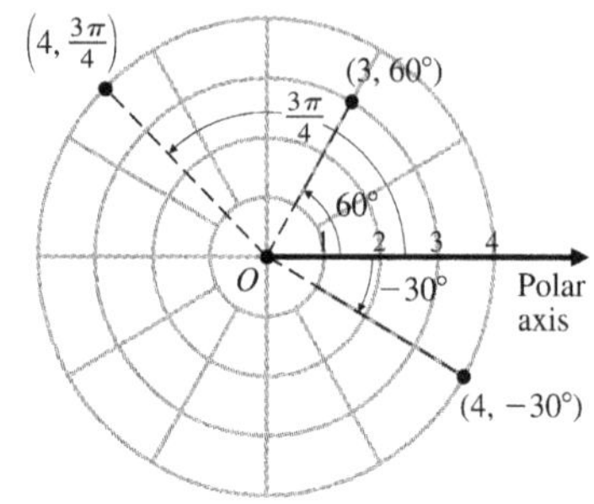

Figure 4

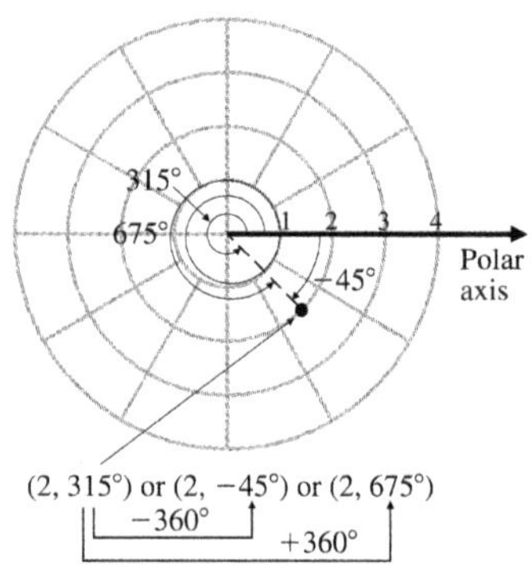

Note that in the Cartesian system each point has a *unique* number pair representation, whereas in the polar system, a point has *infinitely* many possible number pair representations because of periodicity. For example, the polar coordinates (2, 315°), (2, −45°), and (2, 675°) all represent the same point (Figure 4).

In general, if a point P has polar coordinates (r, θ), then for any integer n,

$$(r, \theta + 2\pi n) \quad \text{or} \quad (r, \theta + 360°n)$$

are also polar coordinates of P.

The value of r for polar coordinates (r, θ) may be positive, negative, or 0. If r is negative, then the location of the point (r, θ) is the same as the point

$$(|r|, \theta + \pi) \quad \text{or} \quad (|r|, \theta + 180°)$$

For example, the point $(-3, \pi/4)$ has the same location as $(3, 5\pi/4)$ (Figure 5).

In other words, a negative value for r indicates moving $|r|$ units in the direction opposite that of the terminal side of θ in order to locate the point (r, θ). Also, if $r = 0$, then θ can be any angle.

Figure 5

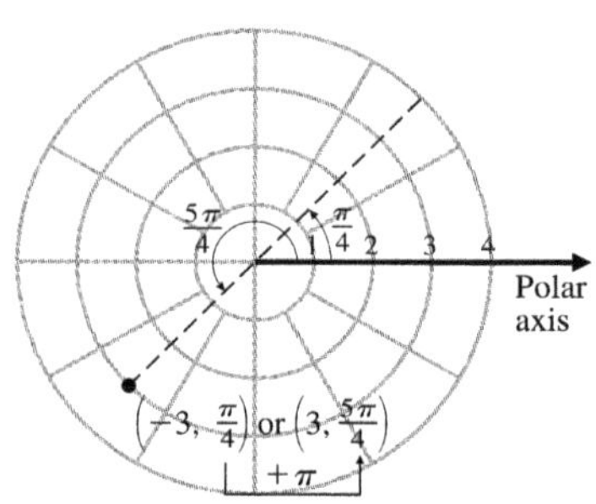

Converting Points from Polar to Cartesian Form and Vice Versa

To establish the relationship between polar coordinates and Cartesian coordinates, we draw the polar axis of a polar coordinate system so that it coincides with the positive x axis of a Cartesian coordinate system.

Any point P in the plane can be located either by polar coordinates (r, θ) or by Cartesian coordinates (x, y).

For instance, suppose that P is in quadrant I and $r > 0$ (Figure 6). Using the right triangle in Figure 6, we obtain the following relationships between the coordinates:

Conversion of Coordinates

Polar or Cartesian

1. To convert a point P from polar coordinates (r, θ) to Cartesian coordinates (x, y), use
$$x = r\cos\theta \quad \text{and} \quad y = r\sin\theta.$$
2. To convert a point P from Cartesian coordinates (x, y) to polar coordinates (r, θ), use
$$r^2 = x^2 + y^2$$
so that if $r > 0$, then $r = \sqrt{x^2 + y^2}$; also use
$$\tan\theta = \frac{y}{x}.$$

Figure 6

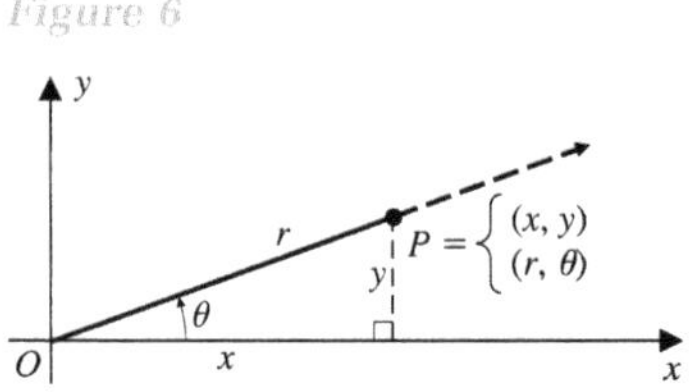

These relationships are true regardless of the quadrant location of P, even when $r \leq 0$.

EXAMPLE 2

Converting Polar Coordinates to Cartesian Coordinates

Convert the given polar coordinates to Cartesian coordinates (x, y). Plot the point.

(a) $(3, 60°)$ (b) $\left(4, -\dfrac{5\pi}{6}\right)$

Solution To convert these points from polar to Cartesian coordinates, we proceed as follows:

(a) For the point $(3, 60°)$, we have

$$x = r\cos\theta = 3\cos 60° = 3 \cdot \frac{1}{2} = \frac{3}{2}$$

$$y = r\sin\theta = 3\sin 60° = 3 \cdot \left(\frac{\sqrt{3}}{2}\right) = \frac{3\sqrt{3}}{2}.$$

So the Cartesian coordinates are

$$\left(\frac{3}{2}, \frac{3\sqrt{3}}{2}\right) \quad \text{(Figure 7a).}$$

(b) For the point $\left(4, -\dfrac{5\pi}{6}\right)$, we have

$$\begin{aligned} x = r\cos\theta &= 4\cos\left(-\frac{5\pi}{6}\right) \\ &= 4\left(-\frac{\sqrt{3}}{2}\right) \\ &= -2\sqrt{3} \\ y = r\sin\theta &= 4\sin\left(-\frac{5\pi}{6}\right) \\ &= 4\left(-\frac{1}{2}\right) \\ &= -2. \end{aligned}$$

So the Cartesian coordinates are

$$(-2\sqrt{3}, -2) \quad \text{(Figure 7b).}$$

Figure 7

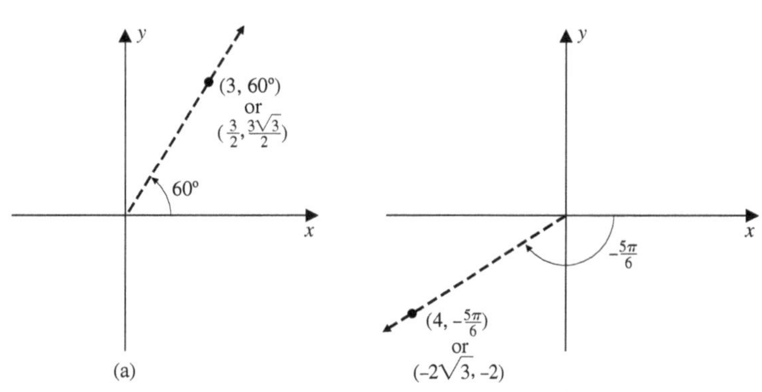

It should be noted that the conversion equation $\tan\theta = y/x$ does not define θ uniquely. To determine the polar coordinates of P, it is helpful to plot the given point because we must pay attention to the quadrant in which P lies and the restrictions on θ, as the next example illustrates.

EXAMPLE 3

Converting Cartesian Coordinates to Polar Coordinates

Convert the given Cartesian coordinates to polar coordinates $(r, 0)$, where $r \geq 0$ and $0° \leq \theta < 360°$.

(a) $(-1, 1)$ (b) $(5.03, -2.76)$

Round off the answers to two decimal places.

Solution Here we use the conversion equations $r = \sqrt{x^2 + y^2}$ and $\tan\theta = y/x$, keeping in mind that the polar coordinates of a given point are not unique.

(a) For the point $(-1, 1)$ (Figure 8a), we have

$$r = \sqrt{x^2 + y^2} = \sqrt{(-1)^2 + 1^2} = \sqrt{2} \quad \text{and} \quad \tan\theta = -1.$$

Since the point $(-1, 1)$ lies in quadrant II, $90° < \theta < 180°$, so we select $\theta = 135°$ as the solution of $\tan\theta = -1$. So polar coordinates of the point are $(\sqrt{2}, 135°)$.

(b) For the point $(5.03, -2.76)$ (Figure 8b), we have

$$r = \sqrt{(5.03)^2 + (-2.76)^2} = 5.74 \text{ (approx.)} \quad \text{and} \quad \tan\theta = \frac{-2.76}{5.03}.$$

Since $\tan\theta$ is negative and θ is restricted to $0° \leq \theta < 360°$, it follows that θ is a positive angle in quadrant IV with reference angle θ_R given by

$$\theta_R = \tan^{-1}\frac{2.76}{5.03} = 28.75° \text{ (approx.)}.$$

So, θ is approximately $360° - \theta_R = 331.25°$, and the polar coordinates are $(5.74, 331.25°)$.

Figure 8

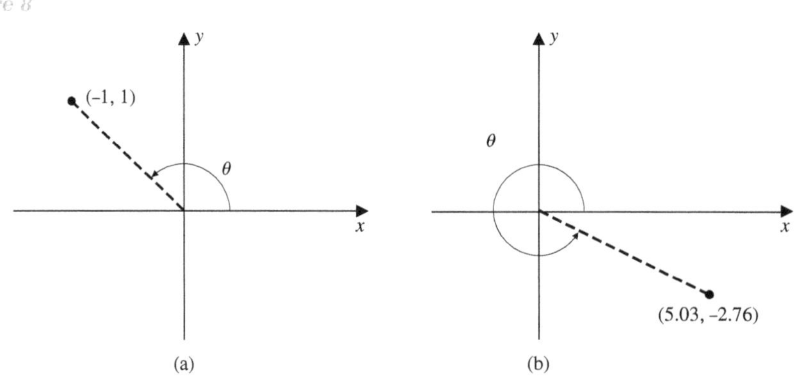

Converting Equations from Cartesian to Polar Form and Vice Versa

We know that an equation in x and y may be represented by a graph in the Cartesian coordinate system. Suppose we superimpose on the xy system a polar system so that the polar axis coincides with the positive x axis. Clearly, the graph maintains its shape. By using the conversion equations we can determine an equation for the graph in terms of polar coordinates r and θ.

EXAMPLE 4 **Converting a Cartesian Equation to Polar Form**

Sketch the graph of the Cartesian equation

$$(x - 2)^2 + y^2 = 4$$

and then find an equation of the graph in polar form.

Solution We recognize the graph of this equation as a circle with radius 2 and center at (2, 0) (Figure 9). To convert the given equation to polar form, we proceed as follows:

Figure 9

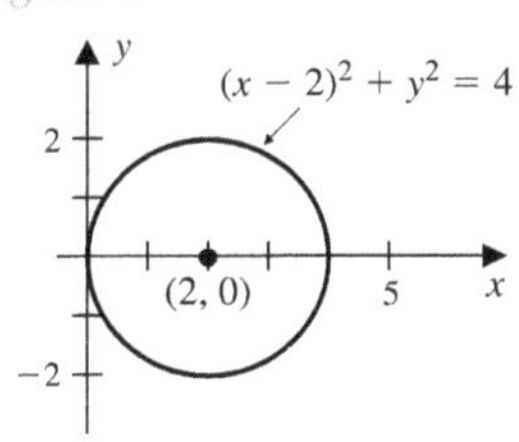

$$(x - 2)^2 + y^2 = 4 \qquad \text{Given}$$
$$x^2 - 4x + 4 + y^2 = 4 \qquad \text{Multiply}$$
$$x^2 + y^2 - 4x = 0 \qquad \text{Simplify and rewrite}$$
$$r^2 - 4r\cos\theta = 0 \qquad x^2 + y^2 = r^2 \text{ and } x = r\cos\theta$$
$$r^2 = 4r\cos\theta \qquad \text{Add } 4r\cos\theta \text{ to each side}$$
$$r = 4\cos\theta \qquad \text{Divide each side by } r$$

Note that the resulting equation is valid even when $r = 0$.

We can also convert an equation from polar form to Cartesian form. Once again, note that even though the equations are given in different coordinate systems, the graphs are the same.

EXAMPLE 5 **Converting a Polar Equation to Cartesian Form**

Convert the polar equation $\theta = \pi/3$ to Cartesian form, and then sketch the graph.

Solution Since $\theta = \pi/3$, it follows from the conversion equation $\tan\theta = y/x$ that

$$\tan\frac{\pi}{3} = \frac{y}{x} \quad \text{or} \quad \sqrt{3} = \frac{y}{x}.$$

Figure 10

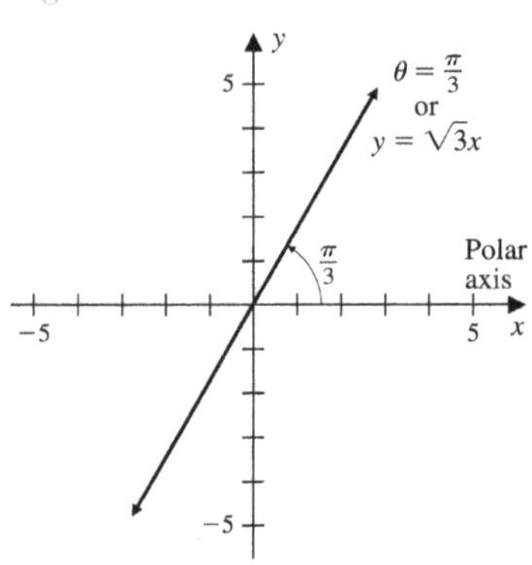

That is,

$$y = \sqrt{3}x.$$

This last equation is the equation of a line with slope $\sqrt{3}$ and containing the origin (Figure 10).

Graphing Polar Equations

By the **graph of a polar equation,** we mean the set of all points (r, θ) that satisfy an equation that relates r and θ. In Example 5, we graphed the polar equation by first converting it to Cartesian form, and then we used the Cartesian coordinate system to sketch the graph. Polar equations can also be graphed directly by using the point-plotting method.

EXAMPLE 6

Graphing Polar Equations Using Point-Plotting

Use point-plotting to sketch the graph of each equation.

(a) $r = 3$ (b) $r = \theta$, where $\theta \geq 0$ (c) $r = 1 + \sin\theta$

Solution (a) The equation $r = 3$ indicates that a point is on the graph if and only if it is of the form $(3, \theta)$, where θ is any angle.

After plotting a few points of this form and then connecting them with a smooth curve, we see that the graph consists of all points that are 3 units away from the pole. The graph is the circle with center at the pole and radius 3 (Figure 11).

Figure 11

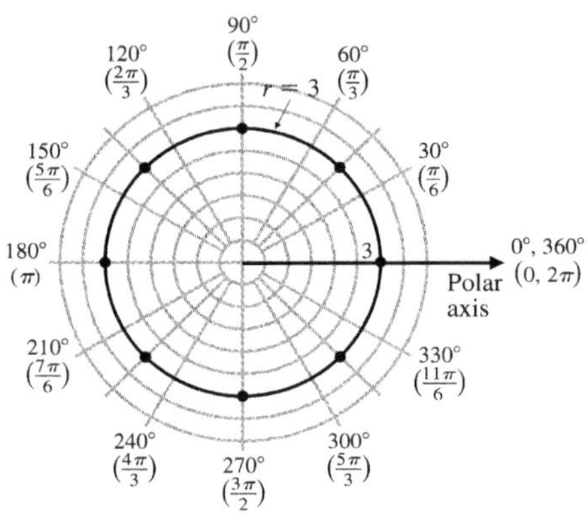

The graph of an equation of the form $r = k\theta$ is called an Archimedean spiral.

(b) Since r represents distances (real numbers) and $r = \theta$ (for $\theta \geq 0$), it is implied that θ represents real numbers or radian measures of angles. Table 1 lists some coordinates that satisfy the equation, and Figure 12 shows the graph resulting when these points are plotted and connected with a smooth curve. Note that as the angle θ is allowed to increase (rotate counterclockwise), the distances r correspondingly get longer. This results in a spiral-shaped curve.

TABLE 1

θ	0	$\frac{\pi}{2}$	π	$\frac{4\pi}{3}$	$\frac{3\pi}{2}$	2π
r	0	1.57 (approx.)	3.14 (approx.)	4.19 (approx.)	4.71 (approx.)	6.28 (approx.)

Figure 12

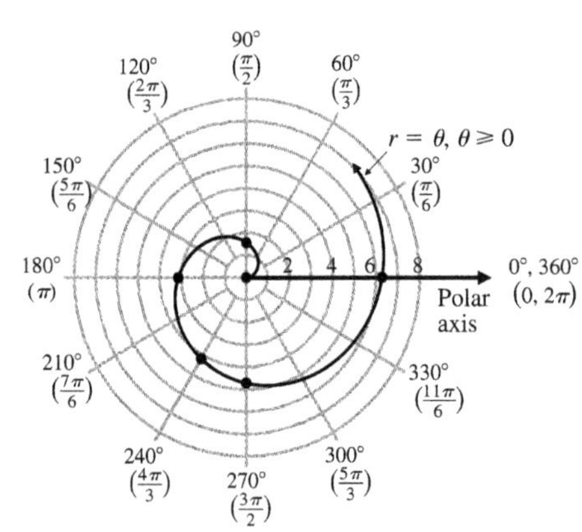

(c) To graph

$$r = 1 + \sin\theta,$$

we first build a table of coordinates that satisfy the equation (Table 2), then plot the points, and finally connect them with a smooth curve. The resulting graph is shown in Figure 13.

TABLE 2

θ	0	$\frac{\pi}{6}$	$\frac{\pi}{3}$	$\frac{\pi}{2}$	$\frac{2\pi}{3}$	$\frac{5\pi}{6}$	π	$\frac{3\pi}{2}$	2π
r	1	$\frac{3}{2} = 1.5$	1.87 (approx.)	2	1.87 (approx.)	$\frac{3}{2} = 1.5$	1	0	1

Figure 13

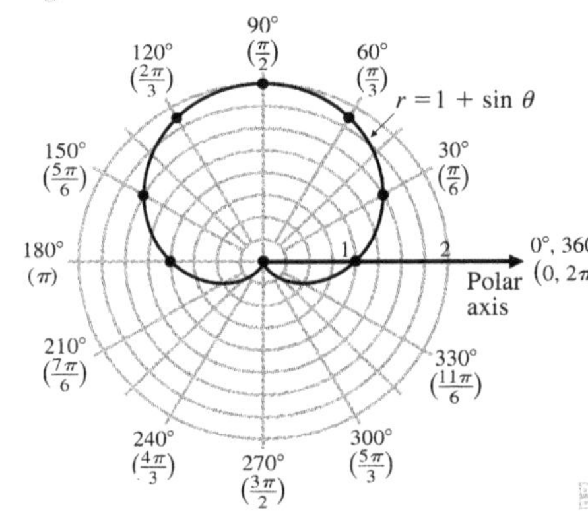

This graph is called a cardioid because of its heart shape.

Graphers have a polar graphing mode that enables us to sketch the graphs of polar equations of the form $r = f(\theta)$. Care must be taken when setting the range for θ, according to whether the degree mode or radian mode has been selected.

EXAMPLE 7

Graphing a Polar Equation

Use a grapher to sketch the graph of each polar equation.

(a) $r = 2\cos 2\theta$ (use radian mode)

(b) $r = 1 + 2\sin\theta$ (use degree mode)

Solution

(a) A viewing window of the graph of $r = 2\cos 2\theta$ is shown in Figure 14a where the radian mode was used.

(b) Figure 14b displays a viewing window of the graph of $r = 1 + 2\sin\theta$, in which degree mode was used.

Figure 14

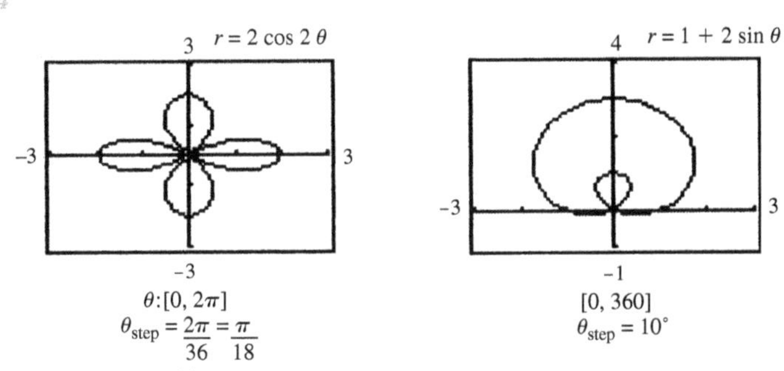

Note that because of the aspect ratio, graphers can produce distorted polar graphs. To improve the accuracy of a distorted graph, most graphers have features that correct the distortion. For instance, the graph of $r = 3$ is a circle, centered at the pole with radius 3 (see Example 6a). Figure 15a shows a viewing window in which the graph is distorted. After applying the distortion correction feature of the grapher, we obtain a more accurate graph (Figure 15b). Both are graphed using degree mode.

Figure 15

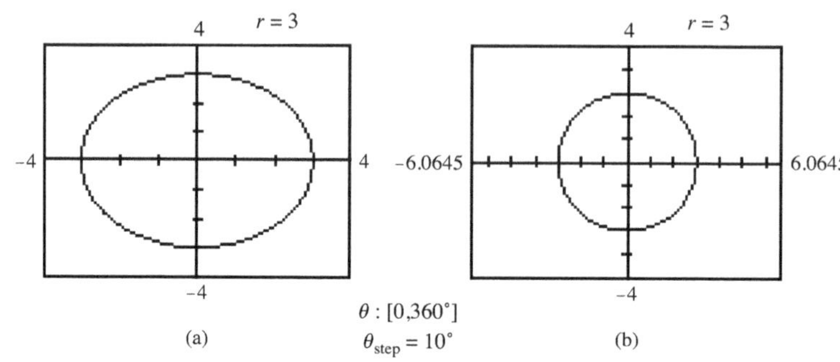

PROBLEM SET 3.2

Mastering the Concepts

In problems 1–6, plot points P and Q on the same polar coordinate system.

1. $P = (2, -45°)$; $Q = (2, 315°)$

2. $P = (3, -60°)$; $Q = (3, 300°)$

3. $P = \left(3, \frac{\pi}{6}\right)$; $Q = \left(-3, \frac{\pi}{6}\right)$

4. $P = \left(4, \frac{7\pi}{6}\right)$; $Q = \left(-4, \frac{7\pi}{6}\right)$

5. $P = \left(4, \frac{2\pi}{3}\right)$, $Q = \left(4, -\frac{2\pi}{3}\right)$

6. $P = \left(-5, \frac{\pi}{4}\right)$; $Q = \left(-5, -\frac{\pi}{4}\right)$

In problems 7 and 8, plot the given point P in the polar coordinate system, and then find three additional representations of P such that:

(i) $r > 0$ (ii) $r < 0$

7. (a) $P = (3, 100°)$
(b) $P = \left(-3, \frac{\pi}{4}\right)$

8. (a) $P = (4, 60°)$
(b) $P = \left(4, -\frac{\pi}{3}\right)$

In problems 9–14, determine the Cartesian coordinates of the points with the given polar coordinates. Round off the answers to two decimal places in problems 11–14.

9. (a) $(6, 30°)$
(b) $\left(4, -\frac{\pi}{4}\right)$

10. (a) $(-8, 45°)$
(b) $\left(4, -\frac{\pi}{6}\right)$

11. (a) $\left(3, -\frac{\pi}{7}\right)$
(b) $(4, -110°)$

12. (a) $\left(-3, \frac{5\pi}{13}\right)$
(b) $(-6, 213°)$

13. (a) $(2, -3)$
(b) $\left(-3, \frac{5\pi}{2}\right)$

14. (a) $(-2, 4)$
(b) $\left(4, -\frac{\pi}{9}\right)$

In problems 15–18, determine the polar coordinates of the points with the given Cartesian coordinates satisfying the conditions that $r > 0$ and $0 \le \theta < 2\pi$. Round off the answers to two decimal places in problems 17 and 18.

15. (a) $(-1, \sqrt{3})$
(b) $(-6, 6\sqrt{3})$

16. (a) $(5, 5)$
(b) $(2\sqrt{3}, -2)$

17. (a) $(-3, 5.1)$
(b) $(-6, -7)$

18. (a) $(4\sqrt{2}, -\sqrt{3})$
(b) $(-4.3, 1.4)$

In problems 19–32, sketch the graph of each Cartesian equation. Then find an equation of the graph in polar form.

19. $x = 2$

20. $y = 3$

21. $y = -3$

22. $x = -2$

23. $y = 3x - 2$

24. $x - y = 5$

25. $x^2 + y^2 = 25$

26. $x^2 + y^2 = 7$

27. $(x + 4)^2 + y^2 = 16$

28. $x^2 + y^2 - 6y = 0$

29. $y = 4x^2$

30. $xy = 4$

31. $y^2 = 8x$

32. $y = \sqrt{x^2 + 4}$

In problems 33–46, find an equation in terms of Cartesian coordinates that corresponds to each polar equation. Then sketch the graph.

33. $r = 3$ **34.** $r = -2$

35. $r = 2 \csc \theta$ **36.** $r = -4 \sec \theta$

37. $\theta = \dfrac{\pi}{4}$ **38.** $\theta = \dfrac{5\pi}{3}$

39. $r = -4 \cos \theta$ **40.** $r = 2 \sin \theta$

41. $r = 4 \tan \theta \sec \theta$ **42.** $r^2 \sin 2\theta = 4$

43. $4r \cos \theta + 3r \sin \theta = 12$

44. $2r \sin \theta - 3r \cos \theta = 6$

45. $r = 4 \cos \theta + 2 \sin \theta$

46. $r = \cos \theta - \sin \theta$

In problems 47–54, use point-plotting to sketch the graph of each polar equation.

47. $r = 5$ **48.** $r = 1$

49. $\theta = -\dfrac{\pi}{3}$ **50.** $\theta = \dfrac{\pi}{6}$

51. $r = \cos \theta$ **52.** $r = -2 \sin \theta$

53. $r = 1 + \cos \theta$ **54.** $r = 1 - \cos \theta$

Developing and Extending the Concepts

55. Show that the graph of the polar equation

$$r = \frac{c}{a \cos \theta + b \sin \theta}$$

where a, b, and c are constants, and a and b are not both 0, is a straight line. What is the slope of the line?

56. Use the law of cosines to show that the distance d between the polar points $P = (r_1, \theta_1)$ and $Q = (r_2, \theta_2)$ satisfies the equation

$$d^2 = r_1^2 + r_2^2 - 2r_1r_2 \cos(\theta_2 - \theta_1).$$

57. Polar Coordinates of a Tile: Find the polar coordinates of the corners of a tile in the shape of a regular hexagon (a polygon with six equal sides) if the center is at the pole, one vertex lies on the polar axis, and the distance from the center to any vertex is a units (Figure 16).

Figure 16

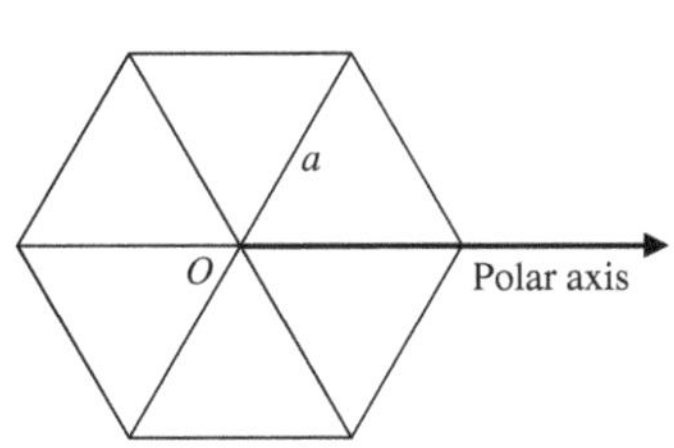

58. Give a polar equation whose graph is a circle with radius 5 and has the given point as the center in the Cartesian coordinate system.

(a) The origin (b) $(-5, 0)$

(c) $(0, 5)$ (d) $(4, 3)$

Problems 59–70 refer to the following: Some graphs of polar equations are symmetric. Table 3 indicates how to recognize certain types of symmetry: apply the symmetry tests in Table 3 to determine the symmetry (if any) of each graph. Then use a grapher to graph each polar equation to demonstrate the symmetry.

TABLE 3 Testing a Polar Equation for Symmetry If (r, θ) Lies on the Graph:

Type of Symmetry	Then These Points Lie on the Graph
1. The polar axis	$(-r, \pi - \theta)$ or $(r, -\theta)$
2. The line $\theta = \pi/2$	$(-r, -\theta)$ or $(r, \pi - \theta)$
3. The pole	$(-r, \theta)$ or $(r, \pi + \theta)$

59. $r = 1 - \sin \theta$ (cardioid)

60. $r = 1 + 2 \cos \theta$ (limaçon)

61. $r = 2 + 3 \cos \theta$ (limaçon)

62. $r = 1 - 2 \sin \theta$ (limaçon)

63. $r = 4 + 3 \sin \theta$ (limaçon)

64. $r = 5 + 3 \cos \theta$ (limaçon)

65. $r = -2 \cos 2\theta$ (four-petaled rose)

66. $r = 4 \sin 2\theta$ (four-petaled rose)

67. $r = 2 \sin 3\theta$ (three-petaled rose)

68. $r = 2 \cos 3\theta$ (three-petaled rose)

69. $r^2 = 4 \cos 2\theta$ (lemniscate)

70. $r^2 = 4 \sin 2\theta$ (lemniscate)

71. (a) Sketch the graph of each polar equation:

$$r = 2 \cos 3\theta$$

$$r = 2 \cos\left[3\left(\theta - \frac{\pi}{6}\right)\right]$$

and $$r = 2 \cos\left[3\left(\theta + \frac{\pi}{4}\right)\right]$$

(b) Describe how the graphs in part (a) are related to each other.

72. (a) The graphs of $r = 2 \sin[(3\theta)/2]$ and $r = 2 \sin[(5\theta)/2]$ are called *rose curves*. Sketch a complete graph of each equation.

(b) How many petals does each graph in part (a) have?

Interlude: Practice Problems 3.2

1. Convert the polar equation $r + r\cos\theta = 3$ to an equation in terms of Cartesian coordinates and sketch the graph of the polar equation (in the polar coordinate plane).

2. Convert the Cartesian equation $3x^2 - 9x = 12y - 3y^2$ to a polar equation and sketch the graph (in the Cartesian coordinate plane).

3. Convert the polar equation $3 \cos \theta - 4 \sin \theta = -\frac{1}{r}$ to an equation in terms of Cartesian coordinates and sketch the graph of the polar equation (in the polar coordinate plane).

4. Find all solutions to the equation $2 - 2 \sin \theta = 3$ for $0 \leqslant \theta < 2\pi$ by using graphs of polar equations. Show your graphs.

Objectives

1. Describe Vector Operations Geometrically
2. Perform Vector Operations Analytically
3. Find Unit Vectors and Direction Angles
4. Solve Applied Problems

3.3 Vectors in the Plane

Until now, we have dealt with quantities that can be measured or represented by single real numbers, such as area, volume, time, temperature, and speed. In this section, we deal with quantities called *vectors,* which cannot be described or represented by a single real number. Vectors are important tools in applications involving force, velocity, acceleration, and displacement. We'll see how trigonometry is used when working with vectors. Here we restrict our study to vectors in a two dimensional setting.

Describing Vector Operations Geometrically

Geometrically, a **vector** in the plane is a line segment with a direction usually denoted by an arrowhead at one end of the segment. The end point that contains the arrowhead is called the **terminal point,** and the other end point is called the **initial point** of the vector. The length of a vector is called its **magnitude.** The vector in Figure 1, denoted by $\overrightarrow{AB}$, has initial point A and terminal point B, and its magnitude is denoted by $|\overrightarrow{AB}|$.

Figure 1

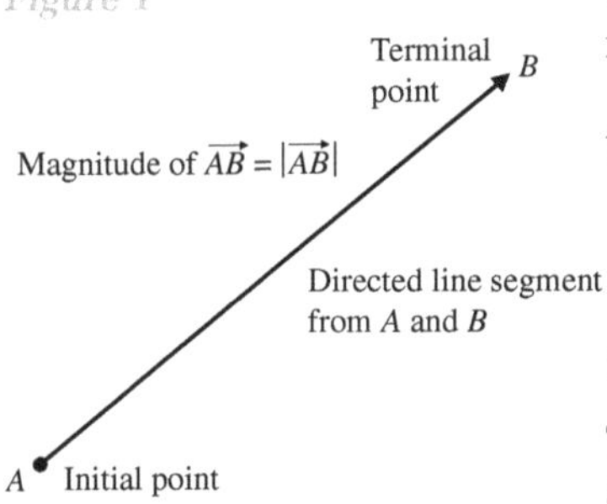

The **zero vector,** written **0**, is a vector whose initial and terminal points are the same. This magnitude is given by

$$|\mathbf{0}| = 0.$$

1. Geometry of Equal Vectors

Two vectors are considered to be **equal** if they have the same magnitude and the same direction, such as vectors $\overrightarrow{PQ}$ and $\overrightarrow{RS}$ in Figure 2a. Note that $\overrightarrow{AB}$ and $\overrightarrow{BA}$ in Figure 2b are not equal, even though their lengths are the same, because they have opposite directions. That is,

$$\overrightarrow{AB} \neq \overrightarrow{BA}, \quad \text{even though} \quad |\overrightarrow{AB}| = |\overrightarrow{BA}|.$$

Figure 2

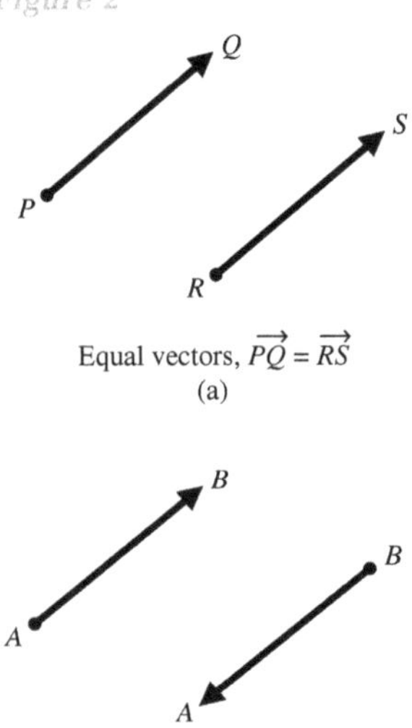

Equal vectors, $\overrightarrow{PQ} = \overrightarrow{RS}$
(a)

Unequal vectors, $\overrightarrow{AB} \neq \overrightarrow{BA}$
(b)

Special notation is used for vectors so that they can be distinguished from real numbers. In this book, we normally use lowercase boldface letters to denote vectors. Hence, we speak of "vector **u**" and just write boldface **u.**

2. Geometry of Vector Addition

Two vectors **u** and **v** (Figure 3a) may be added to form the *sum* **u** + **v** as follows:

First, we shift **v** so that its initial point coincides with the terminal point of **u.** The vector having the same initial point as **u** and the same terminal point as **v** is defined to be the **sum (resultant vector) u + v** of **u** and **v** (Figure 3b). As Figure 3b shows, the resultant vector is a diagonal vector of the parallelogram determined by vectors **u** and **v.** This description of addition is called the **parallelogram law of addition.**

Figure 3

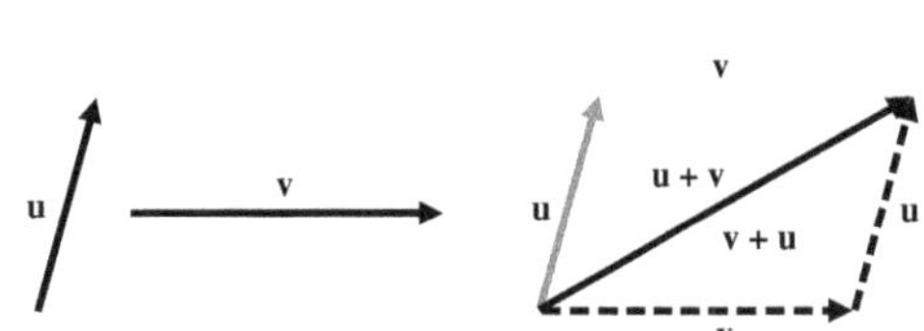

(a) (b)

Figure 3b displays the fact that **u** + **v** = **v** + **u**; *that is, vector addition is commutative.*

3. Geometry of Scalar Multiplication

Vectors can be multiplied by real numbers, called **scalars** in this context, to form other vectors. We refer to this operation as **scalar multiplication,** and we describe it as follows.

If $a \neq 0$ is a scalar (real number) and $\mathbf{u}$ is a vector, then the vector $a\mathbf{u}$ is a vector that satisfies the following conditions:

(i) **The Magnitude of au:**
$|a\mathbf{u}| = |a||\mathbf{u}|$; that is, the magnitude of $a\mathbf{u}$ is the product of the absolute value of a and the magnitude of $\mathbf{u}$.

(ii) **The Direction of au:**
If $a > 0$, then the direction of $a\mathbf{u}$ is the same as the direction of $\mathbf{u}$.
If $a < 0$, then the direction of $a\mathbf{u}$ is opposite the direction of $\mathbf{u}$.

Figure 4 displays some illustrations of scalar multiplication. Note that $2\mathbf{u}$ is a vector in the same direction as $\mathbf{u}$, with magnitude twice that of $\mathbf{u}$. And $(-1/2)\mathbf{u}$ is a vector opposite in direction to $\mathbf{u}$, with a length half that of $\mathbf{u}$.

If $a = 0$, then $a\mathbf{u} = 0\mathbf{u} = \mathbf{0}$.

Figure 4

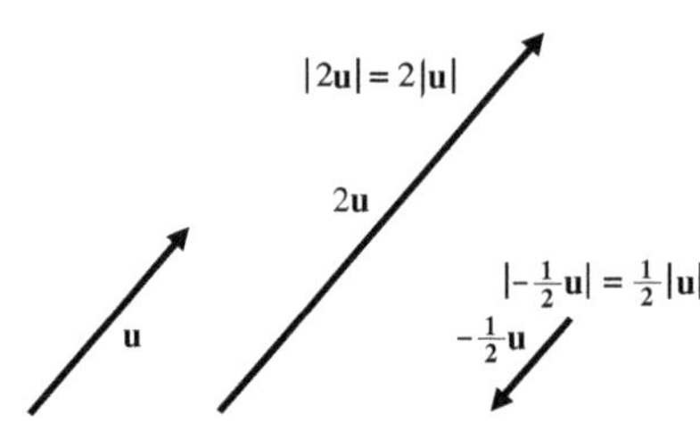

4. Geometry of Vector Subtraction

Now we can use vector addition and scalar multiplication to define vector **subtraction** as

$$\mathbf{u} - \mathbf{v} = \mathbf{u} + (-\mathbf{v})$$

where $-\mathbf{v} = (-1)\,\mathbf{v}$ (Figure 5a).

Figure 5

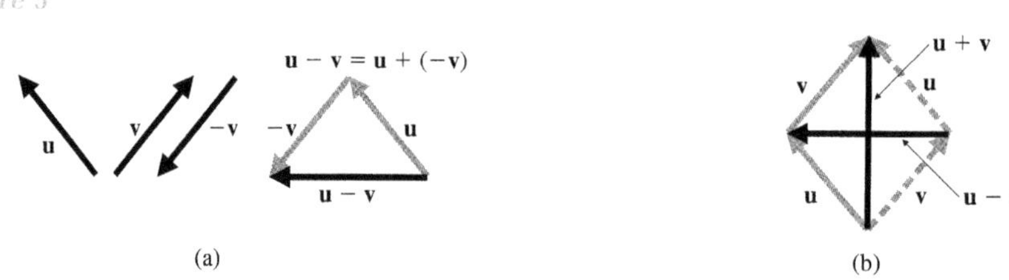

Observe that if we form the parallelogram defined by $\mathbf{u}$ and $\mathbf{v}$ (Figure 5b), one of the diagonal vectors is $\mathbf{u} + \mathbf{v}$ and the other is $\mathbf{u} - \mathbf{v}$.

EXAMPLE 1

Describing Vector Operations Geometrically

Figure 6

Given vectors $\mathbf{u}$ and $\mathbf{v}$ as shown in Figure 6, display the vector $3\mathbf{u} - 2\mathbf{v}$.

Solution To form $3\mathbf{u} - 2\mathbf{v}$, we first form $3\mathbf{u}$ and $2\mathbf{v}$ (Figure 7a). Next we form the parallelogram defined by $3\mathbf{u}$ and $2\mathbf{v}$. The difference $3\mathbf{u} - 2\mathbf{v}$ is the diagonal vector shown in Figure 7b.

Figure 7

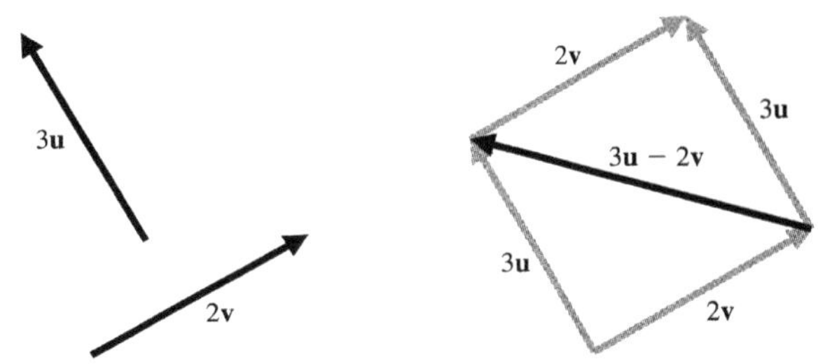

Performing Vector Operations Analytically

Figure 8

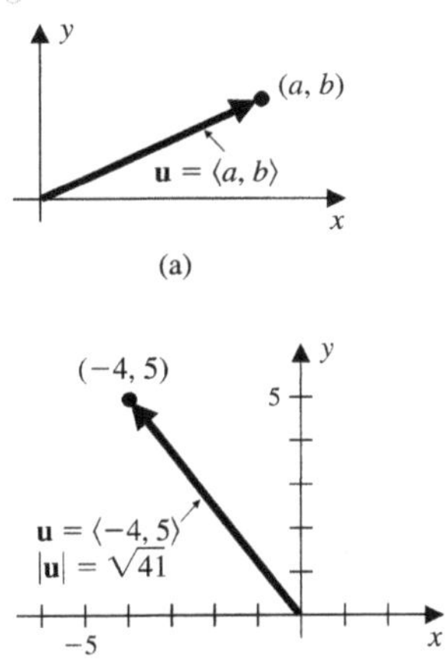

When a vector is positioned in a Cartesian plane so that its initial point is at the origin, it is called a **radius vector** or **standard position vector.** Such a vector can be uniquely represented by the coordinates of its terminal point. For instance, Figure 8a shows a radius vector **u** with terminal point (a, b). We call a the ***x* component** and b the ***y* component** of **u,** and we write **u** in the **component form**

$$\mathbf{u} = \langle a, b \rangle.$$

From the Pythagorean theorem, the **magnitude** $|\mathbf{u}|$ of vector $\mathbf{u} = \langle a, b \rangle$ is given by

$$|\mathbf{u}| = \sqrt{a^2 + b^2}.$$

For example, the vector $\mathbf{u} = \langle -4, 5 \rangle$ has x component -4, y component 5, and $|\mathbf{u}| = \sqrt{(-4)^2 + 5^2} = \sqrt{41}$ (Figure 8b).

1. Equal Vectors

If **u** and **v** are equal position vectors, they have the same initial point (0, 0), so the components must also be equal; that is:

Component Rule
Equal Vectors

If $\mathbf{u} = \langle x_1, y_1 \rangle$ and $\mathbf{v} = \langle x_2, y_2 \rangle$, then

$$\mathbf{u} = \mathbf{v} \quad \text{whenever} \quad x_1 = x_2 \quad \text{and} \quad y_1 = y_2.$$

For instance, if $\mathbf{u} = \mathbf{w}$, where $\mathbf{u} = \langle 1, -3 \rangle$ and $\mathbf{w} = \langle 1, b \rangle$, then $b = -3$.

Suppose that **w** is a vector in a Cartesian plane and **w** is not a radius vector. Assume that the initial point of **w** is (x_1, y_1) and its terminal point is (x_2, y_2), as shown in Figure 9. To find the radius vector representation of **w,** we shift **w** from its given position in such a way that the initial point (x_1, y_1) is positioned at the origin to obtain the component form

$$\mathbf{w} = \langle x_2 - x_1, y_2 - y_1 \rangle.$$

Figure 9

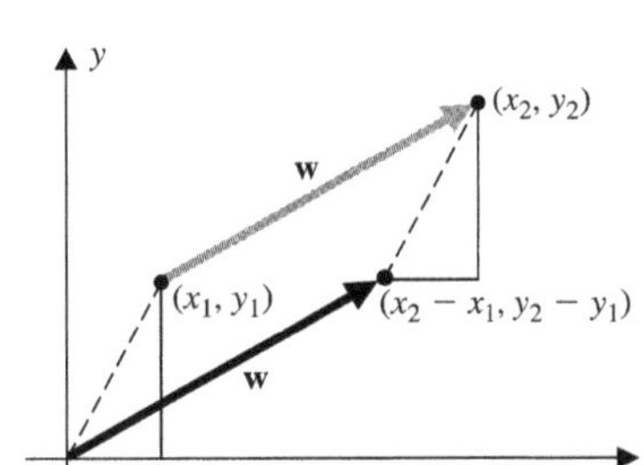

EXAMPLE 2

Finding the Component Form

Let $P = (3, -2)$ be the initial point and $Q = (-4, 1)$ be the terminal point of the vector

$$\mathbf{u} = \overrightarrow{PQ}.$$

Find its component form, and display $\mathbf{u}$ in both its given position and its standard position.

Solution Here

$$P = (3, -2) = (x_1, y_1) \quad \text{and} \quad Q = (-4, 1) = (x_2, y_2).$$

The x and y components of $\mathbf{u} = \overrightarrow{PQ}$ are given, respectively, by

$$x_2 - x_1 = -4 - 3 = -7 \quad \text{and} \quad y_2 - y_1 = 1 - (-2) = 3.$$

So $\mathbf{u} = \langle -7, 3 \rangle$. Figure 10 displays the vector in both its given location and its standard position.

Figure 10

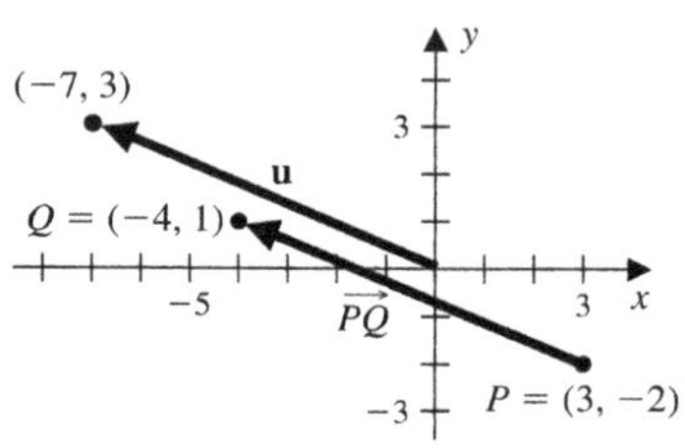

2. Vector Addition

When vectors are written in component form, vector addition becomes very simple. Suppose that $\mathbf{u}$ and $\mathbf{v}$ are in standard position with

$$\mathbf{u} = \langle x_1, y_1 \rangle \quad \text{and} \quad \mathbf{v} = \langle x_2, y_2 \rangle.$$

As displayed in Figure 11, moving the initial point of $\mathbf{v}$ to the terminal point of $\mathbf{u}$ shows that the x component of $\mathbf{u} + \mathbf{v}$ is obtained by adding the x components of $\mathbf{u}$ and $\mathbf{v}$. Similarly, the y component of $\mathbf{u} + \mathbf{v}$ is obtained by adding the y components of $\mathbf{u}$ and $\mathbf{v}$.

Figure 11

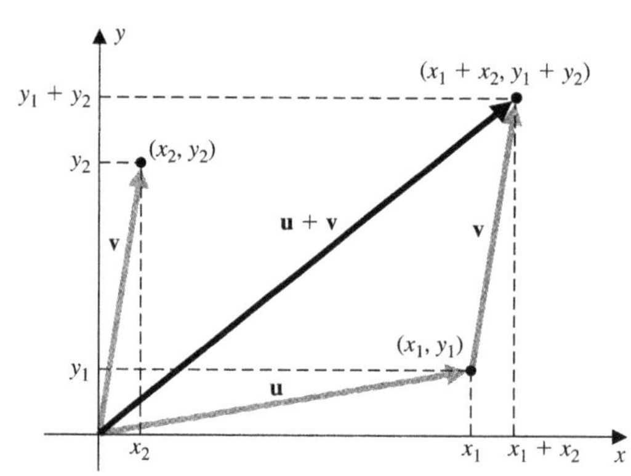

Thus we have the following property:

Component Rule Addition of Vectors

If $\mathbf{u} = \langle x_1, y_1 \rangle$ and $\mathbf{v} = \langle x_2, y_2 \rangle$, then

$$\mathbf{u} + \mathbf{v} = \langle x_1 + x_2, y_1 + y_2 \rangle.$$

3. Scalar Multiplication

As illustrated in Figure 12, we have the following rule:

Component Rule
Scalar Multiplication

If $\mathbf{u} = \langle x_1, y_1 \rangle$ and c is a real number, then

$$c\mathbf{u} = c\langle x_1, y_1 \rangle = \langle cx_1, cy_1 \rangle.$$

Figure 12

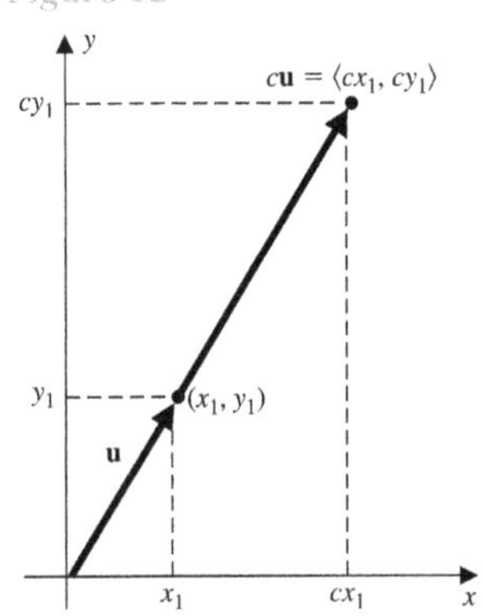

4. Vector Subtraction

Finally, we derive the subtraction rule as follows:
If $\mathbf{u} = \langle x_1, y_1 \rangle$ and $\mathbf{v} = \langle x_2, y_2 \rangle$, then $-\mathbf{v} = -1 \cdot \mathbf{v} = \langle -x_2, -y_2 \rangle$.
So

$$\begin{aligned} \mathbf{u} - \mathbf{v} &= \mathbf{u} + (-\mathbf{v}) \\ &= \langle x_1, y_1 \rangle + \langle -x_2, -y_2 \rangle \\ &= \langle x_1 - x_2, y_1 - y_2 \rangle. \end{aligned}$$

Thus we are led to the following property:

Component Rule
Subtraction of Vectors

If $\mathbf{u} = \langle x_1, y_1 \rangle$ and $\mathbf{v} = \langle x_2, y_2 \rangle$, then

$$\mathbf{u} - \mathbf{v} = \langle x_1 - x_2, y_1 - y_2 \rangle.$$

EXAMPLE 3 **Performing Vector Operations**

Let $\mathbf{u} = \langle 3, 4 \rangle$ and $\mathbf{v} = \langle -5, 6 \rangle$. Find:

(a) $\mathbf{u} + \mathbf{v}$ (b) $\mathbf{u} - \mathbf{v}$

(c) $4\mathbf{u} - 3\mathbf{v}$ (d) $|4\mathbf{u} - 3\mathbf{v}|$

Solution

(a) $$\begin{aligned} \mathbf{u} + \mathbf{v} &= \langle 3, 4 \rangle + \langle -5, 6 \rangle \\ &= \langle 3 - 5, 4 + 6 \rangle \\ &= \langle -2, 10 \rangle \end{aligned}$$

(b) $$\begin{aligned} \mathbf{u} - \mathbf{v} &= \langle 3, 4 \rangle - \langle -5, 6 \rangle \\ &= \langle 3 - (-5), 4 - 6 \rangle \\ &= \langle 8, -2 \rangle \end{aligned}$$

(c) $$\begin{aligned} 4\mathbf{u} - 3\mathbf{v} &= 4\langle 3, 4 \rangle - 3\langle -5, 6 \rangle \\ &= \langle 12, 16 \rangle + \langle 15, -18 \rangle \\ &= \langle 27, -2 \rangle \end{aligned}$$

(d) Using the result from part (c),

$$|4\mathbf{u} - 3\mathbf{v}| = \sqrt{(27)^2 + (-2)^2} = \sqrt{729 + 4} = \sqrt{733}.$$

Finding Unit Vectors and Direction Angles

A vector of magnitude 1 is called a **unit vector.** There are two special unit vectors, called **basis vectors,** defined by

$$\mathbf{i} = \langle 1, 0\rangle \quad \text{and} \quad \mathbf{j} = \langle 0, 1\rangle.$$

Any vector $\mathbf{u} = \langle a, b\rangle$ can be written in terms of **i** and **j** as follows:

$$\langle a, b\rangle = \langle a, 0\rangle + \langle 0, b\rangle = a\langle 1, 0\rangle + b\langle 0, 1\rangle = a\mathbf{i} + b\mathbf{j} \quad \text{(Figure 13)}$$

Figure 13

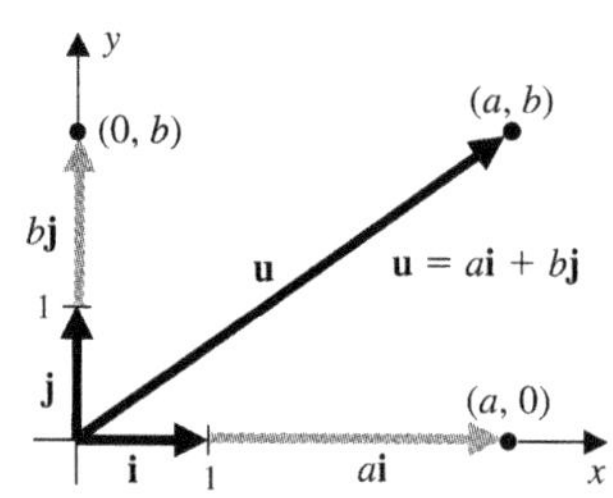

Thus we get the *component form* of the vector.

$$\mathbf{u} = \langle a, b\rangle = a\mathbf{i} + b\mathbf{j}.$$

The component rules for addition, subtraction, and scalar multiplication apply to vectors written in terms of **i** and **j.** For example,

$$(3\mathbf{i} + 5\mathbf{j}) + (2\mathbf{i} - 7\mathbf{j}) = 5\mathbf{i} - 2\mathbf{j}$$
$$-2(3\mathbf{i} + 2\mathbf{j}) = -6\mathbf{i} - 4\mathbf{j}$$
$$\text{and} \quad (2\mathbf{i} - 3\mathbf{j}) - (\mathbf{i} + 5\mathbf{j}) = \mathbf{i} - 8\mathbf{j}$$

Consider the vector $\frac{1}{|\mathbf{u}|}\mathbf{u}$, where $\mathbf{u} \neq \mathbf{0}$. Since $\frac{1}{|\mathbf{u}|} > 0$, the vector $\frac{1}{|\mathbf{u}|}\mathbf{u}$ has the same direction as **u.** Also, the magnitude of $\frac{1}{|\mathbf{u}|}\mathbf{u}$ is given by

$$\left|\frac{1}{|\mathbf{u}|}\mathbf{u}\right| = \frac{1}{|\mathbf{u}|}|\mathbf{u}| = 1.$$

Thus we are led to the following notion:

Definition
Unit Vector

A **unit vector** in the direction of $\mathbf{u} = \langle a, b\rangle$, where $\mathbf{u} \neq \mathbf{0}$, is given by

$$\frac{1}{|\mathbf{u}|}\mathbf{u} = \left\langle \frac{a}{\sqrt{a^2 + b^2}}, \frac{b}{\sqrt{a^2 + b^2}} \right\rangle.$$

The process of forming this vector is referred to as **normalizing the vector u.**

EXAMPLE 4

Normalizing a Vector

Find a unit vector in the same direction as $\mathbf{u} = 4\mathbf{i} - 3\mathbf{j}$.

Solution The magnitude of **u** is given by

$$|\mathbf{u}| = \sqrt{4^2 + (-3)^2} = 5.$$

Thus a unit vector in the same direction as **u** is given by

$$\frac{1}{|\mathbf{u}|}\mathbf{u} = \frac{1}{5}(4\mathbf{i} - 3\mathbf{j}) = \frac{4}{5}\mathbf{i} - \frac{3}{5}\mathbf{j}.$$

Figure 14

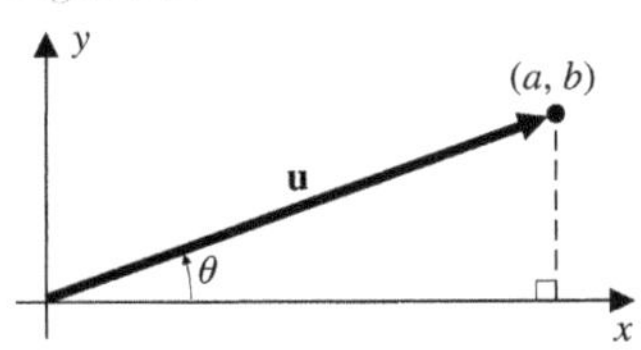

An angle θ formed by the vector $\mathbf{u} = \langle a, b\rangle$ and the positive x axis is called a **direction angle** of the vector **u** (Figure 14).

Using trigonometry, we have

$$\cos\theta = \frac{a}{|\mathbf{u}|} \quad\text{and}\quad \sin\theta = \frac{b}{|\mathbf{u}|}.$$

Rewriting these equations as $a = |\mathbf{u}|\cos\theta$ and $b = |\mathbf{u}|\sin\theta$, respectively, we get a trigonometric form of the vector.

$$\mathbf{u} = a\mathbf{i} + b\mathbf{j} = |\mathbf{u}|(\cos\theta)\mathbf{i} + |\mathbf{u}|(\sin\theta)\mathbf{j}$$

Also, if $a \neq 0$, it follows that

$$\frac{b}{a} = \frac{|\mathbf{u}|\sin\theta}{|\mathbf{u}|\cos\theta} = \frac{\sin\theta}{\cos\theta} = \tan\theta.$$

Thus a direction angle θ for $\mathbf{u} = a\mathbf{i} + b\mathbf{j}$ can be found by solving

$$\tan\theta = \frac{b}{a}$$

where θ is selected according to the quadrant location of (a, b).

For example, the direction angle of $\mathbf{u} = 5\mathbf{i} + 12\mathbf{j}$ satisfies

$$\tan\theta = \frac{b}{a} = \frac{12}{5} = 2.4.$$

Since (5, 12) is in quadrant I, it follows that

$$\theta = \tan^{-1} 2.4 \quad \text{or approx. } 67.4° \quad \text{(Figure 15)}$$

is a direction angle.

Figure 15

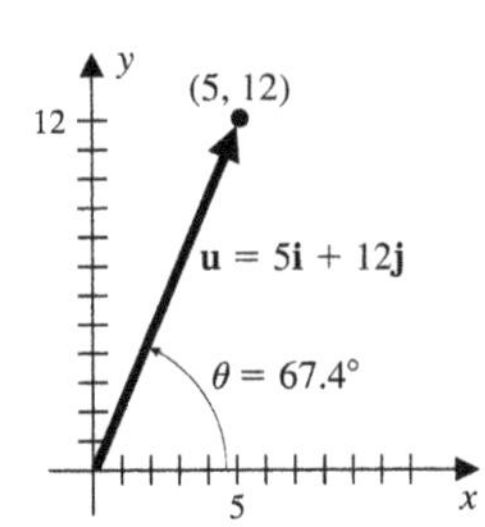

EXAMPLE 5 **Finding the Components of a Vector**

Find the component form of the vector **u** if $|\mathbf{u}| = 2$ and a direction angle $\theta = (5\pi)/4$.

Solution The vector **u** has magnitude 2 and direction angle $\theta = (5\pi)/4$ (Figure 16). So **u** can be written in component form as

Figure 16

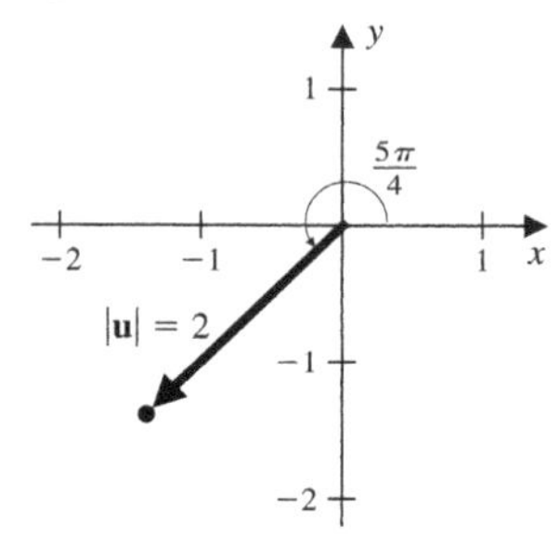

$$\begin{aligned}\mathbf{u} = a\mathbf{i} + b\mathbf{j} &= |\mathbf{u}|\left(\cos\frac{5\pi}{4}\right)\mathbf{i} + |\mathbf{u}|\left(\sin\frac{5\pi}{4}\right)\mathbf{j}\\ &= 2\left(-\frac{\sqrt{2}}{2}\right)\mathbf{i} + 2\left(-\frac{\sqrt{2}}{2}\right)\mathbf{j}\\ &= -\sqrt{2}\,\mathbf{i} - \sqrt{2}\,\mathbf{j}.\end{aligned}$$

Clearly,

the x component of **u** is $-\sqrt{2}$

and

the y component is also $-\sqrt{2}$.

Solving Applied Problems

Vectors are used to model applications in science and engineering.

EXAMPLE 6 **Finding the Magnitude of a Resultant Force**

A boat is pulled along a canal by two ropes on opposite sides of the canal, as shown in Figure 17. The first rope exerts a force of 350 pounds and makes an angle of 25° with respect to the axis of the boat. The other exerts a force of 200 pounds and makes an angle of 32° with respect to the axis of the boat.

Figure 17

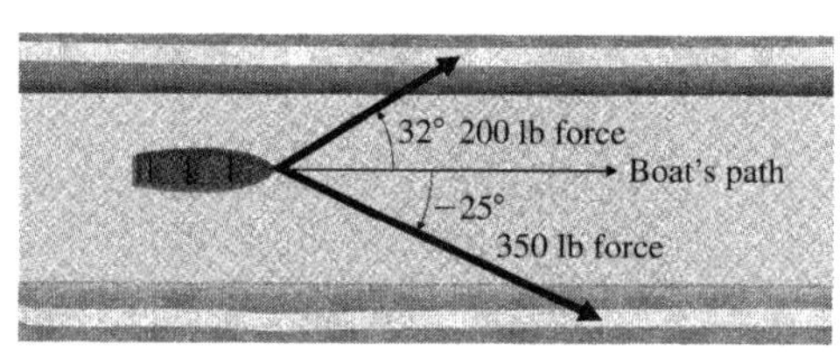

(a) What is the magnitude of the resultant force vector (to two decimal places)?

(b) What is the angle that the resultant vector makes with the axis of the boat (to the nearest hundredth of a degree)?

Solution (a) Let $\mathbf{F}_1$ be the vector representing the 350 pound force and $\mathbf{F}_2$ be the vector representing the 200 pound force.
So we have

$$|\mathbf{F}_1| = 350$$

and

$$|\mathbf{F}_2| = 200.$$

Figure 18

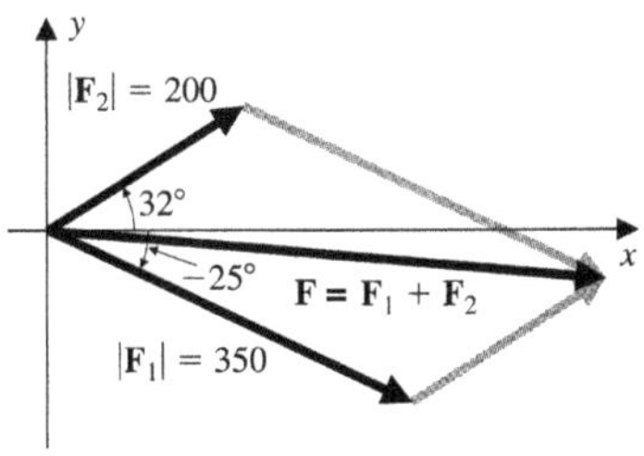

Next we set up a Cartesian coordinate system in such a way that the origin is at the point on which both $\mathbf{F}_1$ and $\mathbf{F}_2$ act. The resultant vector that is given by

$$\mathbf{F} = \mathbf{F}_1 + \mathbf{F}_2$$

describes the force on the boat (Figure 18).

Since $\mathbf{F}_1$ makes an angle of $-25°$ with the positive x axis, it follows that the components of $\mathbf{F}_1$ are given by

$$\begin{aligned} \mathbf{F}_1 &= |\mathbf{F}_1| \cos(-25°)\mathbf{i} + |\mathbf{F}_1| \sin(-25°)\mathbf{j} \\ &= 350(\cos 25°)\mathbf{i} - 350(\sin 25°)\mathbf{j} \\ &= 317.21\mathbf{i} - 147.92\mathbf{j} \text{ (approx.)} \end{aligned}$$

Similarly, $\mathbf{F}_2$ makes an angle of 32° with the positive x axis, so its components are given by

$$\begin{aligned} \mathbf{F}_2 &= |\mathbf{F}_2| (\cos 32°)\mathbf{i} + |\mathbf{F}_2| (\sin 32°)\mathbf{j} \\ &= 200(\cos 32°)\mathbf{i} + 200(\sin 32°)\mathbf{j} \\ &= 169.61\mathbf{i} + 105.98\mathbf{j} \text{ (approx.).} \end{aligned}$$

The resultant vector is given by

$$\begin{aligned} \mathbf{F} &= \mathbf{F}_1 + \mathbf{F}_2 \\ &= (317.21\mathbf{i} - 147.92\mathbf{j}) + (169.61\mathbf{i} + 105.98\mathbf{j}) \\ &= 486.82\mathbf{i} - 41.94\mathbf{j} \text{ (approx.).} \end{aligned}$$

and its magnitude is

$$\begin{aligned} |\mathbf{F}| &= \sqrt{(486.82)^2 + (-41.94)^2} \\ &= 488.62 \text{ (approx.).} \end{aligned}$$

Thus the resultant force has a magnitude of about 488.62 pounds.

(b) The angle θ between $\mathbf{F}$ and the axis of the boat (positive x axis) is given by

$$\tan \theta = \frac{-41.94}{486.82}$$

that is,

$$\theta = \tan^{-1}\left(\frac{-41.94}{486.82}\right) = -4.92° \text{ (approx.).}$$

Vectors are also used in aeronautical navigation, where the following terminology is commonly used: The **heading** of an airplane is the direction in which it is pointed; its **airspeed** is its speed relative to the air. As shown in Figure 19, the vector $\mathbf{v}_1$, whose magnitude is the airspeed and whose direction is the heading, represents the **velocity of the airplane relative to the air.** The vector $\mathbf{v}_2$ is the **velocity vector** for the wind; that is, $|\mathbf{v}_2|$ is the speed of the wind, and the direction angle of $\mathbf{v}_2$ is the direction of the wind. The **course,** or **track,** of the airplane is the direction in which it is actually moving over the ground, and its **ground speed** is its speed relative to the ground. The magnitude of the resultant vector

Figure 19

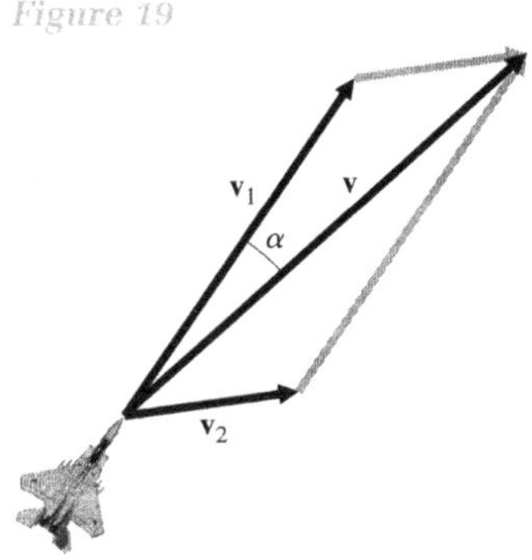

$$\mathbf{v} = \mathbf{v}_1 + \mathbf{v}_2$$

is the ground speed or velocity of the airplane, and the direction of $\mathbf{v}$ is the **course.** In this context, vector $\mathbf{v}$ is called the **velocity vector** of the airplane. The angle α between the vectors $\mathbf{v}$ and $\mathbf{v}_1$ is called the **drift angle.**

EXAMPLE 7

Finding the Course and Speed of an Airplane

An airplane is headed N30°E with an airspeed of 500 miles per hour. The wind is blowing S29°E at a speed of 50 miles per hour. Find:

(a) The ground speed (to two decimal places)

(b) The course of the airplane—that is, the direction angle of the velocity vector relative to the ground (to the nearest hundredth of a degree)

(c) The drift angle (to the nearest hundredth of a degree)

Solution Let $\mathbf{v}_1$ represent the velocity and direction of the airplane relative to the air; let $\mathbf{v}_2$ represent the velocity and direction of the wind relative to the ground; and let $\mathbf{v}$ represent the velocity vector of the airplane (Figure 20). Then

$$\mathbf{v}_1 = 500(\cos 60°)\mathbf{i} + 500(\sin 60°)\mathbf{j}$$
$$= 250.00\mathbf{i} + 433.01\mathbf{j}$$
$$\mathbf{v}_2 = 50[\cos(-61°)]\mathbf{i} + 50[\sin(-61°)]\mathbf{j}$$
$$= 24.24\mathbf{i} + (-43.73)\mathbf{j} \text{ (approx.)}.$$

Figure 20

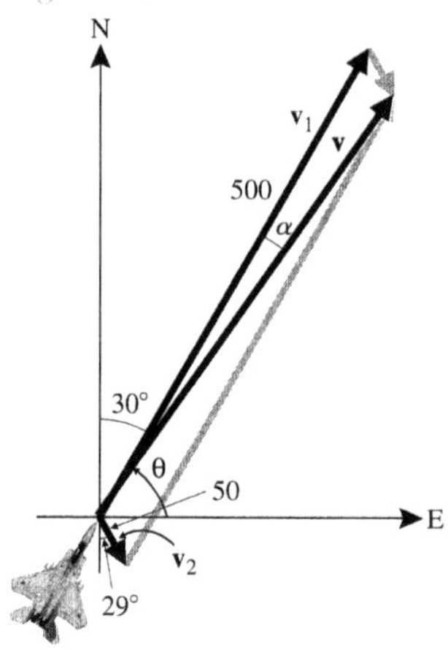

(a) Thus

$$\mathbf{v} = \mathbf{v}_1 + \mathbf{v}_2$$
$$= (250.00\mathbf{i} + 433.01\mathbf{j}) + (24.24\mathbf{i} - 43.73\mathbf{j})$$
$$= 274.24\mathbf{i} + 389.28\mathbf{j}.$$

So

$$|\mathbf{v}| = \sqrt{(274.24)^2 + (389.28)^2}$$
$$= 476.18 \text{ (approx.)}.$$

Therefore, the ground speed is approximately 476.18 miles per hour.

(b) The direction angle θ of the velocity vector $\mathbf{v}$ is given by $\tan\theta = 389.28/274.24$. So one possibility is

$$\theta = \tan^{-1}\frac{389.28}{274.24} = 54.84° \text{ (approx.)}.$$

(c) Upon examining Figure 20 and using the result from part (b), we find that the drift angle $\alpha = 60° - 54.84° = 5.16°$ (approx.).

PROBLEM SET 3.3

Mastering the Concepts

In problems 1 and 2, let $O = (0, 0)$, $P = (-2, 1)$, $Q = (3, -2)$, $R = (4, 3)$, and $S = (-3, -5)$. Sketch each vector.

1. (a) $\mathbf{u} = \overrightarrow{OP}$
(b) $\mathbf{v} = \overrightarrow{OQ}$
(c) $2\mathbf{u}$
(d) $3\mathbf{u} - 4\mathbf{v}$

2. (a) $\mathbf{u} = \overrightarrow{OS}$
(b) $\mathbf{v} = \overrightarrow{OR}$
(c) $-\frac{1}{5}\mathbf{v}$
(d) $-2\mathbf{u} + 5\mathbf{v}$

In problems 3–8, let $\mathbf{u}$ be a vector whose initial point is P and whose terminal point is Q. Draw $\mathbf{u} = \overrightarrow{PQ}$ in standard position. Also, find the component form and the magnitude of $\mathbf{u}$.

3. $P = (8, 6)$, $Q = (3, 4)$

4. $P = (-7, -6)$, $Q = (2, 3)$

5. $P = (-2, 6)$, $Q = (3, -5)$

6. $P = (-3, 2)$, $Q = (1, -3)$
7. $P = (3, 7)$, $Q = (-3, 1)$
8. $P = (1, -3)$, $Q = (5, -1)$

In problems 9–14, use $\mathbf{u} = \langle 3, 4\rangle$, $\mathbf{v} = \langle -2, 4\rangle$, and $\mathbf{w} = \langle 7, 8\rangle$ to find each expression. Display part (a) of each problem graphically.

9. (a) $\mathbf{u} + \mathbf{v}$
   (b) $|\mathbf{u} + \mathbf{v}|$
10. (a) $\mathbf{v} - \mathbf{w}$
   (b) $|\mathbf{v} - \mathbf{w}|$
11. (a) $3\mathbf{u} - 4\mathbf{w}$
   (b) $|3\mathbf{u} - 4\mathbf{w}|$
12. (a) $5\mathbf{w} + 3\mathbf{u}$
   (b) $|5\mathbf{w} + 3\mathbf{u}|$
13. (a) $-6\mathbf{v} - \mathbf{u}$
   (b) $|-6\mathbf{v} - \mathbf{u}|$
14. (a) $-(2\mathbf{u} + \mathbf{v})$
   (b) $|-(2\mathbf{u} + \mathbf{v})|$

In problems 15 and 16, express each vector in the component form of $a\mathbf{i} + b\mathbf{j}$.

15. (a) $\langle -2, 4\rangle$
   (b) $\langle 3, 0\rangle$
16. (a) $\langle 0, 2\rangle$
   (b) $\langle -1, -5\rangle$
17. Let $\mathbf{w} = \mathbf{u} + \mathbf{v}$. Find $\mathbf{v}$ in terms of $\mathbf{i}$ and $\mathbf{j}$, and sketch the vector $\mathbf{v}$ in each case.
   (a) $\mathbf{u} = 3\mathbf{i} - 2\mathbf{j}$ and $\mathbf{w} = 5\mathbf{i} + 3\mathbf{j}$
   (b) $\mathbf{u} = 2\mathbf{i} - \mathbf{j}$ and $\mathbf{w} = -2\mathbf{i} + 3\mathbf{j}$
18. Let $\mathbf{u} = \langle -2, 2\rangle$ and $\mathbf{v} = \langle -5, 0\rangle$. Suppose $\mathbf{w} = \mathbf{u} - \mathbf{v}$. Express $\mathbf{w}$ in terms of $\mathbf{i}$ and $\mathbf{j}$.

In problems 19 and 20, sketch the resultant vector $\mathbf{w} = \mathbf{u} + \mathbf{v}$. Find the length of $\mathbf{w}$ and the angle that $\mathbf{w}$ makes with $\mathbf{v}$.

19.
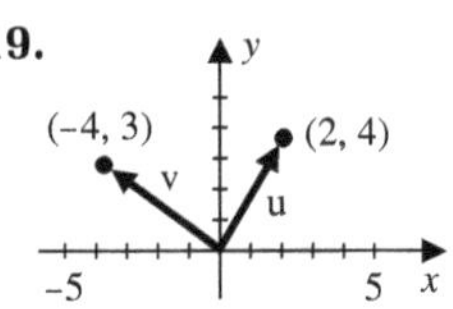

20.
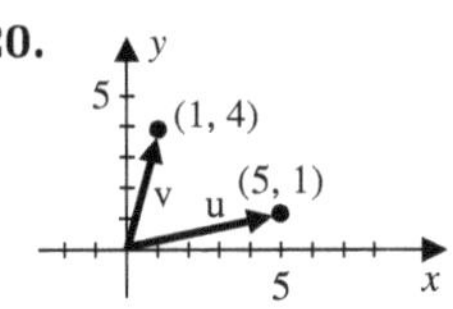

In problems 21–28, normalize each vector. Also, find the direction angle of $\mathbf{u}$ (rounded off to two decimal places).

21. $\mathbf{u} = -\sqrt{3}\mathbf{i} + \mathbf{j}$
22. $\mathbf{u} = -2\mathbf{i} - 2\sqrt{3}\mathbf{j}$
23. $\mathbf{u} = 5\mathbf{i} + 12\mathbf{j}$
24. $\mathbf{u} = 3\mathbf{i} - 3\mathbf{j}$
25. $\mathbf{u} = \sqrt{2}\mathbf{i} - \sqrt{2}\mathbf{j}$
26. $\mathbf{u} = 3\mathbf{i} + 7\mathbf{j}$
27. $\mathbf{u} = (3\mathbf{i} + 8\mathbf{j}) - (\mathbf{i} - 2\mathbf{j})$
28. $\mathbf{u} = \langle 3, -2\rangle - 2\langle 4, 1\rangle$

In problems 29–36, find the x and y components of the vector $\mathbf{u} = x\mathbf{i} + y\mathbf{i}$, if θ represents a direction angle of $\mathbf{u}$. Round off the answers to two decimal places.

29. $|\mathbf{u}| = 3$; $\theta = 0°$
30. $|\mathbf{u}| = 2$; $\theta = 180°$
31. $|\mathbf{u}| = 5$; $\theta = \frac{\pi}{4}$
32. $|\mathbf{u}| = 1$; $\theta = \frac{5\pi}{3}$
33. $|\mathbf{u}| = 2$; $\theta = -150°$
34. $|\mathbf{u}| = 4$; $\theta = -75°$
35. $|\mathbf{u}| = \frac{1}{2}$; $\theta = 100.2°$
36. $|\mathbf{u}| = 7$; $\theta = 253.3°$

Applying the Concepts

In problems 37–44, compute each scalar value to two decimal places and each angle to the nearest hundredth of a degree.

37. **Resultant Force:** Two forces $\mathbf{F}_1$ and $\mathbf{F}_2$ act on a point. If $|\mathbf{F}_1| = 40$ pounds, $|\mathbf{F}_2| = 23$ pounds, and the angle between $\mathbf{F}_1$ and $\mathbf{F}_2$ is 58°, find:
   (a) The magnitude of the resultant force $\mathbf{F}$
   (b) The angle between $\mathbf{F}_1$ and $\mathbf{F}$
38. **Resultant Force:** Two forces acting on the same object produce a resultant force of 17 pounds with an angle of 20° relative to a horizontal axis. Assume that one of the two forces is 20 pounds and has an angle of 50° relative to the horizontal axis. Find the other force and the direction angle relative to the horizontal axis.
39. **Pulling a Sled:** Two children are pulling a third child across the ice on a sled. The first child pulls a rope with a force of 5 newtons and the second child pulls a rope with a force of 8 newtons. If the angle between the ropes is 28° (Figure 21), find:
   (a) The magnitude of the resultant force
   (b) The angle the resultant force vector makes with the first child's rope

Figure 21

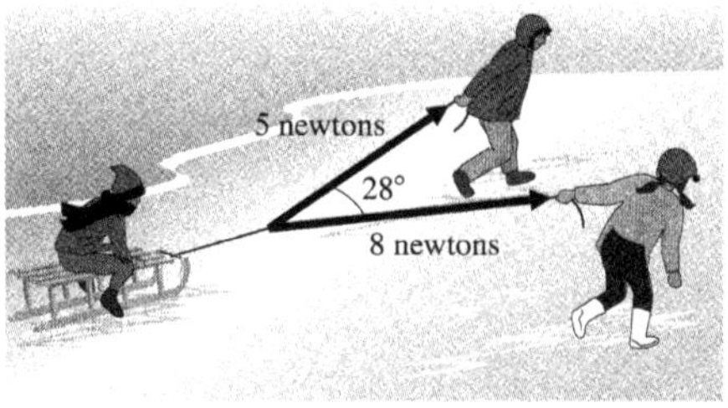

40. **Helicopter Aviation:** A helicopter is flying in the direction N50°W with an airspeed of 165 miles per hour. The wind is blowing at 35 miles per hour in the direction S60°E. Find the course and the ground speed of the helicopter.
41. **Airplane Aviation:** A commercial airplane with an airspeed of 510 miles per hour is headed N85°E. The wind is blowing N35°E at 45 miles per hour. Find:
   (a) The ground speed
   (b) The course of the airplane
   (c) The drift angle

42. Airplane Aviation: An airplane is flying in the direction N25°E with an airspeed of 450 kilometers per hour. Its ground speed is 500 kilometers per hour, and its course is N40°E. Find:
   (a) The speed of the wind
   (b) The direction of the wind

43. Boat Navigation: A boat heading S40°E at a still water speed of 40 kilometers per hour is pushed off course by a current of 32 kilometers per hour flowing in the direction S50°W. Find:
   (a) The speed of the boat
   (b) The course the boat is traveling
   (c) The drift angle

44. Motorboat Navigation: Suppose a motorboat that has a speed in still water of 12 miles per hour heads due west across a river that is flowing due south at a speed of 3 miles per hour.
   (a) What is the speed of the boat?
   (b) What angle does the path of the boat make with the vector representing the current?

Developing and Extending the Concepts

45. Find the terminal point of

$$\mathbf{u} = \langle 10, -4 \rangle$$

if the initial point P is $(3, 5)$.

46. Find the initial point of

$$\mathbf{u} = \langle -4, 7 \rangle$$

if the terminal point Q is $(2, -1)$.

Problems 47 and 48 pertain to the following properties:

> **Vector Addition and Scalar Multiplication Properties**
>
> Let $\mathbf{u}$, $\mathbf{v}$, and $\mathbf{w}$ be vectors, and let t and s be scalars; then the following relationships hold:
>
> **1.** $\mathbf{u} + \mathbf{v} = \mathbf{v} + \mathbf{u}$
> **2.** $\mathbf{u} + (\mathbf{v} + \mathbf{w}) = (\mathbf{u} + \mathbf{v}) + \mathbf{w}$
> **3.** $\mathbf{u} + \mathbf{0} = \mathbf{u}$
> **4.** $\mathbf{u} + (-\mathbf{u}) = \mathbf{0}$
> **5.** $s(\mathbf{u} + \mathbf{v}) = s\mathbf{u} + s\mathbf{v}$
> **6.** $(s + t)\mathbf{u} = s\mathbf{u} + t\mathbf{u}$
> **7.** $(st)\mathbf{u} = s(t\mathbf{u})$
> $\quad = t(s\mathbf{u})$
> **8.** $1\mathbf{u} = \mathbf{u}$
> and
> $0\mathbf{u} = \mathbf{0}$

47. Illustrate properties 1–4 for $\mathbf{u} = \langle 1, 2 \rangle$, $\mathbf{v} = \langle -3, 2 \rangle$, and $\mathbf{w} = \langle -5, 4 \rangle$.

48. Illustrate properties 5–8 for $\mathbf{u} = \langle -2, 5 \rangle$, $\mathbf{v} = \langle 4, 3 \rangle$, $s = 4$, and $t = -3$.

The **dot product,** or **inner product,** of two vectors $\mathbf{u} = \langle a, b \rangle$ and $\mathbf{v} = \langle c, d \rangle$ is given by

$$\mathbf{u} \cdot \mathbf{v} = ac + bd.$$

The angle θ between the vectors $\mathbf{u}$ and $\mathbf{v}$ is given by

$$\cos \theta = \frac{\mathbf{u} \cdot \mathbf{v}}{|\mathbf{u}||\mathbf{v}|}, \text{ where } 0 \leq \theta \leq 180°.$$

In problems 49–54, find:
   (a) The dot product of the two vectors $\mathbf{u}$ and $\mathbf{v}$
   (b) The angle θ between $\mathbf{u}$ and $\mathbf{v}$ (round off θ to the nearest tenth of a degree)

49. $\mathbf{u} = \langle -3, 2 \rangle$; $\mathbf{v} = \langle 1, -3 \rangle$
50. $\mathbf{u} = \langle 1, -3 \rangle$; $\mathbf{v} = \langle 5, -1 \rangle$
51. $\mathbf{u} = \langle -3, 6 \rangle$; $\mathbf{v} = \langle 2, -1 \rangle$
52. $\mathbf{u} = \langle -2, 6 \rangle$; $\mathbf{v} = \langle 3, -5 \rangle$
53. $\mathbf{u} = 5\mathbf{i} - 2\mathbf{j}$; $\mathbf{v} = 4\mathbf{i} + 3\mathbf{j}$
54. $\mathbf{u} = -4\mathbf{i}$; $\mathbf{v} = 4\mathbf{i} + 5\mathbf{j}$

55. Explain why two nonzero vectors $\mathbf{u}$ and $\mathbf{v}$ are **perpendicular** (or **orthogonal**) if and only if

$$\mathbf{u} \cdot \mathbf{v} = 0.$$

[*Hint:* Use the formulas given for problems 49–54.]

In problems 56–58, use the result from problem 55.

56. Find the values of the constant k so that $\mathbf{u}$ and $\mathbf{v}$ are orthogonal. Sketch $\mathbf{u}$ and $\mathbf{v}$ in the same coordinate system.
   (a) $\mathbf{u} = \langle k, 3 \rangle$; $\mathbf{v} = \langle -4, 2 \rangle$
   (b) $\mathbf{u} = \langle 3k, -1 \rangle$; $\mathbf{v} = \langle k, 4 \rangle$

57. Use vectors and the dot product to determine whether the triangle with vertices

$$A = (6, 1), B = (4, 3), \text{ and } C = (2, 1)$$

is a right triangle. If it is a right triangle, which vertex angle is the right angle?

58. Suppose that vector $\mathbf{u}$ has initial point $(2, 1)$ and terminal point $(5, -4)$,

and

$$\mathbf{v} = \langle 5, 3 \rangle.$$

Use the dot product to determine whether $\mathbf{u}$ and $\mathbf{v}$ are orthogonal.

Interlude: Practice Problems 3.3, Part I

1. A vector is drawn from point $P(4,-3)$ to point $Q(-6,10)$. Call this vector $\mathbf{u}$.

(a) Give the standard representation of this vector.

(b) Find the direction angle of this vector, in degrees, accurate to two decimal places.

(c) Find the unit vector in the same direction as $\mathbf{u}$.

2. Let $\mathbf{a} = 8\mathbf{i} + 6\mathbf{j}$ and $\mathbf{b} = -9\mathbf{i} + 2\mathbf{j}$.

(a) Fill in the blanks: $\mathbf{a} = \langle$________, ________$\rangle$ and $\mathbf{b} = \langle$________, ________$\rangle$

(b) Calculate the magnitude of each vector.

(c) Find the angle between the vectors, in radians, correct to three decimal places.

3. The captain of a boat is steering at a heading of N33°W at a speed of 16 mph. There is a 5 mph current at a heading of N48°E

(a) Calculate the true heading of the boat, accurate to two decimal places.

(b) If the captain wishes to actually travel due north, then at what speed should the boat be operated? Give your result accurate to two decimal places.

Interlude: Practice Problems 3.3, Part II

Given two vectors $\mathbf{u} = \langle a, b \rangle$ and $\mathbf{v} = \langle c, d \rangle$, the **dot product** of **u** and **v** is the real number given by $\mathbf{u} \bullet \mathbf{v} = ac + bd$.

1. Find nonzero vectors **u**, **v**, and **w** so that $\mathbf{u} \bullet \mathbf{v} = \mathbf{u} \bullet \mathbf{w}$ but $\mathbf{v} \neq \mathbf{w}$. Verify your work.

It can be shown that **u** and **v** are **perpendicular** (or **orthogonal**) if and only if $\mathbf{u} \bullet \mathbf{v} = 0$.

2. Let $\mathbf{v} = \langle -8, 9 \rangle$. Find the unit vector orthogonal to **v**.

If a constant force **F** is applied to an object moving along vector **D**, then the **work** W done is defined as $W = \mathbf{F} \bullet \mathbf{D}$.

3. A force of 140 pounds is used to pull a sled using a rope that makes an angle of 32° with the horizontal. If the sled is moved 200 feet, then how much work is done? Label your response appropriately and give your result accurate to two decimal places.

4. A 475-pound box lies on a ramp that makes an angle of 18° with the horizontal. A rope is tied to the box from a post at the top of the ramp to keep the box in position. Ignoring friction, how much force is being exerted on the rope by the box?

3.4 Trigonometric Forms of Complex Numbers

Objectives

1. Define the Complex Plane
2. Write a Complex Number in Trigonometric Form
3. Find the Powers of Complex Numbers
4. Find the Roots of Complex Numbers

We have already considered complex numbers and we have seen how to perform basic operations on them. Now we'll discuss how to represent these numbers in trigonometric form. Such representations make multiplication and division of complex numbers much easier to perform. In this section, we'll also see the usefulness of trigonometric forms in finding powers and roots of complex numbers.

Defining the Complex Plane

We begin by representing complex numbers as points in the plane. Each ordered pair of real numbers (a, b) can be associated with the complex number $z = a + bi$, and each complex number $z = a + bi$ can be associated with the ordered pair of real numbers (a, b).

Because of this one-to-one correspondence between the complex numbers and the ordered pairs of real numbers, we can use points in the Cartesian plane to represent the complex numbers.

The plane in which the complex numbers are represented is called the **complex plane;** the horizontal axis (x axis) is called the **real axis** and the vertical axis (y axis) is called the **imaginary axis** (Figure 1a).

For example, the ordered pairs $(2, -3)$ and $(5, 2)$ are used to represent the complex numbers $z_1 = 2 - 3i$ and $z_2 = 5 + 2i$, respectively (Figure 1b). Complex numbers of the form $z = bi$ are represented by points of the form $(0, b)$, that is, as points on the imaginary axis. For example, the complex number $3i$ is represented by the point $(0, 3)$. Complex numbers of the form $z = a$ are represented by points of the form $(a, 0)$, that is, as points on the real axis. Thus, the number 5 is represented by the point $(5, 0)$ (Figure 1c).

Figure 1

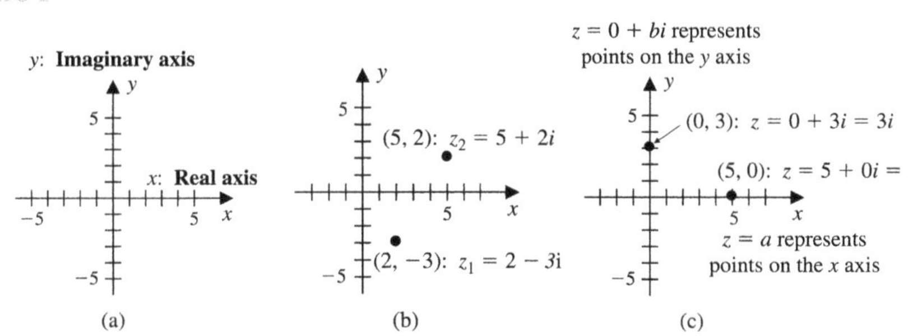

Geometrically, we interpret the *absolute value,* or *magnitude,* of a complex number $a + bi$ as the distance between the origin of the complex plane and the point (a, b). We usually denote it by the absolute value $|z| = |a + bi|$ and refer to it as the **modulus** of the complex number $a + bi$.

So, by the Pythagorean theorem, if $z = a + bi$, then $|z|$ is given by

$$|z| = \sqrt{a^2 + b^2} \quad \text{(Figure 2).}$$

Figure 2

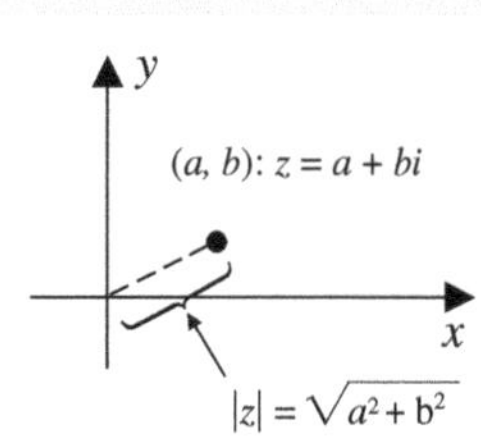

EXAMPLE 1

Finding Absolute Values

Let $z_1 = 4 + 3i$ and $z_2 = \sqrt{3} - i$. Find:

(a) $|z_1|$ (b) $|z_2|$ (c) $|z_1 z_2|$

Solution

(a) $|z_1| = \sqrt{4^2 + 3^2} = \sqrt{16 + 9} = \sqrt{25} = 5$

(b) $|z_2| = \sqrt{(\sqrt{3})^2 + (-1)^2} = \sqrt{4} = 2$

(c) $|z_1 z_2| = |(4 + 3i)(\sqrt{3} - i)|$

$= |(4\sqrt{3} + 3) + (3\sqrt{3} - 4)i|$

$= \sqrt{(4\sqrt{3} + 3)^2 + (3\sqrt{3} - 4)^2}$

$= \sqrt{100} = 10$

Example 1 leads us to the following general results (problem 60):

$$\text{(i) } |z_1 z_2| = |z_1||z_2| \qquad \text{(ii) } \left|\frac{z_1}{z_2}\right| = \frac{|z_1|}{|z_2|}, \quad z_2 \neq 0$$

Writing a Complex Number in Trigonometric Form

Figure 3

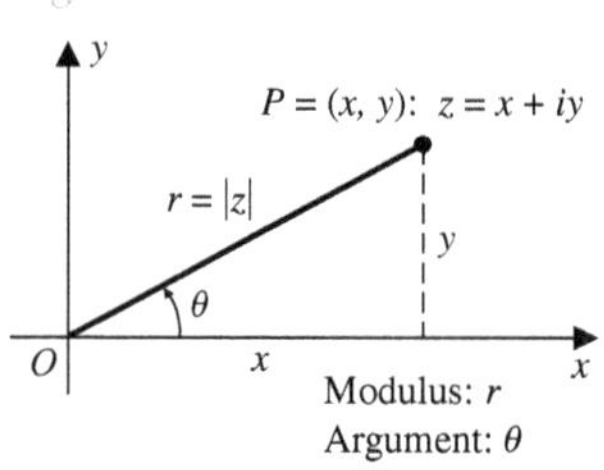

It is possible to write complex numbers in trigonometric form. Suppose that $z = x + yi$ is a nonzero complex number; then its graphical representation is the point $P = (x, y)$. Let θ be any angle in standard position whose terminal side lies on the line segment $\overline{OP}$ and $r = |z| = \sqrt{x^2 + y^2}$. The angle θ is usually chosen in the interval $[0, 2\pi)$ and is called an **argument** of z (Figure 3).

Using trigonometry, we obtain

$$\cos\theta = \frac{x}{r} \quad \text{and} \quad \sin\theta = \frac{y}{r}$$

or

$$x = r\cos\theta \quad \text{and} \quad y = r\sin\theta.$$

Thus we can rewrite any complex number $z = x + yi$ in the form

$$z = r\cos\theta + (r\sin\theta)i = r(\cos\theta + i\sin\theta)$$

where $r = \sqrt{x^2 + y^2}$ is the modulus and θ is an argument.

This form is called the **trigonometric form,** or **polar form,** of a complex number.

Notice that θ is not unique, since

$$r(\cos\theta + i\sin\theta) = r(\cos\theta_1 + i\sin\theta_1)$$

holds whenever $\theta - \theta_1$ is an integer multiple of 2π. Consequently:

Two complex numbers are equal if and only if their moduli are equal and their arguments differ by an integer multiple of 2π.

Thus if z has a modulus r and argument θ, then

$$z = x + iy = r[\cos(\theta + 2\pi k) + i\sin(\theta + 2\pi k)]$$

where k is any integer. If we use degree measure instead of radians to measure θ, then

$$z = r[\cos(\theta + 360°k) + i\sin(\theta + 360°k)]$$

where k is any integer.

EXAMPLE 2

Writing a Complex Number in Cartesian Form

Express the complex number

$$z = 2\left(\cos\frac{\pi}{3} + i\sin\frac{\pi}{3}\right)$$

in the Cartesian form $x + yi$.

Solution The given complex number is in the form $z = r(\cos\theta + i\sin\theta)$, with $r = 2$ and $\theta = \pi/3$.

Figure 4

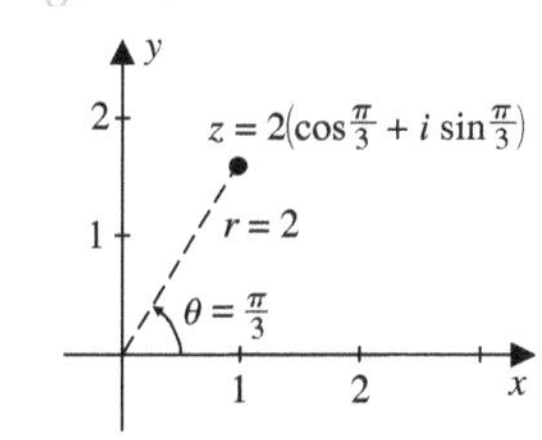

So

$$\begin{aligned} z &= x + yi \\ &= 2\cos\frac{\pi}{3} + \left(2\sin\frac{\pi}{3}\right)i \\ &= 2\left(\frac{1}{2}\right) + 2\left(\frac{\sqrt{3}}{2}\right)i \\ &= 1 + \sqrt{3}i \text{ (Figure 4).} \end{aligned}$$

EXAMPLE 3

Writing a Complex Number in Trigonometric Form

Express $z = -\sqrt{3} - i$ in trigonometric form with $0 \le \theta < 2\pi$.

Solution First we plot the point $(-\sqrt{3}, -1)$ corresponding to $z = -\sqrt{3} - i$ (Figure 5). Then we find the modulus r and angle θ.

From Figure 5, we see that the modulus r is given by

$$r = \sqrt{(-\sqrt{3})^2 + (-1)^2} = 2.$$

Figure 5

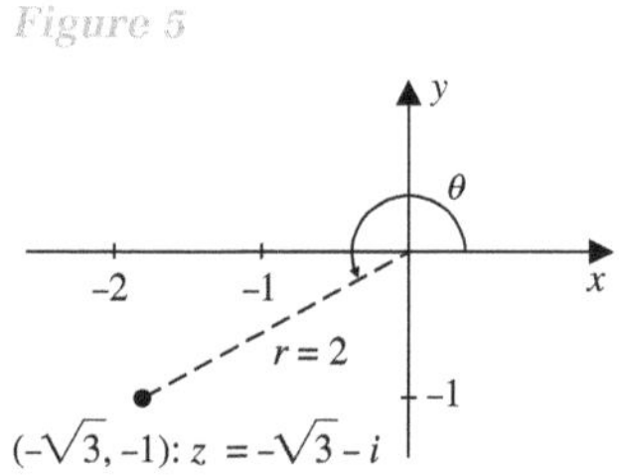

The argument θ satisfies the equation

$$\tan\theta = \frac{(-1)}{-\sqrt{3}} = \frac{1}{\sqrt{3}}$$

and its terminal side is in quadrant III. It follows from trigonometry that $\theta = (7\pi)/6$. Therefore,

$$z = 2\left(\cos\frac{7\pi}{6} + i\sin\frac{7\pi}{6}\right).$$

Finding the Powers of Complex Numbers

Trigonometric forms can be used to multiply and divide complex numbers. Let's examine the product of z_1 and z_2, where

$$z_1 = r_1(\cos\theta_1 + i\sin\theta_1) \quad \text{and} \quad z_2 = r_2(\cos\theta_2 + i\sin\theta_2)$$

We have

$$\begin{aligned} z_1z_2 &= r_1[(\cos\theta_1 + i\sin\theta_1)]\cdot r_2[(\cos\theta_2 + i\sin\theta_2)] \\ &= r_1r_2[(\cos\theta_1\cos\theta_2 - \sin\theta_1\sin\theta_2) + i(\cos\theta_1\sin\theta_2 + \cos\theta_2\sin\theta_1)] \\ &= r_1r_2[\cos(\theta_1+\theta_2) + i\sin(\theta_1+\theta_2)] \quad \text{Sum identities for the cosine and sine} \end{aligned}$$

This result gives us the first of the following two formulas:

Multiplication and Division of Complex Numbers

Suppose that trigonometric forms of the complex numbers z_1 and z_2 are given by

$$z_1 = r_1(\cos\theta_1 + i\sin\theta_1) \quad \text{and} \quad z_2 = r_2(\cos\theta_2 + i\sin\theta_2)$$

Then:

(i) $z_1z_2 = r_1r_2[\cos(\theta_1+\theta_2) + i\sin(\theta_1+\theta_2)]$

(ii) $\dfrac{z_1}{z_2} = \dfrac{r_1}{r_2}[\cos(\theta_1-\theta_2) + i\sin(\theta_1-\theta_2)],\quad z_2 \neq 0$

The derivation of the second formula is left as an exercise (problem 62).
In words, the multiplication and division formulas are stated as follows:

The product of two complex numbers in trigonometric form is obtained by multiplying their moduli and adding their arguments. Also, the quotient of two complex numbers is obtained by dividing their moduli and subtracting their arguments.

EXAMPLE 4 **Multiplying and Dividing Complex Numbers**

Let

$$z_1 = \sqrt{2}\left(\cos\frac{3\pi}{4} + i\sin\frac{3\pi}{4}\right) \quad \text{and} \quad z_2 = 4\left(\cos\frac{3\pi}{2} + i\sin\frac{3\pi}{2}\right).$$

Write each given expression in both trigonometric and Cartesian form:

(a) z_1z_2 (b) $\dfrac{z_1}{z_2}$

Solution (a) Using the multiplication formula, we have

$$\begin{aligned} z_1z_2 &= 4\sqrt{2}\left[\cos\left(\frac{3\pi}{4}+\frac{3\pi}{2}\right) + i\sin\left(\frac{3\pi}{4}+\frac{3\pi}{2}\right)\right] && \text{Multiplication rule} \\ &= 4\sqrt{2}\left(\cos\frac{9\pi}{4} + i\sin\frac{9\pi}{4}\right) && \text{Trigonometric form} \\ &= 4\sqrt{2}\left(\frac{\sqrt{2}}{2} + i\frac{\sqrt{2}}{2}\right) && \text{Evaluate} \\ &= 4 + 4i && \text{Cartesian form} \end{aligned}$$

(b) Using the division formula, we have

$$\frac{z_1}{z_2} = \frac{\sqrt{2}}{4}\left[\cos\left(\frac{3\pi}{4} - \frac{3\pi}{2}\right) + i\sin\left(\frac{3\pi}{4} - \frac{3\pi}{2}\right)\right] \quad \text{Division rule}$$

$$= \frac{\sqrt{2}}{4}\left[\cos\left(-\frac{3\pi}{4}\right) + i\sin\left(-\frac{3\pi}{4}\right)\right] \quad \text{Trigonometric form}$$

$$= \frac{\sqrt{2}}{4}\left(\cos\frac{3\pi}{4} - i\sin\frac{3\pi}{4}\right)$$

$$= \frac{\sqrt{2}}{4}\left(-\frac{\sqrt{2}}{2} - i\frac{\sqrt{2}}{2}\right) \quad \text{Evaluate}$$

$$= -\frac{1}{4} - \frac{1}{4}i \quad \text{Cartesian form}$$

Repeated use of the multiplication rule for complex numbers in trigonometric form allows us to compute powers of a complex number. For instance, if $z = r(\cos\theta + i\sin\theta)$, then

$$z^2 = z \cdot z = r \cdot r[\cos(\theta + \theta) + \mathrm{i}\sin(\theta + \theta)] = r^2(\cos 2\theta + i\sin 2\theta).$$

Since $z^3 = z^2 \cdot z$, then

$$z^3 = r^2 \cdot r[\cos(2\theta + \theta) + \mathrm{i}\sin(2\theta + \theta)] = r^3(\cos 3\theta + i\sin 3\theta).$$

If we repeat the process one more time, we get

$$z^4 = z^3 \cdot z = r^4(\cos 4\theta + i\sin 4\theta).$$

This pattern is expressed in the following theorem, which is attributed to Abraham DeMoivre (1667–1754).

DeMoivre's Theorem

Let $z = r(\cos\theta + i\sin\theta)$. Then for any positive integer n,

$$z^n = [r(\cos\theta + i\sin\theta)]^n = r^n(\cos n\theta + i\sin n\theta).$$

EXAMPLE 5

Finding Powers of Complex Numbers

Use DeMoivre's theorem to determine each of the given powers. Express the answer in Cartesian form.

(a) $[3(\cos 60° + i\sin 60°)]^4$ (b) $(1 + i)^{20}$

Solution

(a) By DeMoivre's theorem,

$$[3(\cos 60° + i\sin 60°)]^4 = 3^4(\cos 240° + i\sin 240°)$$

$$= 81\left(-\frac{1}{2} - i\frac{\sqrt{3}}{2}\right) = -\frac{81}{2} - \frac{81\sqrt{3}}{2}i$$

(b) To use DeMoivre's theorem, we first convert $1 + i$ to trigonometric form:

$$1 + i = \sqrt{2}\left(\cos\frac{\pi}{4} + i\sin\frac{\pi}{4}\right)$$

So the modulus of $1 + i$ is $\sqrt{2}$ and an argument is $\pi/4$. Thus

$$(1 + i)^{20} = \left[\sqrt{2}\left(\cos\frac{\pi}{4} + i\sin\frac{\pi}{4}\right)\right]^{20} = 2^{10}(\cos 5\pi + i\sin 5\pi)$$

$$= 1024(-1 + 0i) = -1024 + 0i.$$

Finding the Roots of Complex Numbers

Recall that a polynomial equation of degree n has n roots (or solutions) in the complex number system. Hence the equation $z^4 = 16$ has four roots. One way to find these roots is to write the equation in the form

$$z^4 - 16 = 0$$

so that

$$(z^2 - 4)(z^2 + 4) = 0$$

$$(z - 2)(z + 2)(z + 2i)(z - 2i) = 0.$$

It follows that the solutions are

$$2, -2, -2i, \text{ and } 2i.$$

It was easy to find the roots for this equation because of the factorization. In more complicated situations, we can find all the roots of a complex number by using DeMoivre's theorem as follows:

Suppose we want to determine all the nth roots of a complex number w. That is, suppose we want to solve the equation

$$z^n = w \quad \text{where } n \text{ is a positive integer.}$$

Assume that trigonometric forms for w and z are given, respectively, by

$$w = R(\cos \phi + i \sin \phi) \quad \text{and} \quad z = r(\cos \theta + i \sin \theta).$$

Then applying DeMoivre's theorem to $z^n = w$ yields

$$\begin{aligned} z^n &= [r(\cos \theta + i \sin \theta)]^n \\ &= r^n(\cos n\theta + i \sin n\theta) \\ &= R(\cos \phi + i \sin \phi). \end{aligned}$$

So

$$r^n = R \quad \text{and} \quad n\theta = \phi + 2\pi k \quad (\text{or } n\theta = \phi + 360°k \text{ if degrees are used}).$$

Hence

$$z = r(\cos \theta + i \sin \theta) \text{ is a root of } z^n = w$$

whenever

$$r = \sqrt[n]{R} \quad \text{and} \quad \theta = \frac{\phi}{n} + \frac{2\pi k}{n} \quad \left(\text{or } \theta = \frac{\phi}{n} + \frac{360°k}{n} \text{ if degrees are used}\right)$$

where

$$k = 0, \pm 1, \pm 2, \pm 3, \ldots.$$

At first glance, it appears that there are an infinite number of roots for the equation $z^n = w$. However, there are only n roots. These n roots can be determined by letting k take on the values $0, 1, 2, 3, 4, \ldots, n - 1$. If we let $k = n$, then

$$\theta = \frac{\phi}{n} + \frac{2\pi n}{n} = \frac{\phi}{n} + 2\pi.$$

This angle has the same terminal side as ϕ/n, so we get the same value of z as when $k = 0$. Similarly, the value of θ obtained by letting $k = n + 1$ gives an angle

$$\theta = \frac{\phi}{n} + \frac{2\pi}{n} + 2\pi$$

and the resulting value of z is the same as when $k = 1$ and so on.

Thus, we are led to the following result.

nth Roots of a Complex Number

If $w = R(\cos\phi + i\sin\phi)$ is any nonzero complex number, and if n is any positive integer, then the distinct nth roots of w are $z_0, z_1, z_2, \ldots, z_{n-1}$, where

$$z_k = \sqrt[n]{R}\left[\cos\left(\frac{\phi}{n} + \frac{2\pi k}{n}\right) + i\sin\left(\frac{\phi}{n} + \frac{2\pi k}{n}\right)\right] \qquad k = 0, 1, 2, \ldots, n-1.$$

If we use degree measure for ϕ, the nth roots of w are given by

$$\sqrt[n]{R}\left[\cos\left(\frac{\phi}{n} + \frac{360^\circ k}{n}\right) + i\sin\left(\frac{\phi}{n} + \frac{360^\circ k}{n}\right)\right] \qquad k = 0, 1, 2, \ldots, n-1.$$

EXAMPLE 6

Finding the Fourth Roots of a Complex Number

Find the four fourth roots of $1 + i$; that is, solve $z^4 = 1 + i$. Express the answers in both trigonometric form and Cartesian form (rounded to two decimal places). Represent the roots graphically.

Solution First, we determine $R(\cos\phi + i\sin\phi)$, a trigonometric representation of $1 + i$.

$$\text{Here } R = \sqrt{1+1} = \sqrt{2} \quad \text{and} \quad \text{we use } \phi = \frac{\pi}{4}$$

So,
$$1 + i = \sqrt{2}\left(\cos\frac{\pi}{4} + i\sin\frac{\pi}{4}\right).$$

Since $n = 4$, the fourth roots are given by

$$z_k = \sqrt[4]{\sqrt{2}}\left[\cos\left(\frac{\pi/4}{4} + \frac{2\pi k}{4}\right) + \mathrm{i}\sin\left(\frac{\pi/4}{4} + \frac{2\pi k}{4}\right)\right] \quad \text{where } k = 0, 1, 2, 3.$$

Using $\sqrt[4]{\sqrt{2}} = \sqrt[8]{2}$, we substitute each value of k into this expression to obtain

$$z_0 = \sqrt[8]{2}\left(\cos\frac{\pi}{16} + i\sin\frac{\pi}{16}\right) \qquad \text{for } k = 0$$

$$z_1 = \sqrt[8]{2}\left(\cos\frac{9\pi}{16} + i\sin\frac{9\pi}{16}\right) \qquad \text{for } k = 1$$

$$z_2 = \sqrt[8]{2}\left(\cos\frac{17\pi}{16} + i\sin\frac{17\pi}{16}\right) \qquad \text{for } k = 2$$

$$z_3 = \sqrt[8]{2}\left(\cos\frac{25\pi}{16} + i\sin\frac{25\pi}{16}\right) \qquad \text{for } k = 3.$$

Figure 6

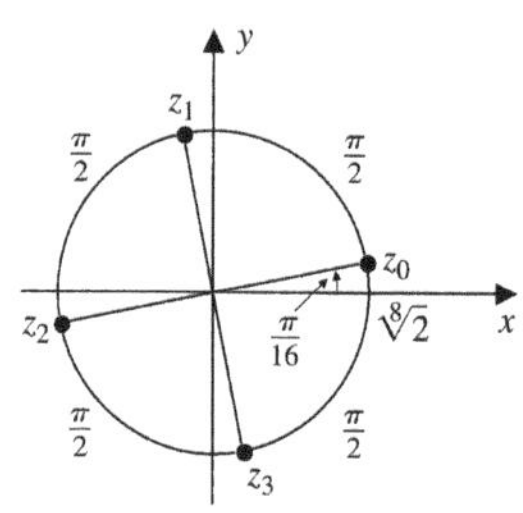

By using a calculator and rounding off the answers to two decimal places, we get the approximations

$$z_0 = 1.07 + 0.21i$$
$$z_1 = -0.21 + 1.07i$$
$$z_2 = -1.07 - 0.21i$$
$$z_3 = 0.21 - 1.07i.$$

These four fourth roots of $1 + i$ are equally spaced on the circle of radius $\sqrt[8]{2}$ centered at the origin; their arguments differ by $\pi/2$ (Figure 6).

The solutions of $z^n = 1$ are called the ***n*th roots of unity.** In the *real* number system, we find the solution of $x^3 = 1$ by taking the cube root of each side of the equation to get $x = 1$. There is only one real number solution. In the *complex* number system, we know that the equation $z^3 = 1$ has three possible solutions. We can now find them by using trigonometric forms of complex numbers.

EXAMPLE 7

Finding the Cube Roots of Unity

Find the cube roots of unity in the form $x + yi$; that is, find all three roots of the equation $z^3 = 1$.

Solution

The number 1 can be written in trigonometric form as $1(\cos 0° + i \sin 0°)$, so $R = 1$, $\phi = 0°$, and $n = 3$. The three roots are given by

$$z_k = \sqrt[3]{1}\left[\cos\left(\frac{0° + 360°k}{3}\right) + i \sin\left(\frac{0° + 360°k}{3}\right)\right] \quad \text{where } k = 0, 1, 2.$$

Substituting these values for k yields the following three roots:

$$z_0 = 1(\cos 0° + i \sin 0°) = 1 + 0i = 1 \qquad \text{for } k = 0$$
$$z_1 = 1(\cos 120° + i \sin 120°) = -\frac{1}{2} + \frac{\sqrt{3}}{2}i \qquad \text{for } k = 1$$
$$z_2 = 1(\cos 240° + i \sin 240°) = -\frac{1}{2} - \frac{\sqrt{3}}{2}i \qquad \text{for } k = 2$$

Figure 7

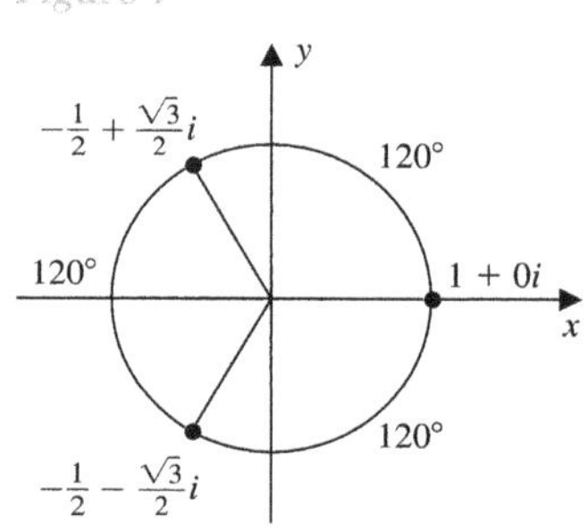

Figure 7 illustrates the three cube roots of 1 graphically. They are equally spaced on the circle of radius 1 with center at the origin; their arguments differ by 120°.

PROBLEM SET 3.4

Mastering the Concepts

In problems 1–4, represent each complex number in the complex plane, and find its modulus.

1. (a) $z = 2 + 7i$
   (b) $z = 5 - 2i$
2. (a) $z = \sqrt{3} - i$
   (b) $z = -4 - 4i$
3. (a) $z = 3 + 4i$
   (b) $z = -5 + 3i$
4. (a) $z = -4$
   (b) $z = 2i$

In problems 5–8, find the value of each expression, where $\bar{z} = a - bi$ is the conjugate of $z = a + bi$.

(a) $|zw|$ (b) $|z\bar{z}|$ (c) $|z||w|$ (d) $\left|\frac{z}{w}\right|$ (e) $\frac{|z|}{|w|}$

5. $z = -1 - i;\quad w = 2 - 2i$
6. $z = \sqrt{3}i;\quad w = -2 - 2i$

7. $z = -4 + 3i; \quad w = -2 - 5i$

8. $z = 4 + 2i; \quad w = 3 - 5i$

In problems 9–16, express each complex number in the Cartesian form $x + yi$, and graph it. Round off to two decimal places in problems 15 and 16.

9. $z = 2(\cos 30° + i \sin 30°)$

10. $z = 10\left(\cos \frac{3\pi}{4} + i \sin \frac{3\pi}{4}\right)$

11. $z = 2\left(\cos \frac{\pi}{2} + i \sin \frac{\pi}{2}\right)$

12. $z = 7\left[\cos\left(-\frac{3\pi}{2}\right) + i \sin\left(-\frac{3\pi}{2}\right)\right]$

13. $z = 8(\cos 900° + i \sin 900°)$

14. $z = 6(\cos 120° + i \sin 120°)$

15. $z = 2(\cos 10° + i \sin 10°)$

16. $z = 2[\cos(-75°) + i \sin(-75°)]$

In problems 17–22, express each complex number in trigonometric form with $0 \le \theta < 2\pi$. Round off to two decimal places when necessary.

17. (a) $z = -1 - i$ (b) $z = -\sqrt{3} - i$

18. (a) $z = -3 + 4i$ (b) $z = 8 + 7i$

19. (a) $z = \sqrt{2} + i$ (b) $z = -5$

20. (a) $z = 3 + 4i$ (b) $z = -3i$

21. (a) $z = -\frac{1}{2} + \frac{\sqrt{3}}{2}i$ (b) $z = \frac{\sqrt{3}}{2} + \frac{1}{2}i$

22. (a) $z = \frac{\sqrt{2}}{2} - \frac{\sqrt{2}}{2}i$ (b) $z = -\frac{\sqrt{2}}{2} + \frac{\sqrt{2}}{2}i$

In problems 23–30, find:

(a) $z_1 z_2$ (b) $\frac{z_1}{z_2}$

Express the answers in both trigonometric form and Cartesian form. In problems 29 and 30, round off answers to two decimal places.

23. $z_1 = 4\left(\cos \frac{5\pi}{6} + i \sin \frac{5\pi}{6}\right)$; $z_2 = 2\left(\cos \frac{\pi}{3} + i \sin \frac{\pi}{3}\right)$

24. $z_1 = 2\left[\cos\left(-\frac{\pi}{6}\right) + i \sin\left(-\frac{\pi}{6}\right)\right]$; $z_2 = 2\left[\cos\left(-\frac{5\pi}{6}\right) + i \sin\left(-\frac{5\pi}{6}\right)\right]$

25. $z_1 = \cos 30° + i \sin 30°$; $z_2 = \cos 60° + i \sin 60°$

26. $z_1 = 5(\cos 30° + i \sin 30°)$; $z_2 = 6(\cos 240° + i \sin 240°)$

27. $z_1 = \sqrt{2}(\cos 45° + i \sin 45°)$; $z_2 = \sqrt{2}(\cos 135° + i \sin 135°)$

28. $z_1 = 4\left(\cos \frac{3\pi}{4} + i \sin \frac{3\pi}{4}\right)$; $z_2 = 2(\cos \pi + i \sin \pi)$

29. $z_1 = 2(\cos 50° + i \sin 50°)$; $z_2 = 3(\cos 40° + i \sin 40°)$

30. $z_1 = 5(\cos 170° + i \sin 170°)$; $z_2 = \cos 55° + i \sin 55°$

In problems 31–44, use DeMoivre's theorem to compute each of the given powers. Express the answer in Cartesian form.

31. $(\cos 30° + i \sin 30°)^7$

32. $(\cos 15° + i \sin 15°)^8$

33. $\left[2\left(\cos \frac{\pi}{6} + i \sin \frac{\pi}{6}\right)\right]^{10}$

34. $\left[3\left(\cos \frac{\pi}{18} + i \sin \frac{\pi}{18}\right)\right]^6$

35. $\left[2\left(\cos \frac{5\pi}{4} + i \sin \frac{5\pi}{4}\right)\right]^8$

36. $[4(\cos 36° + i \sin 36°)]^5$

37. $(5 + 5i)^6$

38. $(1 + \sqrt{3}i)^5$

39. $(\sqrt{3} - i)^4$

40. $\left(-\frac{1}{2} - \frac{\sqrt{3}}{2}i\right)^8$

41. $(\sqrt{3} + i)^{30}$

42. $(1 + i)^{50}$

43. $\left(\frac{1}{\sqrt{2}} + \frac{1}{\sqrt{2}}i\right)^{100}$

44. $\left(\frac{1}{2} + \frac{\sqrt{3}}{2}i\right)^{30}$

In problems 45–52, find all the indicated roots of each complex number in the form $x + yi$. Round off answers to two decimal places. Represent the roots graphically.

45. Square roots of i

46. Square roots of $3 - 3i$

47. Cube roots of 8

48. Cube roots of i

49. Fifth roots of $32(\cos 315° + i \sin 315°)$

50. Fifth roots of unity

51. Fourth roots of $-8 - 3\sqrt{3}i$

52. Fourth roots of -16

In problems 53–58, find all the roots of each equation in both trigonometric and Cartesian form. Round off answers to two decimal places.

53. $z^3 + 8 = 0$

54. $z^3 + 8i = 0$

55. $z^4 + 81 = 0$

56. $z^5 + 1 = 0$

57. $z^6 + 64 = 0$

58. $z^5 - i = 1$

Developing and Extending the Concepts

59. Prove that $|z| = \sqrt{z \cdot \bar{z}}$, where $z = a + bi$ and $\bar{z} = a - bi$ is the conjugate of z.

60. Prove each of the following statements, where $z_1 = a_1 + b_1 i$ and $z_2 = a_2 + b_2 i$:

(a) $|z_1 z_2| = |z_1||z_2|$

(b) $\left|\dfrac{z_1}{z_2}\right| = \dfrac{|z_1|}{|z_2|}$, $z_2 \neq 0$

61. Describe geometrically the set of complex numbers $z = x + yi$ that satisfy each equation. Sketch the graph.

(a) $|z| = 2$

(b) $|z - i| = 1$

62. Let

$$z_1 = r_1(\cos \theta_1 + i \sin \theta_1) \text{ and}$$
$$z_2 = r_2(\cos \theta_2 + i \sin \theta_2).$$

Show that

$$\frac{z_1}{z_2} = \frac{r_1}{r_2}[\cos(\theta_1 - \theta_2) + i \sin(\theta_1 - \theta_2)] \quad z_2 \neq 0.$$

63. Let $z = r(\cos \theta + i \sin \theta)$.
Show that

(a) $\bar{z} = r(\cos \theta - i \sin \theta)$
$= r[\cos(-\theta) + i \sin(-\theta)]$

(b) $z^{-1} = r^{-1}(\cos \theta - i \sin \theta)$
$= r^{-1}[\cos(-\theta) + i \sin(-\theta)]$

64. If $z = r(\cos \theta + i \sin \theta)$, show that

$$z^{-n} = r^{-n}(\cos n\theta - i \sin n\theta)$$

where n is a positive integer. [*Hint:* Use the result from problem 63(b).]

65. Use the result from Problem 64 to write each expression in the form $x + yi$.

(a) $\left(\dfrac{\sqrt{3}}{2} + \dfrac{1}{2}i\right)^{-5}$

(b) $(-2 + 2i)^{-3}$

66. Use DeMoivre's theorem to derive formulas for $\cos 3\theta$ in terms of $\cos \theta$, and $\sin 3\theta$ in terms of $\sin \theta$. [*Hint:* Use the identity $\cos 3\theta + i \sin 3\theta = (\cos \theta + i \sin \theta)^3$.]

Chapter 3 Readiness, Part I

Provide complete solutions in the space provided.

1. The tallest monument in the world is the Gateway to the West Arch in St. Louis. From a point A on the ground in line with the middle of the arch (not directly under the arch) the angle of elevation to the top of the arch is $64.91°$. Moving away from the arch another 57 feet (but staying in line with the middle of the arch) changes the angle of elevation to $60.81°$. Given this, estimate the height of the top of the arch, accurate to two decimal places.

2. Two observation points, A and B, are 1500 feet apart. From these points, the angles of elevation of a hovering hot-air balloon are $43°$ and $47°$, respectively. Determine the height of the balloon, accurate to two decimal places.

3. A church is located at the top of a hill. When the top of the steeple is viewed from the base of the hill, the angle of elevation is 50°. When the top of the steeple is viewed from a point 220 feet out from the base of the hill, the angle of elevation is 40°. The hill rises at a pitch of 32°. Estimate the height of the top of the steeple from the base of the church.

Chapter 3 Readiness, Part II

Provide complete solutions in the space provided.

1. Suppose that a pizza with a diameter of 20 inches is cut into 12 equal-sized slices. If the center of the pizza is the origin, find a polar equation $r = f(\theta)$ along with restrictions on r and θ to describe a slice of the pizza. Show the graph of $r = f(\theta)$ subject to your restrictions on r and θ. Hint: the pizza is circular—the graph of $r = f(\theta)$ should trace the "crust" side of the pizza slice.

2. A plane takes off with a heading of S40°W and an air speed of 600 knots. The wind is blowing in the direction S10°W at 40 knots.

(a) Find the true heading of the airplane, accurate to two decimal places.

(b) Determine the ground speed of the airplane, accurate to two decimal places.

3. Find the fifth roots of $1 + \sqrt{3}i$. Give the exact roots and use radian measure.

4. Consider triangle ABC in Figure 1. Find a formula for the area of the triangle that involves only $\sin \alpha$, b, and c.

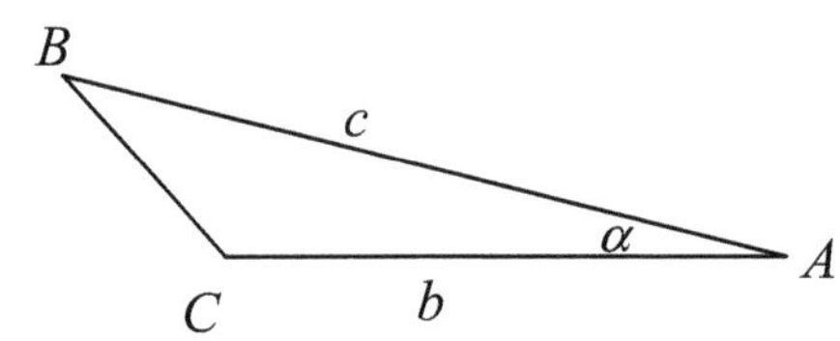

Figure 1

CHAPTER 3 REVIEW PROBLEM SET

In problems 1 and 2, solve triangle ABC. Round off each angle to the nearest hundredth of a degree and each length to two decimal places.

1. (a) $a = 5$, $b = 7$, $\gamma = 30°$
(b) $c = 10$, $\alpha = 45°$, $\beta = 75°$

2. (a) $\alpha = 63.3°$, $\gamma = 81.6°$, $c = 20.7$
(b) $a = 36$, $b = 47$, $c = 41$

3. Locate the point with the given polar coordinates. Then find the corresponding Cartesian coordinates of the point.
(a) $\left(5, \frac{\pi}{4}\right)$ (b) $\left(2, \frac{\pi}{6}\right)$ (c) $(\sqrt{2}, -135°)$

4. Locate the point with the given Cartesian coordinates. Then find polar coordinates that correspond to the point, where $r > 0$ and $0° \le \theta < 360°$.
(a) $(-3, 0)$ (b) $(-2, 2\sqrt{3})$ (c) $(-5\sqrt{3}, -5)$

5. Convert each polar equation to a corresponding equation in Cartesian form. Identify the curve and graph it.
(a) $r + 4 \sin \theta = 0$ (b) $r = -3 \cos \theta$

6. Convert each equation to a corresponding equation in polar form. Identify the curve and graph it.
(a) $3x^2 + 3y^2 = 48$ (b) $y^2 = 4x$

7. Sketch the graph of each polar equation by the point-plotting method. Then use a grapher to demonstrate the validity of the result.
(a) $r = 4 \cos \theta$ (b) $r = 1 + 3 \cos \theta$

8. [G] Use a grapher to graph each equation.
(a) $r = 2 \sin 2\theta$ (b) $r = -2 \cos 3\theta$

9. Let $\mathbf{u} = \langle 3, 4 \rangle$ and $\mathbf{v} = \langle 4, 3 \rangle$. Find each of the following vectors and represent them graphically:
(a) $3\mathbf{u}$ (b) $\mathbf{v} - \mathbf{u}$
(c) $-2\mathbf{u} + \mathbf{v}$ (d) $3\mathbf{u} + 2\mathbf{v}$

10. Determine the components of $\mathbf{u}$ if θ is a direction angle of $\mathbf{u}$.
(a) $|\mathbf{u}| = 5$ and $\theta = 30°$
(b) $|\mathbf{u}| = 6$ and $\theta = 45°$
(c) $|\mathbf{u}| = 8$ and $\theta = \frac{5\pi}{6}$

In problems 11 and 12, find the magnitude and direction of each vector. Round off the answers to two decimal places.

11. (a) $\mathbf{u} = 12\mathbf{i} + 5\mathbf{j}$
(b) $\mathbf{u} = 7\mathbf{i} - 2\mathbf{j}$

12. (a) $\mathbf{u} = 6\mathbf{i} + 3\mathbf{j}$
(b) $\mathbf{u} = \sqrt{3}\mathbf{i} - 2\mathbf{j}$

13. If $\mathbf{u}$ is a vector whose initial point is the first point given and whose terminal point is the second point given, write $\mathbf{u}$ in the form $\mathbf{u} = \langle a, b \rangle$ and find $|\mathbf{u}|$.
(a) $(-6, 8)$; $(4, 3)$
(b) $(-7, 6)$; $(3, -1)$

14. Suppose that vector $\mathbf{u}$ has a magnitude of 10 and a direction of 135°. Find the components of $\mathbf{u}$.

In problems 15–18, express each complex number in trigonometric form. Give the exact angle in problems 15 and 16; in problems 17 and 18, round off the angle measure to two decimal places.

15. (a) $z = -1 + \sqrt{3}i$
(b) $z = -1 + i$

16. (a) $z = 5 + 5i$
(b) $z = 4\sqrt{3} - 12i$

17. (a) $z = 5 + 12i$
(b) $z = 7 - 6i$

18. (a) $z = -12 + 5i$
(b) $z = -7 - 8i$

In problems 19–22, find both z_1z_2 and z_1/z_2. Write the answers in both trigonometric and Cartesian form. In problems 21 and 22, round off the answers to two decimal places.

19. $z_1 = 2(\cos \pi + i \sin \pi)$;
$z_2 = 3\left(\cos \frac{\pi}{2} + i \sin \frac{\pi}{2}\right)$

20. $z_1 = \sqrt{2}(\cos 315° + i \sin 315°)$;
$z_2 = 2\sqrt{2}(\cos 135° + i \sin 135°)$

21. $z_1 = 6(\cos 230° + i \sin 230°)$;
$z_2 = 3(\cos 75° + i \sin 75°)$

22. $z_1 = \sqrt{2}\left(\cos \frac{\pi}{4} + i \sin \frac{\pi}{4}\right)$;
$z_2 = 2\left(\cos \frac{2\pi}{3} + i \sin \frac{2\pi}{3}\right)$

In problems 23–26, use DeMoivre's theorem to write each expression in both trigonometric and Cartesian form. In problems 25 and 26, round off the answers to two decimal places.

23. (a) $(\cos 60° + i \sin 60°)^6$

(b) $\left[\sqrt{6}\left(\cos \frac{\pi}{5} + i \sin \frac{\pi}{5}\right)\right]^5$

24. (a) $(1 + i)^{40}$

(b) $\left(\frac{\sqrt{2}}{2} + \frac{\sqrt{2}}{2}i\right)^{100}$

25. (a) $[3(\cos 27° + i \sin 27°)]^4$

(b) $\left[\sqrt{5}\left(\cos \frac{\pi}{8} + i \sin \frac{\pi}{8}\right)\right]^5$

26. (a) $[2(\cos 44° + i \sin 44°)]^6$

(b) $\left[\sqrt{6}\left(\cos \frac{\pi}{7} + i \sin \frac{\pi}{7}\right)\right]^4$

In problems 27 and 28, write all the roots of each equation in trigonometric form.

27. (a) $z^3 = 8i$

(b) $z^4 = 1 + i$

28. (a) $z^4 = 8\sqrt{2} - 8\sqrt{2}i$

(b) $z^5 = 243i$

In problems 29–32, round off the answers to two decimal places.

29. Surveying: Two points A and B are 50 feet apart on one bank of a straight river. A point C on the bank across the river is located so that angle CAB is 70° and angle ABC is 80°. How wide is the river?

30. Geometry: A diagonal of a parallelogram is 16 inches long and forms angles of 43° and 15°, respectively, with the two sides. How long are the sides of the parallelogram?

31. Engineering: A guy wire attached to the top of a pole is 40 feet long and forms a 50° angle with the ground. How tall is the pole if it is tilted 15° from the vertical directly away from the guy wire?

32. Telephone Pole: A telephone pole AB is 30 feet tall and is located on a 10° slope from a point C on the horizontal plane (Figure 1). If angle ACB is 31°, find the length of the guy wire CB and the distance from C to the base of the pole.

Figure 1

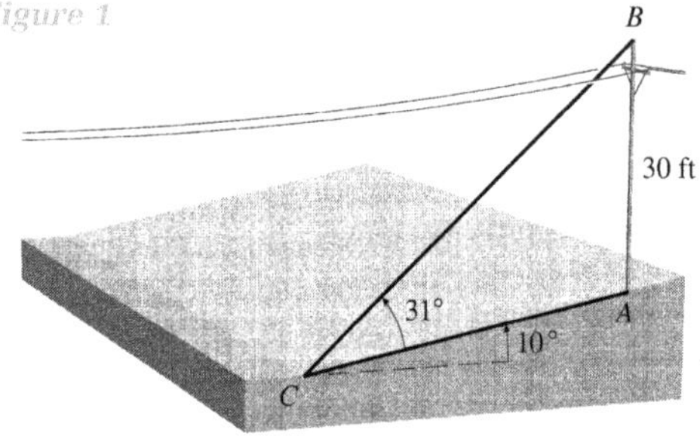

33. Engineering: Two forces, one of 63 pounds and one of 45 pounds, yield a resultant force of 75 pounds. Find the angle between the two forces (to the nearest hundredth of a degree).

34. Navigation: An airplane, which has an airspeed of 550 miles per hour, heading N30°W, is affected by a wind blowing at 45 miles per hour from S45°W. Find the velocity vector of the airplane's path. What is the speed of the airplane, and in what direction is it traveling, relative to the direction of the wind (to the nearest hundredth of a degree)?

CHAPTER 3 TEST

1. Use the law of sines to solve triangle ABC, with

$$a = 70, \quad \beta = 36.65°, \quad \text{and} \quad \alpha = 40°.$$

Round off the answers to two decimal places.

2. Use the law of cosines to find b in triangle ABC, if

$$a = 5.47, \quad c = 8.94, \quad \text{and} \quad \beta = 37.2°.$$

Round off the answer to two decimal places.

3. (a) Write the polar point $(4, 5\pi/3)$ in Cartesian form.
   (b) Write the Cartesian point $(-3, 3)$ in polar form.

4. Sketch the graph of each polar equation. Convert to rectangular form in each case.
   (a) $r = 5$
   (b) $r = 4 \sin \theta$
   (c) $\theta = \dfrac{\pi}{4}$

5. Find the polar form of each equation, and graph the equation.
   (a) $3x + 2y = 5$
   (b) $x^2 + y^2 = 16$

6. Find $4\mathbf{u} - 8\mathbf{v}$ if $\mathbf{u} = 2\mathbf{i} + 3\mathbf{j}$ and $\mathbf{v} = 4\mathbf{i} - \mathbf{j}$.

7. Find a direction angle of the vector $\mathbf{u} = 2\mathbf{i} + 3\mathbf{j}$. Round off the answer to two decimal places.

8. Let $\mathbf{u}$ and $\mathbf{v}$ be vectors such that $|\mathbf{u}| = 6$ and $|\mathbf{v}| = 8$. Assume that angles 110° and 56° are direction angles of $\mathbf{u}$ and $\mathbf{v}$, respectively. Determine $\mathbf{u} + \mathbf{v}$. Round off to two decimal places.

9. Write $z = 4\,(\cos 315° + i \sin 315°)$ in Cartesian form.

10. Write $z = -1 - \sqrt{3}i$ in trigonometric form.

11. For z_1 and z_2 given below, find the following in both the Cartesian and trigonometric forms:

   (a) $z_1 z_2$ (b) $\dfrac{z_1}{z_2}$

$$z_1 = \cos 139° + i \sin 139°$$

and

$$z_2 = 2(\cos 71° + i \sin 71°)$$

Round off each answer to four decimal places.

12. Use DeMoivre's theorem to determine $[2(\cos 60° + i \sin 60°)]^{12}$. Express the answer in the form $a + bi$.

13. Find the three roots of $z^3 = i$. Write the answers in Cartesian form.

14. An airplane flies 40 miles due east, then changes direction and flies 60 miles N20°E. How far is the airplane from its starting point? Round off the answer to one decimal place.

Chapter 3: Answers to Selected Problems

Problem Set 3.1 (on page 240)

1. SAS; $a = 2.05$ **3.** SAS; $b = 5.59$ **5.** SAS; $c = 3.35$ **7.** SSS; $\gamma = 104.48°$ **9.** SSS; $\beta = 50.50°$
11. SSS; $\alpha = 14.15°$ **13.** AAS; $\gamma = 15°$, $b = 28.66$, $c = 10.68$ **15.** ASA; $\beta = 59°$, $a = 34.47$, $c = 22.96$
17. ASA; $\beta = 76.5°$, $a = 13.60$, $c = 17.78$ **19.** AAS; $\alpha = 58.67°$, $a = 4.89$, $b = 4.44$
21. ASA; $\beta = 78°$, $a = 12.49$, $c = 17.15$ **23.** No possible triangle **25.** No possible triangle
27. Case 1: $\alpha = 58.34°$, $\gamma = 84.99°$, $c = 14.51$; Case 2: $\alpha = 121.66°$, $\gamma = 21.67°$, $c = 5.38$
29. $\beta = 53.65°$, $\gamma = 61.17°$, $c = 20.56$
31. Case 1: $\beta = 54.59°$, $\gamma = 92.08°$, $c = 54.56$; Case 2: $\beta = 125.41°$, $\gamma = 21.26°$, $c = 19.80$
33. SSA; no possible triangle **35.** SSS; $\alpha = 41.41°$, $\beta = 41.41°$, $\gamma = 97.18°$
37. AAS; $\gamma = 71°$, $b = 66.83$, $c = 63.67$ **39.** AAS; $\gamma = 74.2°$, $a = 37.98$, $c = 37.15$ **41.** 143.96 ft
43. 343.68 m **45.** 8.23 mi **47.** 3.52 km **49.** 1.08 km; 0.81 km **51.** 1.88 mi **53.** 1.88 m
55. 42.19 nautical mi **57.** **(a)** $d = \sqrt{11943.64t^2 + 3107.75t + 232.44}$
(b) $\frac{1}{2}$ hr., 69.08 km; 1 hr, 123.63 km; 75 mins, 150.93 km
(c) $d = 50$ km, $t \approx 0.32$ hr; $d = 100$ km, $t \approx 0.78$ hr; $d = 150$ km, $t \approx 1.24$ hrs.
59. Angle *PQR:* 52.70°; angle *PRQ:* 50.51°; angle *QPR:* 76.79° **61.** 20.68 in.; 14.91 in.
63. Because of the triangle inequality; $|\cos\theta| > 1$

Problem Set 3.2 (on page 254)

7. Some points are: **(a)** (i) $(3, -260°)$, $(3, 460°)$, $(3, 820°)$ (ii) $(-3, -80°)$, $(-3, 280°)$, $(-3, 640°)$
(b) (i) $\left(3, -\frac{3\pi}{4}\right)$, $\left(3, \frac{5\pi}{4}\right)$, $\left(3, \frac{13\pi}{4}\right)$ (ii) $\left(-3, -\frac{7\pi}{4}\right)$, $\left(-3, \frac{9\pi}{4}\right)$, $\left(-3, \frac{17\pi}{4}\right)$
9. **(a)** $(3\sqrt{3}, 3)$ **(b)** $(2\sqrt{2}, -2\sqrt{2})$ **11.** **(a)** $(2.70, -1.30)$ **(b)** $(-1.37, -3.76)$
13. **(a)** $(-1.98, -0.28)$ **(b)** $(0, -3)$ **15.** **(a)** $\left(2, \frac{2\pi}{3}\right)$ **(b)** $\left(12, \frac{2\pi}{3}\right)$
17. **(a)** $(5.92, 2.10)$ **(b)** $(9.22, 4.00)$ **19.** $r = 2\sec\theta$ **21.** $r = -3\csc\theta$ **23.** $r = \dfrac{2}{3\cos\theta - \sin\theta}$
25. $r = 5$ **27.** $r = -8\cos\theta$ **29.** $r = \frac{1}{4}\tan\theta\sec\theta$ **31.** $r = 8\cot\theta\csc\theta$ **33.** $x^2 + y^2 = 9$
35. $y = 2$ **37.** $y = x$ **39.** $(x + 2)^2 + y^2 = 4$ **41.** $y = \frac{1}{4}x^2$ **43.** $4x + 3y = 12$
45. $(x - 2)^2 + (y - 1)^2 = 5$

47. Circle

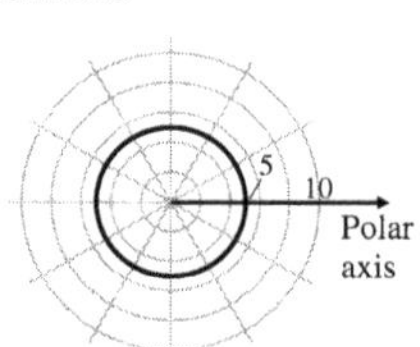

49. Line

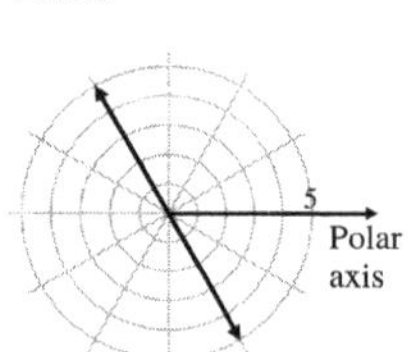

51. Circle

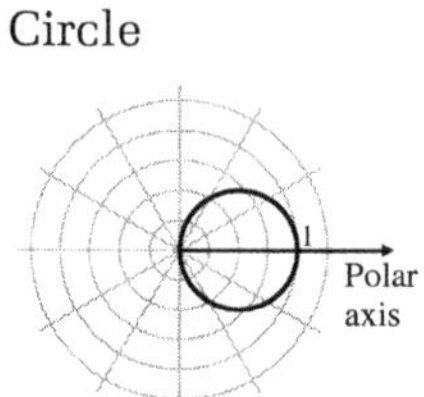

53. Cardioid

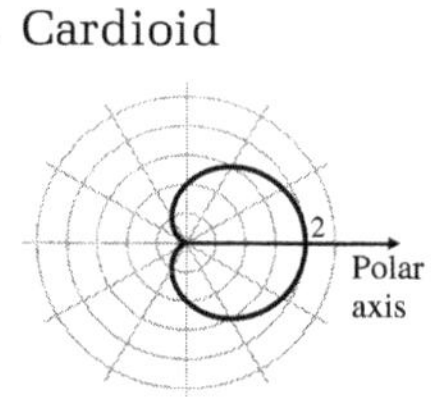

57. $(a, 0)$, $\left(a, \frac{\pi}{3}\right)$, $\left(a, \frac{2\pi}{3}\right)$, (a, π), $\left(a, \frac{4\pi}{3}\right)$, $\left(a, \frac{5\pi}{3}\right)$

59. Symmetry: line $\theta = \frac{\pi}{2}$

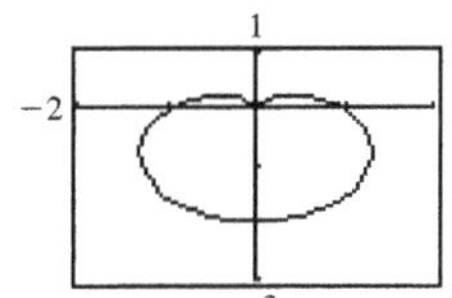

61. Symmetry: polar axis

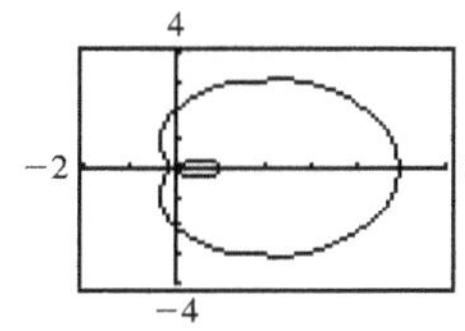

63. Symmetry: line $\theta = \frac{\pi}{2}$

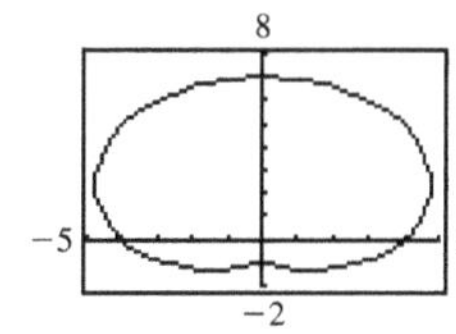

65. Symmetry: pole (and both axes)

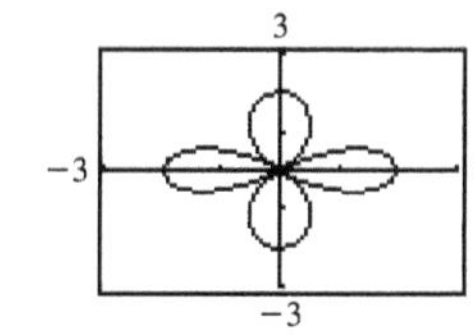

67. Symmetry: line $\theta = \frac{\pi}{2}$

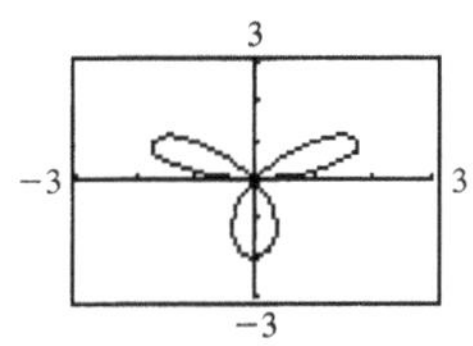

69. Symmetry: polar axis, line $\theta = \frac{\pi}{2}$, and the pole

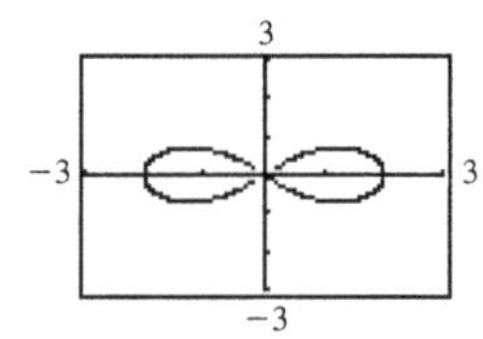

71. **(a)** $r = 2 \cos 3\theta$

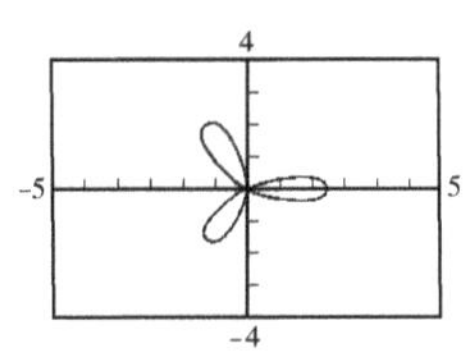

$r = 2 \cos\left[3\left(\theta - \frac{\pi}{6}\right)\right]$

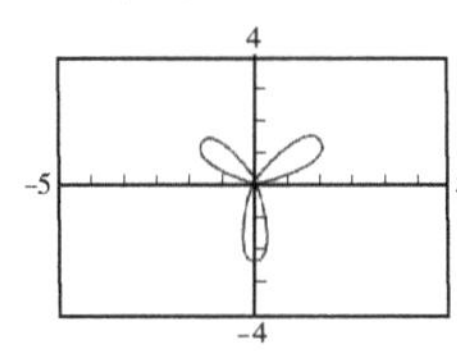

$r = 2 \cos\left[3\left(\theta + \frac{\pi}{4}\right)\right]$

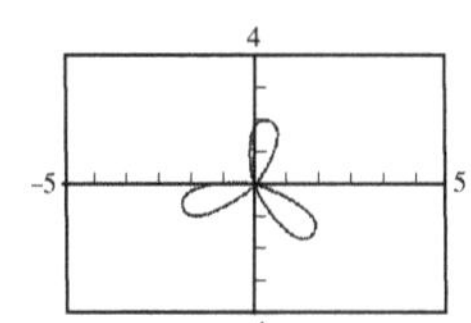

(b) Rotate $r = 2 \cos 3\theta$, $\frac{\pi}{6}$ (30°) counterclockwise to get $r = 2 \cos\left[3\left(\theta - \frac{\pi}{6}\right)\right]$; rotate $r = 2 \cos 3\theta$, $\frac{\pi}{4}$ (45°) clockwise to get $r = 2 \cos\left[3\left(\theta + \frac{\pi}{4}\right)\right]$.

Problem Set 3.3 (on page 268)

3. $\mathbf{u} = \langle -5, -2 \rangle$; $|\mathbf{u}| = \sqrt{29}$ **5.** $\mathbf{u} = \langle 5, -11 \rangle$; $|\mathbf{u}| = \sqrt{146}$ **7.** $\mathbf{u} = \langle -6, -6 \rangle$; $|\mathbf{u}| = 6\sqrt{2}$
9. **(a)** $\langle 1, 8 \rangle$ **(b)** $\sqrt{65}$ **11.** **(a)** $\langle -19, -20 \rangle$ **(b)** $\sqrt{761}$ **13.** **(a)** $\langle 9, -28 \rangle$ **(b)** $\sqrt{865}$
15. **(a)** $-2\mathbf{i} + 4\mathbf{j}$ **(b)** $3\mathbf{i}$ **17.** **(a)** $2\mathbf{i} + 5\mathbf{j}$ **(b)** $-4\mathbf{i} + 4\mathbf{j}$ **19.** $|\mathbf{w}| = \sqrt{53}$; approx. 37.18°
21. $-\frac{\sqrt{3}}{2}\mathbf{i} + \frac{1}{2}\mathbf{j}$; 150° **23.** $\frac{5}{13}\mathbf{i} + \frac{12}{13}\mathbf{j}$; 67.38° **25.** $\frac{1}{\sqrt{2}}\mathbf{i} - \frac{1}{\sqrt{2}}\mathbf{j}$; −45° **27.** $\frac{1}{\sqrt{26}}\mathbf{i} + \frac{5}{\sqrt{26}}\mathbf{j}$; 78.69°
29. $x = 3, y = 0$ **31.** $x = 3.54, y = 3.54$ **33.** $x = -1.73, y = -1$ **35.** $x = -0.09, y = 0.49$
37. **(a)** 55.71 lb **(b)** 20.49° **39.** **(a)** 12.63 N **(b)** 17.29°
41. **(a)** 540.03 mph **(b)** N81.34°E **(c)** 3.66°
43. **(a)** 51.22 kph **(b)** S1.34°E **(c)** 38.66° **45.** (13, 1) **49.** **(a)** −9 **(b)** 142.1°
51. **(a)** −12 **(b)** 143.1° **53.** **(a)** 14 **(b)** 58.7° **57.** Yes; vertex angle B is the right angle.

Problem Set 3.4 (on page 282)

1. **(a)** $|z| = \sqrt{53}$ **(b)** $|z| = \sqrt{29}$ **3.** **(a)** $|z| = 5$ **(b)** $|z| = \sqrt{34}$

5. **(a)** 4 **(b)** 2 **(c)** 4 **(d)** $\frac{1}{2}$ **(e)** $\frac{1}{2}$

7. **(a)** $5\sqrt{29}$ **(b)** 25 **(c)** $5\sqrt{29}$ **(d)** $\frac{5}{\sqrt{29}}$ **(e)** $\frac{5}{\sqrt{29}}$

9. $\sqrt{3} + i$ **11.** $2i$ **13.** -8 **15.** $1.97 + 0.35i$

17. **(a)** $\sqrt{2}\left(\cos\frac{5\pi}{4} + i\sin\frac{5\pi}{4}\right)$ **(b)** $2\left(\cos\frac{7\pi}{6} + i\sin\frac{7\pi}{6}\right)$

19. **(a)** $\sqrt{3}(\cos 0.62 + i\sin 0.62)$ **(b)** $5(\cos\pi + i\sin\pi)$

21. **(a)** $1\left(\cos\frac{2\pi}{3} + i\sin\frac{2\pi}{3}\right)$ **(b)** $1\left(\cos\frac{\pi}{6} + i\sin\frac{\pi}{6}\right)$

23. **(a)** $8\left(\cos\frac{7\pi}{6} + i\sin\frac{7\pi}{6}\right) = -4\sqrt{3} - 4i$ **(b)** $2\left(\cos\frac{\pi}{2} + i\sin\frac{\pi}{2}\right) = 2i$

25. **(a)** $1(\cos 90° + i\sin 90°) = i$ **(b)** $1[\cos(-30°) + i\sin(-30°)] = \frac{\sqrt{3}}{2} - \frac{1}{2}i$

27. **(a)** $2(\cos 180° + i\sin 180°) = -2$ **(b)** $1[\cos(-90°) + i\sin(-90°)] = -i$

29. **(a)** $6(\cos 90° + i\sin 90°) = 6i$ **(b)** $\frac{2}{3}(\cos 10° + i\sin 10°) = 0.66 + 0.12i$

31. $-\frac{\sqrt{3}}{2} - \frac{1}{2}i$ **33.** $512 - 512\sqrt{3}\,i$ **35.** 256 **37.** $-125{,}000i$ **39.** $-8 - 8\sqrt{3}\,i$

41. $-2^{30} = -1{,}073{,}741{,}824$ **43.** -1 **45.** $0.71 + 0.71i, -0.71 - 0.71i$

47. $2, -1 + 1.73i, -1 - 1.73i$ **49.** $0.91 + 1.78i, -1.41 + 1.41i, -1.78 - 0.91i, 0.31 - 1.98i, 1.98 - 0.31i$

51. $1 + 1.73i, -1.73 + i, -1 - 1.73i, 1.73 - i$

53. $2(\cos 60° + i\sin 60°) = 1 + 1.73i$; $2(\cos 180° + i\sin 180°) = -2$; $2(\cos 300° + i\sin 300°) = 1 - 1.73i$

55. $3(\cos 45° + i\sin 45°) = 2.12 + 2.12i$; $3(\cos 135° + i\sin 135°) = -2.12 + 2.12i$; $3(\cos 225° + i\sin 225°) = -2.12 - 2.12i$; $3(\cos 315° + i\sin 315°) = 2.12 - 2.12i$

57. $2(\cos 30° + i\sin 30°) = 1.73 + i$; $2(\cos 90° + i\sin 90°) = 2i$; $2(\cos 150° + i\sin 150°) = -1.73 + i$; $2(\cos 210° + i\sin 210°) = -1.73 - i$; $2(\cos 270° + i\sin 270°) = -2i$; $2(\cos 330° + i\sin 330°) = 1.73 - i$

61. **(a)** On the circle $x^2 + y^2 = 4$ **(b)** On the circle $x^2 + (y - 1)^2 = 1$

65. **(a)** $-\frac{\sqrt{3}}{2} - \frac{1}{2}i$ **(b)** $\frac{1}{32} - \frac{1}{32}i$

Chapter 3 Review Problem Set (on page 289)

1. **(a)** $c = 3.66, \alpha = 43.12°, \beta = 106.88°$ **(b)** $\gamma = 60°, a = 8.16, b = 11.15$

3. **(a)** $\left(\frac{5\sqrt{2}}{2}, \frac{5\sqrt{2}}{2}\right)$ **(b)** $(\sqrt{3}, 1)$ **(c)** $(-1, -1)$

5. **(a)** $x^2 + (y + 2)^2 = 4$; circle **(b)** $\left(x + \frac{3}{2}\right)^2 + y^2 = \frac{9}{4}$; circle

7. **(a)**

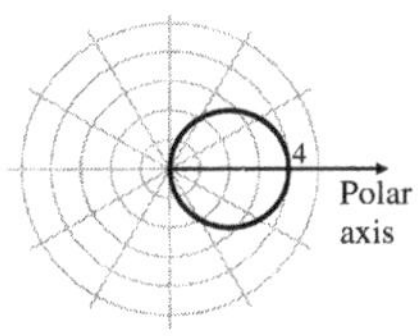

(b)

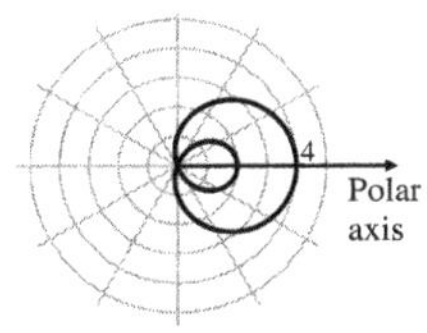

9. **(a)** $\langle 9, 12\rangle$ **(b)** $\langle 1, -1\rangle$ **(c)** $\langle -2, -5\rangle$ **(d)** $\langle 17, 18\rangle$

11. **(a)** $|\mathbf{u}| = 13$; $\theta = 22.62°$ **(b)** $|\mathbf{u}| = 7.28$; $\theta = 344.05°$

13. **(a)** $\mathbf{u} = \langle 10, -5\rangle$; $|\mathbf{u}| = 5\sqrt{5}$ **(b)** $\mathbf{u} = \langle 10, -7\rangle$; $|\mathbf{u}| = \sqrt{149}$

15. **(a)** $2\left(\cos\frac{2\pi}{3} + i\sin\frac{2\pi}{3}\right)$ **(b)** $\sqrt{2}\left(\cos\frac{3\pi}{4} + i\sin\frac{3\pi}{4}\right)$

17. **(a)** $13(\cos 1.18 + i\sin 1.18)$ **(b)** $\sqrt{85}(\cos 319.40° + i\sin 319.40°)$

19. $z_1z_2 = 6\left(\cos\frac{3\pi}{2} + i\sin\frac{3\pi}{2}\right) = -6i; \frac{z_1}{z_2} = \frac{2}{3}\left(\cos\frac{\pi}{2} + i\sin\frac{\pi}{2}\right) = \frac{2}{3}i$

21. $z_1z_2 = 18(\cos 305° + i\sin 305°) = 10.32 - 14.74i; \frac{z_1}{z_2} = 2(\cos 155° + i\sin 155°) = -1.81 + 0.85i$

23. **(a)** $1(\cos 360° + i\sin 360°) = 1$ **(b)** $36\sqrt{6}(\cos\pi + i\sin\pi) = -36\sqrt{6}$

25. **(a)** $81(\cos 108° + i\sin 108°) = -25.03 + 77.04i$ **(b)** $25\sqrt{5}\left(\cos\frac{5\pi}{8} + i\sin\frac{5\pi}{8}\right) = -21.39 + 51.65i$

27. **(a)** $2\left(\cos\frac{\pi}{6} + i\sin\frac{\pi}{6}\right); 2\left(\cos\frac{5\pi}{6} + i\sin\frac{5\pi}{6}\right); 2\left(\cos\frac{3\pi}{2} + i\sin\frac{3\pi}{2}\right)$

(b) $\sqrt[8]{2}\left(\cos\frac{\pi}{16} + i\sin\frac{\pi}{16}\right); \sqrt[8]{2}\left(\cos\frac{9\pi}{16} + i\sin\frac{9\pi}{16}\right);$

$\sqrt[8]{2}\left(\cos\frac{17\pi}{16} + i\sin\frac{17\pi}{16}\right); \sqrt[8]{2}\left(\cos\frac{25\pi}{16} + i\sin\frac{15\pi}{16}\right)$

29. 92.54 ft **31.** 31.72 ft **33.** 93.73°

Chapter 3 Test (on page 291)

1. $b = 65.01, c = 105.96, \gamma = 103.35°$ **2.** $b = 5.65$ **3.** **(a)** $\left(2, -2\sqrt{3}\right)$ **(b)** $\left(3\sqrt{2}, \frac{3\pi}{4}\right)$

4. **(a)** $x^2 + y^2 = 25$ **(b)** $x^2 + (y-2)^2 = 4$ **(c)** $y = x$

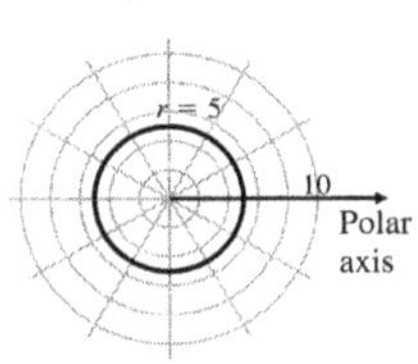

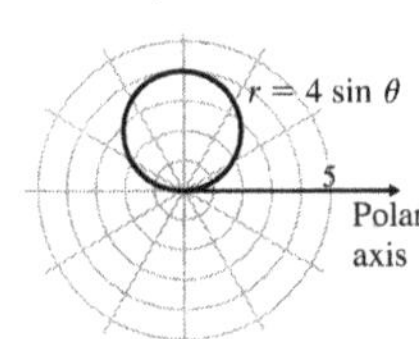

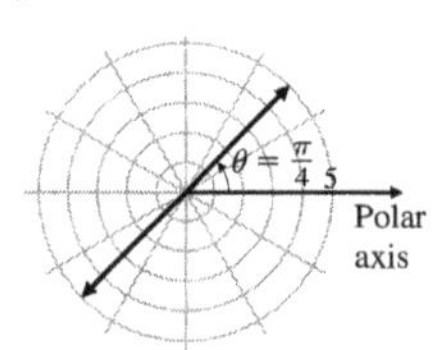

5. **(a)** $r = \dfrac{5}{3\cos\theta + 2\sin\theta}$ **(b)** $r = 4$ **6.** $24i + 20j$ **7.** 56.31° **8.** $2.42i + 12.27j$

9. $2\sqrt{2} - 2\sqrt{2}\,i$ **10.** $2(\cos 240° + i\sin 240°)$

11. **(a)** $2(\cos 210° + i\sin 210°) = \sqrt{3} - i \approx -1.732 - i$ **(b)** $\frac{1}{2}(\cos 68° + i\sin 68°) \approx 0.1873 + 0.4636i$

12. $4096 + 0i$ **13.** $\frac{\sqrt{3}}{2} + \frac{1}{2}i; \frac{-\sqrt{3}}{2} + \frac{1}{2}i; -i$ **14.** 82.7 mi

CHAPTER 4

Systems of Equations and Inequalities

Chapter Contents

In the state of Michigan, most gas stations offer three grades of gasoline with the following ratings on the pumps: low grade (87 octane), middle grade (89 octane) and high grade (93 octane). However, some of these gas stations are equipped with only two large underground tanks—one for the lowest grade (87 octane) and the other for the highest grade (93 octane). To produce the middle grade (89 octane), the pump's computer mixes the other two grades from the underground tanks according to a programmed formula. How much of the 87 octane grade should be blended with the 93 octane grade to obtain 36 gallons of the 89 octane grade? The solution of this problem is given in Example 5 on page 304.

© JupiterImages, Inc.

In science, business, and economics, we often encounter mathematical models that involve more than one equation and contain two or more variables. Such a set of equations is referred to as a *system of equations*. In this chapter, we study several different techniques for solving systems of equations.

Objectives

1. Solve a System by Substitution or by Elimination
2. Solve a System by Gaussian Elimination
3. Solve Applied Problems

4.1 Solutions of Linear Systems by Substitution and Elimination

In this section we investigate systems that contain only linear equations. Such systems are called **linear systems.** In general, a **solution** of a linear system of equations is a collection of particular values for the variables that satisfy all equations in the system.

For example,

$$\begin{cases} 3x + 4y = 12 \\ 3x - 8y = 0 \end{cases}$$

We use a brace to group together the equations in the system.

is a linear system of two equations that contains the two variables x and y.

If we substitute 8/3 for x and 1 for y into each equation in the first system, we find that these values satisfy both equations:

$$\begin{cases} 3\left(\dfrac{8}{3}\right) + 4(1) = 8 + 4 = 12 \\ 3\left(\dfrac{8}{3}\right) - 8(1) = 8 - 8 = 0 \end{cases}$$

We say that the pair of values 8/3 and 1 for x and y, respectively, is a *solution* of the system. A solution of a system of two equations in two variables is often written as an ordered pair of numbers, so we express the solution of the first system in the form

$$\left(\frac{8}{3}, 1\right).$$

The solution of the system

$$\begin{cases} 2x - 3y + 4z = 17 \\ 4x + 5y - z = 5 \\ -x + 2y + z = -3 \end{cases}$$

is given by the ordered triple

$$(3, -1, 2).$$

Again, if we substitute 3 for x, -1 for y, and 2 for z, we find that these values satisfy each of the three equations.

Solving a System by Substitution or by Elimination

The process of finding all solutions of a system is called **solving the system.** To solve a linear system, we can use either *substitution* or *elimination.*

To solve a linear system of two equations in two variables by *substitution,* we use one of the equations to express one of the variables in terms of the other. Then we substitute this expression for the variable in the other equation to obtain an equation in one variable. This procedure is illustrated in the next example.

EXAMPLE 1 **Solving a System by Substitution**

Solve the system by substitution: $\begin{cases} 4x - y = 2 \\ 3x + y = 5 \end{cases}$ and check the solution.

Solution We begin by solving the first equation for y in terms of x to get

$$y = 4x - 2.$$

Replacing y with $4x - 2$ in the second equation of the system, gives us

$$3x + (4x - 2) = 5, \quad 7x = 7 \quad \text{or} \quad x = 1.$$

The value of y that solves the system is obtained by substituting 1 for x in $y = 4x - 2$ to get

$$y = 4(1) - 2 = 2.$$

Thus, the solution is the pair of numbers $x = 1$ and $y = 2$, or (1, 2).

Check To check the solution, replace x by 1 and y by 2 in the original system to get

Graphically, this means that the two lines with equations $4x - y = 2$ and $3x + y = 5$ intersect at the point (1, 2).

$$\begin{cases} 4(1) - 2 = 2 \\ 3(1) + 2 = 5. \end{cases}$$

The substitution method is efficient for solving certain systems with two equations. However, it tends to become cumbersome when more than two equations are involved. An alternate method, called the *elimination method,* enables us to solve systems of linear equations in any number of variables more efficiently. The elimination method is based on the idea of *equivalent systems* of equations.

Two systems of equations are said to be **equivalent** if they have the same solutions.

For example, the systems

$$\begin{cases} 2x + y = 1 \\ -x + 3y = -4 \end{cases} \quad \text{and} \quad \begin{cases} 2x + y = 1 \\ y = -1 \end{cases}$$

are equivalent, since they have the same and only solution: $(1, -1)$.

A linear system can be converted to an equivalent form by performing one or more of the following three elementary operations:

Elementary Operations Yielding Equivalent Systems

1. Interchange the positions of two equations in the system.
2. Replace an equation in the system by a nonzero multiple of itself by multiplying each side of the equation by the same nonzero number.
3. Replace an equation in the system by the sum of a nonzero multiple of that equation and another equation in the system.

The basic strategy behind the elimination method is to use these operations to convert a system to an equivalent one that contains fewer variables and is easier to solve. For instance, consider the system

$$\begin{cases} x + y = 6 \\ x - y = 2. \end{cases}$$

If we replace the second equation by the sum of itself and the first equation, we get the equivalent system

$$\begin{cases} x + y = 6 \\ 2x = 8. \end{cases}$$

This form of the system is easier to solve than the original. By inspection, we see from the second equation that $x = 4$. Substituting this value into the first equation, we get

$$4 + y = 6 \quad \text{or} \quad y = 2.$$

So the solution of the system is (4, 2).

EXAMPLE 2

Solving a System by Elimination

Solve each system by elimination. Then represent the system graphically; that is, graph both equations in the same coordinate system. Relate the solution to the graph of the system.

(a) $\begin{cases} x - y = 1 \\ 3x - y = -1 \end{cases}$ (b) $\begin{cases} x + 3y = 4 \\ 2x + 6y = -6 \end{cases}$ (c) $\begin{cases} 3x - y = 1 \\ 6x - 2y = 2 \end{cases}$

Solution (a) For

$$\begin{cases} x - y = 1 \\ 3x - y = -1 \end{cases}$$

we'll start by using the third elementary operation to convert to an equivalent system in which the x variable is eliminated in the second equation. This is accomplished by replacing the second equation by the sum of -3 times the first equation and the second to obtain

$$\begin{cases} x - y = 1 \\ 3x - y = -1 \end{cases} \quad \rightarrow \quad \begin{cases} x - y = 1 \\ \quad 2y = -4. \end{cases}$$

Next we multiply both sides of the second equation by 1/2 to obtain an equivalent system that is simpler to solve:

Figure 1

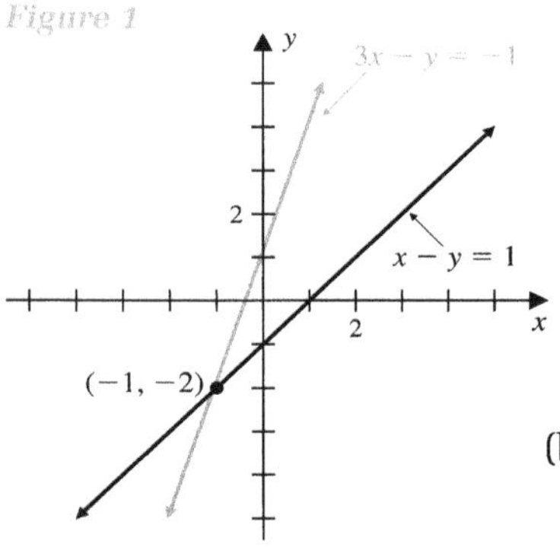

(a)

$$\begin{cases} x - y = 1 \\ \quad y = -2 \end{cases}$$

By substituting -2 for y into the first equation, we get $x - y = 1$, or $x - (-2) = 1$, that is, $x = -1$. Thus the solution is $(-1, -2)$.

The graph of the system is shown in Figure 1a. Clearly, the graphs intersect at one point. Since the coordinates of this point must satisfy both equations, it follows that the solution of the system and the point of intersection are the same: $(-1, -2)$.

(b) For

$$\begin{cases} x + 3y = 4 \\ 2x + 6y = -6 \end{cases}$$

our strategy is to produce an equivalent system by eliminating x in the second equation. To do this, we replace the second equation by the sum of -2 times the first equation and the second equation to obtain

$$\begin{cases} x + 3y = 4 \\ 2x + 6y = -6 \end{cases} \quad \rightarrow \quad \begin{cases} x + 3y = 4 \\ \quad 0 = -14. \end{cases}$$

This leads to a contradiction, so the system has no solution.

Figure 1

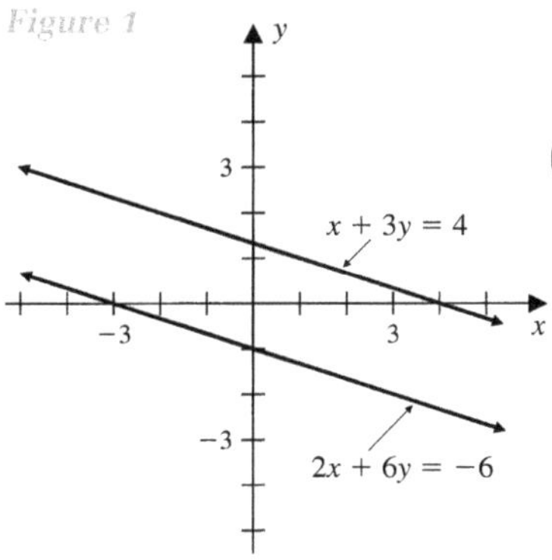

(b)

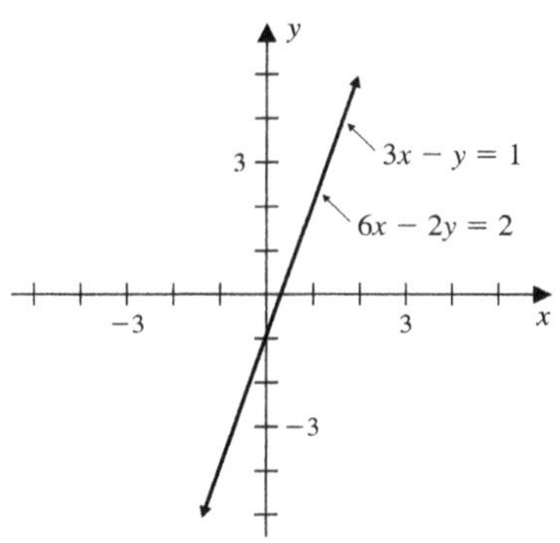

(c)

As Figure 1b shows, the two lines are parallel (the lines have the same slope $m = -1/3$), so there is no point common to the two lines. Consequently, no pair of numbers can satisfy both equations simultaneously.

(c) For

$$\begin{cases} 3x - y = 1 \\ 6x - 2y = 2 \end{cases}$$

we eliminate the x term in the second equation by replacing the second equation by the sum of -2 times the first equation and the second to get

$$\begin{cases} 3x - y = 1 \\ 6x - 2y = 2 \end{cases} \rightarrow \begin{cases} 3x - y = 1 \\ 0 = 0. \end{cases}$$

We conclude that the solution consists of all pairs of numbers (x, y) that satisfy the equation

$$3x - y = 1 \quad \text{or} \quad y = 3x - 1.$$

It follows that the solution consists of ordered pairs of the form $(x, 3x - 1)$, where x is a real number. In Figure 1c, we can see that the two equations have the same graph, so there are infinitely many solutions (one for each point on the graph).

Referring to the three systems in Example 2, the system in part (a) is said to be **consistent and independent,** with the unique solution $x = -1$ and $y = -2$ (Figure 1a). The system in part (b) is said to be **inconsistent** with no solution (Figure 1b). And the system in part (c) is said to be **consistent and dependent** with an infinite number of solutions (Figure 1c).

Solving a System by Gaussian Elimination

In Example 2a, we solved the given system by using the equivalent form

$$\begin{cases} x - y = 1 \\ \quad\; y = -2. \end{cases}$$

This latter system is said to be in *triangular form.* Other examples of triangular form are:

$$\begin{cases} x + 2y = 7 \\ \quad\;\; y = 1 \end{cases} \quad \text{and} \quad \begin{cases} x + 2y - 3z = 1 \\ \quad\;\; y + 2z = 4 \\ \qquad\quad\;\; z = 7 \end{cases}$$

Note the staggered pattern of a triangular form. The leading coefficient is 1 for each equation, and each equation after the first one has one less variable than the one above it. When a system is written in triangular form, we can find its solution by using *back-substitution.*

For instance, the solution of

$$\begin{cases} x + 2y - 3z = 1 \\ \quad\quad y + 2z = 4 \\ \quad\quad\quad\quad z = 7 \end{cases}$$

is found as follows:

Clearly, $z = 7$ by inspection. Substituting this value into the second equation, we get

$$y + 2(7) = 4, \quad y + 14 = 4, \quad \text{or} \quad y = -10.$$

Finally, we substitute -10 for y and 7 for z into the first equation to get

$$x + 2(-10) - 3(7) = 1, \quad x - 20 - 21 = 1, \quad x - 41 = 1, \quad \text{or} \quad x = 42.$$

Thus the solution is $(42, -10, 7)$.

The process of using elimination to convert a given linear system to an equivalent triangular one and then using back-substitution to find the solution is referred to as the **Gaussian elimination method.**

EXAMPLE 3

Subscripts are often used to denote variables in linear systems.

Solving a System by Gaussian Elimination

Solve system (A) by using the Gaussian elimination method.

$$\begin{cases} 2x_1 + x_2 - 2x_3 = 10 \\ 3x_1 + 2x_2 + 2x_3 = 1 \\ 5x_1 + 4x_2 + 3x_3 = 4 \end{cases} \quad \text{(A)}$$

Solution First we eliminate the x_1 term in the second equation by replacing the second equation in system (A) by the sum of -3 times the first equation and 2 times the second equation to obtain the equivalent system (B):

$$\begin{cases} 2x_1 + x_2 - 2x_3 = 10 \\ \quad\quad x_2 + 10x_3 = -28 \\ 5x_1 + 4x_2 + 3x_3 = 4 \end{cases} \quad \text{(B)}$$

In the next step, we eliminate the x_1 term in the third equation by replacing the third equation of (B) with the sum of -5 times the first equation and 2 times the third equation to obtain the equivalent system (C):

$$\begin{cases} 2x_1 + x_2 - 2x_3 = 10 \\ \quad\quad x_2 + 10x_3 = -28 \\ \quad\quad 3x_2 + 16x_3 = -42 \end{cases} \quad \text{(C)}$$

Now we eliminate the x_2 term in the third equation by replacing the third equation of (C) by the sum of the third equation and -3 times the second equation to get the equivalent system (D):

$$\begin{cases} 2x_1 + x_2 - 2x_3 = 10 \\ \quad\quad x_2 + 10x_3 = -28 \\ \quad\quad\quad - 14x_3 = 42 \end{cases} \quad \text{(D)}$$

Next we multiply the third equation by $-1/14$ to obtain the equivalent system (E), in which the coefficient of x_3 is 1:

$$\begin{cases} 2x_1 + x_2 - 2x_3 = 10 \\ \quad\;\; x_2 + 10x_3 = -28 \\ \qquad\qquad x_3 = -3 \end{cases} \tag{E}$$

Finally, in order to change the coefficient of x_1 in the first equation to 1, we multiply the first equation by $1/2$ to obtain the equivalent triangular system (F):

$$\begin{cases} x_1 + \frac{1}{2}x_2 - x_3 = 5 \\ \quad\;\; x_2 + 10x_3 = -28 \\ \qquad\qquad x_3 = -3 \end{cases} \tag{F}$$

Now we use back-substitution to solve this system. Substituting $x_3 = -3$ into the second equation, we get $x_2 = 2$; then substituting $x_2 = 2$ and $x_3 = -3$ into the first equation, we get $x_1 = 1$. Thus the solution is the triple of numbers $x_1 = 1$, $x_2 = 2$, and $x_3 = -3$, or $(1, 2, -3)$.

Some graphers have a special feature that finds the coordinates of a point of intersection of two graphs. This feature can be used to approximate the values of the solution of a system of two linear equations in two variables. For example, to solve the following system graphically

$$\begin{cases} 0.3x - 2.7y = -0.82 \\ 0.2x + 5.4y = 1.83 \end{cases}$$

we graph the two equations in the some viewing window of Figure 2. By using the INTERSECT -finding feature of a grapher, we approximate the location of the point of intersection to be $(0.24, 0.33)$. So the solution of the system is approximately $x = 0.24$ and $y = 0.33$.

Figure 2

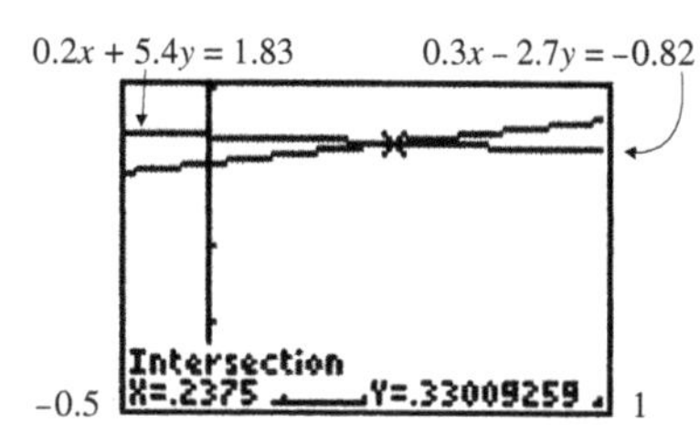

Although graphs are excellent tools for visualizing solutions of systems of equations in two variables, they do not always give the exact solution, especially if the point of intersection is not located at integer values of x and y. For this reason we normally rely on algebraic methods for solving these systems.

Solving Applied Problems

Systems of linear equations are often used to model real-life situations, as the next example shows.

EXAMPLE 4

Modeling a Physics Problem

Figure 3

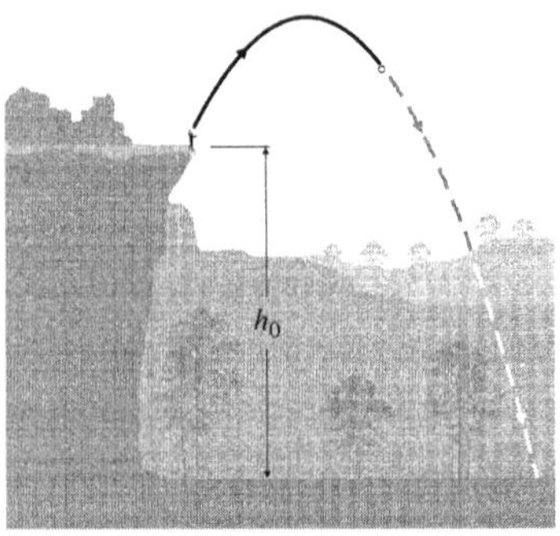

Suppose that a ball is thrown up in the air from a cliff h_0 feet above ground level with an initial upward speed of v_0 (in feet per second). Its height h (in feet) above ground level t seconds after it was thrown is given by the model

$$h = -16t^2 + v_0t + h_0 \quad \text{(Figure 3).}$$

(a) Suppose that the ball is 67 feet above the ground after 1 second and 61 feet above the ground after 1.5 seconds. Use a linear system to determine the initial upward speed v_0 and the height of the cliff h_0.

(b) Use the results from part (a) to determine the specific model that expresses the height h as a function of time t under the given conditions.

(c) How high above ground level is the ball after 2.5 seconds?

Solution

(a) Since the ball is 67 feet above the ground after 1 second and 61 feet after 1.5 seconds, we substitute 1 for t when $h = 67$ and 1.5 for t when $h = 61$ in the equation of the model to get the system

$$\begin{cases} 67 = -16(1)^2 + v_0(1) + h_0 \\ 61 = -16(1.5)^2 + v_0(1.5) + h_0 \end{cases} \quad \text{or} \quad \begin{cases} v_0 + h_0 = 83 \\ 1.5v_0 + h_0 = 97. \end{cases}$$

Solving this system, we get $v_0 = 28$ and $h_0 = 55$. Therefore, the initial speed is 28 feet per second and the height of the cliff is 55 feet.

(b) The general model for this situation is given by

$$h = -16t^2 + v_0t + h_0.$$

Under the given conditions $v_0 = 28$ and $h_0 = 55$, so we get the specific model

$$h = -16t^2 + 28t + 55.$$

(c) When $t = 2.5$, $h = -16(2.5)^2 + 28(2.5) + 55 = 25$. So the height of the ball above ground level is 25 feet after 2.5 seconds.

EXAMPLE 5

Solving a Gasoline Mixture Problem

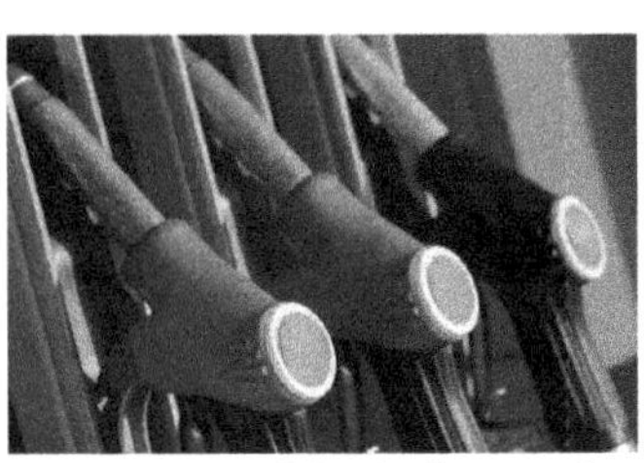

In the state of Michigan, most gas stations offer three grades of gasoline with the following ratings on the pumps: low grade (87 octane), middle grade (89 octane) and high grade (93 octane). However, some of these gas stations are equipped with only two large underground tanks—one for the lowest grade (87 octane) and the other for the highest grade (93 octane). To produce the middle grade (89 octane), the pump's computer mixes the other two grades from the underground tanks according to a programmed formula. How much of the 87 octane grade should be blended with the 93 octane grade to obtain 36 gallons of the 89 octane grade?

Solution

Let x represent the number of gallons of the 87 octane grade and y represent the number of gallons of the 93 octane grade needed to obtain a mix of 36 gallons of the 89 octane grade.

We translate the problem to the following system of equations:

$$\begin{cases} x + y = 36 & \text{Total number of gallons} \\ 87x + 93y = 36\,(89) & \text{Octane mixture} \end{cases}$$

Solving this system of equations, we get $x = 24$ and $y = 12$.

Therefore, 24 gallons of the 87 octane grade should be blended with 12 gallons of the 93 octane grade to obtain 36 gallons of the 89 octane grade.

EXAMPLE 6

Modeling a Manufacturing Problem

A manufacturer of radios produces three types of radios, model *A*, model *B*, and model *C*. The manufacturer determines that the total times allocated for production, assembly, and testing are 638 hours, 253 hours, and 159 hours, respectively. Table 1 indicates the production, assembly, and testing times required for a radio of each type. Use a linear system to determine how many of each type can be produced if all the allocated time is used up.

TABLE 1

Model	Production Hours *Per radio*	Assembly Hours *Per radio*	Testing Hours *Per radio*
A	1.4	0.5	0.3
B	1.6	0.6	0.4
C	1.8	0.8	0.5

Solution Let

x represent the number of radios of model *A* produced,
y the number of model *B*, and
z the number of model *C*.

Now we use the data in Table 1. Since the production hours add up to 638, the assembly hours add up to 253, and the testing hours add up to 159, we obtain the system

$$\begin{cases} 1.4x + 1.6y + 1.8z = 638 \\ 0.5x + 0.6y + 0.8z = 253 \\ 0.3x + 0.4y + 0.5z = 159. \end{cases}$$

Solving this system, we get

$$x = 150,\ y = 110,\ \text{and } z = 140.$$

Therefore, the manufacturer should be able to produce

150 radios of model *A*, 110 of model *B*, and 140 of model *C*

in order to use up all the allocated times.

PROBLEM SET 4.1

Mastering the Concepts

In problems 1–12, solve each system by both substitution and elimination. Represent the system graphically, and then relate the solution to the graph of the system.

1. $\begin{cases} 3x - 2y = 1 \\ x + y = 5 \end{cases}$

2. $\begin{cases} x - y = 2 \\ 4x - 2y = 6 \end{cases}$

3. $\begin{cases} 2x + 6y = 7 \\ x + 3y = 1 \end{cases}$

4. $\begin{cases} 2x - y = 1 \\ 4x - 2y = 5 \end{cases}$

5. $\begin{cases} x + 2y = 3 \\ 2x - 3y = 1 \end{cases}$

6. $\begin{cases} 3x + 2y = 11 \\ -2x + y = 2 \end{cases}$

7. $\begin{cases} 2x - y = 5 \\ -6x + 3y = 13 \end{cases}$

8. $\begin{cases} 3x + y = 4 \\ 7x - y = 6 \end{cases}$

9. $\begin{cases} y = x - 5 \\ 2y = 2x - 10 \end{cases}$

10. $\begin{cases} 3x + 6y = 1 \\ x - 2y = -6 \end{cases}$

11. $\begin{cases} y = 1 - x \\ 5x - y = 13 \end{cases}$

12. $\begin{cases} 3x - 4y = -5 \\ x = 3 - y \end{cases}$

In problems 13–16, solve each system by elimination.

13. $\begin{cases} x + y = 5 \\ x + z = 1 \\ y + z = 2 \end{cases}$

14. $\begin{cases} x - 3y = -11 \\ 2y - 5z = 26 \\ 7x - 3z = -2 \end{cases}$

15. $\begin{cases} x + y + 2z = 11 \\ x - y + z = 3 \\ 2x + y + 3z = 17 \end{cases}$

16. $\begin{cases} x + 3y + z = 4 \\ 3x - 2y + 4z = 11 \\ 2x + y + 3z = 13 \end{cases}$

In problems 17–34, solve each system by the Gaussian elimination method.

17. $\begin{cases} x + 3y = 9 \\ x - y = 1 \end{cases}$

18. $\begin{cases} x + y = 1 \\ y - 3x = 3 \end{cases}$

19. $\begin{cases} 5x - y = 7 \\ 10x - 4y = 18 \end{cases}$

20. $\begin{cases} 2x + 4y = 3 \\ -x + y = 3 \end{cases}$

21. $\begin{cases} 3y + 2z = 5 \\ 2y = -3z + 1 \end{cases}$

22. $\begin{cases} 4x - 3y = 11 \\ 3x + 4y = 2 \end{cases}$

23. $\begin{cases} 3x + 2y = 4 \\ 5x + 3y = 7 \end{cases}$

24. $\begin{cases} 2x - 3y = 1 \\ 5x + 2y = 12 \end{cases}$

25. $\begin{cases} 2x - 7y = -5 \\ 4x + 3y = 7 \end{cases}$

26. $\begin{cases} x + y + 2z = 4 \\ x + y - 2z = 0 \\ x - y = 0 \end{cases}$

27. $\begin{cases} x + y + z = 2 \\ x + 2y - z = 4 \\ 2x + y + z = 0 \end{cases}$

28. $\begin{cases} x_1 + 2x_2 + 4x_3 = 12 \\ 2x_1 - 3x_2 + x_3 = 10 \\ 3x_1 - x_2 - 2x_3 = 1 \end{cases}$

29. $\begin{cases} x + y + z = 6 \\ x - y + 2z = 12 \\ 2x + y + z = 1 \end{cases}$

30. $\begin{cases} x + y + 2z = 4 \\ x - 5y + z = 5 \\ 3x - 4y + 7z = 24 \end{cases}$

31. $\begin{cases} x_1 + 2x_2 + 5x_3 = 4 \\ 4x_1 + x_2 + 3x_3 = 9 \\ 6x_1 + 9x_2 + x_3 = 21 \end{cases}$

32. $\begin{cases} 2x + y - 3z = 9 \\ x - 2y + 4z = 5 \\ 3x + y - 2z = 15 \end{cases}$

33. $\begin{cases} x + 3y - 2z = -21 \\ 7x - 5y + 4z = 31 \\ 2x + y + 3z = 17 \end{cases}$

34. $\begin{cases} x_1 - 5x_2 + 4x_3 = 8 \\ 3x_1 + x_2 - 2x_3 = 4 \\ 9x_1 - 3x_2 + 6x_3 = 6 \end{cases}$

G In problems 35 and 36, solve each system graphically by finding the location of the point of intersection by using a grapher. Round off the answers to two decimal places.

35. $\begin{cases} 0.2x + 0.3y = 0.7 \\ 0.4x - 0.5y = 0.3 \end{cases}$

36. $\begin{cases} 4.01x + 6.07y = 2.33 \\ 7.57x - 3.11y = 13.81 \end{cases}$

Applying the Concepts

In problems 37–52, use a linear system to solve each problem.

37. **Grading:** Suppose that three-fifths of the male students and two-thirds of the female students have A or B averages. If there are 120 students in the class and 46 have an average of C or lower, how many male and how many female students are in the class?

38. **Recreation:** A theater group plans to sell 500 tickets for a play. They charge \$45 for each orchestra seat and \$30 for each balcony seat, and they plan to collect \$18,000 for a performance. How many of each type of tickets do they expect to sell?

39. **Physics:** An arrow is shot straight up in the air from a height of h_0 feet above the ground with an initial upward speed of v_0 (in feet per second). Its height h (in feet) above the ground t seconds after it was shot is given by

$$h = -16t^2 + v_0 t + h_0.$$

(a) If the arrow is 22 feet above the ground after 1.5 seconds and 10 feet above the ground after 2 seconds, find its initial speed and its initial height.

(b) Use the results from part (a) to find the model that expresses the height h as a function of t.

(c) How high above ground level is the arrow after 1 second? When does the arrow hit the ground? Round off the answers to two decimal places.

40. **Physics:** A stone is thrown vertically upward from a height of h_0 (in feet) above the ground with an initial upward speed of v_0 (in feet per second). Its height h (in feet) above the ground t seconds after it was thrown is given by

$$h = -16t^2 + v_0 t + h_0.$$

(a) If the stone is 40 feet above the ground after 1/2 second and 48 feet above the ground after 1.5 seconds, find its initial speed v_0 and its initial height h_0.

(b) Use the results from part (a) to express h as a function of t.

(c) How high is the stone after 2 seconds? When does the stone hit the ground? Round off the answers to two decimal places.

41. **Chemistry:** A chemist has one solution that is 30% acid and another that is 20% acid. How much of each should be used to make 10 liters of a solution that is 22% acid?

42. **Chemistry:** A chemist prepares two acid solutions. The first solution is 20% acid and the second is 50% acid. How many milliliters of each solution should be mixed to obtain 12 milliliters of a 30% acid solution?

43. **Aviation:** An airplane flying with the wind takes 3.75 hours to fly the 2500 miles from Los Angeles to New York, but 4.4 hours to fly the same distance from New York to Los Angeles against the wind. Assuming that the airplane travels at a constant airspeed in still air and the wind blows at a constant rate, find the speed of the airplane in still air and the wind speed.

44. **Water Speed:** It takes a powerboat traveling upstream (against the current) 3 hours to make a 36 mile trip on a river. Returning downstream (with the current), the same trip takes 2 hours. Assuming that the boat travels at a constant speed in still water and the rate of the current is constant, find the speed of the boat in still water and the rate of the current.

45. **Shoe Sales:** A national sporting gear store sells running shoes and tennis shoes. A pair of tennis shoes sells for \$46 and a pair of running shoes sells for \$79. During a 1 day sale, a total of 260 pairs of tennis and running shoes were sold. The total receipts were \$14,435. How many pairs of each kind were sold?

46. **Carpooling:** To encourage carpooling, a city parking lot charges \$7 a day per single driver vehicle or \$4 per day for each carpool vehicle with two or more people. If the city parking lot took in \$649 and 115 vehicles used the lot for 1 day, how many of each kind of car used the lot that day?

47. **Investment Portfolio:** An investment club invested \$42,000 in three accounts. During one year, the first account earned 5% simple interest, the second earned 7% simple interest, and the third earned 9% simple interest. The total interest from the three investments for the year was \$2600. The interest in the first account was \$200 less than the total interest from the other two accounts. How much was invested in each type of account?

48. **Landscape Engineering:** A landscape architect wishes to construct a sprinkler system by locating the sprinklers at the centers of three circular regions that are mutually tangent to each other. The center of each circle is labeled A, B, and C in Figure 4. If the distances $|\overline{AB}| = 15$ meters, $|\overline{BC}| = 13$ meters, and $|\overline{CA}| = 16$ meters, find the radii of the circles.

Figure 4

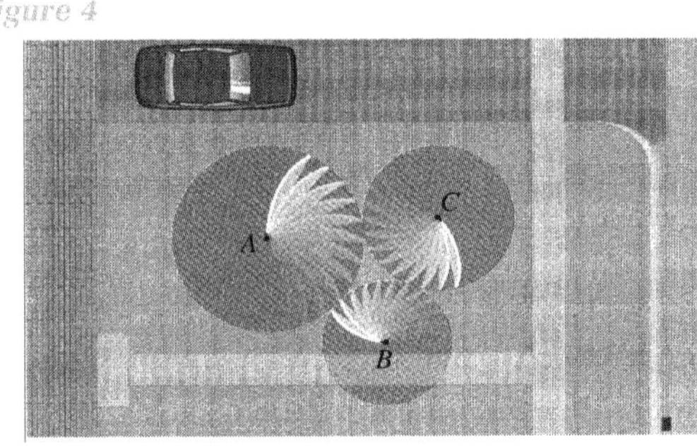

49. **Airport Shuttle:** An airport shuttle service has three sizes of vans. The biggest van holds 12 passengers, the next holds 10 passengers, and the smallest holds 6 passengers. The service manager has 15 vans available to accommodate 152 passengers. However, to reduce fuel costs, the manager wants to use 2 more of the available 12 passenger vans than the 6 passenger vans. How many vans of each kind should be used?

50. **Sports:** The total number of seats in a basketball sports arena is 18,000. The arena is divided into three sections: courtside, balcony, and end zone. There are three times as many balcony seats as courtside seats. For the conference championship game, ticket prices are \$25 for courtside, \$15 for balcony, and \$10 for end zone. If the arena is sold out and the total receipts are \$270,000, how many seats are courtside?

51. **Irrigation:** A grower uses three pumps to provide water to irrigate a grove. When pumps A, B, and C are used for 2 hours, they can pump 74,000 gallons. If the grower runs pump A for 4 hours and pump B for 2 hours, then 64,000 gallons of water are pumped. If pump B is used for 5 hours and pump C for 4 hours, then 120,000 gallons are pumped. How many gallons per hour can each pump handle?

52. Manufacturing: An electronics manufacturer places three wires next to each other inside a radio transmitter. The cross section of each wire is circular, and these circles are tangent to each other. Figure 5 shows the centers of these circles labeled as A, B, and C. If $|\overline{AB}| = 13$ millimeters, $|\overline{BC}| = 22$ millimeters, and $|\overline{DE}| = 52$ millimeters, find the radius of each wire.

Figure 5

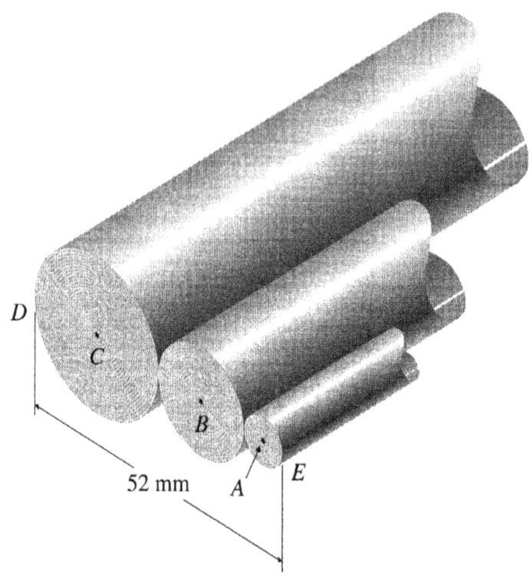

Developing and Extending the Concepts

53. Solve each system by using the substitutions

$$u = 1/x \quad \text{and} \quad v = 1/y.$$

(a) $\begin{cases} \dfrac{5}{x} - \dfrac{6}{y} = 21 \\ \dfrac{1}{x} - \dfrac{2}{y} = 5 \end{cases}$ (b) $\begin{cases} \dfrac{4}{x} + \dfrac{5}{y} = 6 \\ \dfrac{2}{x} - \dfrac{1}{y} = 10 \end{cases}$

54. Solve the system by using substitutions $u = 1/x$, $v = 1/y$, and $w = 1/z$.

$$\begin{cases} \dfrac{3}{x} + \dfrac{4}{y} - \dfrac{1}{z} = -7 \\ \dfrac{1}{x} - \dfrac{1}{y} + \dfrac{2}{z} = 0 \\ \dfrac{4}{x} + \dfrac{1}{y} - \dfrac{3}{z} = -17 \end{cases}$$

55. Give an example of a linear system that has no solution if the system has:

(a) Two variables (b) Three variables

56. Give an example of a linear system that has infinitely many solutions if the system has:

(a) Two variables (b) Three variables

Interlude: Practice Problems 4.1

1. Consider the system $\begin{cases} u \sin x + v \cos x &= 0 \\ u \cos x - v \sin x &= \sec x \end{cases}$. Solve the system for u and v (treat the trigonometric functions as constants).

2. Solve the system $\begin{cases} x - 3y = 8 \\ -2x + y = -10 \\ 5x - y = 23 \end{cases}$ graphically. Show all graphs you use and clearly indicate your conclusion.

3. Suppose that a car gets 18 miles per gallon in city driving and 24 miles per gallon in highway driving. A total of 465 miles are driven using 23 gallons of gas. It is deemed desirable to determine how many miles of city driving and how many miles of highway driving composed the entire 465 miles.

(a) Set up a system of linear equations that can be used to solve the problem. Be sure to indicate what your variables represent.

(b) Solve the system of equations from **(a)** graphically. Show the graphs you used, label the point of intersection appropriately, and indicate how many miles of each type have been driven.

(c) Solve the system of equations from **(a)** algebraically. Show your method, and indicate how many miles of each type have been driven.

Objectives

1. Write Linear Systems in Matrix Notation
2. Solve Linear Systems Using Row-Reduction
3. Decompose Fractions
4. Solve Applied Problems

4.2 Solutions of Linear Systems Using Augmented Matrices

In carrying out the Gaussian elimination method, we wrote the equations in the system in such a way that the same variables were aligned vertically in each step of the process. In this section we'll develop a procedure based on the Gaussian method in which we solve a system of linear equations by keeping track of the changing coefficients without writing down the variables. This procedure makes use of *matrices.*

A **matrix** is a rectangular array of numbers. The numbers in the matrix are called the *entries* or *elements* of the matrix. Traditionally, we enclose the elements of a matrix in brackets. Matrices are usually denoted by capital letters.

For example,

$$A = \begin{bmatrix} 2 & -1 & 3 \\ 5 & 4 & 7 \end{bmatrix}, \quad B = \begin{bmatrix} 1 & 3 \\ 2 & -1 \end{bmatrix}, \quad \text{and} \quad C = \begin{bmatrix} 4 & 3 & -1 \end{bmatrix}$$

are all matrices.

Each horizontal line of numbers is called a *row* of the matrix; each vertical line of numbers is called a *column.* When we refer to a matrix, we usually refer to its *size* or *dimension* by specifying the numbers of rows and the number of columns in that order.

In the given examples, matrix A is a 2 by 3 matrix (2 rows and 3 columns), which is usually written as 2×3. Matrix B is 2×2, and matrix C is 1×3.

An entry or element of a matrix can be identified by subscripts to indicate its row and column position. For instance, in matrix A, we write

$$a_{11} = 2, \quad a_{12} = -1, \quad a_{13} = 3$$
$$a_{21} = 5, \quad a_{22} = 4, \quad a_{23} = 7.$$

In general

a_{mn} is the entry in the mth row and nth column of matrix A.

Writing Linear Systems in Matrix Notation

Consider the linear system of equations

$$\begin{cases} 3x - y = 7 \\ 4x + 7y = 15. \end{cases}$$

It is customary to draw a dashed vertical line to separate the last column (the column that corresponds to the constants).

If we list the coefficients and constants of the system, we obtain the rectangular array of numbers

$$\left[\begin{array}{cc:c} 3 & -1 & 7 \\ 4 & 7 & 15 \end{array}\right].$$

This matrix is called the **augmented matrix** form of the system. The matrix of the coefficients of the variables of the system,

$$\begin{bmatrix} 3 & -1 \\ 4 & 7 \end{bmatrix}$$

is called the **coefficient matrix.** In this example, the first column of this matrix lists the coefficients of the x variables and the second column lists the coefficients of the y variables.

EXAMPLE 1

Writing a System in Matrix Notation

Write the system in matrix notation:

$$\begin{cases} 4x - 3y = 5 \\ x + 7y = 8 \end{cases}$$

Solution The augmented matrix form of the system is

$$\left[\begin{array}{cc|c} 4 & -3 & 5 \\ 1 & 7 & 8 \end{array}\right].$$

If any variable does not appear in an equation, a 0 is inserted in the appropriate position in the array of coefficients.

EXAMPLE 2

Writing a System Corresponding to a Matrix

Write the system in terms of x and y corresponding to the following augmented matrix, and then solve it.

$$\left[\begin{array}{cc|c} 1 & -2 & 7 \\ 0 & 1 & -3 \end{array}\right]$$

Solution Here we used a 0 to denote the coefficient of x in the second equation. So the system is

$$\begin{cases} x - 2y = 7 \\ \quad\;\; y = -3. \end{cases}$$

Using back-substitution, we see that

$$\begin{aligned} x - 2(-3) &= 7 \\ x + 6 &= 7 \\ x &= 1. \end{aligned}$$

Therefore, the solution is $(1, -3)$.

Solving Linear Systems Using Row-Reduction

In Section 4.1 we used elementary operations to transform a linear system of equations into an equivalent system in triangular form, because the latter form is easier to solve. In this section, we perform the same process, but rather than working with the equations in the system, we work with the rows of the corresponding augmented matrix form. Each row serves as an abbreviation of the resulting equation when an elementary operation is applied.

In the following illustration we solve a system of linear equations in two ways:

1. On the left, the elementary operations are used to convert the given system to an equivalent triangular form.
2. On the right, we show the augmented matrix corresponding to each equivalent linear system and describe the elementary operation in terms of a row manipulation. A vertical arrow is used to indicate that each matrix is *row-equivalent* to the one below it in the sense that they both represent equivalent systems.

System of Equations	Augmented Matrix of System	Strategy
(A) $\begin{cases} 3x - y = 15 \\ x + 2y = -2 \end{cases}$	(A) $\left[\begin{array}{cc\|c} 3 & -1 & 15 \\ 1 & 2 & -2 \end{array}\right]$	Given
↓ First interchange equations	↓ First interchange rows	We want the first equation to have a leading coefficient of 1.
(B) $\begin{cases} x + 2y = -2 \\ 3x - y = 15 \end{cases}$	(B) $\left[\begin{array}{cc\|c} 1 & 2 & -2 \\ 3 & -1 & 15 \end{array}\right]$	
↓ Next replace the second equation with the sum of −3 times the first equation and the second equation	↓ Next we replace the second row with the sum of −3 times the first row and the second row	We want to eliminate the x term in the second equation.
(C) $\begin{cases} x + 2y = -2 \\ 0 - 7y = 21 \end{cases}$	(C) $\left[\begin{array}{cc\|c} 1 & 2 & -2 \\ 0 & -7 & 21 \end{array}\right]$	
↓ Multiply the second equation by $-\frac{1}{7}$	↓ Multiply the second row by $-\frac{1}{7}$	We want the second equation to have a leading coefficient of 1.
(D) $\begin{cases} x + 2y = -2 \\ y = -3 \end{cases}$	(D) $\left[\begin{array}{cc\|c} 1 & 2 & -2 \\ 0 & 1 & -3 \end{array}\right]$	
↓ The system is in triangular form	↓ The matrix is in triangular form	

Thus the last matrix (D) is an abbreviated form of the last system form (D). Clearly,

$$y = -3.$$

To get x, we substitute -3 for y in the first equation of D to obtain

$$x + 2(-3) = -2,$$
$$\text{or} \quad x = 4.$$

So the solution is

$$(4, -3).$$

As this illustration shows, the two approaches for solving a linear system are equivalent. However, with some practice, we can work more efficiently with augmented matrices than with the actual systems of equations.

In our earlier discussion of using elimination by addition, we said that two systems were equivalent if they have the same solution. We used the elementary operations on page 297 to transform a system into an equivalent form. Analogous to this approach, we now say that two augmented matrices are **row-equivalent,** if they are augmented matrices of equivalent systems. We use the row operations listed in Table 1 to transform augmented matrices into row-equivalent matrices.

TABLE 1 Elementary Row Operations

Row Operation Performed on a Matrix	Explanation of Operation	Symbolic Representation of Operation
1. Interchange two rows	Interchange row i and row j	$R_i \leftrightarrow R_j$
2. Multiply a row by a nonzero constant	Multiply each entry in row i by a constant k to get new row i, where $k \neq 0$	$kR_i \rightarrow R_i$
3. Add a constant multiple of a row to another row	Multiply each entry of row i by the constant k and add the resulting entries to each entry in row j to get a new row j	$kR_i + R_j \rightarrow R_j$

Table 2 gives illustrations of the elementary row operations.

TABLE 2 Examples of Elementary Row Operations

Original Matrix	Row Operation	Symbolic Representation	Resulting Row-Equivalent Matrix
1. $\begin{bmatrix} 1 & 2 & -3 \\ 4 & 5 & 6 \end{bmatrix}$	Interchange row 1 and row 2	$R_1 \leftrightarrow R_2$	$\begin{bmatrix} 4 & 5 & 6 \\ 1 & 2 & -3 \end{bmatrix}$
2. $\begin{bmatrix} 1 & 2 & -3 & 5 \\ 4 & 5 & 6 & -1 \end{bmatrix}$	Multiply each entry of row 2 by -3 to get a new row 2	$-3R_2 \rightarrow R_2$	$\begin{bmatrix} 1 & 2 & -3 & 5 \\ -12 & -15 & -18 & 3 \end{bmatrix}$
3. $\begin{bmatrix} 2 & 4 & -2 \\ 3 & 1 & 5 \\ 1 & -6 & -1 \end{bmatrix}$	Multiply each entry in row 1 by 2 and add the resulting entries to the entries in row 3 to get a new row 3	$2R_1 + R_3 \rightarrow R_3$	$\begin{bmatrix} 2 & 4 & -2 \\ 3 & 1 & 5 \\ 5 & 2 & -5 \end{bmatrix}$

EXAMPLE 3 **Solving a System Using Matrix Row Operations**

Write the associated augmented matrix of the system and solve it using elementary row operations.

$$\begin{cases} 4x - 3y = 15 \\ x + 2y = 1 \end{cases}$$

Solution We first write the augmented matrix of the system:

$$\left[\begin{array}{cc|c} 4 & -3 & 15 \\ 1 & 2 & 1 \end{array}\right]$$

Next we convert it into an equivalent matrix in triangular form. The easiest way to get a 1 in the upper left corner is to interchange rows 1 and 2:

$$\left[\begin{array}{cc|c} 4 & -3 & 15 \\ 1 & 2 & 1 \end{array}\right] \xrightarrow{R_1 \leftrightarrow R_2} \left[\begin{array}{cc|c} 1 & 2 & 1 \\ 4 & -3 & 15 \end{array}\right]$$

We want a 0 to be the first entry in row 2, so we multiply row 1 by -4 and then add the result to row 2:

$$\left[\begin{array}{cc|c} 1 & 2 & 1 \\ 4 & -3 & 15 \end{array}\right] \xrightarrow{-4R_1 + R_2 \rightarrow R_2} \left[\begin{array}{cc|c} 1 & 2 & 1 \\ 0 & -11 & 11 \end{array}\right]$$

Finally, to get 1 as the first nonzero entry in row 2, we multiply row 2 by $-1/11$:

$$\left[\begin{array}{cc|c} 1 & 2 & 1 \\ 0 & -11 & 11 \end{array}\right] \xrightarrow{-\frac{1}{11}R_2 \rightarrow R_2} \left[\begin{array}{cc|c} 1 & 2 & 1 \\ 0 & 1 & -1 \end{array}\right]$$

The last matrix is in the desired triangular form. By writing the associated system, we have

$$\begin{cases} x + 2y = 1 \\ \quad\;\; y = -1. \end{cases}$$

Using back-substitution, we replace y by -1 in the first equation to obtain

$$x + 2(-1) = 1 \quad \text{or} \quad x - 2 = 1, \quad \text{so} \quad x = 3.$$

Therefore, the solution is $(3, -1)$.

We can avoid the necessity of back-substitution in Example 3 by reducing the last matrix even further. To do this, multiply row 2 by -2 and add the result to row 1 to obtain:

$$\left[\begin{array}{cc|c} 1 & 2 & 1 \\ 0 & 1 & -1 \end{array}\right] \xrightarrow{-2R_2 + R_1 \to R_1} \left[\begin{array}{cc|c} 1 & 0 & 3 \\ 0 & 1 & -1 \end{array}\right].$$

This final matrix is an especially simple one to interpret because the solution,

$$x = 3 \quad \text{and} \quad y = -1,$$

can be read directly from the entries in the last column.

This latter matrix is said to be in *row-reduced echelon form.*

Definition

Row-Reduced Echelon Form of a Matrix

A matrix is said to be in **row-reduced echelon form** if it satisfies the following conditions:

1. The first nonzero entry or leading term in each row is 1.
2. Any rows that consist entirely of 0s are below all rows that do not consist entirely of 0s.
3. The first nonzero entry in each row is to the right of the first nonzero entry in the preceding row.
4. In each column that contains a leading 1 of some row, all other entries are 0.

For example, the matrices

$$\begin{bmatrix} 1 & -2 & 0 \\ 0 & 0 & 1 \end{bmatrix}, \quad \begin{bmatrix} 1 & -1 & 0 & -3 \\ 0 & 0 & 1 & 4 \end{bmatrix},$$

and

$$\begin{bmatrix} 1 & -4 & 0 & 3 & 5 \\ 0 & 0 & 1 & -2 & 4 \\ 0 & 0 & 0 & 0 & 0 \end{bmatrix}$$

are in row-reduced echelon form, because each satisfies the conditions of the above definition.

However,

$$A = \begin{bmatrix} 0 & 1 \\ 0 & -2 \end{bmatrix}, \quad B = \begin{bmatrix} 0 & 0 & 0 \\ 1 & 0 & 0 \end{bmatrix},$$

and

$$C = \begin{bmatrix} 1 & 1 & 0 \\ 0 & 1 & 0 \\ 0 & 0 & 1 \end{bmatrix}$$

are not in row-reduced echelon form. In matrix A, condition 1 is violated, because the leading term in the second row is not 1. In matrix B, condition 2 is violated; and in matrix C, condition 4 is violated, because the entry in row 1, column 2 is 1 rather than 0.

EXAMPLE 4 **Interpreting Row-Reduced Echelon Matrices**

Each of the following augmented matrices is the row-reduced echelon matrix form corresponding to a linear system of equations. Solve each system.

(a) $\left[\begin{array}{cc:c} 1 & 0 & 2 \\ 0 & 1 & 3 \end{array}\right]$ (b) $\left[\begin{array}{cc:c} 1 & 0 & 5 \\ 0 & 0 & -1 \end{array}\right]$ (c) $\left[\begin{array}{ccc:c} 1 & 0 & 0 & 2 \\ 0 & 1 & 2 & 5 \\ 0 & 0 & 0 & 0 \end{array}\right]$

Solution (a) The matrix

$$\left[\begin{array}{cc:c} 1 & 0 & 2 \\ 0 & 1 & 3 \end{array}\right]$$

corresponds to the system

$$\begin{cases} x = 2 \\ y = 3 \end{cases} \text{so the system has one solution, } (2, 3).$$

(b) The matrix

$$\left[\begin{array}{cc:c} 1 & 0 & 5 \\ 0 & 0 & -1 \end{array}\right] \text{corresponds to the system} \begin{cases} x = 5 \\ 0 = -1. \end{cases}$$

Because of the contradiction in the second equation, we conclude that the system has no solution.

(c) The matrix

$$\left[\begin{array}{ccc:c} 1 & 0 & 0 & 2 \\ 0 & 1 & 2 & 5 \\ 0 & 0 & 0 & 0 \end{array}\right]$$

corresponds to the system

$$\begin{cases} x = 2 \\ y + 2z = 5. \end{cases}$$

This implies that there are infinitely many solutions of the form (x, y, z), where

$$x = 2, y = 5 - 2z,$$

and z is any real number.

That is, the system has infinitely many solutions of the form

$$(2, 5 - 2z, z),$$

where z is any real number.

For instance,

if we let $z = 0$, we get the solution $(2, 5, 0)$;

if $z = 1$, we get $(2, 3, 1)$,

and so forth.

We now can describe a matrix method for solving linear systems, called the *Gauss–Jordan elimination method.*

Gauss–Jordan Elimination Method

Step 1 Represent the linear system with an augmented matrix.

Step 2 Use elementary row operations to transform the original matrix to an equivalent row-reduced echelon form.

Step 3 Change the reduced matrix from step 2 to the equation form of the linear system.

Step 4 Interpret the result in step 3 to get the solution of the system, if there is one.

EXAMPLE 5

Solving a System by Gauss–Jordan Elimination

Solve the system below by using Gauss–Jordan elimination.

$$\begin{cases} x_1 - 2x_2 + 3x_3 = -20 \\ 2x_1 - x_2 + 2x_3 = -14 \\ 3x_1 + x_2 + 2x_3 = -11 \end{cases}$$

Solution First we write the augmented matrix of the system; then we transform the matrix into an equivalent matrix in row-reduced echelon form as follows:

$$\left[\begin{array}{rrr|r} 1 & -2 & 3 & -20 \\ 2 & -1 & 2 & -14 \\ 3 & 1 & 2 & -11 \end{array}\right]$$

$\downarrow -2R_1 + R_2 \to R_2$

$$\left[\begin{array}{rrr|r} 1 & -2 & 3 & -20 \\ 0 & 3 & -4 & 26 \\ 3 & 1 & 2 & -11 \end{array}\right]$$

$\downarrow -3R_1 + R_3 \to R_3$

$$\left[\begin{array}{rrr|r} 1 & -2 & 3 & -20 \\ 0 & 3 & -4 & 26 \\ 0 & 7 & -7 & 49 \end{array}\right]$$

These operations are performed in order to obtain a matrix in which there are 0 entries in the column below the leading entry of 1 in row 1.

$\downarrow \frac{1}{3}R_2$ to R_3

$$\left[\begin{array}{rrr|r} 1 & -2 & 3 & -20 \\ 0 & 1 & -\frac{4}{3} & \frac{26}{3} \\ 0 & 7 & -7 & 49 \end{array}\right]$$

$\downarrow \frac{1}{7}R_3 \to R_3$

$$\left[\begin{array}{rrr|r} 1 & -2 & 3 & -20 \\ 0 & 1 & -\frac{4}{3} & \frac{26}{3} \\ 0 & 1 & -1 & 7 \end{array}\right]$$

These operations produce an equivalent matrix in which the leading entries in rows 2 and 3 are 1's.

$\downarrow R_2 \leftrightarrow R_3$

$$\left[\begin{array}{rrr|r} 1 & -2 & 3 & -20 \\ 0 & 1 & -1 & 7 \\ 0 & 1 & -\frac{4}{3} & \frac{26}{3} \end{array}\right]$$

This operation puts the row with fractions on the bottom to simplify our work.

$\downarrow 2R_2 + R_1 \to R_1$

(Continued in next column)

$$\left[\begin{array}{rrr|r} 1 & 0 & 1 & -6 \\ 0 & 1 & -1 & 7 \\ 0 & 1 & -\frac{4}{3} & \frac{26}{3} \end{array}\right]$$

$\downarrow -R_2 + R_3 \to R_3$

$$\left[\begin{array}{rrr|r} 1 & 0 & 1 & -6 \\ 0 & 1 & -1 & 7 \\ 0 & 0 & -\frac{1}{3} & \frac{5}{3} \end{array}\right]$$

These operations lead to a matrix form in which the entries in column 2 above and below the 1 in row 2 are 0s.

$\downarrow -3R_3 \to R_3$

$$\left[\begin{array}{rrr|r} 1 & 0 & 1 & -6 \\ 0 & 1 & -1 & 7 \\ 0 & 0 & 1 & -5 \end{array}\right]$$

This operation puts a 1 in the leading entry in row 3.

$\downarrow R_3 + R_2 \to R_2$

$$\left[\begin{array}{rrr|r} 1 & 0 & 1 & -6 \\ 0 & 1 & 0 & 2 \\ 0 & 0 & 1 & -5 \end{array}\right]$$

$\downarrow -R_3 + R_1 \to R_1$

$$\left[\begin{array}{rrr|r} 1 & 0 & 0 & -1 \\ 0 & 1 & 0 & 2 \\ 0 & 0 & 1 & -5 \end{array}\right]$$

These operations produce an equivalent matrix in which the entries in column 3 above the 1 in row 3 are 0s.

This is the row-reduced echelon form.

The last matrix corresponds to the system

$$\begin{cases} x_1 = -1 \\ x_2 = 2 \\ x_3 = -5. \end{cases}$$

So the solution is $(-1, 2, -5)$.

Decomposing Fractions

A rational expression is said to be *proper* if the degree of its numerator is less than the degree of its denominator. Examples of proper fractions are

$$\frac{5x}{(x+1)(x-2)}, \quad \frac{3}{x^2+5}, \quad \text{and} \quad \frac{5x^2+7x+3}{x^3-5x-1}.$$

In calculus it is necessary at times to rewrite a rational expression as a sum (or difference) of fractions that contain linear or quadratic factors in the denominators according to the factorization of the denominator of the given rational expression. Such a representation is referred to as a ***partial fraction expansion*** or ***decomposition.*** One process of decomposing a proper fraction into partial fractions involves solving linear systems, as illustrated in the next example.

EXAMPLE 6 **Decomposing a Fraction**

Given that the fraction

$$\frac{2x-2}{(x+2)(x+5)}$$

has a partial fraction expansion of the form

$$\frac{2x-2}{(x+2)(x+5)} = \frac{A}{x+2} + \frac{B}{x+5}$$

find the constants A and B.

Solution Given that

$$\frac{2x-2}{(x+2)(x+5)} = \frac{A}{x+2} + \frac{B}{x+5}$$

we proceed to find A and B as follows:

$2x - 2 = A(x + 5) + B(x + 2)$ Multiply each side by the LCD $(x + 2)(x + 5)$

$2x - 2 = Ax + 5A + Bx + 2B$ Multiply

$2x - 2 = (A + B)x + (5A + 2B)$ Rewrite the right side

Equating the coefficients of the two sides of the latter equation results in

$$\begin{cases} A + B = 2 \\ 5A + 2B = -2. \end{cases}$$

The associated augmented matrix of the system is

$$\left[\begin{array}{cc|c} 1 & 1 & 2 \\ 5 & 2 & -2 \end{array}\right].$$

The equivalent matrix in row-reduced echelon form is

$$\left[\begin{array}{cc|c} 1 & 0 & -2 \\ 0 & 1 & 4 \end{array}\right]$$

so $A = -2$ and $B = 4$. Thus

$$\frac{2x - 2}{(x + 2)(x + 5)} = \frac{-2}{x + 2} + \frac{4}{x + 5}.$$

Figure 1

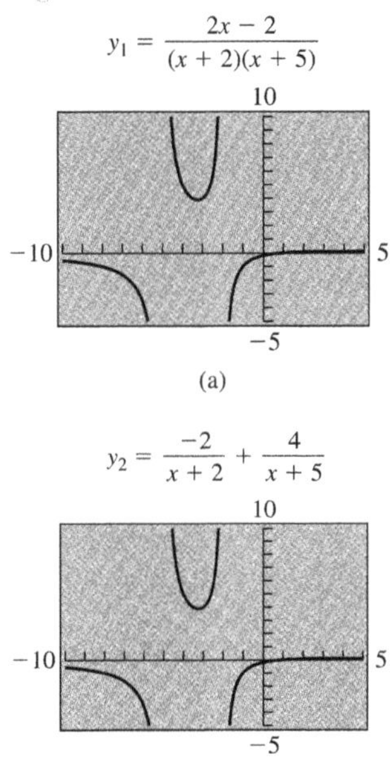

(a) (b)

The Gauss–Jordan elimination method for solving linear systems often involves tedious numerical computations. However, some graphers are equipped with special keys that transform a matrix to an equivalent row-reduced echelon form. Also, a grapher can be used to demonstrate the validity of partial fraction decompositions graphically.

For instance, to demonstrate the validity of the result in Example 6 above graphically, we graph each side of the equation as a function of x. Figure 1a shows the graph of

$$y_1 = \frac{2x - 2}{(x + 2)(x + 5)}$$

whereas, Figure 1b shows the graph of

$$y_2 = \frac{-2}{x + 2} + \frac{4}{x + 5}$$

for the same window settings.

Figures 1a and 1b suggest that the two graphs are the same, thus demonstrating the validity of the given decomposition.

Solving Applied Problems

The following example uses row-reduction to solve a linear system of equations that models a real-life situation.

EXAMPLE 7

Modeling a Diet Problem

A dietician prepares a special diet that consists of iron, carbohydrates, and protein. Table 3 lists the information about the content of iron, carbohydrates, and protein in an 8 ounce glass of skim milk, 1/4 pound of lean red meat, and 2 slices of whole-grain bread. If a person on a special diet must have 10.5 milligrams of iron, 61.0 grams of carbohydrates, and 94.5 grams of protein, how many 8 ounce glasses of skim milk, 1/4 pound servings of lean red meat, and 2-slice servings of whole-grain bread will supply this diet?

TABLE 3

Food Source	Iron Milligrams Per Serving	Carbohydrates Grams Per Serving	Protein Grams Per Serving
8 ounce glass of skim milk	0.1	1.0	8.5
1/4 pound of lean red meat	3.4	20.0	22.0
2 slices of whole-grain bread	2.2	12.0	10.0

Solution Let
x represent the number of 8 ounce glasses of skim milk,
y represent the number of 1/4 pound servings of lean red meat, and
z represent the number of 2-slice servings of whole-grain bread.

The system of linear equations that provides the special diet is given as follows:

$$\begin{cases} 0.1x + 3.4y + 2.2z = 10.5 \\ x + 20.0y + 12.0z = 61.0 \\ 8.5x + 22.0y + 10.0z = 94.5 \end{cases}$$

The augmented matrix that corresponds to this system is

$$\left[\begin{array}{ccc|c} 0.1 & 3.4 & 2.2 & 10.5 \\ 1.0 & 20.0 & 12.0 & 61.0 \\ 8.5 & 22.0 & 10.0 & 94.5 \end{array}\right].$$

By using the Gauss–Jordan elimination method, we get the equivalent row-reduced echelon matrix.

$$\left[\begin{array}{ccc|c} 1 & 0 & 0 & 5 \\ 0 & 1 & 0 & 1 \\ 0 & 0 & 1 & 3 \end{array}\right]$$

The associated system for the transformed matrix is

$$\begin{cases} x = 5 \\ y = 1 \\ z = 3. \end{cases}$$

So it takes 5 (8 ounce) glasses of skim milk,
1 serving (1/4 pound) of lean red meat,
and 3 servings (2 slices each) of whole-grain bread to supply this diet.

PROBLEM SET 4.2

Mastering the Concepts

In problems 1–4, write the augmented matrix of each linear system.

1. $\begin{cases} x + 6y = 7 \\ 3x + 7y = 10 \end{cases}$

2. $\begin{cases} 4x + y = 7 \\ 8x - 3y = 11 \end{cases}$

3. $\begin{cases} 2x + y = 2 \\ x - y = 4 \\ 3x + z = 6 \end{cases}$

4. $\begin{cases} x_1 + 2x_2 - x_3 = 4 \\ -x_1 + 3x_2 - 2x_3 = 5 \\ 2x_1 - x_2 + x_3 = 0 \end{cases}$

In problems 5–10, each matrix is in triangular form and is the augmented matrix form of a corresponding linear system of equations. Solve each system.

5. $\left[\begin{array}{cc|c} 1 & -2 & -5 \\ 0 & 1 & 3 \end{array}\right]$

6. $\left[\begin{array}{cc|c} 1 & 3 & 4 \\ 0 & 1 & 1 \end{array}\right]$

7. $\left[\begin{array}{ccc|c} 1 & 2 & 3 & 6 \\ 0 & 1 & -4 & -3 \\ 0 & 0 & 1 & 1 \end{array}\right]$

8. $\left[\begin{array}{ccc|c} 1 & 8 & 7 & -22 \\ 0 & 1 & 4 & -7 \\ 0 & 0 & 1 & -1 \end{array}\right]$

9. $\left[\begin{array}{ccc|c} 1 & 0 & -1 & 7 \\ 0 & 1 & 3 & -3 \\ 0 & 0 & 1 & -2 \end{array}\right]$

10. $\left[\begin{array}{ccc|c} 1 & 1 & 2 & 4 \\ 0 & 1 & 2 & 3 \\ 0 & 0 & 1 & 1 \end{array}\right]$

11. Let

$$A = \begin{bmatrix} 1 & 0 & 3 \\ -3 & 1 & 4 \\ 5 & -1 & 5 \end{bmatrix}.$$

Find the matrix obtained by performing the following elementary row operations on A:

(a) Interchange the second and third rows.

(b) Multiply the third row of A by 3.

(c) Add 3 times the first row of A to the third row.

12. Let

$$A = \begin{bmatrix} 2 & 0 & 4 & 2 \\ 3 & -2 & 5 & 6 \\ -1 & 3 & 1 & 1 \end{bmatrix}.$$

Find the matrix obtained by performing the following elementary row operations on A:

(a) Interchange the second and third rows.

(b) Multiply the first row of A by 1/2.

(c) Add 2 times the third row of A to the first row.

In problems 13–16, which of the matrices are in row-reduced echelon form?

13. (a) $\begin{bmatrix} 1 & -2 & 0 & 1 \\ 0 & 0 & 1 & -4 \end{bmatrix}$

(b) $\begin{bmatrix} 1 & -1 & 0 & 4 \\ 0 & 0 & 1 & 2 \\ 0 & 0 & 0 & 0 \end{bmatrix}$

14. (a) $\begin{bmatrix} 1 & 0 & 0 & -1 \\ 0 & 1 & 0 & -3 \\ 0 & 0 & 1 & 2 \end{bmatrix}$

(b) $\begin{bmatrix} 1 & 0 & 0 \\ 0 & 0 & 1 \\ 0 & 1 & 0 \end{bmatrix}$

15. (a) $\begin{bmatrix} 1 & 0 & 1 \\ 0 & 2 & -3 \end{bmatrix}$

(b) $\begin{bmatrix} 1 & 2 & -1 & 0 & 2 \\ 0 & 0 & 0 & 1 & 4 \\ 0 & 0 & 0 & 0 & 0 \end{bmatrix}$

16. (a) $\begin{bmatrix} 1 & 2 & 0 & 2 & 5 \\ 0 & 0 & 1 & -3 & 4 \\ 0 & 0 & 0 & 0 & 0 \end{bmatrix}$

(b) $\begin{bmatrix} 1 & 0 & -2 & -2 & 3 \\ 0 & 1 & 4 & 3 & 1 \\ 0 & 0 & 1 & -1 & 5 \end{bmatrix}$

In problems 17–22, each row-reduced echelon matrix is the augmented matrix form of a corresponding linear system of equations. Solve each system.

17. $\left[\begin{array}{cc|c} 1 & 0 & 1 \\ 0 & 1 & 1 \end{array}\right]$

18. $\left[\begin{array}{ccc|c} 1 & 0 & 0 & 0 \\ 0 & 1 & 0 & 0 \\ 0 & 0 & 1 & 2 \end{array}\right]$

19. $\left[\begin{array}{ccc|c} 1 & 0 & -1 & 2 \\ 0 & 1 & -2 & 3 \end{array}\right]$

20. $\left[\begin{array}{ccc|c} 1 & 0 & 0 & 3 \\ 0 & 1 & 0 & 0 \\ 0 & 0 & 1 & 5 \end{array}\right]$

21. $\left[\begin{array}{ccc|c} 1 & 0 & 0 & -3 \\ 0 & 1 & 0 & 4 \\ 0 & 0 & 0 & 7 \end{array}\right]$

22. $\left[\begin{array}{ccc|c} 1 & 0 & 0 & 3 \\ 0 & 1 & 0 & -2 \\ 0 & 0 & 0 & 4 \end{array}\right]$

In problems 23–40, solve each system by the Gauss–Jordan elimination method.

23. $\begin{cases} 3x + y = 14 \\ 2x - y = 1 \end{cases}$

24. $\begin{cases} 4x + 3y = 15 \\ 3x + 5y = 14 \end{cases}$

25. $\begin{cases} -2x + 3y = 8 \\ 2x - y = 5 \end{cases}$

26. $\begin{cases} x - 2y = 5 \\ 3x - 6y = 4 \end{cases}$

27. $\begin{cases} x + y + z = 6 \\ 3x - y + 2z = 7 \\ 2x + 3y - z = 5 \end{cases}$

28. $\begin{cases} 2x + 3y + z = 6 \\ x - 2y + 3z = -3 \\ 3x + y - z = 8 \end{cases}$

29. $\begin{cases} x + y + 2z = 4 \\ x + y - 2z = 0 \\ x - y = 0 \end{cases}$

30. $\begin{cases} x + y + z = 4 \\ x - y + 2z = 8 \\ 2x + y - z = 3 \end{cases}$

31. $\begin{cases} 2x + y - z = 7 \\ y - x = 1 \\ z - y = 1 \end{cases}$

32. $\begin{cases} x_1 + 2x_2 + 3x_3 = 0 \\ x_1 - 2x_3 = 1 \\ x_2 + x_3 = 1 \end{cases}$

33. $\begin{cases} x - 2y + z = -1 \\ 3x + y - 2z = 4 \\ y - z = 1 \end{cases}$

34. $\begin{cases} x + y - 2z = 3 \\ 3x - y + z = 5 \\ 3x + 3y - 6z = 9 \end{cases}$

35. $\begin{cases} 2x + y + z = 1 \\ 4x + 2y + 3z = 1 \\ -2x - y + z = 2 \end{cases}$

36. $\begin{cases} x + y + z = 0 \\ 2x - y - 4z = 15 \\ x - 2y - z = 7 \end{cases}$

37. $\begin{cases} 2x - 3y + z = 4 \\ x - 4y - z = 3 \end{cases}$

38. $\begin{cases} 2x + 3y - z = -2 \\ x - y + 2z = 4 \end{cases}$

39. $\begin{cases} 2x_1 - 3x_3 = 5 \\ 4x_2 + 2x_3 = -20 \\ 5x_1 + 2x_2 = -16 \end{cases}$

40. $\begin{cases} x_1 + x_2 = -1 \\ x_1 - 2x_3 = -11 \\ x_2 - 3x_3 = -15 \end{cases}$

In problems 41–50, find the partial fraction expansion.

41. $\dfrac{3x - 5}{(x - 1)(x - 2)} = \dfrac{A}{x - 1} + \dfrac{B}{x - 2}$

42. $\dfrac{x + 14}{(x + 4)(x + 12)} = \dfrac{A}{x + 4} + \dfrac{B}{x + 12}$

43. $\dfrac{-9x - 7}{(3x + 1)(x + 1)} = \dfrac{A}{3x + 1} + \dfrac{B}{x + 1}$

44. $\dfrac{13x - 24}{(3x - 5)(x - 2)} = \dfrac{A}{3x - 5} + \dfrac{B}{x - 2}$

45. $\dfrac{-21x + 11}{(x - 1)(x - 2)(x - 3)} = \dfrac{A}{x - 1} + \dfrac{B}{x - 2} + \dfrac{C}{x - 3}$

46. $\dfrac{x^2 - 3x + 4}{(x - 1)(x + 1)(x + 2)} = \dfrac{A}{x - 1} + \dfrac{B}{x + 1} + \dfrac{C}{x + 2}$

47. $\dfrac{5x^2 - 21x + 13}{(x - 3)^2(x + 2)} = \dfrac{A}{x - 3} + \dfrac{B}{(x - 3)^2} + \dfrac{C}{x + 2}$

48. $\dfrac{-x^2 + 13x - 26}{(x + 1)^2(x - 4)} = \dfrac{A}{x + 1} + \dfrac{B}{(x + 1)^2} + \dfrac{C}{x - 4}$

49. $\dfrac{6}{x(x^2 + 1)} = \dfrac{A}{x} + \dfrac{Bx + C}{x^2 + 1}$

50. $\dfrac{30x}{(x - 6)(x^2 + 4)} = \dfrac{A}{x - 6} + \dfrac{Bx + C}{x^2 + 4}$

In problems 51 and 52, use a grapher to demonstrate the validity of the partial fraction decomposition found in the problem.

51. Problem 41
52. Problem 42

Applying the Concepts

In problems 53–58, set up an appropriate linear system, and then use the Gauss–Jordan method to solve each system.

53. **Nutrition:** A nutritionist wants to create a special diet out of three substances *A, B,* and *C.* The requirements in the diet are 480 units of calcium, 190 units of iron, and 360 units of vitamins. The calcium, iron, and vitamin contents in substances *A, B,* and *C* are listed in Table 4. How many ounces of each substance should be used to prepare this diet?

TABLE 4

Substance	Calcium *Units Per Ounce*	Iron *Units Per Ounce*	Vitamin *Units Per Ounce*
A	30	10	15
B	20	10	10
C	10	5	20

54. **Production:** Three assembly lines, *A, B,* and *C,* can produce 8400 TV dinners per day. Together, lines *A* and *B* can produce 4900 TV dinners, whereas lines *B* and *C* together can produce 5600 TV dinners. Find the number of TV dinners each assembly line can produce alone.
55. **Horticulture:** A horticulturist wishes to mix three types of fertilizer which contain 25%, 35%, and 40% nitrogen, respectively, in order to get a mixture of 4000 pounds of $35\frac{5}{8}\%$ nitrogen. The final mixture should contain three times as much of the 40% type as the 25% type. How much of each type is in the final mixture?
56. **Fast-Food Chain:** A fast-food chain sells three types of franchises, *A, B,* and *C.* Franchise *A* sells for \$20,000, franchise *B* for \$25,000, and franchise *C* for \$30,000. In one year, the company sold 18 franchises for \$430,000. If the number sold of franchise *A* is twice the number sold of franchise *C,* how many of each type did the company sell that year?

57. Electrical Engineering: In electronics, the analysis of a circuit leads naturally to a system of linear equations. Suppose an electrical network contains currents I_1, I_2, and I_3 in the circuit, resistors R_1, R_2, and R_3; a 6-volt battery ($E_1 = 6$) and a 12-volt battery ($E_2 = 12$), as shown in Figure 2.

Figure 2

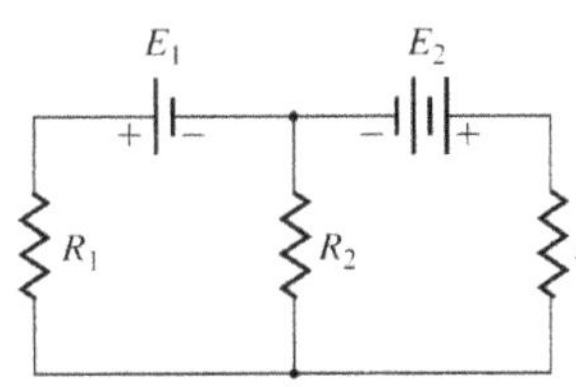

Assume the network satisfies the system given by

$$\begin{cases} I_1 - I_2 + I_3 = 0 \\ R_1 I_1 + R_2 I_2 = E_1 \\ R_2 I_2 + R_3 I_3 = E_2. \end{cases}$$

If $R_1 = 4$ ohms, $R_2 = 10$ ohms, and $R_3 = 21$ ohms, find the currents I_1, I_2, and I_3 (in amperes).

58. Irrigation Network: Figure 3 is a diagram of a system of irrigation canals that shows the flow of water into the system at junction A and out at junctions B, C, and D. The flows f_1, f_2, f_3, f_4, and f_5 (in gallons per minute) are solutions to the system

$$\begin{cases} f_1 + f_2 + f_3 = 500 \text{ at junction } A \\ f_1 - f_5 = 100 \text{ at junction } B \\ f_2 - f_4 + f_5 = 100 \text{ at junction } C \\ f_3 + f_4 = 300 \text{ at junction } D. \end{cases}$$

Figure 3

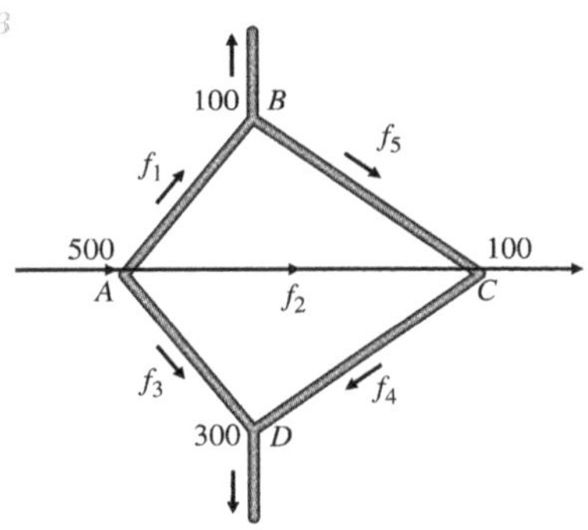

(a) Solve the system for f_1, f_2, f_3, f_4, and f_5.

(b) Discuss some choices of flows for this network.

Developing and Extending the Concepts

In problems 59–62, use matrices to solve each system. In each case, write the augmented matrix and reduce it to row-reduced echelon form.

59. $\begin{cases} x + y + 2z + 3w = 5 \\ x - 2y + z + w = 5 \\ 3x + y + z - w = 2 \\ 2x - y + z + 4w = 8 \end{cases}$

60. $\begin{cases} x_1 - x_2 + 2x_3 + x_4 = 8 \\ -x_1 + 2x_2 + 3x_3 - x_4 = 4 \\ x_1 - x_2 + x_3 - 2x_4 = -6 \\ -x_1 + x_2 + x_3 + x_4 = 7 \end{cases}$

61. $\begin{cases} 3x + y - 2z + 2w = 7 \\ 2x + z + 4w = 3 \\ y + 3z + 5w = 0 \\ -x + 2y - 3w = 4 \end{cases}$

62. $\begin{cases} x_1 + 4x_2 - x_3 + x_4 = 11 \\ 3x_1 + 2x_2 + x_3 + 2x_4 = 9 \\ x_1 - 6x_2 + 3x_4 = 23 \\ x_1 + 14x_2 - 5x_3 + 2x_4 = -13 \end{cases}$

In problems 63–68, find (if possible) conditions on a, b, and c so that the system has:

(a) No solution **(b)** One solution

(c) Infinite number of solutions

63. $\begin{cases} 3x_1 + x_2 - x_3 = a \\ x_1 - x_2 + 2x_3 = b \\ 5x_1 + 3x_2 - 4x_3 = c \end{cases}$

64. $\begin{cases} x_1 - x_3 = a \\ 2x_1 + x_2 - x_3 = b \\ 2x_2 + 3x_3 = c \end{cases}$

65. $\begin{cases} x_1 + x_2 + x_3 = 2 \\ 2x_1 + 3x_2 + 2x_3 = 5 \\ 2x_1 + 3x_2 + (a^2 - 1)x_3 = a \end{cases}$

66. $\begin{cases} x_1 + x_2 + x_3 = 2 \\ x_1 + 2x_2 + x_3 = 3 \\ x_1 + x_2 + (c^2 - 5)x_3 = c \end{cases}$

67. Find a, b, and c, so that (1, 2, 3) is a solution of the system

$$\begin{cases} ax + by + cz = 6 \\ 2bx - cy + 3az = 9 \\ -cx + 2by + 2az = 9. \end{cases}$$

68. Find the three angles of a triangle, α, β, and γ, if

$$\begin{cases} \alpha + \beta = 75° \\ \gamma - \beta = 52°. \end{cases}$$

Interlude: Practice Problems 4.2, Part I

1. Find appropriate constants A and B so that $\dfrac{4x - 88}{(x + 4)(x - 9)} = \dfrac{A}{x + 4} + \dfrac{B}{x - 9}$.

2. Write a system of linear equations in three variables containing all nonzero coefficients so that $(-1, 2, 5)$ is the unique solution to the system. Show verification that your system is valid.

3. On an 18-hole golf course, there are par-3 holes, par-4 holes, and par-5 holes. A golfer who shoots par on every hole has a score of 70. The course has twice as many par-4 holes as par-5 holes. How many holes of each type are there on the course? Set up an appropriate model, declare what all variables represent, and solve the problem. Clearly state the conclusion.

Interlude: Practice Problems 4.2, Part II

1. Use augmented matrices to solve the system $\begin{cases} w - 3x + 2y - z = 10 \\ 3w + x - 4y + 8z = -24 \\ x - y + z = -5 \\ 2w - 3z = 18 \end{cases}$. Show all steps and indicate the row operations you use to pass between row-equivalent matrices.

2. A store owner wishes to create an 800-pound salt mixture that is 30% salt. He has four types of salt compounds to use. Compound *A* contains 15% salt. Compound *B* contains 25% salt. Compound *C* contains 35% salt. Compound *D* contains 40% salt. The resulting mixture must contain twice as much of compound *B* salt as compound *C* salt and three times as much compound *A* salt as compound *C* salt. Given this, you are to determine how many pounds of each compound he should use. Set up an appropriate model, declare what all variables represent, and solve the problem. Clearly state your conclusion.

Interlude: Practice Problems 4.2, Part III

Consider the matrix $A = \begin{bmatrix} 2 & 4 & -8 \\ 7 & 10 & 0 \\ -4 & 6 & 1 \\ -3 & 5 & 3 \end{bmatrix}$.

1. I performed this sequence of row operations on A: $-5R_4 + R_2 \to R_2$, $2R_3 + R_1 \to R_1$, $4R_3 \to R_3$. Show the resulting matrix after each row operation.

2. List appropriate row operations to reduce the first column of A to the form $\begin{bmatrix} 1 \\ 0 \\ 0 \\ 0 \end{bmatrix}$ and show the resulting matrix after each operation.

3. Consider the matrix $B = \begin{bmatrix} 1 & 3 & -11 \\ 0 & 1 & -7 \\ 0 & 0 & 1 \\ 0 & 0 & 1 \end{bmatrix}$. Is B row-equivalent to A? Explain.

Objectives

1. Perform Matrix Arithmetic
2. Find the Inverse of a Square Matrix
3. Solve Linear Systems Using Inverse Matrices
4. Solve Applied Problems

4.3 Solutions of Linear Systems Using Inverse Matrices

In Section 4.2 we used elementary row operations on augmented matrices to solve linear systems of equations. In this section we examine another method for solving linear systems that involves the algebra of matrices.

We say that **two matrices are equal** if and only if they are the same size and the corresponding elements in each matrix are equal. For example,

$$\begin{bmatrix} 2 & 3 & -1 \\ 5 & 8 & 3 \end{bmatrix} = \begin{bmatrix} 2 & x & y \\ a & b & c \end{bmatrix}.$$

if and only if $x = 3$, $y = -1$, $a = 5$, $b = 8$, and $c = 3$.

A matrix whose elements are all 0s is called a **zero matrix** and is denoted by **0**. For example, the zero matrix of size 2×3 is given by

$$\mathbf{0} = \begin{bmatrix} 0 & 0 & 0 \\ 0 & 0 & 0 \end{bmatrix}.$$

Performing Matrix Arithmetic

The addition (or subtraction) of two matrices can be performed *only* if the matrices are the same size. Then, to **add** (or **subtract**) the matrices, we add (or subtract) the corresponding entries.

EXAMPLE 1

Adding and Subtracting Matrices

Let

$$A = \begin{bmatrix} 3 & 5 & 1 \\ 6 & 2 & -3 \end{bmatrix} \quad \text{and} \quad B = \begin{bmatrix} 1 & 3 & 7 \\ 5 & -2 & 4 \end{bmatrix}.$$

Determine:

(a) $A + B$ (b) $A - B$

Solution

(a) $$A + B = \begin{bmatrix} 3 & 5 & 1 \\ 6 & 2 & -3 \end{bmatrix} + \begin{bmatrix} 1 & 3 & 7 \\ 5 & -2 & 4 \end{bmatrix}$$
$$= \begin{bmatrix} 3+1 & 5+3 & 1+7 \\ 6+5 & 2+(-2) & -3+4 \end{bmatrix} = \begin{bmatrix} 4 & 8 & 8 \\ 11 & 0 & 1 \end{bmatrix}$$

(b) $$A - B = \begin{bmatrix} 3 & 5 & 1 \\ 6 & 2 & -3 \end{bmatrix} - \begin{bmatrix} 1 & 3 & 7 \\ 5 & -2 & 4 \end{bmatrix}$$
$$= \begin{bmatrix} 3-1 & 5-3 & 1-7 \\ 6-5 & 2-(-2) & -3-4 \end{bmatrix} = \begin{bmatrix} 2 & 2 & -6 \\ 1 & 4 & -7 \end{bmatrix}$$

If A is a matrix and k is a real number, we define kA to be the matrix obtained by multiplying each entry of A by k. This process is referred to as **multiplication by a scalar.**

EXAMPLE 2

Multiplying a Matrix by a Real Number

Let

$$A = \begin{bmatrix} -5 & 1 \\ 3 & 2 \end{bmatrix}.$$

Find each matrix:

(a) $4A$ (b) $-2A$

Solution

(a) $$4A = 4\begin{bmatrix} -5 & 1 \\ 3 & 2 \end{bmatrix} = \begin{bmatrix} 4(-5) & 4(1) \\ 4(3) & 4(2) \end{bmatrix} = \begin{bmatrix} -20 & 4 \\ 12 & 8 \end{bmatrix}$$

(b) $$-2A = -2\begin{bmatrix} -5 & 1 \\ 3 & 2 \end{bmatrix} = \begin{bmatrix} (-2)(-5) & (-2)(1) \\ (-2)(3) & (-2)(2) \end{bmatrix} = \begin{bmatrix} 10 & -2 \\ -6 & -4 \end{bmatrix}$$

Because matrices are added by adding their corresponding elements, it follows from the properties of real numbers that the addition of matrices obeys both the *commutative* and *associative properties.* That is, if A, B, and C are matrices of the same size, then we have the following properties:

1. **Commutative Property:** $A + B = B + A$
2. **Associative Property:** $A + (B + C) = (A + B) + C$

A zero matrix acts as the identity for the operation of addition. That is, if A and $\mathbf{0}$ are of the same size, then we have:

3. **Zero Matrix Property:** $A + \mathbf{0} = \mathbf{0} + A = A$

For example,

$$\begin{bmatrix} 1 & 2 \\ 3 & 8 \end{bmatrix} + \begin{bmatrix} 0 & 0 \\ 0 & 0 \end{bmatrix} = \begin{bmatrix} 1 & 2 \\ 3 & 8 \end{bmatrix}.$$

The matrix $-A$ is the same as $(-1)A$. Also, $A - B$ can be interpreted as $A - B = A + (-B)$.

If A is a matrix, the **additive inverse of** A, denoted by $-A$, is the matrix consisting of entries that are opposite in sign to the entries of A. Thus, we have:

4. **Additive Inverse Property:** $A + (-A) = (-A) + A = \mathbf{0}$

The following properties involving multiplication by a scalar, and addition, can be easily verified.

Scalar Multiplication Properties

If A and B are matrices of the same size, and c and d are real numbers, then

5. $(cd)A = c(dA)$
6. $(c + d)A = cA + dA$
7. $c(A + B) = cA + cB$

To understand how to *multiply two matrices,* we need some background. A matrix consisting of one row is called a **row matrix,** and a matrix consisting of a single column is called a **column matrix.**

For example,

$$\begin{bmatrix} 2 & -1 & -4 \end{bmatrix} \quad \text{and} \quad \begin{bmatrix} 1 & 0 & -1 & 5 \end{bmatrix}$$

are row matrices, whereas

$$\begin{bmatrix} 3 \\ 4 \\ -1 \end{bmatrix} \quad \text{and} \quad \begin{bmatrix} 1 \\ 0 \\ -1 \\ 5 \end{bmatrix}$$

are column matrices.

Suppose that R is a row matrix, C is a column matrix, and the number of columns in R is the same as the number of rows in C. Then the **product** of R and C is a matrix formed by multiplying the corresponding entries of the two matrices and adding the resulting products.

For instance, if

$$R = \begin{bmatrix} 2 & -1 & 3 & 4 \end{bmatrix} \quad \text{and} \quad C = \begin{bmatrix} 5 \\ 2 \\ -3 \\ 1 \end{bmatrix}$$

then

$$\begin{aligned} RC &= \begin{bmatrix} 2 & -1 & 3 & 4 \end{bmatrix} \begin{bmatrix} 5 \\ 2 \\ -3 \\ 1 \end{bmatrix} \\ &= 2(5) + (-1)(2) + 3(-3) + 4(1) \\ &= [3]. \end{aligned}$$

Unlike addition and subtraction of matrices, the product of two matrices cannot be determined by multiplying corresponding elements.

Note that in order to multiply two matrices they do not have to be of the same size. However, in order to form the product AB of two matrices A and B, the number of columns of A must be the same as the number of rows of B. The product AB will have as many rows as A and as many columns as B. That is:

If A is an $m \times k$ matrix and B is a $k \times n$ matrix, then the product AB is an $m \times n$ matrix.

The following diagram illustrates these restrictions:

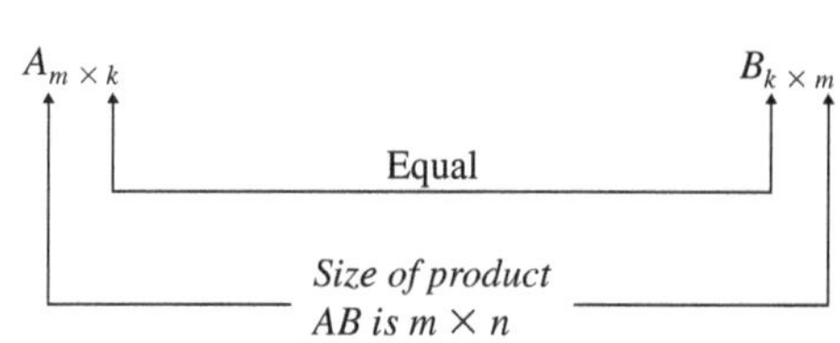

The entry in the ith row and jth column of the product AB is found by multiplying the ith row of a matrix A and the jth column of matrix B, as we did for the row and column matrices. For example, to find the product AB if

$$A = \begin{bmatrix} 1 & 3 & 2 \\ -2 & 4 & 5 \end{bmatrix} \quad \text{and} \quad B = \begin{bmatrix} -4 & 7 \\ 3 & -5 \\ 1 & 6 \end{bmatrix}.$$

we first observe that

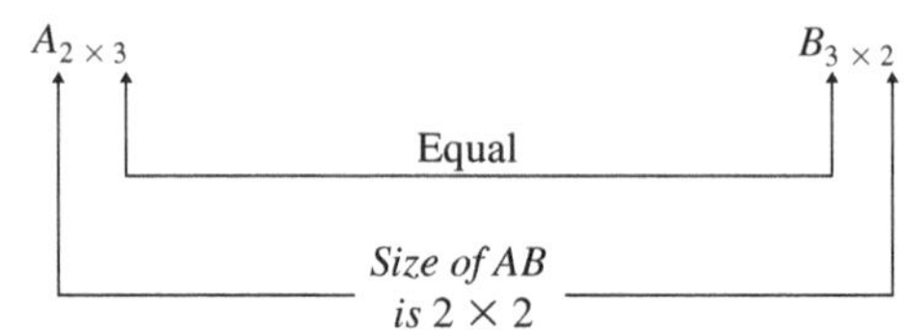

The product AB is determined as follows:

Row 1 in A times column 1 in B ↓   Row 1 in A times column 2 in B ↓

$$AB = \begin{bmatrix} 1 & 3 & 2 \\ -2 & 4 & 5 \end{bmatrix}\begin{bmatrix} -4 & 7 \\ 3 & -5 \\ 1 & 6 \end{bmatrix} = \begin{bmatrix} [1 \quad 3 \quad 2]\begin{bmatrix} -4 \\ 3 \\ 1 \end{bmatrix} & [1 \quad 3 \quad 2]\begin{bmatrix} 7 \\ -5 \\ 6 \end{bmatrix} \\ [-2 \quad 4 \quad 5]\begin{bmatrix} -4 \\ 3 \\ 1 \end{bmatrix} & [-2 \quad 4 \quad 5]\begin{bmatrix} 7 \\ -5 \\ 6 \end{bmatrix} \end{bmatrix}$$

↑ Row 2 in A times column 1 in B ↑ Row 2 in A times column 2 in B

$$= \begin{bmatrix} 1(-4) + 3(3) + 2(1) & 1(7) + 3(-5) + 2(6) \\ -2(-4) + 4(3) + 5(1) & -2(7) + 4(-5) + 5(6) \end{bmatrix}$$

$$= \begin{bmatrix} 7 & 4 \\ 25 & -4 \end{bmatrix}$$

EXAMPLE 3 Multiplying Matrices

Find AB if $A = \begin{bmatrix} 3 & -2 \\ 1 & 5 \\ 4 & 6 \end{bmatrix}$ and $B = \begin{bmatrix} 7 & -2 \\ 3 & 0 \end{bmatrix}$.

Solution Since A is a 3×2 matrix and B is a 2×2 matrix, then AB is a 3×2 matrix. The product AB is given by

$$AB = \begin{bmatrix} 3 & -2 \\ 1 & 5 \\ 4 & 6 \end{bmatrix} \begin{bmatrix} 7 & -2 \\ 3 & 0 \end{bmatrix}$$

$$= \begin{bmatrix} 3(7) + (-2)(3) & 3(-2) + (-2)(0) \\ 1(7) + 5(3) & 1(-2) + 5(0) \\ 4(7) + 6(3) & 4(-2) + 6(0) \end{bmatrix}$$

$$= \begin{bmatrix} 15 & -6 \\ 22 & -2 \\ 46 & -8 \end{bmatrix}$$

The product AB may be defined even when BA is not. For instance, the product BA is not defined in Example 3 because the number of rows of A is not the same as the number of columns of B.

Matrix multiplication is *not* commutative; that is, it is *not* generally true that matrix product AB is the same as matrix product BA. For instance,

$$AB = \begin{bmatrix} 2 & 1 \\ 4 & 2 \end{bmatrix} \begin{bmatrix} -1 & 2 \\ 3 & 0 \end{bmatrix} = \begin{bmatrix} 1 & 4 \\ 2 & 8 \end{bmatrix}$$

whereas

$$BA = \begin{bmatrix} -1 & 2 \\ 3 & 0 \end{bmatrix} \begin{bmatrix} 2 & 1 \\ 4 & 2 \end{bmatrix} = \begin{bmatrix} 6 & 3 \\ 6 & 3 \end{bmatrix}.$$

Although matrix multiplication is not commutative, matrix multiplication obeys the associative and distributive properties. That is, if A, B, and C are matrices where multiplication is defined, and if k is a constant real number, then:

1. **Associative Properties:**
   (a) $A(BC) = (AB)C$
   (b) $(kA)B = k(AB)$
2. **Distributive Properties:**
   (a) $A(B + C) = AB + AC$
   (b) $(B + C)A = BA + CA$

Finding the Inverse of a Square Matrix

The 2×2 matrix

$$\begin{bmatrix} 1 & 0 \\ 0 & 1 \end{bmatrix}$$

and the 3×3 matrix

$$\begin{bmatrix} 1 & 0 & 0 \\ 0 & 1 & 0 \\ 0 & 0 & 1 \end{bmatrix}$$

are referred to as **identity matrices.** If I is an identity matrix and A is an $n \times n$ matrix, so that AI and IA are defined, then

$$AI = IA = A.$$

For example,

$$\begin{bmatrix} 1 & -3 \\ 2 & 4 \end{bmatrix}\begin{bmatrix} 1 & 0 \\ 0 & 1 \end{bmatrix} = \begin{bmatrix} 1 & -3 \\ 2 & 4 \end{bmatrix} \quad \text{and} \quad \begin{bmatrix} 1 & 0 \\ 0 & 1 \end{bmatrix}\begin{bmatrix} 1 & -3 \\ 2 & 4 \end{bmatrix} = \begin{bmatrix} 1 & -3 \\ 2 & 4 \end{bmatrix}.$$

An $n \times n$ matrix A is said to be **invertible** if there is an $n \times n$ matrix B such that

$$AB = BA = I.$$

The matrix B is called the **inverse matrix** of A and is denoted by A^{-1}; that is,

$$AA^{-1} = A^{-1}A = I.$$

EXAMPLE 4

Verifying That One Matrix Is an Inverse of Another

Show that the matrix

$$B = \begin{bmatrix} -5 & 2 \\ 3 & -1 \end{bmatrix}$$

is the inverse of the matrix

$$A = \begin{bmatrix} 1 & 2 \\ 3 & 5 \end{bmatrix}.$$

Solution Since

$$AB = \begin{bmatrix} 1 & 2 \\ 3 & 5 \end{bmatrix}\begin{bmatrix} -5 & 2 \\ 3 & -1 \end{bmatrix} = \begin{bmatrix} 1 & 0 \\ 0 & 1 \end{bmatrix}$$

and

$$BA = \begin{bmatrix} -5 & 2 \\ 3 & -1 \end{bmatrix}\begin{bmatrix} 1 & 2 \\ 3 & 5 \end{bmatrix} = \begin{bmatrix} 1 & 0 \\ 0 & 1 \end{bmatrix}$$

it follows that $AB = I = BA$, so B is indeed the inverse of A; that is, $B = A^{-1}$. Similarly, $A = B^{-1}$.

We know from algebra that every nonzero real number a has a multiplicative inverse $a^{-1} = 1/a$ such that

$$aa^{-1} = a^{-1}a = 1.$$

The analogous statement for matrix multiplication does *not* always hold, that is, not every square nonzero matrix has an inverse.

For instance, consider the matrix

$$A = \begin{bmatrix} 1 & 1 \\ 1 & 1 \end{bmatrix}.$$

If A has an inverse, say

$$B = \begin{bmatrix} a & b \\ c & d \end{bmatrix}$$

then the product AB would have to equal the 2×2 identity matrix. However,

$$AB = \begin{bmatrix} 1 & 1 \\ 1 & 1 \end{bmatrix}\begin{bmatrix} a & b \\ c & d \end{bmatrix} = \begin{bmatrix} a+c & b+d \\ a+c & b+d \end{bmatrix}.$$

Clearly,

$$\begin{bmatrix} a+c & b+d \\ a+c & b+d \end{bmatrix} \quad \text{cannot equal} \quad \begin{bmatrix} 1 & 0 \\ 0 & 1 \end{bmatrix}$$

since it is impossible to have both $a + c = 1$ and $a + c = 0$. Thus A cannot have an inverse.

Elementary row operations can be used to find the inverse of a square matrix if it exists. For example, to find the inverse of the matrix

$$A = \begin{bmatrix} 3 & 5 \\ 1 & 2 \end{bmatrix}$$

we need to find a matrix

$$B = \begin{bmatrix} x & u \\ y & v \end{bmatrix}$$

with entries x, y, u, and v so that

$$AB = \begin{bmatrix} 3 & 5 \\ 1 & 2 \end{bmatrix}\begin{bmatrix} x & u \\ y & v \end{bmatrix} = \begin{bmatrix} 1 & 0 \\ 0 & 1 \end{bmatrix}.$$

Multiplying the left side, we get

$$\begin{bmatrix} 3x+5y & 3u+5v \\ x+2y & u+2v \end{bmatrix} = \begin{bmatrix} 1 & 0 \\ 0 & 1 \end{bmatrix}.$$

Equating the corresponding entries of the two matrices, we have

$$\begin{cases} 3x + 5y = 1 \\ x + 2y = 0 \end{cases} \quad \text{and} \quad \begin{cases} 3u + 5v = 0 \\ u + 2v = 1. \end{cases}$$

Now we can use augmented matrices and row-reduction to solve these two systems. Notice that the coefficient matrices of both systems are the same. Therefore, we can write the matrix so that the first three columns form the augmented matrix of the first system, and the first two columns and last column form the augmented matrix of the second system. That is, we write the matrix

$$\left[\begin{array}{cc|cc} 3 & 5 & 1 & 0 \\ 1 & 2 & 0 & 1 \end{array}\right].$$

Next we perform elementary row operations to express this matrix in a row-reduced echelon form:

$$\left[\begin{array}{cc|cc} 3 & 5 & 1 & 0 \\ 1 & 2 & 0 & 1 \end{array}\right]$$

$$\xrightarrow{R_1 \leftrightarrow R_2} \left[\begin{array}{cc|cc} 1 & 2 & 0 & 1 \\ 3 & 5 & 1 & 0 \end{array}\right]$$

$$\xrightarrow{-3R_1 + R_2 \to R_2} \left[\begin{array}{cc|cc} 1 & 2 & 0 & 1 \\ 0 & -1 & 1 & -3 \end{array}\right]$$

$$\xrightarrow{(-1)R_2 \to R_2} \left[\begin{array}{cc|cc} 1 & 2 & 0 & 1 \\ 0 & 1 & -1 & 3 \end{array}\right]$$

$$\xrightarrow{-2R_2 + R_1 \to R_1} \left[\begin{array}{cc|cc} 1 & 0 & 2 & -5 \\ 0 & 1 & -1 & 3 \end{array}\right]$$

From the first three columns of the last matrix, we see that $x = 2$ and $y = -1$. Similarly, from the first two columns and the last column, it follows that $u = -5$ and $v = 3$. Therefore,

$$B = A^{-1} = \begin{bmatrix} 2 & -5 \\ -1 & 3 \end{bmatrix}$$

Notice that A^{-1} is the right half of the augmented matrix we found above, and that the left half is the identity matrix I.

This illustration leads us to the following general procedure for finding the inverse of a square matrix A:

Procedure
Finding the Inverse of an $n \times n$ Matrix A

1. Form the $n \times 2n$ matrix.

$$[A \vdots I]$$

$n \times n$ matrix ↑ (under A) ↑ $n \times n$ identity matrix (under I)

2. Perform elementary row operations on this matrix to transform it, if possible, to a matrix of the form.

$$[I \vdots B]$$

$n \times n$ identity matrix ↑ (under I)

3. The matrix B is A^{-1}; that is, $A^{-1} = B$.

If the above procedure does not lead to the identity matrix on the left, then the matrix A has no inverse and we say A^{-1} does not exist.

EXAMPLE 5 **Finding the Inverse of a 3 × 3 Matrix**

Find A^{-1} where $A = \begin{bmatrix} 1 & 2 & 3 \\ 1 & 1 & 2 \\ 0 & 1 & 2 \end{bmatrix}$.

Solution We begin by forming the matrix $[A \vdots I]$:

$$\left[\begin{array}{ccc|ccc} 1 & 2 & 3 & 1 & 0 & 0 \\ 1 & 1 & 2 & 0 & 1 & 0 \\ 0 & 1 & 2 & 0 & 0 & 1 \end{array}\right]$$

Next we perform elementary row operations on the entire matrix until the left half is transformed into the identity matrix:

$$\left[\begin{array}{ccc|ccc} 1 & 2 & 3 & 1 & 0 & 0 \\ 1 & 1 & 2 & 0 & 1 & 0 \\ 0 & 1 & 2 & 0 & 0 & 1 \end{array}\right] \xrightarrow{-R_1 + R_2 \to R_2} \left[\begin{array}{ccc|ccc} 1 & 2 & 3 & 1 & 0 & 0 \\ 0 & -1 & -1 & -1 & 1 & 0 \\ 0 & 1 & 2 & 0 & 0 & 1 \end{array}\right]$$

$$\xrightarrow{R_2 + R_3 \to R_3} \left[\begin{array}{ccc|ccc} 1 & 2 & 3 & 1 & 0 & 0 \\ 0 & -1 & -1 & -1 & 1 & 0 \\ 0 & 0 & 1 & -1 & 1 & 1 \end{array}\right]$$

$$\xrightarrow{2R_2 + R_1 \rightarrow R_1} \left[\begin{array}{ccc:ccc} 1 & 0 & 1 & -1 & 2 & 0 \\ 0 & -1 & -1 & -1 & 1 & 0 \\ 0 & 0 & 1 & -1 & 1 & 1 \end{array}\right]$$

$$\xrightarrow{(-1)R_2 \rightarrow R_2} \left[\begin{array}{ccc:ccc} 1 & 0 & 1 & -1 & 2 & 0 \\ 0 & 1 & 1 & 1 & -1 & 0 \\ 0 & 0 & 1 & -1 & 1 & 1 \end{array}\right]$$

$$\xrightarrow{(-1)R_3 + R_2 \rightarrow R_2} \left[\begin{array}{ccc:ccc} 1 & 0 & 1 & -1 & 2 & 0 \\ 0 & 1 & 0 & 2 & -2 & -1 \\ 0 & 0 & 1 & -1 & 1 & 1 \end{array}\right]$$

$$\xrightarrow{(-1)R_3 + R_1 \rightarrow R_1} \left[\begin{array}{ccc:ccc} 1 & 0 & 0 & 0 & 1 & -1 \\ 0 & 1 & 0 & 2 & -2 & -1 \\ 0 & 0 & 1 & -1 & 1 & 1 \end{array}\right]$$

This result can be confirmed by verifying that $AA^{-1} = A^{-1}A = I$.

Therefore,
$$A^{-1} = \begin{bmatrix} 0 & 1 & -1 \\ 2 & -2 & -1 \\ -1 & 1 & 1 \end{bmatrix}.$$

Solving Linear Systems Using Inverse Matrices

Inverse matrices can also be used to solve linear systems of equations when the system has the same number of equations as variables—provided that the coefficient matrix has an inverse. Let's consider a linear system consisting of two equations in two variables x and y:

$$\begin{cases} a_1x + a_2y = k_1 \\ b_1x + b_2y = k_2 \end{cases}$$

We can express this system as a product of matrices as follows: Let A be the 2×2 coefficient matrix of the system of equations, let K be the column matrix of the constants on the right side, and let X be the column matrix of unknowns. That is, let

$$A = \begin{bmatrix} a_1 & a_2 \\ b_1 & b_2 \end{bmatrix}, \quad X = \begin{bmatrix} x \\ y \end{bmatrix}, \quad \text{and} \quad K = \begin{bmatrix} k_1 \\ k_2 \end{bmatrix}.$$

Then AX is the matrix with 2 rows and 1 column given by

$$AX = \begin{bmatrix} a_1 & a_2 \\ b_1 & b_2 \end{bmatrix}\begin{bmatrix} x \\ y \end{bmatrix} = \begin{bmatrix} a_1x + a_2y \\ b_1x + b_2y \end{bmatrix}.$$

Since $a_1x + a_2y = k_1$ and $b_1x + b_2y = k_2$, it follows that $AX = K$. The latter equation is called the **matrix equation form** of the system. To solve this equation, we multiply both sides *on the left* by A^{-1}, assuming it exists, to isolate X on the left side:

Do not change the order of multiplying A^{-1} on each side of the equation $AX = K$, since $A^{-1}(AX)$ and KA^{-1} are not necessarily the same.

$$AX = K$$
$$A^{-1}(AX) = A^{-1}K$$
$$(A^{-1}A)X = A^{-1}K$$
$$IX = A^{-1}K$$

so

$$X = A^{-1}K.$$

The entries in the matrix product $A^{-1}K$ make up the solution of the original system.

EXAMPLE 6 **Solving a 2 × 2 Linear System by Using the Inverse of a Matrix**

Solve the system by using the inverse of the coefficient matrix.

$$\begin{cases} 3x + 5y = 11 \\ x + 2y = 4 \end{cases}$$

Solution The given system of equations is equivalent to the matrix equation $AX = K$, where

$$A = \begin{bmatrix} 3 & 5 \\ 1 & 2 \end{bmatrix}, \quad X = \begin{bmatrix} x \\ y \end{bmatrix}, \quad \text{and} \quad K = \begin{bmatrix} 11 \\ 4 \end{bmatrix}.$$

From the illustration on pages 337 and 338, we know that A^{-1} exists and that

$$A^{-1} = \begin{bmatrix} 2 & -5 \\ -1 & 3 \end{bmatrix}.$$

It follows that the solution X is given by

$$\begin{aligned} X &= A^{-1}K \\ &= \begin{bmatrix} 2 & -5 \\ -1 & 3 \end{bmatrix}\begin{bmatrix} 11 \\ 4 \end{bmatrix} = \begin{bmatrix} 2 \\ 1 \end{bmatrix}. \end{aligned}$$

Therefore, $x = 2$ and $y = 1$, and the solution is $(2, 1)$.

EXAMPLE 7 **Solving a 3 × 3 Linear System by Using the Inverse of a Matrix**

Solve the system by using the inverse of the coefficient matrix.

$$\begin{cases} x + 2y + 3z = 4 \\ x + y + 2z = 5 \\ \quad\;\; y + 2z = 4 \end{cases}$$

Solution If we let

$$A = \begin{bmatrix} 1 & 2 & 3 \\ 1 & 1 & 2 \\ 0 & 1 & 2 \end{bmatrix}, \quad X = \begin{bmatrix} x \\ y \\ z \end{bmatrix}, \quad \text{and} \quad K = \begin{bmatrix} 4 \\ 5 \\ 4 \end{bmatrix}$$

we can write the system as $AX = K$. From Example 5, we know that

$$A^{-1} = \begin{bmatrix} 0 & 1 & -1 \\ 2 & -2 & -1 \\ -1 & 1 & 1 \end{bmatrix} \text{ so that } X = A^{-1}K = \begin{bmatrix} 0 & 1 & -1 \\ 2 & -2 & -1 \\ -1 & 1 & 1 \end{bmatrix}\begin{bmatrix} 4 \\ 5 \\ 4 \end{bmatrix} = \begin{bmatrix} 1 \\ -6 \\ 5 \end{bmatrix}.$$

Thus $x = 1$, $y = -6$, and $z = 5$, and the solution is $(1, -6, 5)$.

G Most graphers can perform matrix operations, including matrix multiplication and matrix inversion. For example, to solve the following system by using a grapher

$$\begin{cases} 0.1s + 0.2t + 0.1u = 27 \\ 0.5s + 0.2t + 0.3u = 9 \\ 0.4s + 0.8t + 0.1u = 36 \end{cases}$$

we first find the coefficient matrix A and the matrix equation

$$A = \begin{bmatrix} 0.1 & 0.2 & 0.1 \\ 0.5 & 0.2 & 0.3 \\ 0.4 & 0.8 & 0.1 \end{bmatrix} \quad \text{and} \quad AX = K,$$

$$\text{where} \quad X = \begin{bmatrix} s \\ t \\ u \end{bmatrix} \quad \text{and} \quad K = \begin{bmatrix} 27 \\ 9 \\ 36 \end{bmatrix},$$

Here we took advantage of the grapher's numerical conversion feature to display the entries of A^{-1} in fraction form rather than in decimal form.

then we input matrix A and find

$$A^{-1} = \begin{bmatrix} -\frac{55}{6} & \frac{5}{2} & \frac{5}{3} \\ \frac{35}{12} & -\frac{5}{4} & \frac{5}{6} \\ \frac{40}{3} & 0 & \frac{-10}{3} \end{bmatrix}$$

Next we use matrix multplication to obtain

$$X = A^{-1}K = \begin{bmatrix} -165 \\ 97.5 \\ 240 \end{bmatrix}.$$

So $\quad s = -165, \quad t = 97.5 \quad \text{and} \quad u = 240.$

Solving Applied Problems

The next example illustrates a way in which matrix multiplication is used to interpret data.

EXAMPLE 8

Solving a Business Model

A fruit market packages fruit in three different ways for gift baskets. The economy basket E has 5 apples, 3 grapefruits, and 4 pears. The standard basket S has 4 apples, 4 grapefruits, and 5 pears. The luxury basket L has 6 apples, 5 grapefruits, and 5 pears. The cost is \$0.40, \$0.70 and \$0.50 for each apple, grapefruit, and pear, respectively. What is the cost of preparing each basket of fruit?

Solution We arrange the given data in matrix form as follows:

Cost per fruit

$$A = \begin{array}{c} \\ \left[\right. \end{array}\begin{array}{ccc} \text{Apple} & \text{Grapefruit} & \text{Pear} \\ \$0.40 & \$0.70 & \$0.50 \end{array}\begin{array}{c} \\ \left. \right] \end{array}$$

Number of items in each basket type

$$B = \begin{array}{l} \text{Apples} \\ \text{Grapefruits} \\ \text{Pears} \end{array} \overset{\begin{array}{ccc} E & S & L \end{array}}{\begin{bmatrix} 5 & 4 & 6 \\ 3 & 4 & 5 \\ 4 & 5 & 5 \end{bmatrix}}$$

To find the cost of each basket, multiply matrix A by matrix B to get

$$AB = [0.40 \quad 0.70 \quad 0.50]\begin{bmatrix} 5 & 4 & 6 \\ 3 & 4 & 5 \\ 4 & 5 & 5 \end{bmatrix}$$

$$= [6.10 \quad 6.90 \quad 8.40].$$

Therefore, the cost of each economy, standard, and luxury basket are

$6.10, $6.90, and $8.40, respectively.

PROBLEM SET 4.3

Mastering the Concepts

In problems 1–6, perform each operation on the given matrices A and B

(a) $A + B$ (b) $A - B$ (c) $4A$
(d) $-5B$ (e) $4A - 5B$

1. $A = \begin{bmatrix} 3 & 5 \\ -1 & 4 \end{bmatrix}$; $B = \begin{bmatrix} 5 & 6 \\ -1 & 3 \end{bmatrix}$

2. $A = \begin{bmatrix} -1 & 2 \\ 3 & 0 \end{bmatrix}$; $B = \begin{bmatrix} 2 & -1 \\ 3 & -4 \end{bmatrix}$

3. $A = \begin{bmatrix} 3 & 1 & 2 \\ -1 & 3 & 1 \end{bmatrix}$; $B = \begin{bmatrix} 5 & 11 & 6 \\ 3 & 0 & -1 \end{bmatrix}$

4. $A = \begin{bmatrix} 0 & 1 & 4 & 2 \\ -5 & 6 & 1 & 3 \end{bmatrix}$; $B = \begin{bmatrix} 6 & -1 & 3 & 1 \\ -2 & 0 & 1 & 4 \end{bmatrix}$

5. $A = \begin{bmatrix} -5 & 2 & 1 \\ -2 & 1 & 2 \\ 3 & -1 & 2 \end{bmatrix}$; $B = \begin{bmatrix} -4 & 6 & 2 \\ -1 & 2 & 5 \\ 3 & -1 & 2 \end{bmatrix}$

6. $A = \begin{bmatrix} 1 & 0 & 0 & 1 & -1 \\ 2 & 1 & 3 & 5 & 7 \\ 3 & 6 & 2 & -1 & 4 \end{bmatrix}$;

$B = \begin{bmatrix} 6 & 1 & -1 & 3 & 5 \\ 7 & 1 & 2 & 8 & -1 \\ 0 & 1 & 2 & 0 & 1 \end{bmatrix}$

In problems 7–12, if possible, find:

(a) AB (b) BA

7. $A = \begin{bmatrix} 1 & 2 \\ -1 & 1 \end{bmatrix}$; $B = \begin{bmatrix} 5 & -1 \\ 7 & 0 \end{bmatrix}$

8. $A = \begin{bmatrix} 1 & -1 \\ 0 & 2 \end{bmatrix}$; $B = \begin{bmatrix} 3 & 1 & -1 \\ -2 & 0 & 1 \end{bmatrix}$

9. $A = \begin{bmatrix} 2 & -3 & 5 \\ -1 & 1 & 3 \end{bmatrix}$; $B = \begin{bmatrix} -3 & 1 \\ 1 & 2 \\ 0 & -5 \end{bmatrix}$

10. $A = \begin{bmatrix} 3 & 2 & 1 \\ 1 & 4 & -1 \\ 2 & 1 & -3 \end{bmatrix}$; $B = \begin{bmatrix} -3 & 0 & 0 \\ 0 & -3 & 0 \\ 0 & 0 & -3 \end{bmatrix}$

11. $A = \begin{bmatrix} 4 & 3 & -1 \\ 8 & -2 & 3 \\ 6 & 5 & 2 \end{bmatrix}$; $B = \begin{bmatrix} 1 & 0 & 0 \\ 0 & 1 & 0 \\ 0 & 0 & 1 \end{bmatrix}$

12. $A = \begin{bmatrix} -1 & 1 & 2 & 0 \\ -2 & 3 & 5 & 1 \\ 2 & -1 & 3 & 2 \end{bmatrix}$; $B = \begin{bmatrix} 1 & -3 & 4 \\ 2 & 1 & 0 \\ -1 & 0 & 2 \\ 0 & -1 & 3 \end{bmatrix}$

In problems 13–18, determine whether A and B are inverses of each other.

13. $A = \begin{bmatrix} 1 & -1 \\ 2 & 1 \end{bmatrix}$; $B = \begin{bmatrix} \frac{1}{3} & \frac{1}{3} \\ -\frac{2}{3} & \frac{1}{3} \end{bmatrix}$

14. $A = \begin{bmatrix} 2 & 3 \\ 5 & 1 \end{bmatrix}$; $B = \begin{bmatrix} -\frac{1}{13} & \frac{3}{13} \\ \frac{5}{13} & -\frac{2}{13} \end{bmatrix}$

15. $A = \begin{bmatrix} 3 & 2 \\ 2 & -3 \end{bmatrix}$; $B = \begin{bmatrix} \frac{3}{13} & \frac{2}{13} \\ \frac{2}{13} & -\frac{3}{13} \end{bmatrix}$

16. $A = \begin{bmatrix} 1 & 1 & 1 \\ 2 & -1 & -1 \\ 1 & -1 & 2 \end{bmatrix}$; $B = \begin{bmatrix} \frac{1}{3} & \frac{1}{3} & 0 \\ \frac{5}{9} & -\frac{1}{9} & -\frac{1}{3} \\ \frac{1}{9} & -\frac{2}{9} & \frac{1}{3} \end{bmatrix}$

17. $A = \begin{bmatrix} 1 & 2 & 4 \\ 2 & -3 & 1 \\ 3 & -1 & -2 \end{bmatrix}$; $B = \begin{bmatrix} \frac{1}{7} & 0 & \frac{2}{7} \\ \frac{1}{7} & -\frac{2}{7} & \frac{1}{7} \\ \frac{1}{7} & \frac{1}{7} & -\frac{1}{7} \end{bmatrix}$

18. $A = \begin{bmatrix} 1 & 1 & 1 \\ 2 & 3 & -1 \\ 3 & 5 & 1 \end{bmatrix}$; $B = \begin{bmatrix} 2 & 1 & -1 \\ -\frac{5}{4} & -\frac{1}{2} & \frac{3}{4} \\ \frac{1}{4} & -\frac{1}{2} & \frac{1}{4} \end{bmatrix}$

In problems 19–26, use elementary row operations to find the inverse of the given matrix, if it exists.

19. $\begin{bmatrix} 3 & 5 \\ 1 & 2 \end{bmatrix}$

20. $\begin{bmatrix} 2 & -5 \\ -1 & 3 \end{bmatrix}$

21. $\begin{bmatrix} -1 & 1 \\ 1 & 0 \end{bmatrix}$

22. $\begin{bmatrix} 1 & 2 \\ \frac{1}{4} & \frac{3}{4} \end{bmatrix}$

23. $\begin{bmatrix} 1 & 2 & 0 \\ 0 & 2 & 3 \\ 1 & 3 & 1 \end{bmatrix}$

24. $\begin{bmatrix} 2 & 7 & 1 \\ 1 & 4 & -1 \\ 1 & 3 & 0 \end{bmatrix}$

25. $\begin{bmatrix} 7 & 2 & -6 \\ -3 & -1 & 3 \\ 2 & 1 & -2 \end{bmatrix}$

26. $\begin{bmatrix} -1 & 0 & 2 \\ 3 & 1 & 0 \\ 0 & 2 & -3 \end{bmatrix}$

In problems 27–36, use an inverse matrix to solve each system.

27. $\begin{cases} x - y = 1 \\ 2x + y = 5 \end{cases}$ (see problem 13)

28. $\begin{cases} 2x_1 + 3x_2 = 7 \\ 5x_1 + x_2 = -2 \end{cases}$ (see problem 14)

29. $\begin{cases} 3x + 2y = 8 \\ 2x - 3y = 14 \end{cases}$ (see problem 15)

30. $\begin{cases} 3x + 5y = 13 \\ x + 2y = 5 \end{cases}$

31. $\begin{cases} 2x - 5y = -3 \\ -x + 3y = 2 \end{cases}$

32. $\begin{cases} 4x + y = 17 \\ 3x + 2y = 17 \end{cases}$

33. $\begin{cases} x + y + z = 2 \\ 2x + 3y - z = 3 \\ 3x + 5y + z = 8 \end{cases}$

34. $\begin{cases} x_1 + 2x_2 + 4x_3 = 12 \\ 2x_1 - 3x_2 + x_3 = 10 \\ 3x_1 - x_2 - 2x_3 = 1 \end{cases}$

35. $\begin{cases} x + y + z = 6 \\ 2x - y - z = 0 \\ x - y + 2z = 7 \end{cases}$

36. $\begin{cases} 2x + 7y + z = 10 \\ x + 4y - z = 4 \\ x + 3y = 4 \end{cases}$

In problems 37–40, write each system in the form $AX = K$. Then use a grapher to find A^{-1} and $X = A^{-1}K$, the solution of the system. In problems 38, 39, and 40, round off the entries of A^{-1} to two decimal places.

37. $\begin{cases} x + 3y + 2z + w = 4 \\ x + 2y + 4z = -3 \\ 2x + 6y + 5z + 2w = -1 \\ x + 3y + 2z + 2w = 2 \end{cases}$

38. $\begin{cases} x + 2y + z - w = 3 \\ 2x + 4y + z + w = 2 \\ 3x + 2y - z - 2w = -4 \\ 2x + 5y + 3z = 7 \end{cases}$

39. $\begin{cases} 0.2x + 0.4y - 0.6z = 0.2 \\ -0.2x + 0.1y + 0.3z + 0.2w = 0.7 \\ 0.2x + 0.9y - 0.9z + 0.1w = 1.1 \\ 0.4x + 0.3y - 0.3z = 0.7 \end{cases}$

40. $\begin{cases} 0.3x + 0.1y - 0.1z + 0.1w = 0 \\ 0.6x + 0.4y - 0.1z + 0.4w = 1.5 \\ -0.3x - 0.1y + 0.2z - 0.1w = 1 \\ 0.3x + 0.5y + 0.3w = 2.1 \end{cases}$

Applying the Concepts

41. Crop Shipments: A nursery raises two kinds of flowers that are shipped in bundles to three different florists. The number of bundles of flower k that are shipped to store m is given by entry a_{km} in the matrix

$$A = \begin{array}{c} \text{Store 1} \quad \text{Store 2} \quad \text{Store 3} \\ \begin{bmatrix} 200 & 85 & 80 \\ 250 & 170 & 100 \end{bmatrix} \end{array} \begin{array}{l} \text{Flower 1} \\ \text{Flower 2} \end{array}$$

The profit (in dollars) per bundle is represented by the matrix.

$$T = \begin{array}{c} \text{Flower 1} \quad \text{Flower 2} \\ \begin{bmatrix} 4.50 & 5.70 \end{bmatrix} \end{array}$$

(a) Find the product TA.

(b) Indicate what each entry of the product represents.

42. Inventory Levels: A store sells two brands, B_1 and B_2, of microwave ovens. The following matrices give the sales figures and costs for three months. Use matrix multiplication to determine the total dollar sales and the total dollar costs of these items for these three months.

	May	July	October
Number of brand B_1 ovens sold	16	10	14
Number of brand B_2 ovens sold	20	15	16

	Brand B_1	Brand B_2
Retail price	210	140
Dealer cost	150	90

43. Total Revenue: An automobile manufacturer produces three different new models of cars, A, B, and C, that are shipped to two different locations for distribution. The following matrices give the number of cars that are shipped as well as the price in dollars per car.

Number shipped

$$\begin{array}{l} \text{Model } A \\ \text{Model } B \\ \text{Model } C \end{array} \begin{array}{c} \text{Site 1} \quad \text{Site 2} \\ \begin{bmatrix} 6{,}000 & 7{,}000 \\ 9{,}000 & 10{,}000 \\ 4{,}000 & 8{,}000 \end{bmatrix} \end{array} = S$$

Price

$$\begin{array}{c} \text{Model } A \quad \text{Model } B \quad \text{Model } C \\ \begin{bmatrix} 10{,}000 & 12{,}000 & 14{,}000 \end{bmatrix} \end{array} = P$$

(a) Find the product PS.

(b) Indicate what each entry of the product represents.

44. Manufacturing: A manufacturer of calculators has two plants, each producing scientific calculators and graphing calculators. The manufacturing time requirements (in hours per calculator) and the assembly and packing costs (in dollars per hour) are given by the following matrices:

Manufacturing time (hours per unit)

$$A = \begin{array}{c} \text{Assembly} \quad \text{Packaging} \\ \begin{bmatrix} 0.2 & 0.1 \\ 0.3 & 0.1 \end{bmatrix} \end{array} \begin{array}{l} \text{Scientific calculator} \\ \text{Graphing calculator} \end{array}$$

Costs (dollars per hour)

$$B = \begin{array}{c} \text{First plant} \quad \text{Second plant} \\ \begin{bmatrix} 5 & 6 \\ 4 & 5 \end{bmatrix} \end{array} \begin{array}{l} \text{Assembly} \\ \text{Packaging} \end{array}$$

(a) Compute AB.

(b) Interpret the meaning of each entry in AB.

Developing and Extending the Concepts

45. Let

$$A = \begin{bmatrix} 3 & 2 \\ -1 & 4 \end{bmatrix}, \quad B = \begin{bmatrix} 0 & 1 \\ -1 & 2 \end{bmatrix},$$

and

$$C = \begin{bmatrix} 2 & -3 \\ 1 & -4 \end{bmatrix}.$$

(a) Demonstrate the commutative property:

$$A + B = B + A$$

(b) Demonstrate the associative property:

$$A + (B + C) = (A + B) + C$$

(c) Find $-A$ and $-C$. Then show that

$$A + (-A) = \mathbf{0}$$

and

$$C + (-C) = \mathbf{0},$$

the 2×2 zero matrix.

(d) Find matrix E so that

$$A + E = C.$$

(e) Demonstrate that $A(B - C) = AB - AC$.

46. Find the values of x and y if:

(a) $\begin{bmatrix} 3 & 1 \\ 5 & 7 \end{bmatrix} = \begin{bmatrix} 3 & x \\ -y & 7 \end{bmatrix}$

(b) $\begin{bmatrix} 1 & -1 \\ 2 & 6 \end{bmatrix} + \begin{bmatrix} x & 3 \\ y & 2 \end{bmatrix} = \begin{bmatrix} 8 & 2 \\ 0 & 8 \end{bmatrix}$

47. Find the values of x, y, z, and w if:

$$\begin{bmatrix} x & y \\ z & w \end{bmatrix} \begin{bmatrix} 3 & -5 \\ -1 & 2 \end{bmatrix} = \begin{bmatrix} 1 & -1 \\ 2 & 0 \end{bmatrix}$$

48. Let

$$A = \begin{bmatrix} 3 & -1 \\ 0 & -2 \end{bmatrix}.$$

Show that

$$A^2 - A - 6I = \begin{bmatrix} 0 & 0 \\ 0 & 0 \end{bmatrix}$$

where

$$I = \begin{bmatrix} 1 & 0 \\ 0 & 1 \end{bmatrix}.$$

49. Suppose that A and B are $n \times n$ matrices and

$$AB = BA.$$

(a) Show that

$$(A + B)^2 = A^2 + 2AB + B^2.$$

(b) Show that

$$(A - B)(A + B) = A^2 - B^2.$$

50. If

$$A^{-1} = \begin{bmatrix} 1 & -1 & 3 \\ 2 & 0 & 5 \\ -1 & 1 & 0 \end{bmatrix}$$

find a matrix B such that

$$AB = \begin{bmatrix} 1 & -1 & 2 \\ 0 & 1 & 1 \\ 1 & 0 & 0 \end{bmatrix}.$$

51. If

$$A^{-1} = \begin{bmatrix} 1 & -1 & 3 \\ 2 & 1 & 1 \\ 0 & 2 & -2 \end{bmatrix}$$

find A.

52. Let $A = \begin{bmatrix} 2 & 3 \\ 4 & 5 \end{bmatrix}$ and $B = \begin{bmatrix} 7 & 8 \\ 6 & 7 \end{bmatrix}$.

(a) Compute

$$A^{-1},\ B^{-1}, \text{ and } B^{-1}A^{-1}.$$

(b) Compute

$$(AB)^{-1} \text{ and compare it to } B^{-1}A^{-1}.$$

(c) Can you generalize the result in part (b)? Explain.

Interlude: Practice Problems 4.3, Part I

1. The table below summarizes the closing price per share for Baracus (BA), Hammersmith Labs (HL), and Suncorp (SUN) common stock at the end of the indicated year, as well as the number of shares of the stock owned by Nigel, Bonnie, and Jamie at those times.

$/share	BA	HL	SUN
2011	89.92	47.26	18.55
2012	94.33	46.81	22.14
2013	92.65	49.42	26.77

shares	BA	HL	SUN
Nigel	420	180	300
Bonnie	225	175	450
Jamie	190	360	95

(a) Create a 3×3 matrix P to represent the stock price information, labeling the rows and columns of P appropriately.

(b) Create a 3×3 matrix S to represent the holdings of the three people, labeling the rows and columns of S appropriately.

(c) Show and calculate a matrix product that gives the value of the stock holdings of the three people at the end of the indicated years. Label the rows and columns of the product matrix appropriately.

2. SpecialTees sells specialty T-shirts and sweatshirts, with production plants in Verndale and Menahga. The company offers this apparel in three quality levels: standard, deluxe, and premium. Last fall the Verndale plant produced 2980 standard, 1950 deluxe, and 1220 premium T-shirts, along with 1840 standard, 1020 deluxe, and 760 premium sweatshirts. The Menahga plant produced 3980 standard, 2450 deluxe, and 1330 premium T-shirts, along with 2840 standard, 3650 deluxe, and 990 premium sweatshirts in the same time period.

   (a) Write "production matrices" of the same size for each plant that show how many of each different type of apparel were manufactured. Label your rows and columns appropriately.

   (b) Use the matrices from (a) to determine how many more or less articles of each type were produced in Menahga compared to in Verndale. Be clear about what matrix operation you are using.

   (c) Suppose that both plants expect a 3% increase in business next fall. Using an appropriate matrix operation on the matrices from **(a)**, construct new matrices that show how many items of each type are expected to be produced next fall. Be clear about what matrix operation you are using.

Interlude: Practice Problems 4.3, Part II

1. Let $A = \begin{bmatrix} 1 & 0 & -2 \\ a & b & 4 \\ 2 & 0 & 0 \end{bmatrix}$ and suppose that a and b are appropriate nonzero constants so that A is invertible. Write an appropriate sequence of row operations to calculate the second row of A^{-1} and give that row.

2. Let $A = \begin{bmatrix} 1 & -1 & 3 \\ -1 & 2 & h \\ 2 & 0 & 10 \end{bmatrix}$. Determine all values of h for which A will not be invertible.

3. Suppose that $A = \begin{bmatrix} 1 & 2 \\ 3 & 5 \end{bmatrix}$ and B is a square matrix such that $(AB)^{-1} = \begin{bmatrix} 4 & 5 \\ -3 & -4 \end{bmatrix}$. Find B.

4. Consider this statement: "If A and B are invertible $n \times n$ matrices, then $A + B$ is also invertible." If the statement is true, then explain why. If the statement is false, then give a specific example that shows so, and verify the example.

Interlude: Practice Problems 4.3, Part III

Given a matrix A, the ***transpose of*** A, denoted by A^T, is the matrix whose rows are formed from the columns of A.

1. Let $A = \begin{bmatrix} -2 & 2 \\ 3 & -4 \end{bmatrix}$.

(a) Calculate A^{-1} and A^T.

(b) Calculate $(A^T)^{-1}$ and $(A^{-1})^T$. How are the two related?

2. Let $A = \begin{bmatrix} 2 & -3 & 5 \\ 1 & 0 & -7 \end{bmatrix}$ and let $B = \begin{bmatrix} 8 & 2 \\ -3 & 4 \\ 6 & 9 \end{bmatrix}$.

(a) Calculate A^T and B^T.

(b) Calculate $(AB)^T$.

(c) Calculate $B^T A^T$.

Chapter 4 Readiness, Part I

Provide complete solutions in the space provided.

1. Let $A = \begin{bmatrix} -2 & 3 & -3 \\ 4 & 0 & 4 \\ 1 & 7 & -1 \end{bmatrix}$ and $C = \begin{bmatrix} -25 & 30 \\ 8 & -8 \\ -46 & 58 \end{bmatrix}$. Find a matrix B so that $AB = C$. If you believe no such matrix exists, then explain why.

2. Let $A = \begin{bmatrix} 2 & 3 \\ -1 & 0 \end{bmatrix}$ and $B = \begin{bmatrix} -8 & -12 \\ k & 0 \end{bmatrix}$. What value(s) of k, if any, will make $AB = BA$?

3. If $A = \begin{bmatrix} -1 & 4 \\ 6 & 2 \\ 7 & -3 \end{bmatrix}$ and $AB = \begin{bmatrix} 10 & 37 \\ 70 & 12 \\ 55 & -34 \end{bmatrix}$, then determine B.

4. Let $A = \begin{bmatrix} a & b \\ c & d \end{bmatrix}$ and suppose that $ad - bc \neq 0$. Let $B = \dfrac{1}{ab - bc}\begin{bmatrix} d & -b \\ -c & a \end{bmatrix}$. Show that A and B are inverses of each other.

Chapter 4 Readiness, Part II

Provide complete solutions in the space provided.

1. Consider the system $\begin{cases} -2x + hy = 1 \\ 6x + ky = -2 \end{cases}$.

   (a) Determine all possible values of h and k so that the system has exactly one solution. Show how you arrived at your conclusion.

   (b) Determine all possible values of h and k so that the system has an infinite number of solutions. Show how you arrived at your conclusion.

   (c) Determine all possible values of h and k so that the system has no solutions. Show how you arrived at your conclusion.

2. I intend to fill a 150-gallon tank with a blend of regular gasoline (10% ethanol), E-30, and E-85. I also wish for the filled tank to be 50% ethanol.

   (a) Set up a system of equations that models this problem. Declare what variables represent.

   (b) Solve the system from **(a)**.

   (c) Suppose that 100 gallons of the blend needs to come from some combination of E-30 and E-85. How many gallons of each of the three types of fuel are needed to accomplish this?

CHAPTER 4 REVIEW PROBLEM SET

In problems 1–4, solve each system by elimination.

1. $\begin{cases} x - 2y = 3 \\ x + y = -3 \end{cases}$

2. $\begin{cases} 8r - 7s = 28 \\ 5r + 2s = -8 \end{cases}$

3. $\begin{cases} x - y = 3 \\ 2x + y = 3 \end{cases}$

4. $\begin{cases} x - y + 2z = 0 \\ 3x + y + z = 2 \\ 2x - y + 5z = 5 \end{cases}$

In problems 5–8, each row-reduced echelon matrix is the augmented matrix form of a corresponding system of linear equations. Solve each system.

5. $\left[\begin{array}{cc|c} 1 & 0 & -9 \\ 0 & 1 & 5 \end{array}\right]$

6. $\left[\begin{array}{ccc|c} 1 & 0 & 0 & -2 \\ 0 & 1 & 3 & 7 \\ 0 & 0 & 0 & 0 \end{array}\right]$

7. $\left[\begin{array}{ccc|c} 1 & 0 & 3 & -1 \\ 0 & 1 & 1 & 2 \\ 0 & 0 & 0 & 0 \end{array}\right]$

8. $\left[\begin{array}{ccc|c} 1 & 0 & -4 & 1 \\ 0 & 0 & 0 & 2 \end{array}\right]$

In problems 9–12, use row-reduction to solve each system.

9. $\begin{cases} 4x - y = -4 \\ x + 2y = 6 \end{cases}$

10. $\begin{cases} 2x - 3y = 1 \\ -5x + y = 0 \end{cases}$

11. $\begin{cases} 3x - y + 2z = 1 \\ x - 2y + 4z = 2 \\ x - y + z = 0 \end{cases}$

12. $\begin{cases} 2x - y + z = 11 \\ x + y - z = -2 \\ -3x - 2y = -7 \end{cases}$

In problems 13 and 14, find the partial fraction expansion.

13. $\dfrac{x + 2}{x^2 - 6x - 7} = \dfrac{A}{x - 7} + \dfrac{B}{x + 1}$

14. $\dfrac{3x + 1}{x^2 + x} = \dfrac{A}{x} + \dfrac{B}{x + 1}$

In problems 15–18, use the given matrices to perform the indicated matrix operations, if possible.

15. $A = \begin{bmatrix} 2 & 1 \\ 1 & -1 \end{bmatrix}$; $B = \begin{bmatrix} 5 & -2 \\ 3 & 4 \end{bmatrix}$

(a) $A + 2B$ (b) $-3A + 4B$

16. $A = \begin{bmatrix} 1 & -4 & -8 \\ 5 & 20 & 3 \end{bmatrix}$; $B = \begin{bmatrix} 7 & -2 & 7 \\ 3 & 5 & 3 \end{bmatrix}$

(a) $3A - 2B$ (b) $4A + 2B$

17. $A = \begin{bmatrix} 1 & -2 & 1 \\ 3 & 1 & -2 \\ 0 & 1 & -1 \end{bmatrix}$; $B = \begin{bmatrix} 4 & -1 \\ 3 & 1 \\ 2 & 8 \end{bmatrix}$

(a) AB (b) BA

18. $A = \begin{bmatrix} 1 & 3 \\ -1 & 2 \\ 0 & 1 \end{bmatrix}$; $B = \begin{bmatrix} -2 & 3 & 1 \\ 0 & 2 & -3 \\ 1 & -1 & 4 \end{bmatrix}$

(a) AB (b) BA

In problems 19 and 20, verify that $AA^{-1} = A^{-1}A = I$.

19. $A = \begin{bmatrix} 3 & -2 \\ 8 & -5 \end{bmatrix}$; $A^{-1} = \begin{bmatrix} -5 & 2 \\ -8 & 3 \end{bmatrix}$

20. $A = \begin{bmatrix} 1 & -1 & 1 \\ 0 & 2 & -1 \\ 2 & 3 & 0 \end{bmatrix}$; $A^{-1} = \begin{bmatrix} 3 & 3 & -1 \\ -2 & -2 & 1 \\ -4 & -5 & 2 \end{bmatrix}$

In problems 21 and 22, use the inverse of the matrix of coefficients to solve each system. (Refer to the inverses given in problems 19 and 20.)

21. $\begin{cases} 3x - 2y = -7 \\ 8x - 5y = -18 \end{cases}$

22. $\begin{cases} x - y + z = 7 \\ 2y - z = -7 \\ 2x + 3y = -1 \end{cases}$

In problems 23–26, evaluate each determinant.

23. $\begin{vmatrix} 1 & -1 \\ 3 & 2 \end{vmatrix}$

24. $\begin{vmatrix} -2 & 1 \\ -1 & 7 \end{vmatrix}$

25. $\begin{vmatrix} 2 & 7 & 4 \\ 3 & -1 & 5 \\ 4 & 14 & 8 \end{vmatrix}$

26. $\begin{vmatrix} 2 & -4 & 7 \\ 0 & 1 & 2 \\ 0 & 0 & 5 \end{vmatrix}$

In problems 27–30, use Cramer's rule to solve each system.

27. $\begin{cases} 3x - y = 7 \\ 2x + 3y = 12 \end{cases}$

28. $\begin{cases} 5x + 3y = 13 \\ 7x - 5y = 18 \end{cases}$

29. $\begin{cases} 2x - 3y + z = 10 \\ 5x + y - z = 6 \\ x - y + 2z = 6 \end{cases}$

30. $\begin{cases} 3x - y + 2z = 5 \\ 2x + 3y + z = 1 \\ 5x + y + 4z = 8 \end{cases}$

In problems 31 and 32, sketch the graph of the solution set of each system of inequalities.

31. $\begin{cases} x + 2y > 5 \\ 3x - y < 9 \end{cases}$ **32.** $\begin{cases} x^2 + 2y \geq 0 \\ x + 2y \leq 6 \end{cases}$

In problems 33–36, graph the region defined by each constraint system, locate each corner point, and then find the maximum and minimum values of the given linear expression over the region.

33. $F = 7x + 3y$

$\begin{cases} x \geq 0, y \geq 0 \\ 7x + 2y \leq 14 \end{cases}$

34. $F = 2x + y$

$\begin{cases} x \geq 1 \\ y \geq 2 \\ x + 2y \leq 10 \end{cases}$

35. $F = x + 5y$

$\begin{cases} x \geq 0, y \geq 0 \\ x + y \leq 4 \\ 4x + y \leq 7 \end{cases}$

36. $F = x + y$

$\begin{cases} x \geq 0, y \geq 0 \\ 3y - 2x \leq 6 \\ 3y + 4x \leq 24 \end{cases}$

In problems 37–40, sketch the graphs of both equations in the system in the same viewing window. Use the result to determine the number of solutions. Then solve the system.

37. $\begin{cases} 3x - 4y = 25 \\ x^2 + y^2 = 25 \end{cases}$ **38.** $\begin{cases} 2x - y = 2 \\ x^2 + 2y^2 = 12 \end{cases}$

39. $\begin{cases} x + y^2 = 6 \\ x^2 + y^2 = 36 \end{cases}$ **40.** $\begin{cases} \log_6 x + \log_6 y = 1 \\ x^2 + y^2 = 13 \end{cases}$

41. Income Tax: An income tax service developed the following relationships among taxes for a corporation:

$$\begin{cases} x + 0.095y = 2{,}584.42 \\ 0.01x + 1.01y + 0.01z = 1{,}529.70 \\ 0.48x + 0.48y + z = 12{,}253.24 \end{cases}$$

where x denotes the state income tax owed, y denotes the city income tax owed, and z denotes the federal income tax owed (all in dollars). Solve the system to determine the taxes owed (to the nearest dollar).

In problems 42–45, use a system of equations to solve each problem.

42. Investment Portfolio: Suppose that \$10,000 is invested in two funds paying 13% and 14% annual simple interest rate. If the annual return from both investments is \$1400, what amount was invested at each rate?

43. Puzzle: The sum of the squares of two numbers is 113. When 5 times the square of one number is added to the square of the other, the sum is 309. What are the numbers?

44. Geometry: The perimeter of a rectangular garden is 46 meters and the area of the garden is 60 square meters. Find the length and the width of the garden.

45. Art: An artist determines the monthly cost of manufacturing a total of 190 of types A, B, and C statues is \$910. Suppose that the manufacturing cost of each statue of type A is \$7, of type B is \$6, and of type C is \$3. If statue A sells for \$25, statue B for \$15, and statue C for \$10, how many of each type should be made to produce a monthly revenue of \$2800?

CHAPTER 4 TEST

1. Solve the system

$\begin{cases} 4x - y = 3 \\ -2x + 3y = 1 \end{cases}$

by using:

(a) The elimination method (without matrices)

(b) The elimination method using matrices

2. The row-reduced matrix

$$\left[\begin{array}{ccc|c} 1 & 0 & 0 & 4 \\ 0 & 1 & 0 & -1 \\ 0 & 0 & 1 & 3 \end{array}\right]$$

is the augmented matrix form of a corresponding system of linear equations. Solve the system.

3. Let

$$A = \begin{bmatrix} 3 & -1 \\ 4 & 2 \end{bmatrix} \text{ and } B = \begin{bmatrix} 3 & -2 \\ 2 & 7 \end{bmatrix}$$

Perform each matrix operation:

(a) $A + 2B$

(b) AB

(c) $3A - 2B$

4. (a) Let $A = \begin{bmatrix} 1 & -2 \\ 1 & 1 \end{bmatrix}$.

Verify that $A^{-1} = \begin{bmatrix} \frac{1}{3} & \frac{2}{3} \\ -\frac{1}{3} & \frac{1}{3} \end{bmatrix}$.

(b) Use the inverse of the matrix of coefficients to solve the system

$$\begin{cases} x - 2y = 3 \\ x + y = -3. \end{cases}$$

5. (a) Evaluate $\begin{vmatrix} 5 & 2 \\ 2 & -3 \end{vmatrix}$.

(b) Use Cramer's rule to solve the system

$$\begin{cases} 5x + 2y = 3 \\ 2x - 3y = 5. \end{cases}$$

6. Find the partial fraction decomposition:

$$\frac{x + 4}{x^2 - 3x} = \frac{A}{x} + \frac{B}{x - 3}$$

7. Solve the system: $\begin{cases} x^2 - 2y = 0 \\ x + 2y = 6 \end{cases}$

8. Sketch the region defined by the system

$$\begin{cases} x \geq 0,\ y \geq 0 \\ x + y \leq 4 \\ 2x + y \leq 6. \end{cases}$$

and find the coordinates of the corner points of the boundary of the region.

9. Find the maximum and minimum values of $F = 3x + 4y + 1$ under the constraints

$$\begin{cases} x \geq 0,\ y \geq 0 \\ x + 2y \leq 8 \\ x + y \leq 5. \end{cases}$$

10. A plumber charges a fixed charge plus an hourly rate for service on a house call. The plumber charged \$70 to repair a water tank that required 2 hours of labor and \$100 to repair a water tank that took 3.5 hours of labor. Find the plumber's fixed charge and hourly rate. (Use a system of equations to solve this problem.)

11. From geometry, the sum of the interior angles of a triangle *ABC* is 180°. The measure of angle *A* is 100° less than the measure of the sum of the angles *B* and *C*, and the measure of angle *C* is 40° less than the measure of twice angle *B*. Find the measure of each angle. (Use a system of equations to solve this problem.)

Chapter 4: Answers to Selected Problems

Problem Set 4.1 (on page 305)

1. $\left(\frac{11}{5}, \frac{14}{5}\right)$ **3.** No solution **5.** $\left(\frac{11}{7}, \frac{5}{7}\right)$ **7.** No solution **9.** $(a, a - 5)$, where a is a real number

11. $\left(\frac{7}{3}, -\frac{4}{3}\right)$ **13.** $(2, 3, -1)$ **15.** $(4, 3, 2)$ **17.** $(3, 2)$ **19.** $(1, -2)$ **21.** $(y, z) = \left(\frac{13}{5}, -\frac{7}{5}\right)$

23. $(2, -1)$ **25.** $(1, 1)$ **27.** $\left(-2, \frac{10}{3}, \frac{2}{3}\right)$ **29.** $\left(-5, \frac{5}{3}, \frac{28}{3}\right)$ **31.** $(2, 1, 0)$

33. $(-1, -2, 7)$ **35.** $(2, 1)$ **37.** 30 females, 90 males

39. **(a)** $V_0 = 32$ ft./sec, $h_0 = 10$ ft. **(b)** $h = -16t^2 + 32t + 10$
(c) $h(1) = 26$ ft, $h = 0$ when $t \approx 2.27$ secs. **41.** 2 L 30% solution, 8 L 20% solution

43. Airplane: approx. 617 mph; wind: approx. 49 mph

45. 185 pairs tennis shoes; 75 pairs running shoes

47. \$24,000 at 5% interest; \$11,000 at 7% interest; \$7000 at 9% interest

49. 3 of the 12 passenger vans, 11 of the 10 passenger vans, 1 of the 6 passenger vans

51. A pumps 10,000 gal/hr; B pumps 12,000 gal/hr; C pumps 15,000 gal/hr

53. **(a)** $x = \frac{1}{3}, y = -1$ **(b)** $x = \frac{1}{4}, y = -\frac{1}{2}$

Problem Set 4.2 (on page 320)

1. $\left[\begin{array}{cc|c} 1 & 6 & 7 \\ 3 & 7 & 10 \end{array}\right]$ **3.** $\left[\begin{array}{ccc|c} 2 & 1 & 0 & 2 \\ 1 & -1 & 0 & 4 \\ 3 & 0 & 1 & 6 \end{array}\right]$ **5.** $(1, 3)$ **7.** $(1, 1, 1)$ **9.** $(5, 3, -2)$

11. **(a)** $\begin{bmatrix} 1 & 0 & 3 \\ 5 & -1 & 5 \\ -3 & 1 & 4 \end{bmatrix}$ **(b)** $\begin{bmatrix} 1 & 0 & 3 \\ -3 & 1 & 4 \\ 15 & -3 & 15 \end{bmatrix}$ **(c)** $\begin{bmatrix} 1 & 0 & 3 \\ -3 & 1 & 4 \\ 8 & -1 & 14 \end{bmatrix}$ **13.** Both (a) and (b)

15. (b) **17.** $(1, 1)$ **19.** $(2 + t, 3 + 2t, t)$, where t is a real number **21.** No solution **23.** $(3, 5)$

25. $\left(\frac{23}{4}, \frac{13}{2}\right)$ **27.** $(1, 2, 3)$ **29.** $(1, 1, 1)$ **31.** $(4, 5, 6)$ **33.** $(1, 1, 0)$ **35.** No solution

37. $\left(\frac{7}{5} - \frac{7}{5}t, -\frac{2}{5} - \frac{3}{5}t, t\right)$, where t is a real number **39.** $\left(-\frac{23}{13}, -\frac{93}{26}, -\frac{37}{13}\right)$ **41.** $A = 2, B = 1$

43. $A = -1, B = -6$ **45.** $A = -5, B = 31, C = -26$ **47.** $A = 2, B = -1, C = 3$

49. $A = 6, B = -6, C = 0$ **53.** 10 oz A, 5 oz B, 8 oz C **55.** 500 lb 25%, 2000 lb 35%, 1500 lb 40%

57. $I_1 = \frac{33}{167}A, I_2 = \frac{87}{167}A, I_3 = \frac{54}{167}A$ **59.** $(1, -1, 1, 1)$ **61.** $(3, 2, 1, -1)$

63. **(a)** $2a - b - c \neq 0$ **(b)** Not possible **(c)** $2a - b - c = 0$

65. **(a)** $a = -\sqrt{3}, \sqrt{3}$ **(b)** $a \neq -\sqrt{3}, \sqrt{3}$ **(c)** Not possible **67.** $a = b = c = 1$

Problem Set 4.3 (on page 342)

1. **(a)** $\begin{bmatrix} 8 & 11 \\ -2 & 7 \end{bmatrix}$ **(b)** $\begin{bmatrix} -2 & -1 \\ 0 & 1 \end{bmatrix}$ **(c)** $\begin{bmatrix} 12 & 20 \\ -4 & 16 \end{bmatrix}$ **(d)** $\begin{bmatrix} -25 & -30 \\ 5 & -15 \end{bmatrix}$ **(e)** $\begin{bmatrix} -13 & -10 \\ 1 & 1 \end{bmatrix}$

3. **(a)** $\begin{bmatrix} 8 & 12 & 8 \\ 2 & 3 & 0 \end{bmatrix}$ **(b)** $\begin{bmatrix} -2 & -10 & -4 \\ -4 & 3 & 2 \end{bmatrix}$ **(c)** $\begin{bmatrix} 12 & 4 & 8 \\ -4 & 12 & 4 \end{bmatrix}$

(d) $\begin{bmatrix} -25 & -55 & -30 \\ -15 & 0 & 5 \end{bmatrix}$ **(e)** $\begin{bmatrix} -13 & -51 & -22 \\ -19 & 12 & 9 \end{bmatrix}$

5. (a) $\begin{bmatrix} -9 & 8 & 3 \\ -3 & 3 & 7 \\ 6 & -2 & 4 \end{bmatrix}$ **(b)** $\begin{bmatrix} -1 & -4 & -1 \\ -1 & -1 & -3 \\ 0 & 0 & 0 \end{bmatrix}$ **(c)** $\begin{bmatrix} -20 & 8 & 4 \\ -8 & 4 & 8 \\ 12 & -4 & 8 \end{bmatrix}$

(d) $\begin{bmatrix} 20 & -30 & -10 \\ 5 & -10 & -25 \\ -15 & 5 & -10 \end{bmatrix}$ **(e)** $\begin{bmatrix} 0 & -22 & -6 \\ -3 & -6 & -17 \\ -3 & 1 & -2 \end{bmatrix}$

7. (a) $\begin{bmatrix} 19 & -1 \\ 2 & 1 \end{bmatrix}$ **(b)** $\begin{bmatrix} 6 & 9 \\ 7 & 14 \end{bmatrix}$ **9. (a)** $\begin{bmatrix} -9 & -29 \\ 4 & -14 \end{bmatrix}$ **(b)** $\begin{bmatrix} -7 & 10 & -12 \\ 0 & -1 & 11 \\ 5 & -5 & -15 \end{bmatrix}$

11. (a) $\begin{bmatrix} 4 & 3 & -1 \\ 8 & -2 & 3 \\ 6 & 5 & 2 \end{bmatrix}$ **(b)** $\begin{bmatrix} 4 & 3 & -1 \\ 8 & -2 & 3 \\ 6 & 5 & 2 \end{bmatrix}$ **13.** Inverses **15.** Inverses **17.** Inverses

19. $\begin{bmatrix} 2 & -5 \\ -1 & 3 \end{bmatrix}$ **21.** $\begin{bmatrix} 0 & 1 \\ 1 & 1 \end{bmatrix}$ **23.** $\begin{bmatrix} 7 & 2 & -6 \\ -3 & -1 & 3 \\ 2 & 1 & -2 \end{bmatrix}$ **25.** $\begin{bmatrix} 1 & 2 & 0 \\ 0 & 2 & 3 \\ 1 & 3 & 1 \end{bmatrix}$

27. (2, 1) **29.** (4, −2) **31.** (1, 1) **33.** (−1, 2, 1) **35.** (2, 1, 3)

37. $\begin{bmatrix} 1 & 3 & 2 & 1 \\ 1 & 2 & 4 & 0 \\ 2 & 6 & 5 & 2 \\ 1 & 3 & 2 & 2 \end{bmatrix}\begin{bmatrix} x \\ y \\ z \\ w \end{bmatrix} = \begin{bmatrix} 4 \\ -3 \\ -1 \\ 2 \end{bmatrix}, A^{-1} = \begin{bmatrix} 12 & 3 & -8 & 2 \\ -2 & -1 & -2 & -1 \\ -2 & 0 & 1 & 0 \\ -1 & 0 & 0 & 1 \end{bmatrix}, X = \begin{bmatrix} 51 \\ -9 \\ -9 \\ -2 \end{bmatrix}.$

39. $\begin{bmatrix} 0.2 & 0.4 & -0.6 & 0 \\ -0.2 & 0.1 & 0.3 & 0.2 \\ 0.2 & 0.9 & -0.9 & 0.1 \\ 0.4 & 0.3 & -0.3 & 0 \end{bmatrix}\begin{bmatrix} x \\ y \\ z \\ w \end{bmatrix} = \begin{bmatrix} 0.2 \\ 0.7 \\ 1.1 \\ 0.7 \end{bmatrix}, A^{-1} = \begin{bmatrix} 2 & 1 & -2 & 3 \\ -11 & 3 & 6 & 1 \\ -8.33 & -1.67 & 3.33 & 1.67 \\ 20 & 10 & -10 & 0 \end{bmatrix}, X = \begin{bmatrix} 1 \\ 3 \\ 2 \\ 0 \end{bmatrix}.$

41. (a) [2325 1351.5 930]
(b) Profit from store 1 is \$2325; profit from store 2 is \$1351.50; profit from store 3 is \$930.

43. (a) [224,000,000 302,000,000]
(b) Value of cars shipped to site 1 is \$224,000,000; value of cars shipped to site 2 is \$302,000,000.

47. $x = 1, y = 2, z = 4, w = 10$ **51.** $\begin{bmatrix} -1 & 1 & -1 \\ 1 & -\frac{1}{2} & \frac{5}{4} \\ 1 & -\frac{1}{2} & \frac{3}{4} \end{bmatrix}$

Chapter 4 Review Problem Set (on page 357)

1. (−1, −2) **3.** (2, −1) **5.** (−9, 5) **7.** $(-1 - 3t, 2 - t, t)$, where t is a real number

9. $\left(-\frac{2}{9}, \frac{28}{9}\right)$ **11.** (0, 1, 1) **13.** $A = \frac{9}{8}, B = -\frac{1}{8}$ **15. (a)** $\begin{bmatrix} 12 & -3 \\ 7 & 7 \end{bmatrix}$ **(b)** $\begin{bmatrix} 14 & -11 \\ 9 & 19 \end{bmatrix}$

17. (a) $\begin{bmatrix} 0 & 5 \\ 11 & -18 \\ 1 & -7 \end{bmatrix}$ **(b)** Not possible **21.** (−1, 2) **23.** 5 **25.** 0 **27.** (3, 2) **29.** $\left(\frac{52}{29}, -\frac{50}{29}, \frac{36}{29}\right)$

31. 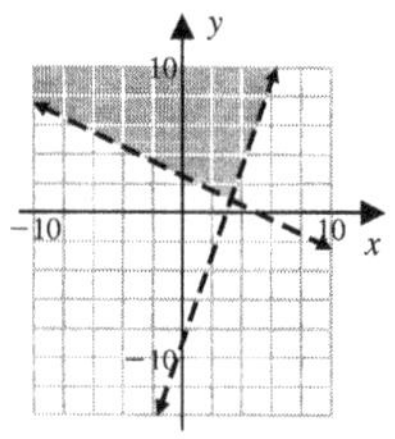

33. Maximum: 21; Minimum: 0 **35.** Maximum: 20; Minimum: 0
37. One solution: (3, −4) **39.** Three solutions: $(-5, \sqrt{11})$, (6, 0), $(-5, -\sqrt{11})$
41. \$2453 in state income tax, \$1387 in city income tax, \$10,410 in federal income tax
43. 7 and 8, 7 and −8, −7 and 8, or −7 and −8 **45.** 40 of A, 60 of B, 90 of C

Chapter 4 Test (on page 358)

1. **(a)–(b)** (1, 1) **2.** (4, −1, 3) **3.** **(a)** $\begin{bmatrix} 9 & -5 \\ 8 & 16 \end{bmatrix}$ **(b)** $\begin{bmatrix} 7 & -13 \\ 16 & 6 \end{bmatrix}$ **(c)** $\begin{bmatrix} 3 & 1 \\ 8 & -8 \end{bmatrix}$

4. **(b)** (−1, −2) **5.** **(a)** −19 **(b)** (1, −1) **6.** $A = -\frac{4}{3}, B = \frac{7}{3}$ **7.** (2, 2), $\left(-3, \frac{9}{2}\right)$

8. (0, 4), (2, 2), (3, 0), (0, 0) **9.** Maximum: 19; Minimum: 1 **10.** Fixed charge = \$30; hourly rate = \$20

11. A = 40°, B = 60°, C = 80°

CHAPTER 5

Additional Topics

Chapter Contents

Parametric equations provide us with a way of modeling the motion of a moving object. For example, suppose a space shuttle rises for its launch pad and reaches a height of 30,000 feet. Because of a computer malfunction, the booster rockets are turned of prematurely, at a moment when the shuttle is moving at a speed of 32,000 feet per second at an agle of 42 degree with an imaginery reference plane tangent to the earth directly below.

What is the maximum height of the space shuttle above the earth and how long will it take the space shuttle to reach its maximum height? The solution of this problem is given in Example 4 on page 367.

Many applications in physical science involve models that use elements of trigonometry or models that have similar characteristics as the trigonometric models we have studied so far. Here we investigate some of these models.

Objectives

1. Graph Parametric Equations by Point-Plotting
2. Eliminate the Parameter
3. Represent Cartesian Equations Parametrically
4. Solve Applied Problems

5.1 Parametric Equations

So far we have generated curves that result from graphing equations either relating x and y (in the Cartesian system) or r and θ (in the polar system). There is another way of generating curves that is used in modeling the path of a moving object. These curves are obtained from equations that express each x and y as functions of a third variable, say t (which represents time in many applications).

For instance, suppose that a particle moves in the xy plane in such a way that its x and y coordinates are functions of time t, say

$$\begin{cases} x = f(t) \\ y = g(t) \end{cases} \quad t \geq 0.$$

Then these functions taken together describe the path that the particle traces out in the xy plane as time t elapses.

In general, we have the following definition:

Definition

Parametric Equations in the Plane

Suppose that the points (x, y) on a curve in the plane are defined by the functions

$$\begin{cases} x = f(t) \\ y = g(t) \end{cases}$$

where t is a real number in a given interval. The functions are called **parametric equations** for the curve, and t is referred to as the **parameter.**

Some examples of parametric equations for curves in the plane are

$$\begin{cases} x = t + 1 \\ y = t^2 \end{cases} \qquad \begin{cases} x = 2\cos t \\ y = 3\sin t \end{cases} \quad 0 \leq t \leq 2\pi \qquad \begin{cases} x = t^2 \\ y = t^2 + 1 \end{cases} \quad 0 \leq t \leq 1.$$

Graphing Parametric Equations by Point-Plotting

We can use point-plotting to graph a curve represented by parametric equations. This is done by evaluating x and y for some values of t, plotting the corresponding points, and then connecting them with a smooth curve, if appropriate.

By monitoring the formation of the curve as the values of t increase, we can see how the curve is traced out, not just its shape.

EXAMPLE 1

Graphing Parametric Equations by Point-Plotting

Sketch the curve represented by the parametric equations by using point-plotting.

$$\begin{cases} x = t + 1 \\ y = t^2 \end{cases} \quad 0 \leq t \leq 5.$$

Explain how the curve is being generated as the values of t increase.

Solution

We begin by making a table of values of t and the corresponding values of x and y (Table 1). By plotting the (x, y) points in Table 1 and then connecting them with a smooth curve, we get a sketch of the graph (Figure 1).

TABLE 1

t	$x = t + 1$	$y = t^2$	(x, y)
0	1	0	(1, 0)
1	2	1	(2, 1)
2	3	4	(3, 4)
3	4	9	(4, 9)
4	5	16	(5, 16)
5	6	25	(6, 25)

Figure 1

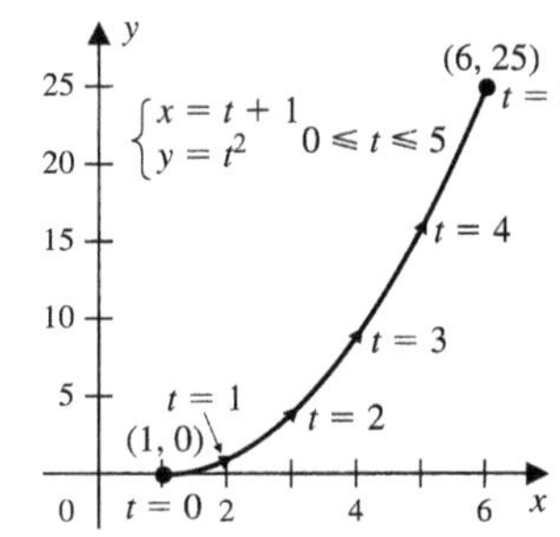

As the values of t increase from 0 to 5, the corresponding points on the graph trace out the curve starting at the point (1, 0) and ending at the point (6, 25). This *orientation* is denoted by the arrowheads in Figure 1.

Eliminating the Parameter

We can often derive an equation in terms of x and y for the curve defined by parametric equations. This process is referred to as **eliminating the parameter.**

For instance, in Example 1 where

$$\begin{cases} x = t + 1 \\ y = t^2 \end{cases} \quad 0 \le t \le 5$$

Eliminating the parameter often helps us to recognize the shape of the curve. However, it is important to examine the parametric form when determining the orientation of the curve.

we can eliminate the parameter algebraically as follows:

First solve the equation $x = t + 1$ for t to get $t = x - 1$.

Then substitute $x - 1$ for t in the second equation, $y = t^2$, to obtain

$$y = (x - 1)^2.$$

The condition $0 \le t \le 5$ requires that $0 \le x - 1 \le 5$, or $1 \le x \le 6$.

So the graph in Figure 1 is actually a portion of the parabola $y = (x - 1)^2$ where $1 \le x \le 6$.

The process of eliminating the parameter sometimes involves trigonometric identities.

EXAMPLE 2

Eliminating the Parameter Using Trigonometry

In many applications, t represents time, but it could denote other quantities such as an angle measure.

Eliminate the parameter and then sketch the curve of the parametric equations

$$\begin{cases} x = 2 \cos t \\ y = 3 \sin t \end{cases} \quad 0 \le t \le 2\pi.$$

Describe the orientation of a point on the curve as t varies from 0 to 2π.

Solution First we solve the equation $x = 2 \cos t$ for $\cos t$ and the second equation for $\sin t$ to obtain

$$\cos t = \frac{x}{2} \quad \text{and} \quad \sin t = \frac{y}{3}.$$

Squaring both sides of each equation produces

$$\cos^2 t = \frac{x^2}{4} \quad \text{and} \quad \sin^2 t = \frac{y^2}{9}.$$

Figure 2

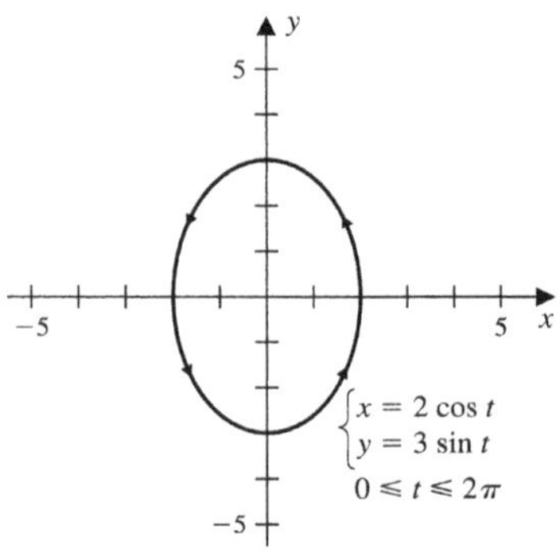

Adding these two equations, we have

$$\cos^2 t + \sin^2 t = \frac{x^2}{4} + \frac{y^2}{9}.$$

Since $\cos^2 t + \sin^2 t = 1$, we get

$$\frac{x^2}{4} + \frac{y^2}{9} = 1$$

which is an equation of an ellipse with center (0, 0), semimajor axis of length 3 on the *y* axis, and semiminor axis of length 2 on the *x* axis (Figure 2).

The restriction $0 \le t \le 2\pi$ suggests that the orientation of a point moving along the elliptical curve, as *t* increases from 0 to 2π,
starts at point

$$(2 \cos 0, 3 \sin 0) = (2, 0)$$

passes through points

$$(2 \cos(\pi/2), 3 \sin(\pi/2)) = (0, 3)$$

and

$$(2 \cos \pi, 3 \sin \pi) = (-2, 0)$$

and continues around the ellipse to end at

$$(2 \cos 2\pi, 3 \sin 3\pi) = (2, 0).$$

In other words, a point moves along the ellipse in a counterclockwise direction and traverses the ellipse once as *t* increases from 0 to 2π (Figure 2).

Representing Cartesian Equations Parametrically

We now consider the reverse problem—that is, finding a parametric representation for the graph of an equation in *x* and *y*.

If a curve is described by the function $y = f(x)$, we can let $x = t$ to obtain the following parametric equations for this curve:

$$\begin{cases} x = t \\ y = f(t) \end{cases} \quad \text{where } t \text{ is in the domain of } f$$

For example, the curve defined by

$$y = 3x^2 + 7,$$

where $0 \le x \le 10$, can be obtained from the following parametric equations:

$$\begin{cases} x = t \\ y = 3t^2 + 7 \end{cases} \quad 0 \le t \le 10$$

However, parametric representations for curves are not unique. For instance, if we let $x = 2t$, then another set of parametric equations for $y = 3x^2 + 7$, where $0 \le x \le 10$, is given by

$$\begin{cases} x = 2t \\ y = 3(2t)^2 + 7 \end{cases} \quad \text{or} \quad \begin{cases} x = 2t \\ y = 12t^2 + 7. \end{cases}$$

Since

$$x = 2t \quad \text{and} \quad 0 \le x \le 10,$$

then

$$0 \le 2t \le 10 \quad \text{or} \quad 0 \le t \le 5.$$

Rewriting the Cartesian equation of a curve in terms of parametric equations may involve trigonometric functions as illustrated in the next example.

EXAMPLE 3

Representing an Equation Parametrically by Using Trigonometry

Represent the equation of the circle

$$x^2 + y^2 = 4$$

parametrically if the orientation of points on this curve starts at (2, 0) and traverses the circle counterclockwise once.

Solution Using Figure 3 and trigonometry, we see that the equations

$$\begin{cases} x = 2\cos t \\ y = 2\sin t \end{cases} \quad 0 \le t \le 2\pi$$

Figure 3

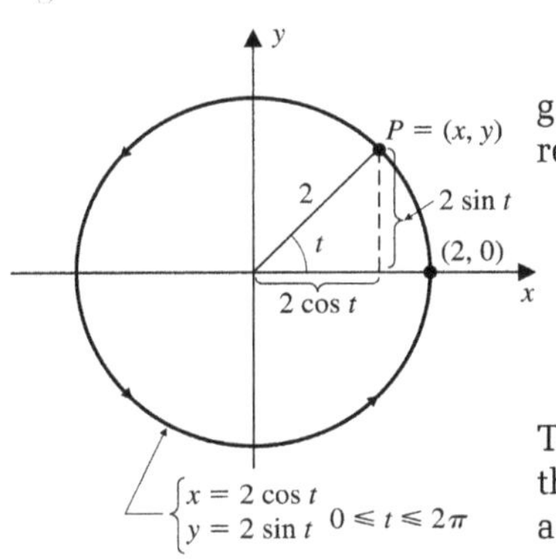

give a parametric representation for every point P on the circle. We verify this result by observing that

$$\begin{aligned} x^2 + y^2 &= 4\cos^2 t + 4\sin^2 t \\ &= 4(\cos^2 t + \sin^2 t) \\ &= 4. \end{aligned}$$

The condition, $0 \le t \le 2\pi$, indicates that a point moves along the circle starting at the point (2, 0), when $t = 0$, and traverses the circle in a counterclockwise direction as t increases from 0 to 2π, ending at point (2, 0), when $t = 2\pi$.

Solving Applied Problems

Figure 4

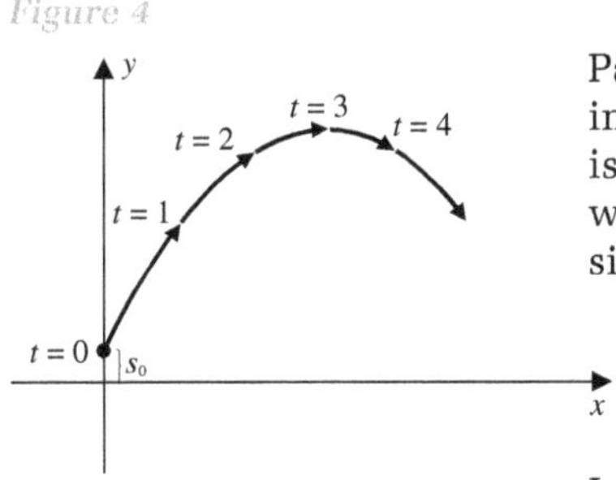

Parametric equations are useful in modeling real-world situations, especially those in which more than one variable is dependent upon time. For example, if an object is launched from a height of s_0 feet into the air at an angle θ with the ground and with an initial speed of v_0 feet per second, then physicists have determined the position of the object after t seconds is modeled by the parametric equations

$$\begin{cases} x = (v_0 \cos\theta)t \\ y = -16t^2 + (v_0 \sin\theta)t + s_0 \end{cases} \quad t \ge 0.$$

In this model, $v_0 \cos\theta$ is the horizontal component of the initial speed and $v_0 \sin\theta$ is the vertical component of the initial speed. The constant s_0 is the vertical distance (in feet) between the ground and the point from which the object is propelled. Figure 4 shows a general graph that describes the path of the object.

EXAMPLE 4

Modeling the Path of a Space Shuttle

A space shuttle rises from its launch pad and reaches a height of 30,000 feet. Because of a computer malfunction, the booster rockets are turned off prematurely, at a moment when the shuttle is moving at a speed of 32,000 feet per second at an angle of 42° with an imaginary reference plane tangent to the earth directly below.

(a) Use the model given above to determine the specific parametric equations that describe the shuttle's flight. Round off the values to the nearest thousands.

(b) Use the result in part (a) to eliminate the parameter.

(c) Find the maximum height of the space shuttle above the earth and how long it will take the space shuttle to reach its maximum height. Round off the time to the nearest 10 seconds and the height to the nearest 1,000 feet.

Solution

(a) We use the parametric equations:

$$\begin{cases} x = (v_0 \cos \theta)t \\ y = -16t^2 + (v_0 \sin \theta)t + s_0 \end{cases}$$

where $v_0 = 32{,}000$, $\theta = 42°$, and $S_0 = 30{,}000$.

The horizontal position of the space shuttle is represented by

$$\begin{aligned} x &= (v_0 \cos \theta)t \\ &= (32{,}000 \cos 42°)t = 24{,}000t \text{ (to the nearest thousands).} \end{aligned}$$

The vertical position is given by

$$\begin{aligned} y &= -16t^2 + (32{,}000 \sin 42°)t + 30{,}000 \\ &= -16t^2 + 21{,}000t + 30{,}000 \text{ (to the nearest thousands).} \end{aligned}$$

Therefore, the specific parametric equations are given by

$$\begin{cases} x = 24{,}000t \\ y = -16t^2 + 21{,}000t + 30{,}000,\ t \geq 0. \end{cases}$$

(b) From part (a) we use the result $x = 24{,}000t$ to write $t = \dfrac{x}{24{,}000}$. Thus to eliminate the parameter we use substitution to get the quadratic function

$$y = -16\left(\frac{x}{24{,}000}\right)^2 + 21{,}000\left(\frac{x}{24{,}000}\right) + 30{,}000;\ x > 0$$

or

$$y = \left(\frac{-1}{36 \cdot 10^6}\right)x^2 + \frac{7}{8}x + 30{,}000;\ x \geq 0.$$

(c) Since the equation found in part (b) is a quadratic function, its graph is a parabola and its maximum occurs when

Figure 5

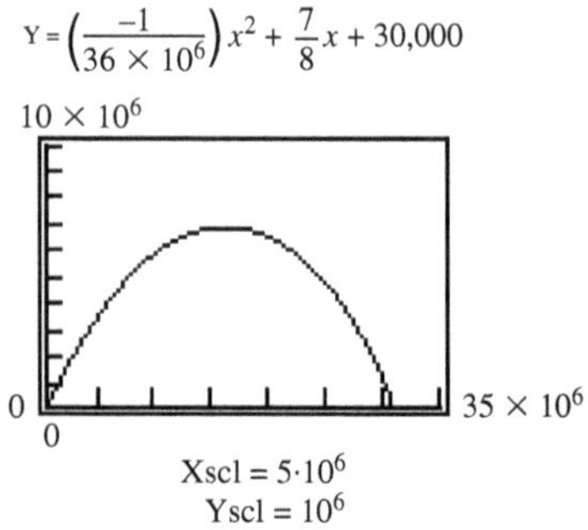

$$x = \frac{-b}{2a} = \frac{\left(-\dfrac{7}{8}\right)}{2\left(\dfrac{-1}{36 \cdot 10^6}\right)} = 15{,}750{,}000$$

that is, when

$$t = \frac{x}{24{,}000} = \frac{15{,}750{,}000}{24{,}000} = 656.25$$

or approximately 660 seconds (about 11 minutes).

The maximum height is found by substituting $x = 15{,}750{,}000$ into the quadratic function found in part (b) to get

$$y = 6{,}921{,}000 \text{ feet (approx.)}$$

or about 1310.8 miles

Figure 6

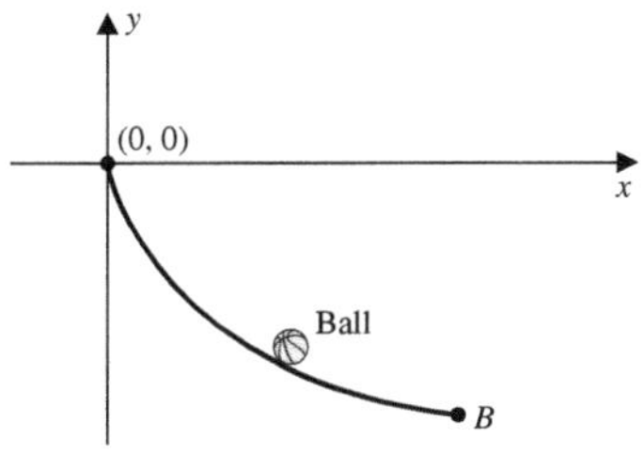

With the aid of a grapher we can graph the function found in part (b) of Example 4 (Figure 5). The graph displays the occurence of a maximum height.

At times, parametric equations represent curves generated by physical motion in cases where it may be difficult to eliminate the parameter to find a Cartesian equation. For instance, suppose a ball at point (0, 0) is allowed to drop and then slide down a curve to point B, not directly under (0, 0) (Figure 6). It can be shown that the ball takes the fastest path along a curve whose parametric equations are

$$\begin{cases} x = a(t - \sin t) \\ y = a(1 - \cos t) \end{cases} \quad \text{where } a \text{ is a constant and } t \text{ represents time.}$$

The curve represented by these equations is called a *cycloid*. Point-plotting is impractical to graph these equations; however, it is easy to generate such a graph by using a grapher programmed to handle parametric equations.

For instance, to graph the cycloid represented by the parametric equations

When using a grapher to graph parametric equations, it is important to take into account the given restrictions on the parameter.

$$\begin{cases} x = 2(t - \sin t) \\ y = 2(1 - \cos t) \end{cases} \quad 0 \le t \le 8\pi$$

we set the grapher in parametric mode, making certain that the restrictions on t are properly entered.

Figure 7 shows a viewing window of the graph of the curve. The curve is periodic with period 4π. Note that by watching certain graphers as the curve is being generated we can observe the orientation of the curve.

Figure 7

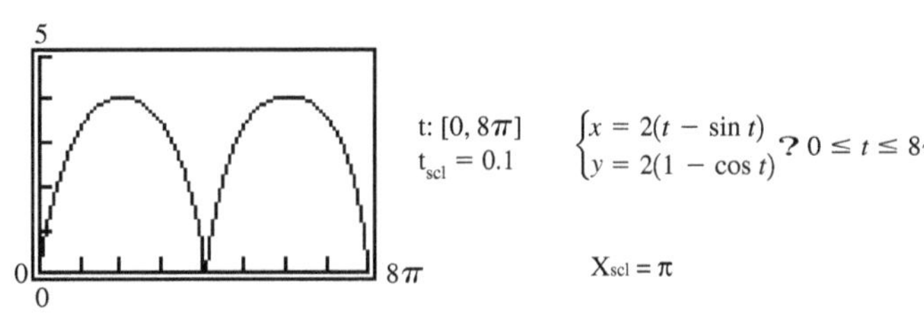

PROBLEM SET 5.1

Mastering the Concepts

In problems 1–4, sketch the curve represented by the parametric equations by using point-plotting. Describe the orientation of a point on the curve as t varies over its values.

1. $\begin{cases} x = 2t + 1 \\ y = t - 1 \end{cases} \quad 0 \le t \le 4$

2. $\begin{cases} x = (1/2)t + 1 \\ y = 3t - 4 \end{cases} \quad 0 \le t \le 5$

3. $\begin{cases} x = \cos t \\ y = 4 \sin t \end{cases} \quad 0 \le t \le 2\pi$

4. $\begin{cases} x = 2 \cos t \\ y = 5 \sin t \end{cases} \quad 0 \le t \le 2\pi$

In problems 5–20, eliminate the parameter and then sketch the curve for the given parametric equations. Describe the orientation of a point on the curve as t varies over its values.

5. $\begin{cases} x = 1 - 3t \\ y = 2 + (5/2)t \end{cases} \quad -3 \le t \le 3$

6. $\begin{cases} x = -2 - 4t \\ y = -5 + 3t \end{cases} \quad 1 \le t \le 8$

7. $\begin{cases} x = 1 + 2t \\ y = -1 + 4t^2 \end{cases} \quad -2 \le t \le 2$

8. $\begin{cases} x = 3t^2 \\ y = 2t \end{cases} \quad -2 \le t \le 2$

9. $\begin{cases} x = t \\ y = \sqrt{1 - t^2} \end{cases} \quad -1 \le t \le 1$

10. $\begin{cases} x = -3t \\ y = \sqrt{4 + 2t} \end{cases} \quad -2 \le t \le 3$

11. $\begin{cases} x = 3 \cos t \\ y = 3 \sin t \end{cases} \quad 0 \le t \le \pi$

12. $\begin{cases} x = 3 \cos t \\ y = -4 \sin t \end{cases} \quad 0 \le t \le 2\pi$

13. $\begin{cases} x = -3 + \cos t \\ y = 4 - \sin t \end{cases} \quad 0 \le t \le 2\pi$

14. $\begin{cases} x = 1 + \sin t \\ y = -1 + 4 \cos t \end{cases} \quad 0 \le t \le 2\pi$

15. $\begin{cases} x = \csc t \\ y = \cot t \end{cases} \quad -\frac{\pi}{2} < t < 0 \quad \text{or} \quad 0 < t < \frac{\pi}{2}$

16. $\begin{cases} x = \sin t \\ y = \csc t \end{cases} \quad 0 < t < \pi$

17. $\begin{cases} x = e^t \\ y = e^{-2t} \end{cases} \quad t \ge 0$

18. $\begin{cases} x = t^2 \\ y = 3 \ln t \end{cases} \quad t > 0$

19. $\begin{cases} x = 2 \sec t \\ y = 3 \tan t \end{cases} \quad -\frac{\pi}{2} < t < \frac{\pi}{2}$

20. $\begin{cases} x = 2 \cos^3 t \\ y = 2 \sin^3 t \end{cases} \quad 0 \le t \le 2\pi$

In problems 21–30, find parametric equations that represent the given oriented curve.

21. $y = 3x^2 + 5$; from $(0, 5)$ to $(2, 17)$
22. $y = x^3$; from $(-1, -1)$ to $(2, 8)$
23. $y = x^2 - 4x + 3$; from $(-1, 8)$ to $(1, 0)$
24. $y^2 = x$; from $(1, -1)$ to $(4, 2)$
25. Along the line segment from $(1, 2)$ to $(3, 5)$
26. Along the line segment from $(0, 2)$ to $(2, 0)$
27. Around the circle $x^2 + y^2 = 9$ once, counterclockwise starting at $(3, 0)$
28. Around the ellipse $(x^2/9) + (y^2/4) = 1$ once, counterclockwise starting at $(0, 2)$
29. Around the circle $x^2 + (y - 1)^2 = 1$ once, counterclockwise starting at $(1, 1)$
30. Along the hyperbola $x^2 - 4y^2 = 4$, where $x > 0$

Applying the Concepts

31. **Path of a Projectile:** A projectile is launched from a height of $s_0 = 10$ feet at an angle $\theta = 50°$ with the horizontal and with an initial velocity $\nu_0 = 200$ feet per second.
    (a) Use the model on page 365 to determine the parametric equations that describe the projectile's flight.
    (b) Determine how long it takes the projectile to return to ground level (rounded off to the nearest tenth of a second). Also, find the horizontal distance the projectile has traveled to the nearest foot when it hits the ground.
    (c) Graph the curve.
    (d) Determine the maximum height reached by the projectile to the nearest foot. When does it obtain that height?

32. **Path of a Projectile:** A projectile is launched from ground level with an initial velocity of 64 feet per second and at an angle of 45° with the horizontal.
    (a) Determine the parametric equations that model the projectile's path.
    (b) Find the time it takes for the projectile to hit the ground (to the nearest hundredth of a second). Also, find the horizontal distance the projectile has traveled when it hits the ground to the nearest foot.
    (c) Graph the curve.
    (d) Determine the maximum height reached by the projectile to the nearest foot. When does it attain that height?

33. **Baseball Path:** A baseball player hits a baseball 4 feet above home plate at an angle of 25° with the horizontal and with an initial velocity of 125 feet per second.
    (a) Find parametric equations that model this path.
    (b) The outfield fence is 380 feet from home plate and the fence is 8 feet high. Determine how high the ball is when it is 380 feet from home plate. Will the ball clear the fence for a home run if it is not caught?

34. **Sports:** The parametric equations

$$\begin{cases} x = 2(t - \sin t) \\ y = -2(1 - \cos t) \end{cases}$$

describe the shape of a skateboard ramp. Locate the low point of the ramp.

Developing and Extending the Concepts

35. Find parametric equations for the curve represented by

$$2x^2 - 3xy + 2y^2 = 6$$

36. (a) Show that

$$\begin{cases} x = a \cos t + h \\ y = b \sin t + k \end{cases} \quad 0 \le t \le 2\pi$$

where $a > 0$, $b > 0$, and $a \ne b$ are parametric equations of an ellipse with center at (h, k), horizontal axis of length $2a$, and vertical axis of length $2b$.

(b) What is the curve if $a = b$?

37. Given parametric equations: $\begin{cases} x = 3 + 5t \\ y = 4 - 2t \end{cases}$

(a) Graph the parametric equations to discover that the graph is a straight line.

(b) Eliminate the parameter to find the equation of the line in terms of x and y.

(c) How does the slope of the line found in part (b) relate to the parametric equations in part (a)?

(d) Can you generalize the above results to relate the parametric equations of a line

$$\begin{cases} x = a + bt \\ y = c + dt \end{cases}$$

to the slope of the line?

38. Show that

$$\begin{cases} x = h + a \sec t \\ y = k + b \tan t \end{cases} \quad -\frac{\pi}{2} < t < \frac{3\pi}{2}, \ t \neq \frac{\pi}{2}$$

where $a > 0$ and $b > 0$ are parametric equations of a hyperbola with center at (h, k), transverse axis of length $2a$, and conjugate axis of length $2b$.

39. Given parametric equations: $\begin{cases} x = t^2 \\ y = t^2 \end{cases}$

(a) Eliminate the parameter for the given system.

(b) Explain why the graph of the given parametric equations is different from the graph of $y = x$.

40. Do the following sets of parametric equations represent the same curve? Explain.

$$\begin{cases} x = \cos t \\ y = \sin t \end{cases} \quad 0 \leq t \leq 2\pi$$

$$\begin{cases} x = \sin t \\ y = \cos t \end{cases} \quad 0 \leq t \leq 2\pi$$

In problems 41–46, use a grapher to graph each curve represented by the parametric equations.

41. $\begin{cases} x = 3t - 2 \\ y = 4t^2 - 3 \end{cases} \quad 0 \leq t \leq 4$

42. $\begin{cases} x = t^2 - 2t \\ y = t^3 - 3t \end{cases} \quad 0 \leq t \leq 4$

43. $\begin{cases} x = 2 - (1/t) \\ y = 2t + (1/t) \end{cases} \quad 1 \leq t \leq 10$

44. $\begin{cases} x = 1/(t^2 - 1) \\ y = -5t + 3 \end{cases} \quad 2 \leq t \leq 10$

45. $\begin{cases} x = 3t - 3 \sin t \\ y = 3 - 3 \cos t \end{cases} \quad 0 \leq t \leq 10\pi$

46. $\begin{cases} x = 2 \cos^2 t \\ y = 4 \sin^2 t \end{cases} \quad 0 \leq t \leq 2\pi$

Interlude: Practice Problems 5.1, Part I

1. Consider the parametric equations $\begin{cases} x = t^2 \\ y = t - 1 \end{cases}, \; -2 \leq t \leq 2.$

   (a) Find a Cartesian equation corresponding to the given parametric equations. Give appropriate restrictions on x and y.

   (b) Sketch the graph in the Cartesian plane corresponding to the given parametric equations. Use arrows to indicate the orientation of a point moving along the curve.

2. Give a pair of parametric equations to represent the graph of the curve with equation $xy^2 + 9x - 6y = 0$ (fill in the blanks)

$$\begin{cases} x = \_\_\_\_\_\_\_\_\_\_\_\_ \\ y = \_\_\_\_\_\_\_\_\_\_\_\_ \end{cases}$$

3. Consider the parametric equations $\begin{cases} x = 5\cos t \\ y = 6\sin t \end{cases}$, $0 \le t \le 2\pi$.

(a) Find a Cartesian equation corresponding to the given parametric equations. Give appropriate restrictions on x and y.

(b) Sketch the graph in the Cartesian plane corresponding to the given parametric equations. Use arrows to indicate the orientation of a point moving along the curve.

4. Consider the parametric equations $\begin{cases} x = 4\cos 2t \\ y = \sin t \end{cases}$, $\frac{\pi}{2} \le t \le \frac{3\pi}{2}$.

(a) Find a Cartesian equation corresponding to the given parametric equations. Give appropriate restrictions on x and y.

(b) Sketch the graph in the Cartesian plane corresponding to the given parametric equations. Use arrows to indicate the orientation of a point moving along the curve.

Interlude: Practice Problems 5.1, Part II

You are standing off to the side of a Ferris wheel with a diameter of 40 feet (pretend you're standing at the origin in the Cartesian plane). The base of the Ferris wheel is 60 feet away at ground level.

1. Use parametric equations to create a graph in the Cartesian plane representing the Ferris wheel. Show that graph.

2. The Ferris wheel is not moving at the moment. On a dare, you decide throw a ball from your position (the origin) up to the top of the Ferris wheel, where your buddy Crazy Eddy is presently sitting. If you throw the ball at an angle of 60°, what initial velocity do you need to get the ball up to Crazy Eddy? Give parametric equations that describe the path of the ball. Show the graphs corresponding to the path of the ball and the Ferris wheel.

3. You have your ball back. Having succeeded in getting the ball to Crazy Eddy (and not getting kicked out of the amusement park), you now decide to throw the ball back to Crazy Eddy when the Ferris wheel starts moving again. The Ferris wheel completes one rotation every minute. If you throw the ball with an initial velocity of 66 feet per second at the moment the ride begins moving again, at what angle do you need to throw in order to give Crazy Eddy a chance to catch the ball? Create parametric equations that describe the path of the ball and Crazy Eddy's path.

Objectives

1. Define Hyperbolic Trigonometric Functions
2. Discover Properties and Identities
3. Establish Inverses For Hyperbolic Sine and Hyperbolic Cosine Functions

5.2 Hyperbolic Trigonometric Functions

Certain combinations of the exponential functions $f(x) = e^x$ and $g(x) = e^{-x}$ arise so frequently in applications of mathematics that they are given special names. In particular, we will consider the six basic **hyperbolic trigonometric functions**. We shall see that these functions have some interesting similarities to the basic trigonometric functions.

Defining the Hyperbolic Trigonometric Functions

The six basic hyperbolic trigonometric functions (with accompanying special notation) are defined as follows:

Hyperbolic Sine Function	$\sinh x = \dfrac{e^x - e^{-x}}{2}$	**Hyperbolic Cosecant Function**	$\operatorname{csch} x = \dfrac{1}{\sinh x}$
Hyperbolic Cosine Function	$\cosh x = \dfrac{e^x + e^{-x}}{2}$	**Hyperbolic Secant Function**	$\operatorname{sech} x = \dfrac{1}{\cosh x}$
Hyperbolic Tangent Function	$\tanh x = \dfrac{\sinh x}{\cosh x}$	**Hyperbolic Cotangent Function**	$\coth x = \dfrac{\cosh x}{\sinh x}$

Sketch graphs of $y = \sinh x$, $y = \cosh x$, and $y = \tanh x$ on separate sets of axes.

Determine the domain and range of each function.

$y = \sinh x$	$y = \cosh x$	$y = \tanh x$
Domain: **Range:**	**Domain:** **Range:**	**Domain:** **Range:**

Discovering Properties of the Hyperbolic Trigonometric Functions

Based on your graphs from above, which of the three functions appear to be odd? Which appear to be even?

Explore/Discuss

By definition, show that the hyperbolic sine and hyperbolic tangent functions are odd, while the hyperbolic cosine function is even.

There are some identities dealing with hyperbolic trigonometric functions that are remarkably similar to identities we have studied involving the basic trigonometric functions.

Explore/Discuss

Verify that $\cosh^2 x - \sinh^2 x = 1$ is an identity.

Explore/Discuss

Verify that $\sinh(x + y) = \sinh x \cosh y + \cosh x \sinh y$ is an identity by transforming the right side of the equation into the left.

Explore/Discuss

Verify that $1 - \tanh^2 x = \operatorname{sech}^2 x$ is an identity by transforming the left side of the equation into the right.

It is natural to wonder why the term "hyperbolic" is applied to these functions.

Explore/Discuss

Graph the curves with parametric equations

$$\begin{cases} x = \cosh t \\ y = \sinh t \end{cases} \text{and} \begin{cases} x = -\cosh t \\ y = -\sinh t \end{cases} \text{for } -2\pi \leq t \leq 2\pi.$$

The basic trigonometric functions are called *circular functions* because of their relationship to the unit circle. The basic hyperbolic trigonometric functions are thus named because of the relationships between these functions and the hyperbola with equation $x^2 - y^2 = 1$. In fact, given any real number t, the point $P(\cosh t, \sinh t)$ lies on the graph of that hyperbola (this can be shown using the identity $\cosh^2 x - \sinh^2 x = 1$).

Applications of hyperbolic functions can be found in many areas, including applications in the study of light, electricity, velocity, and radioactivity. A prominent application of the hyperbolic cosine function is in modeling the shape of a uniform flexible chain or cable suspended on each end at the same height (imagine a telephone wire). While this shape may resemble a parabola, it is actually a *caternary*, which comes from the Latin word for "chain". Caternary curves are modeled by equations of the form $y = a \cosh(x/a)$, for some real number a.

Inverse Hyperbolic Trigonometric Functions

Looking at the graphs of the hyperbolic sine, hyperbolic cosine, and hyperbolic tangent functions, it is clear that all of those functions are one-to-one as their graphs pass the horizontal line test. As this is the case, each has an inverse function. We will consider what these inverse functions look like.

Calculating Inverse Functions

Recall that any function f that is one-to-one has an inverse function, typically denoted by f^{-1}, so that $(f^{-1} \circ f)(x) = x$ for all x in the domain of f. We will find the inverse function for the hyperbolic sine function, which we will denote by $\sinh^{-1}$.

Note that $y = \sinh^{-1} x$ if and only if $x = \sinh y$. From this we will show that

$$\sinh^{-1} x = \ln\left(x + \sqrt{x^2 + 1}\right)$$

Now, if $x = \sinh y$, then we have

$$x = \frac{e^y - e^{-y}}{2}$$

Consequently,

$$2x = e^y - e^{-y}$$

so

$$e^y - 2x - e^{-y} = 0$$

Multiplying both sides of that equation by e^y yields

$$\left(e^y\right)^2 - 2xe^y - 1 = 0$$

We use the quadratic formula to obtain

$$e^y = \frac{2x \pm \sqrt{4x^2 + 4}}{2}$$

Simplifying yields

$$e^y = x + \sqrt{x^2 + 1} \text{ or } e^y = x - \sqrt{x^2 + 1}$$

It can be shown that $\sqrt{x^2 + 1} > x$ for all values of x (we leave this as an exercise for the reader), so $x - \sqrt{x^2 + 1} < 0$. Since e^y cannot be negative, we discard the second solution above. Consequently, we have

$$e^y = x + \sqrt{x^2 + 1}$$

so that, by definition,

$$y = \ln\left(x + \sqrt{x^2 + 1}\right)$$

This means that $\sinh^{-1} x = \ln\left(x + \sqrt{x^2 + 1}\right)$, as proposed.

PROBLEM SET 5.2

In problems 1-8, verify that each equation is an identity by transforming one side of the equation into the other.

1. $\cosh x + \sinh x = e^x$
2. $\cosh x - \sinh x = e^{-x}$
3. $\cosh(x + y) = \cosh x \cosh y + \sin x \sinh y$
4. $\coth^2 x - 1 = \operatorname{csch}^2 x$
5. $\tanh(x + y) = \dfrac{\tanh x + \tanh y}{1 + \tanh x \tanh y}$
6. $\cosh 2x = \cosh^2 x + \sinh^2 x$
7. $\sinh 2x = 2\sinh x \cosh x$
8. show that $\tanh^{-1} x = \dfrac{1}{2} \ln\left(\dfrac{1 + x}{1 - x}\right)$ for $-1 < x < 1$

Interlude: Practice Problems 5.2

Show that $\cosh^{-1} x = \ln\left(x + \sqrt{x^2 - 1}\right)$ for $x \geq 1$.

Chapter 5 Readiness

Provide complete solutions in the space provided.

1. Give a pair of parametric equations to represent the graph of the curve with polar equation $r^2 = \dfrac{144}{16\sin^2\theta + 9\cos^2\theta}$ (fill in the blanks), and then sketch the graph in the Cartesian coordinate plane. Indicate the orientation of a point on the curve with arrows and label all intercepts with exact coordinates.

$$\begin{cases} x = \_\_\_\_\_\_\_\_\_\_\_\_\_\_\_\_ \\ y = \_\_\_\_\_\_\_\_\_\_\_\_\_\_\_\_ \end{cases}$$

2. Consider the curves shown in Figure 1. Create four separate systems of parametric equations that would generate the curves shown in the figure (fill in the blanks). Assume that gridlines each mark one unit. Include appropriate restrictions on t for the curves generating the "mouth" on the figure.

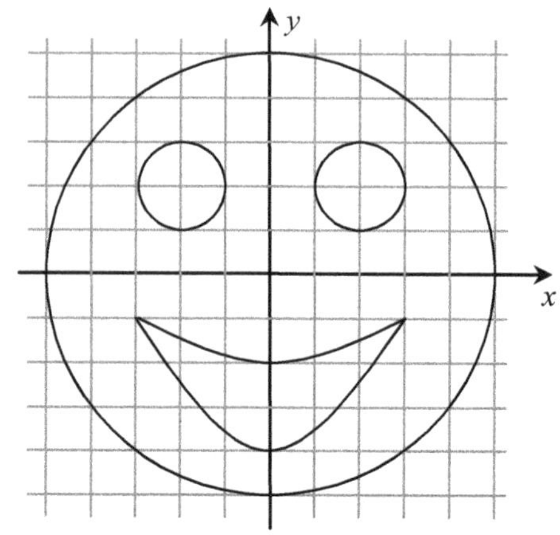

Figure 1

$\begin{cases} x = \_\_\_\_\_\_\_\_\_\_ \\ y = \_\_\_\_\_\_\_\_\_\_ \end{cases}$

$\begin{cases} x = \_\_\_\_\_\_\_\_\_\_ \\ y = \_\_\_\_\_\_\_\_\_\_ \end{cases}$

$\begin{cases} x = \_\_\_\_\_\_\_\_\_\_ \\ y = \_\_\_\_\_\_\_\_\_\_ \end{cases}$

$\begin{cases} x = \_\_\_\_\_\_\_\_\_\_ \\ y = \_\_\_\_\_\_\_\_\_\_ \end{cases}$

$\begin{cases} x = \_\_\_\_\_\_\_\_\_\_ \\ y = \_\_\_\_\_\_\_\_\_\_ \end{cases}$

Chapter 5: Answers to Selected Problems

Problem Set 5.1 (on page 369)

1.

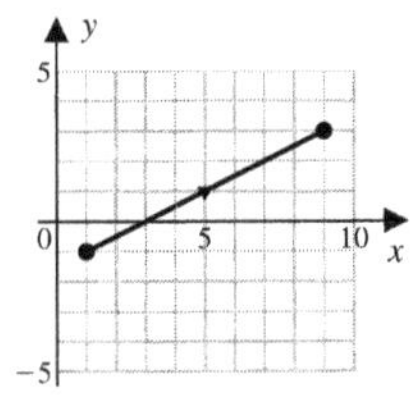

3.

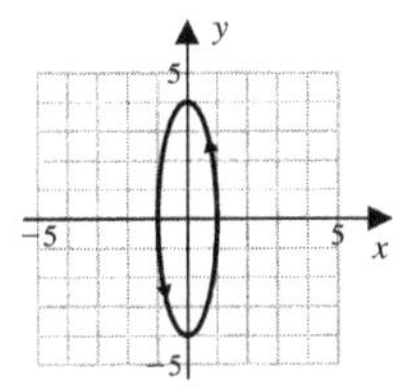

5. $y = \frac{17}{6} - \frac{5}{6}x$

7. $y = x^2 - 2x$

9. $y = \sqrt{1 - x^2}$

11. $x^2 + y^2 = 9,\ y \geq 0$

13. $(x + 3)^2 + (y - 4)^2 = 1$

15. $x^2 - y^2 = 1,\ x \geq 1$ or $x \leq -1$

17. $y = \frac{1}{x^2},\ x \geq 1$

19. $\frac{x^2}{4} - \frac{y^2}{9} = 1,\ x \geq 2$

21. One possibility: $\begin{cases} x = t \\ y = 3t^2 + 5 \end{cases}$ $\quad 0 \leq t \leq 2$

23. One possibility: $\begin{cases} x = t \\ y = t^2 - 4t + 3 \end{cases}$ $\quad -1 \leq t \leq 1$

25. One possibility: $\begin{cases} x = t \\ y = \frac{3}{2}t + \frac{1}{2} \end{cases}$ $\quad 1 \leq t \leq 3$

27. One possibility: $\begin{cases} x = 3 \cos t \\ y = 3 \sin t \end{cases}$ $\quad 0 \leq t \leq 2\pi$

29. One possibility: $\begin{cases} x = \cos t \\ y = 1 + \sin t \end{cases}$ $\quad 0 \leq t \leq 2\pi$

31. **(a)** $\begin{cases} x = (200 \cos 50°)t \\ y = -16t^2 + (200 \sin 50°)t + 10 \end{cases}$ **(b)** 9.6 sec; 1239 ft **(d)** 377 ft; 4.8 sec

33. **(a)** $\begin{cases} x = (125 \cos 25°)t \\ y = -16t^2 + (125 \sin 25°)t + 4 \end{cases}$ **(b)** Approx. 1.2 ft; no

35. One possibility: $\begin{cases} x = \sqrt{6} \cos t - \sqrt{\frac{6}{7}} \sin t \\ y = \sqrt{6} \cos t + \sqrt{\frac{6}{7}} \sin t \end{cases}$ $\quad 0 \leq t \leq 2\pi$

37. **(b)** $y = -\frac{2}{5}x + \frac{26}{5}$ **(c)** Slope $= \frac{\text{Coefficient of } t \text{ values in } y}{\text{Coefficient of } t \text{ values in } x}$ **(d)** The slope is d/b.

39. **(a)** $y = x$ **(b)** Since t^2 cannot be negative, $x \geq 0$ and $y \geq 0$.

41.

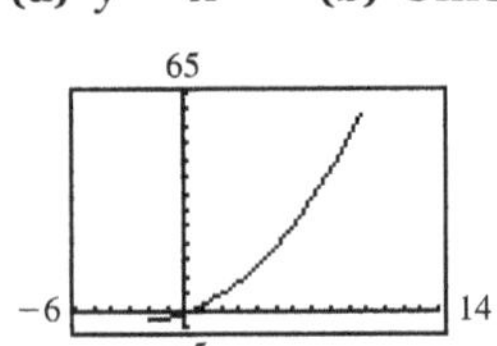

43.

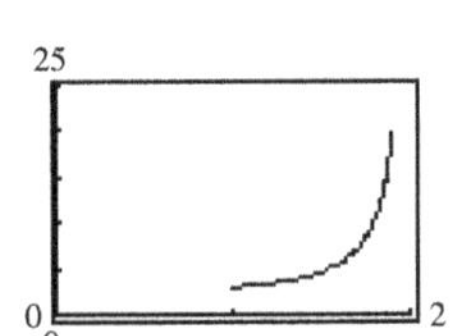

45.

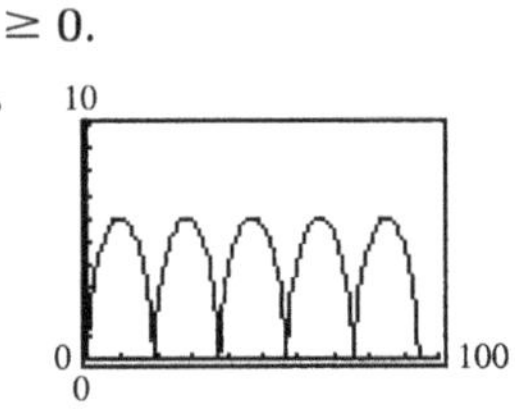

Appendix I

Objectives

1. Define Complex Numbers
2. Add, Subtract, Multiply, and Divide Complex Numbers

Review of Complex Numbers

When restricting our attention to real numbers only, we find that equations like $x^2 = -1$ and $x^2 = -7$ do not have solutions. In order to solve equations like these, it is necessary to expand our number system to allow for the square roots of negative numbers. We will look at such a number system, which mathematicians call the **complex number system**. This system will also include the real number system, with which we are already somewhat familiar, so the arithmetic within this number system will in many ways be similar to the ordinary real number arithmetic we comfortably employ.

Imaginary and Complex Numbers

The equation $x^2 = -1$ has no real number solutions, as there are no real numbers whose squares are negative. However, using the principle of square roots with this equation, we see that

$$x = \sqrt{-1} \text{ or } x = -\sqrt{-1}$$

Substituting into the given equation, we see that these two results do, in fact, satisfy the equation (we leave it as an exercise to the reader to do this). So, it seems reasonable to say that the equation $x^2 = -1$ does have solutions, but the solutions are not real numbers; rather, they involve the square root of a negative number.

> ***The Imaginary Unit and Imaginary Numbers***
> - The **imaginary unit** i is a number such that $i^2 = -1$ (i.e., $i = \sqrt{-1}$)
> - **Imaginary numbers** are numbers of the form $\sqrt{k}$, where $k < 0$; that is, imaginary numbers are square roots of negative numbers

Consequently, an introduction follows to what mathematicians call the **imaginary unit** and the collection of **imaginary numbers**.

Using standard simplification properties with square roots and this i notation, we can simplify imaginary numbers and write them in terms of the imaginary unit. For instance, consider the imaginary number $\sqrt{-24}$. Notice that

$$\sqrt{-24} = \sqrt{4 \cdot 6 \cdot -1} = \sqrt{4} \cdot \sqrt{6} \cdot \sqrt{-1} = 2\sqrt{6} \cdot i$$

It is commonplace to write complex numbers in terms of i, and we will do so for the remainder of this discussion. Consequently, we will refer to all numbers that we can write as bi, where b is a real number, as imaginary numbers.

Now, consider the quadratic equation $x^2 + 2x + 5 = 0$. Using the quadratic formula, we see that the numbers

$$x = \frac{-2+\sqrt{-16}}{2} \text{ and } x = \frac{-2-\sqrt{-16}}{2}$$

are solutions, but are not real numbers. Rewriting these in terms of i, we have

$$x = \frac{-2+\sqrt{16}\cdot i}{2} \text{ and } x = \frac{-2-\sqrt{16}\cdot i}{2}$$

or, simplifying a bit,

$$x = -1 + 2i \text{ and } x = -1 - 2i$$

The numbers $-1 + 2i$ and $-1 - 2i$ are arithmetic combinations of a real number and an imaginary number. We will call numbers with this form **complex numbers**.

> ***Complex Numbers***
>
> **Complex numbers** are numbers that can be written in the form $a + bi$, where a and b are real numbers. Mathematicians call this form the **standard form** of a complex number. The number a is called the **real part** of the complex number, while b is called the **imaginary part** of the number.

From this definition, we see that any real number a is also a complex number, since $a = a + 0i$. Similarly, any imaginary number bi is also a complex number, as $bi = 0 + bi$. Consequently, the collection of all complex numbers includes all real numbers as well as all imaginary numbers.

Complex Number Arithmetic

Since the collection of complex numbers contains the collection of real numbers, it is quite natural to define basic arithmetic operations of addition, subtraction, multiplication and division for complex numbers in such a way that our familiar arithmetic with real numbers still holds. Specifically, when adding, subtracting or multiplying complex numbers, we proceed just as if we were working with polynomials, where we identify and collect like terms, as illustrated in the next two examples.

EXAMPLE 1 **Adding and Subtracting Complex Numbers** Perform the indicated operations and write the results in standard form.

(a) $(3 + 6i) + (5 - 8i)$ (b) $(4 - i) - (6 + 3i)$

Solution As mentioned, we proceed in a manner similar to adding and subtracting binomials.

(a) $(3 + 6i) + (5 - 8i) = (3 + 5) + (6i - 8i) = 8 - 2i$

(b) $(4 - i) - (6 + 3i) = (4 - 6) + (-i - 3i) = -2 - 4i$

EXAMPLE 2 **Multiplying Complex Numbers** Find the following products and write the results in standard form.

(a) $(1 + 3i)(-7 + 10i)$ (b) $12i(6 + 5i)$

Solution We multiply in the same manner as we would multiply polynomials, keeping in mind that $i^2 = -1$.

(a) $$\begin{aligned}(1 + 3i)(-7 + 10i) &= (1)(-7) + (3i)(-7) + (1)(10i) + (3i)(10i)\\ &= -7 - 21i + 10i + 30i^2\\ &= -7 - 11i + 30(-1)\\ &= -37 - 11i\end{aligned}$$

(b) $12i(6 + 5i) = (12i)(6) + (12i)(5i)$
$= 72i + 60i^2$
$= -60 + 72i$

In working with the complex number system, we understand exponentiation with integer exponents in the same way as when we are working with real numbers. When looking at positive integer powers of i, an interesting pattern develops.

$$i^1 = i = \sqrt{-1} \qquad i^5 = i^4 \cdot i = i$$
$$i^2 = -1 \qquad i^6 = i^4 \cdot i^2 = -1$$
$$i^3 = i^2 \cdot i = -i \qquad i^7 = i^4 \cdot i^3 = -i$$
$$i^4 = \left(i^2\right)^2 = (-1)^2 = 1 \qquad i^8 = \left(i^4\right)^2 = 1$$

The positive integer powers of i "cycle through" the four values i, -1, $-i$ and 1. This makes it rather easy to simplify higher positive integer powers of i just by using basic properties of exponents. For instance

$$i^{35} = i^{32} \cdot i^3 = \left(i^4\right)^8 \cdot (-i) = 1^8 \cdot (-i) = -i$$

You should note that some familiar properties of exponents from the real number system do not necessarily always hold in the complex number system. For instance, if x and y are nonnegative real numbers, then we know that $\sqrt{xy} = \sqrt{x} \cdot \sqrt{y}$ (or using equivalent exponential notation, $(xy)^{1/2} = (x^{1/2} \cdot y^{1/2})$. However, if both x and y are negative numbers, that property will not hold. For example,

$$\sqrt{-4} \cdot \sqrt{-25} = 2i \cdot 5i = 10i^2 = -10$$

However,

$$\sqrt{(-4)(-25)} = \sqrt{100} = 10$$

So, $\sqrt{(-4)(-25)} \neq \sqrt{-4} \cdot \sqrt{-25}$.

In general, the properties of real numbers and positive integer exponents do hold within the complex number system, but properties involving fractional exponents may not.

Now, consider the two complex numbers $5 + 9i$ and $5 - 9i$. We say that these numbers are **complex conjugates** of each other. Notice what happens when we multiply these:

$$(5 + 9i)(5 - 9i) = 25 + 45i - 45i - 81i^2 = 25 + 81 = 106$$

In fact, given any complex conjugate pair $a + bi$ and $a - bi$, we see that

$$(a + bi)(a - bi) = a^2 + abi - abi + b^2 = a^2 + b^2$$

This property of conjugate pairs is important for us because we will use it when we perform division with complex numbers.

Since $i = \sqrt{-1}$, expressions of the form $\dfrac{18 + 3i}{2 + 7i}$ (indicating division of one complex number by another) actually have a radical in the denominator, so we proceed as if we are rationalizing the denominator, except this time we will use the complex conjugate of the denominator:

$$\begin{aligned}\frac{18 + 3i}{2 + 7i} &= \frac{18 + 3i}{2 + 7i} \cdot \frac{2 - 7i}{2 - 7i} \\ &= \frac{36 + 6i - 126i - 21i^2}{2^2 + 7^2} \\ &= \frac{57 - 120i}{53} \\ &= \frac{57}{53} - \frac{120}{53}i\end{aligned}$$

In general, then, we have the following summary of addition, subtraction, multiplication and division in the complex number system.

Arithmetic in the Complex Number System

Suppose that $a + bi$ and $c + di$ are complex numbers. Then

- $(a + bi) + (c + di) = (a + c) + (b + d)i$
- $(a + bi) - (c + di) = (a - c) + (b - d)i$
- $(a + bi)(c + di) = (ac - bd) + (bc + ad)i$
- $\dfrac{a + bi}{c + di} = \dfrac{a + bi}{c + di} \cdot \dfrac{c - di}{c - di} = \dfrac{(a + bi)(c - di)}{c^2 + d^2}$, provided $c + di \neq 0$

Exercises

For these exercises, assume that $i = \sqrt{-1}$.

1. Write each expression in the form $a + bi$.

(a) $\sqrt{-16}$

(b) $3 + \sqrt{-8}$

(c) $7 - 3\sqrt{-9}$

(d) $\dfrac{2 - \sqrt{-4}}{6}$

(e) $i^6 + 5$

2. Write each expression in the form $a + bi$.

(a) $\sqrt{-64}$

(b) $-6 - \sqrt{-12}$

(c) $\sqrt{-14} + \sqrt{-21}$

(d) $2i^2 - 6i + 9$

(e) $\dfrac{\sqrt{-36} + \sqrt{-81}}{i^5}$

3. Perform the indicated operations and simplify. Write each result in the form $a + bi$.

(a) $(3 + 2i) + (7 - 3i)$

(b) $(4 - i) - (7 - 3i)$

(c) $(-3 - 6i) - (3 + 7i)$

(d) $(18 + 2i) + (-6 - 5i)$

4. Perform the indicated operations and simplify. Write each result in the form $a + bi$.

(a) $(-3i)(-6i)$

(b) $\sqrt{-36}\sqrt{-9}$

(c) $2i\,(5 + 3i)$

(d) $-7i\,(6 + i)$

5. Perform the indicated operations and simplify. Write each result in the form $a + bi$.

(a) $(2 + 8i)(3 + 5i)$

(b) $(-2 - 5i)(-9 + 4i)$

(c) $(3 + 4i)^2$

(d) $(-5 - 2i)^2$

6. Perform the indicated operations and simplify. Write each result in the form $a + bi$.

(a) $\dfrac{7}{4 + i}$

(b) $\dfrac{6}{3 - 2i}$

(c) $\dfrac{9}{7i}$

(d) $\dfrac{8i + 16}{8i}$

7. Perform the indicated operations and simplify. Write each result in the form $a + bi$.

(a) $\dfrac{3i}{4 + 2i}$

(b) $\dfrac{2 + 7i}{5i}$

(c) $\dfrac{2 + i}{5 + 4i}$

(d) $\dfrac{3 - 6i}{12 + 9i}$

(e) $\dfrac{4 - 8i}{2 - 9i}$

8. Simplify each of the following.

(a) i^7

(b) i^{11}

(c) i^{52}

(d) $(-i)^9$

(e) $(-i)^{34}$

9. Simplify each of the following.

(a) $(3i)^4$

(b) $(-2i)^5$

(c) $5i^5 + 4i^3$

(d) $\dfrac{i^5 + i^6 + i^8}{(1 - i)^4}$

(e) $(1 - i)^3(1 + i)^3$

Answers to Problems

1. (a) $4i$

(b) $3 + 2\sqrt{2i}$

(c) $7 - 9i$

(d) $\frac{1}{3} - \frac{1}{3}i$

(e) 4

3. (a) $10 - i$

(b) $-3 - 4i$

(c) $-6 - 13i$

(d) $12 - 3i$

5. (a) $-34 + 34i$

(b) $38 + 37i$

(c) $-7 + 24i$

(d) $21 + 20i$

7. (a) $\frac{3}{10} + \frac{3}{5}i$

(b) $\frac{7}{5} - \frac{2}{5}i$

(c) $\frac{14}{41} - \frac{3}{41}i$

(d) $-\frac{2}{25} - \frac{11}{25}i$

(e) $\frac{16}{17} + \frac{4}{17}i$

9. (a) 81

(b) $-32i$

(c) i

(d) $-\frac{1}{4}i$

(e) 8

Appendix II

ALGEBRA

SPECIAL PRODUCTS

1. $(a + b)^2 = a^2 + 2ab + b^2$
2. $(a - b)^2 = a^2 - 2ab + b^2$
3. $(a + b)^3 = a^3 + 3a^2b + 3ab^2 + b^3$
4. $(a - b)^3 = a^3 - 3a^2b + 3ab^2 - b^3$

FACTORING

1. $a^2 - b^2 = (a + b)(a - b)$
2. $a^2 \pm 2ab + b^2 = (a \pm b)^2$
3. $a^3 + b^3 = (a + b)(a^2 - ab + b^2)$
4. $a^3 - b^3 = (a - b)(a^2 + ab + b^2)$

EXPONENT PROPERTIES

1. $a^m \cdot a^n = a^{m+n}$
2. $(a^m)^n = a^{mn}$
3. $(ab)^n = a^n b^n$
4. $\left(\frac{a}{b}\right)^n = \frac{a^n}{b^n}, b \neq 0$
5. $\frac{a^m}{a^n} = a^{m-n}, a \neq 0$
6. $a^{-n} = \frac{1}{a^n}, a \neq 0$

RADICAL PROPERTIES

1. $\sqrt[n]{ab} = \sqrt[n]{a} \cdot \sqrt[n]{b}$
2. $\sqrt[n]{\frac{a}{b}} = \frac{\sqrt[n]{a}}{\sqrt[n]{b}}$
3. $\sqrt[n]{a^m} = \left(\sqrt[n]{a}\right)^m = a^{m/n}$
4. $\sqrt[n]{\sqrt[m]{a}} = \sqrt[nm]{a}$

PROPERTIES OF LOGARITHMS

1. $\log_b MN = \log_b M + \log_b N$
2. $\log_b \frac{M}{N} = \log_b M - \log_b N$
3. $\log_b N^y = y \log_b N$
4. $b^{\log_b N} = N$

IMPORTANT FORMULAS

QUADRATIC FORMULA

If $ax^2 + bx + c = 0$, for $a \neq 0$, then

$$x = \frac{-b \pm \sqrt{b^2 - 4ac}}{2a}$$

DISTANCE FORMULA

The distance d between $P_1 = (x_1, y_1)$ and $P_2 = (x_2, y_2)$ is

$$d = \sqrt{(x_2 - x_1)^2 + (y_2 - y_1)^2}$$

MIDPOINT FORMULA

The midpoint m between $P_1 = (x_1, y_1)$ and $P_2 = (x_2, y_2)$ is given by

$$m = \left(\frac{x_1 + x_2}{2}, \frac{y_1 + y_2}{2}\right)$$

GEOMETRY

A = area, C = circumference, V = volume, r = radius, h = altitude, l = length, w = width, b (or a) = length of a base, s = length of a side

1. Rectangle	$A = lw$	**6.** Cube	$V = s^3$
2. Parallelogram	$A = bh$	**7.** Rectangular box	$V = lwh$
3. Triangle	$A = \frac{1}{2}bh$	**8.** Cylinder	$V = \pi r^2 h$
4. Circle	$A = \pi r^2$; $C = 2\pi r$	**9.** Sphere	$V = \frac{4}{3}\pi r^3$
5. Trapezoid	$A = \frac{1}{2}(a + b)h$	**10.** Cone	$V = \frac{1}{3}\pi r^2 h$

STANDARD GRAPHS OF FUNCTIONS

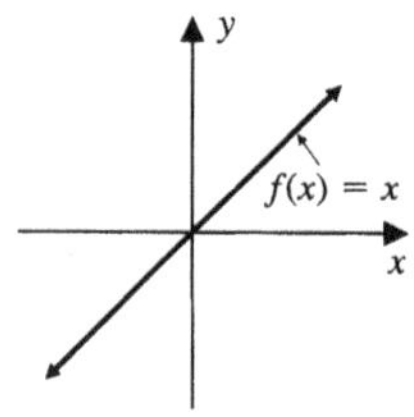

1. Identity function

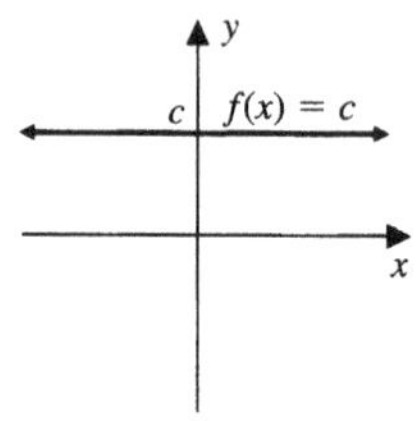

2. Constant function
(c is a constant)

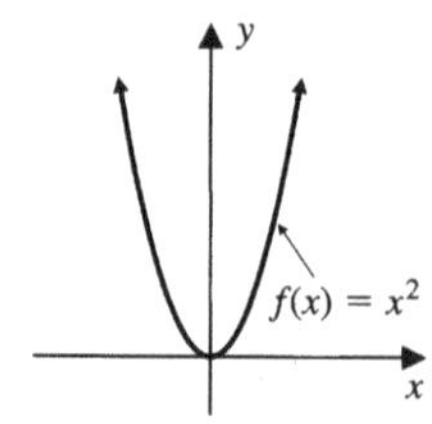

3. Squaring function

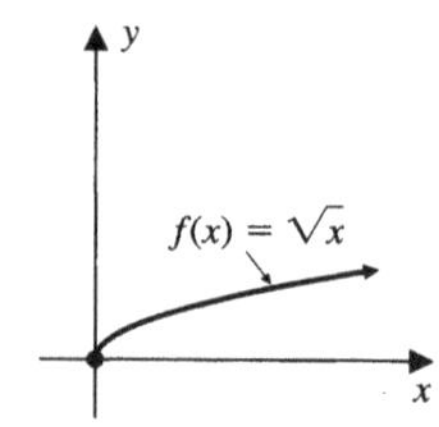

4. Square root function

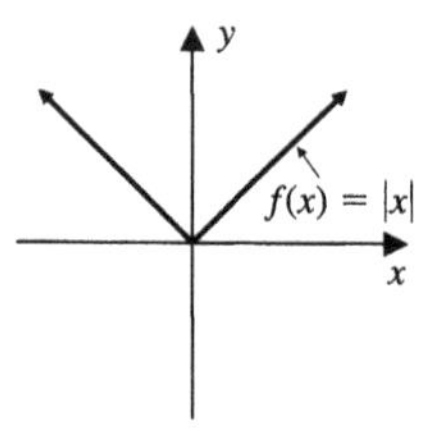

5. Absolute value function

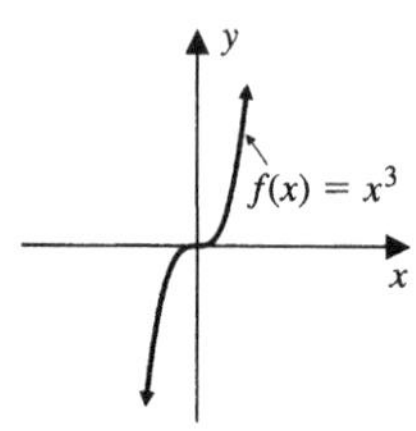

6. Cubing function

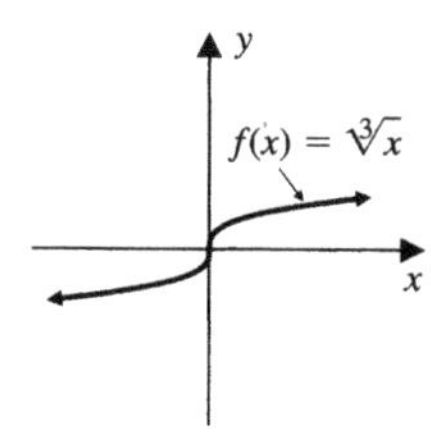

7. Cube root function

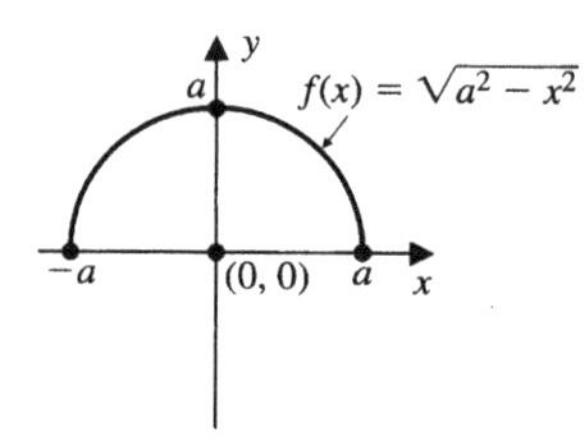

8. Upper half-circle function
(center: (0, 0); radius a)

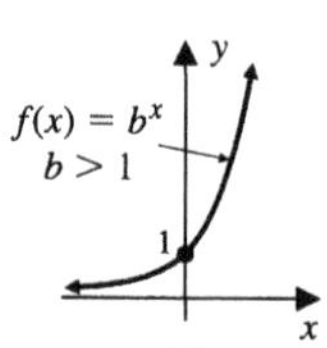

(a)

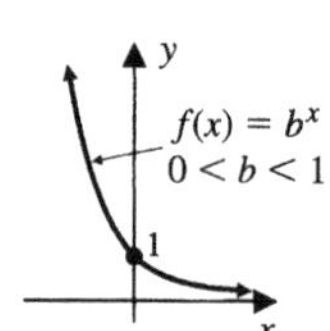

(b)

9. Exponential functions

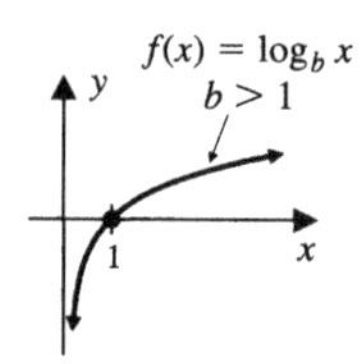

(a)

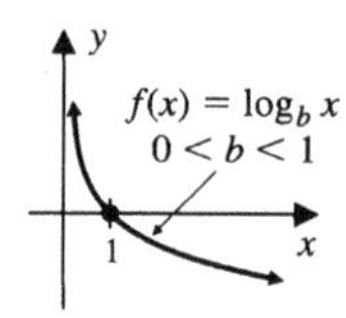

(b)

10. Logarithmic functions

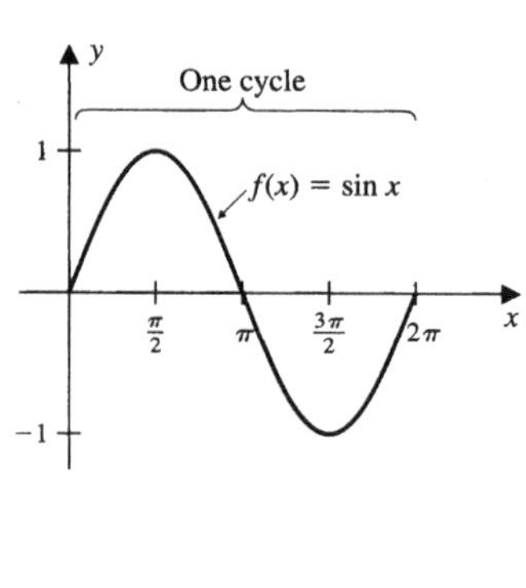

(a)

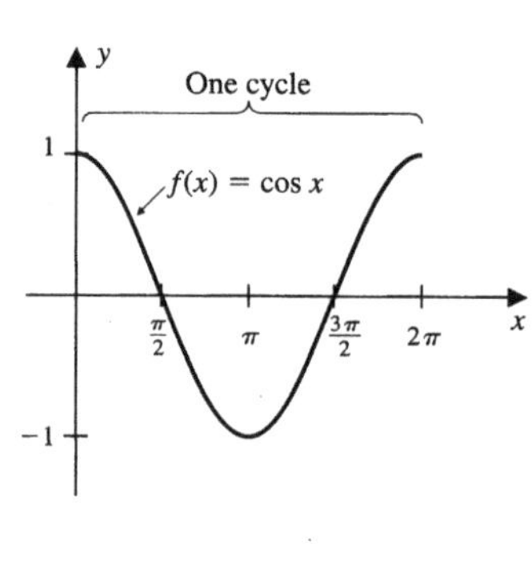

(b)

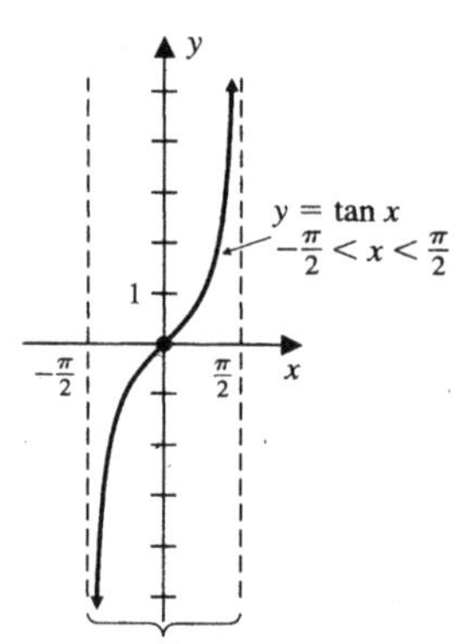

(c)

11. Trigonometric functions

TRIGONOMETRIC FUNCTIONS

ACUTE ANGLES

$\sin\theta = \dfrac{\text{opp}}{\text{hyp}}$ $\csc\theta = \dfrac{\text{hyp}}{\text{opp}}$

$\cos\theta = \dfrac{\text{adj}}{\text{hyp}}$ $\sec\theta = \dfrac{\text{hyp}}{\text{adj}}$

$\tan\theta = \dfrac{\text{opp}}{\text{adj}}$ $\cot\theta = \dfrac{\text{adj}}{\text{opp}}$

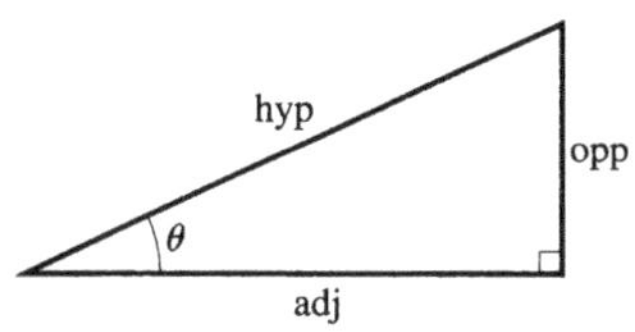

GENERAL ANGLES

$\sin\theta = \dfrac{y}{r}$ $\csc\theta = \dfrac{r}{y}$

$\cos\theta = \dfrac{x}{r}$ $\sec\theta = \dfrac{r}{x}$

$\tan\theta = \dfrac{y}{x}$ $\cot\theta = \dfrac{x}{y}$

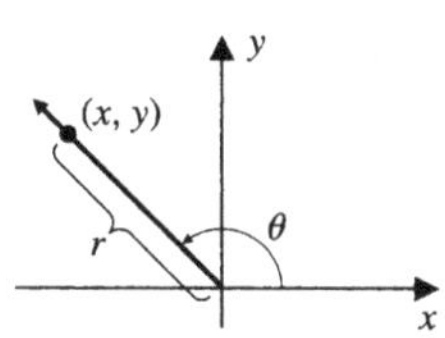

ADDITION FORMULAS

1. $\sin(\alpha+\beta) = \sin\alpha\cos\beta + \sin\beta\cos\alpha$
2. $\cos(\alpha+\beta) = \cos\alpha\cos\beta - \sin\alpha\sin\beta$
3. $\tan(\alpha+\beta) = \dfrac{\tan\alpha + \tan\beta}{1 - \tan\alpha\tan\beta}$

SUBTRACTION FORMULAS

1. $\sin(\alpha-\beta) = \sin\alpha\cos\beta - \sin\beta\cos\alpha$
2. $\cos(\alpha-\beta) = \cos\alpha\cos\beta + \sin\alpha\sin\beta$
3. $\tan(\alpha-\beta) = \dfrac{\tan\alpha - \tan\beta}{1 + \tan\alpha\tan\beta}$

TRIGONOMETRIC IDENTITIES

FUNDAMENTAL IDENTITIES

1. $\csc\theta = \dfrac{1}{\sin\theta}$
2. $\sec\theta = \dfrac{1}{\cos\theta}$
3. $\cot\theta = \dfrac{1}{\tan\theta}$
4. $\tan\theta = \dfrac{\sin\theta}{\cos\theta}$
5. $\cot\theta = \dfrac{\cos\theta}{\sin\theta}$
6. $\cos^2\theta + \sin^2\theta = 1$
7. $1 + \tan^2\theta = \sec^2\theta$
8. $1 + \cot^2\theta = \csc^2\theta$

EVEN–ODD IDENTITIES

1. $\sin(-\theta) = -\sin\theta$
2. $\cos(-\theta) = \cos\theta$
3. $\tan(-\theta) = -\tan\theta$
4. $\cot(-\theta) = -\cot\theta$
5. $\sec(-\theta) = \sec\theta$
6. $\csc(-\theta) = -\csc\theta$

THE UNIT CIRCLE

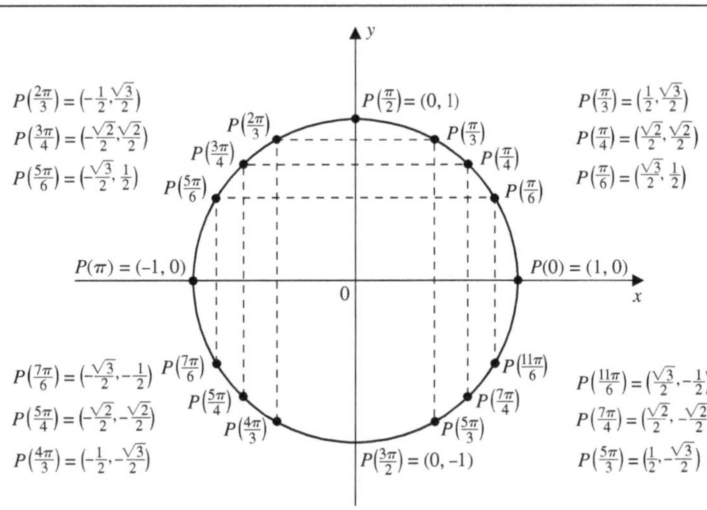

COFUNCTION FORMULAS

1. $\cos\left(\frac{\pi}{2} - \theta\right) = \sin \theta$
2. $\sin\left(\frac{\pi}{2} - \theta\right) = \cos \theta$
3. $\cot\left(\frac{\pi}{2} - \theta\right) = \tan \theta$
4. $\tan\left(\frac{\pi}{2} - \theta\right) = \cot \theta$
5. $\csc\left(\frac{\pi}{2} - \theta\right) = \sec \theta$
6. $\sec\left(\frac{\pi}{2} - \theta\right) = \csc \theta$

DOUBLE-ANGLE FORMULAS

1. $\sin 2\theta = 2 \sin \theta \cos \theta$
2. $\cos 2\theta = \cos^2 \theta - \sin^2 \theta$
   $= 2\cos^2 \theta - 1$
   $= 1 - 2\sin^2 \theta$
3. $\tan 2\theta = \dfrac{2 \tan \theta}{1 - \tan^2 \theta}$

PRODUCT FORMULAS

1. $\sin \alpha \cos \beta = \frac{1}{2}[\sin(\alpha + \beta) + \sin(\alpha - \beta)]$
2. $\cos \alpha \sin \beta = \frac{1}{2}[\sin(\alpha + \beta) - \sin(\alpha - \beta)]$
3. $\cos \alpha \cos \beta = \frac{1}{2}[\cos(\alpha + \beta) + \cos(\alpha - \beta)]$
4. $\sin \alpha \sin \beta = \frac{1}{2}[\cos(\alpha - \beta) - \cos(\alpha + \beta)]$

HALF-ANGLE FORMULAS

1. $\cos \frac{\theta}{2} = \pm\sqrt{\dfrac{1 + \cos \theta}{2}}$
2. $\sin \frac{\theta}{2} = \pm\sqrt{\dfrac{1 - \cos \theta}{2}}$
3. $\cos^2 \theta = \dfrac{1 + \cos 2\theta}{2}$
4. $\sin^2 \theta = \dfrac{1 - \cos 2\theta}{2}$

TABLES OF SPECIAL VALUES OF TRIGONOMETRIC FUNCTIONS

Angle θ in degrees	Angle θ in radians	$\sin \theta$	$\cos \theta$	$\tan \theta$	$\cot \theta$	$\sec \theta$	$\csc \theta$
0°	0	0	1	0	—	1	—
30°	$\pi/6$	1/2	$\sqrt{3}/2$	$\sqrt{3}/3$	$\sqrt{3}$	$2\sqrt{3}/3$	2
45°	$\pi/4$	$\sqrt{2}/2$	$\sqrt{2}/2$	1	1	$\sqrt{2}$	$\sqrt{2}$
60°	$\pi/3$	$\sqrt{3}/2$	1/2	$\sqrt{3}$	$\sqrt{3}/3$	2	$2\sqrt{3}/3$

www.ingramcontent.com/pod-product-compliance
Ingram Content Group UK Ltd.
Pitfield, Milton Keynes, MK11 3LW, UK
UKHW051510290926
15471UKWH00008B/1544

9 781524 928025